EXPLORING PLANT STRESS TOLERANCE

A Metabolomics Approach

Shuvasish Choudhury, PhD
Debojyoti Moulick, PhD
Mirza Hasanuzzaman, PhD
Editors

AAP | APPLE ACADEMIC PRESS

First edition published 2026

Apple Academic Press Inc.
1265 Goldenrod Circle, NE,
Palm Bay, FL 32905 USA

760 Laurentian Drive, Unit 19,
Burlington, ON L7N 0A4, Canada

CRC Press
2385 NW Executive Center Drive,
Suite 320, Boca Raton FL 33431

4 Park Square, Milton Park,
Abingdon, Oxon, OX14 4RN, UK

Library and Archives Canada Cataloguing in Publication

CIP data on file with Canada Library and Archives

Library of Congress Cataloging-in-Publication Data

CIP data on file with US Library of Congress

ISBN: 978-1-77964-172-4 (hbk)
ISBN: 978-1-77964-173-1 (ebk)
DOI: 10.1201/9781779641731

About the Editors

Shuvasish Choudhury, PhD
Associate Professor, Plant Stress Biology, Department of Life Science and Bioinformatics, Assam University, Silchar, Assam, India.

Shuvasish Choudhury, PhD, is an Associate Professor of Plant Stress Biology in the Department of Life Science and Bioinformatics, Assam University, Silchar, Assam, India. He earned his PhD in Life Science from Assam University, Silchar, in 2009 and did his postdoctoral research at the Department of Biochemistry and Molecular Biology, Oklahoma State University, USA. His research areas include plant stress biology and plant metabolomics. He has published more than 80 research papers and related articles, chapters, etc., in international journals of repute indexed by Scopus and Web of Science, as well as several book chapters and edited books from reputed international publishing house. He is also an acting Associate Editor for Plant Abiotic Stress by Frontiers in Plant Science, Switzerland. He has received a Young Botanist Award in 2010 from the Indian Botanical Society, India. He was awarded with Fast Track Young Scientist by the Department of Science and Technology, Govt. of India, and Overseas Associateship by DBT, Government of India. He has successfully completed research projects funded by agencies such as UGC, DBT, SERB, and CSIR, Government of India. He has more than 3900 citations by Google Scholar and 2000 citations indexed by SCOPUS, cited in research works published in top journals in the area. His current research includes biochemistry and metabolomics of abiotic stress in plants, salinity stress, nutrient management strategies for abiotic stress resilience in crops, conversion of waste bioresources to biochar, and climate change impact on agriculture. Dr. Choudhury has been among the 2% most influential scientist list by Stanford University, USA and Elsevier in 2024.

Debojyoti Moulick, PhD
Postdoctoral Fellow, Assam University, Silchar, Assam, India.

Debojyoti Moulick, PhD, is a Postdoctoral Fellow at Assam University, Silchar, India. In 2018, he was awarded the prestigious Research Associateship (DBT-RA) award from the Indian Institute of Science (IISc), Bangalore, and the Department of Biotechnology, Govt. of India. Recently in 2025, Dr. Moulick has secured his place among world's top 2% most influential scientists list, across all fields, prepared by Stanford University, USA, and Elsevier, based on bibliometric data from the Scopus database. He is a graduate in Microbiology and postgraduate in Biotechnology. He was awarded an MPhil and PhD degrees in Environmental Science by the University of Kalyani, West Bengal, India. Dr. Moulick is also a recipient of a Senior Research Fellowship (SRF) from the Ministry of Environment and Forest for supporting his PhD research. In addition, he was also awarded a Senior Research Fellowship (SRF) by the Indian Council for Agricultural Research (ICAR), Govt. of India. He carried out his PhD work under the supervision of Prof. S. C. Santra, a well-recognized environmental biotechnologist and a global expert of arsenic toxicity. Before beginning his MPhil degree, he worked on gas-hydrates and geochemical exploration in the Geochemistry Division of KDMIPE of ONGC in Dehradun, India, under the supervision of Dr. A. K. Jain. Dr. Moulick has gathered extensive experience in biomonitoring, seed priming technology and phytotoxicity, grain quality analysis, etc., in soil–water–crop agroecosystems. In 2018, Dr. Moulick's research work has been recognized by the Department of Science and Technology (Govt. of India) in Vigyan Prasar (India Science Wire) and several media houses as a breakthrough in proposing an alternative, cheap, and farmer-friendly mitigation option for minimizing arsenic-induced phytotoxicity and transmission through rice plant and grain respectively. He has to his credit several edited volumes published by CRC Press/Taylor and Franics, Frontiers Media SA, and Springer Nature (Switzerland), with several in press now. In addition, since 2017 he has reviewed over 350 papers for various international publication houses, served as an editorial board member for *BMC Plant Biology* and *Integrated Environmental Assessment and Management*. As of December 2025, Dr. Moulick's publications have been cited over 2540+ times as per Google Scholar with an h-index of 30 and i10 index of 63.

Mirza Hasanuzzaman, PhD
Professor of Agronomy,
Sher-e-Bangla Agricultural University (SAU),
Dhaka, Bangladesh.

Mirza Hasanuzzaman, PhD, is a Professor of Agronomy at Sher-e-Bangla Agricultural University (SAU), Dhaka, Bangladesh. He completed his PhD from the United Graduate School of Agricultural Sciences, Ehime University, Japan with a Japanese Government (MEXT) Scholarship. Later, he completed his postdoctoral research at the Center of Molecular Biosciences (COMB), University of the Ryukyus, Okinawa, Japan, with a "Japan Society for the Promotion of Science (JSPS)" postdoctoral fellowship. Subsequently, he became an Adjunct Senior Researcher at the University of Tasmania with an Australian Government's Endeavour Research Fellowship. In 2006, Dr. Hasanuzzaman joined as a Lecturer in the Department of Agronomy, SAU. Since then he has received three early promotions and has reached the top level, becoming a full professor in June 2017, all within the same university. Prof. Hasanuzzaman has over 300 Scopus-indexed publications, with more than 250 includedded in the Web of Science. His publications are cited over 36,000 times as per Scopus, with an h-index of 97 (as of December 2025). He is an editor and a reviewer for more than 80 peer-reviewed international journals and the recipient of a "Publons Peer Review Award 2017, 2018, and 2019." He is acting Associate Editor of *Plant Signaling & Behavior* (Taylor & Francis), *Frontiers in Plant Science* (Switzerland), *Phyton* (Tech Science Press, USA) and *Stresses* (MDPI, Switzerland); Academic Editor of *PeerJ* (USA), *Plants* (MDPI), *Open Agriculture* (De Gruyter, Poland); and member of the editorial board of *Plant Physiology and Biochemistry* (Elsevier), *Scientific Reports* (Nature), *Annals of Agricultural Sciences* (Elsevier), *Communicative and Integrative Biology* (Taylor & Francis) and so on. He has received many prestigious awards, such as the World Academy of Science (TWAS) Young Scientist Award 2014; University Grants Commission (UGC) Gold Medal 2018; Bangladesh Academy of Sciences (BAS) Gold Medal Award 2022 (Senior Group); Global Network of Bangladeshi Biotechnologists (GNOBB) Award 2021; Society for Plant Research Young Scientist Award (Agriculture) 2023; Eminent Scientist Award 2022 from the Society for Science of Climate Change and Sustainable Environment; and so on. He is a Fellow of the Bangladesh Academy of Sciences (BAS), the World Academy of Sciences (TWAS); Royal Society of Biology, International Society of Environmental Botanists, and a Foreign Fellow of the Society for Science of Climate Change and Sustainable Environment. He was named a Highly Cited Researcher by Clarivate in 2021 and 2022.

Contents

Contributors

Magdi T. Abdelhamid
Botany Department, National Research Centre, Dokki, Cairo, Egypt; Department of Research Centers, Montana State University, Bozeman, MT, USA

Hammad Ahmed
Seed Biotech Laboratories, Centre of Excellence in Molecular Biology, University of the Punjab, Lahore, Pakistan

Md. Zishan Ansari
Department of Biotechnology, Guru Ghasidas Vishwavidyalaya (A Central University), Bilaspur, Chhattisgarh, India

Usman Arif
Seed Biotech Laboratories, Centre of Excellence in Molecular Biology, University of the Punjab, Lahore, Pakistan

Rabiah Ashraf
Seed Biotech Laboratories, Centre of Excellence in Molecular Biology, University of the Punjab, Lahore, Pakistan

Jay Prakash Awasthi
Department of Botany, Government College Umariyapan, Katni, Madhya Pradesh, India

Allah Bakhsh
Seed Biotech Laboratories, Centre of Excellence in Molecular Biology, University of the Punjab, Lahore, Pakistan

Dharmeswar Barhoi
Department of Zoology, The Assam Royal Global University, Guwahati, Assam, India

Shabana Begum
Department of Life Science and Bioinformatics, Assam University, Silchar, Assam, India

Nikha Borah
Microbiology Laboratory, Department of Life Sciences, Dibrugarh University, Dibrugarh, Assam, India

Uddipta Borthakur
Department of Botany, Gauhati University, Guwahati, India

Kabyashree Buragohain
Microbiology Laboratory, Department of Life Sciences, Dibrugarh University, Dibrugarh, Assam, India

Anwesha Chatterjee
Amity Institute of Biotechnology, Amity University, Kolkata, Kolkata, West Bengal, India

Bhaben Chowardhara
Department of Botany, Faculty of Science and Technology, Department of Botany, Arunachal University of Studies, Namsai, Arunachal Pradesh, India

Umakanta Chowra
Department of Botany, Guwahati College, Guwahati, Assam, India

Anupam Das
Department of Soil Science and Agricultural Chemistry, Bihar Agricultural University, Bhagalpur, Bihar, India

Namita Devi
Biswanath College, Biswanath Chariali, Assam, India

Ahmed H. El-Desoky
Pharmacognosy Department, Institute of Pharmaceutical and Drug Industries Research, National Research Centre (NRC), Dokki, Giza, Egypt

Shaimaa I. M. Elsayed
Horticultural Department, National Research Centre, Giza, Egypt

Hend Fouad
Medicinal and Aromatic Plants Research Department, National Research Centre, Giza, Egypt; Institute for Sustainable Food, University of Sheffield, Western Bank, Sheffield, United Kingdom

Rasha Fouad
Medicinal and Aromatic Plants Research Department, National Research Centre, Giza, Egypt

Proma Ghosh
Amity Institute of Biotechnology, Amity University Kolkata, Kolkata, West Bengal, India

Ankita Gogoi
Microbiology Laboratory, Department of Life Sciences, Dibrugarh University, Dibrugarh, Assam, India

Dikshit Goswami
Department of Botany, Gauhati University, Guwahati, Assam, India

Sunayana Goswami
Biswanath College, Biswanath Chariali, Assam, India

Swati Hazra
School of Agricultural Sciences, Sharda University, Greater Noida, Uttar Pradesh, India

Akbar Hossain
Division of Soil Science, Bangladesh Wheat and Maize Research Institute (BWMRI), Nashipur, Dinajpur, Bangladesh

Peter Jackson
Institute for Sustainable Food, University of Sheffield, Western Bank, Sheffield, United Kingdom

Subrat Kakati
Department of Botany, Gauhati University, Guwahati, Assam, India

Khirod Kalita
Department of Botany, Gauhati University, Guwahati, Assam, India

Sibashish Kityania
Department of Life Science and Bioinformatics, Assam University, Silchar, Cachar, Assam, India

Ashish Kumar
Department of Biotechnology, Guru Ghasidas Vishwavidyalaya (A Central University), Bilaspur, Chhattisgarh, India

Sanjeev Kumar
Department of Botany, Central University of Punjab, Bhatinda, Punjab, India

Bikash Kumar Kundu
Department of Botany, Gauhati University, Guwahati, Assam, India

Kamel Mahmoud
Pharmacognosy Department, College of Pharmaceutical Sciences and Drug Manufacturing, Misr University for Sciences and Technology (MUST), Egypt

Sayanti Mandal
Department of Chemistry and Biochemistry, Sharda School of Engineering & Sciences, Sharda University, Greater Noida, Uttar Pradesh, India

Pranab Behari Mazumder
Department of Biotechnology, Assam University, Silchar, Assam, India

Arkabanee Mukherjee
Indian Institute of Tropical Meteorology, Pune, Maharashtra, India

Abdul Mukti
Department of Botany, Central University, of Punjab, Bhatinda, Punjab, India

Deepa Nath
Department of Botany, Guru Charan University, Silchar, Cachar, Assam, India

Manabendra Nath
Department of Botany, Gauhati University, Guwahati, Assam, India

Rajat Nath
Department of Biotechnology and Microbiology, School of Natural Sciences, Techno India University Tripura (TIUT), Agartala, Tripura, India

Ratul Nath
Microbiology Laboratory, Department of Life Sciences, Dibrugarh University, Dibrugarh, Assam, India

Rupshikha Nath
Department of Life Science and Bioinformatics, Assam University, Silchar, Cachar, Assam, India

Elsayed A. Omar
Medicinal and Aromatic Plants Research Department, National Research Centre, Dokki, Giza, Egypt

Harshata Pal
Amity Institute of Biotechnology, Amity University Kolkata, Kolkata, West Bengal, India

Amit Kumar Pradhan
Department of Botany, Pragjyotish College, Guwahati, Assam, India

Preetom Regon
Department of Entomology, Agricultural Research Organization, The Volcani Institute, Rishon LeZion, Israel

Farid M. Rohim
Horticultural Crops Technology Department, National Research Centre, Giza, Egypt

Bedabrata Saha
Plant Pathology and Weed Research Department, Newe Ya'ar Research Centre, ARO, Ramat, Yishay, Israel

L. R. Saikia
Department of Life Sciences, Dibrugarh University, Dibrugarh, Assam, India

Misbah Saleem
Seed Biotech Laboratories, Centre of Excellence in Molecular Biology, University of the Punjab, Lahore, Pakistan

Zina Moni Shandilya
Department of Botany, Gauhati University, Guwahati, Assam, India

Nibedita Sharma
Department of Botany, Gauhati University, Guwahati, Assam, India

Nilofer Sheikh
Biswanath College, Biswanath Chariali, Assam, India

Utsab Singha
Department of Zoology, The Assam Royal Global University, Guwahati, Assam, India

Sukanya Sonowal
Microbiology Laboratory, Department of Life Sciences, Dibrugarh University, Dibrugarh, Assam, India

Anupam Das Talukdar
Department of Life Science and Bioinformatics, Assam University, Silchar, Cachar, Assam, India

Dulumoni Tamuly
Department of Zoology, MDKG College, Dibrugarh, Assam, India

Bhaben Tanti
Department of Botany, Gauhati University, Guwahati, Assam, India

Ahmed M. Taweela
Pharmacognosy Department, Institute of Pharmaceutical and Drug Industries Research, National Research Centre (NRC), Dokki, Giza, Egypt

Abbreviations

$^1O_2$	singlet oxygen
AAO3	abscisic aldehyde oxidase 3
ABA	abscisic acid
ABC	ATP-binding cassette
ACC	1-aminocyclopropane-1-carboxylic acid
AFPs	antifreeze proteins
AMF	arbuscular mycorrhizal fungi
ANDI-MS	analytical data interchange for mass spectroscopy
ANOVA	Analysis of Variance
AOX	oxidase-like alternative oxidase
AP2/ERF	APETALA2/ethylene response factor
APCI	air pressure chemical ionization
API	application program interface
APS	ATP sulfurylase
APX	ascorbate peroxidase
APX	ascorbate peroxidase
AR	adaptive response
AsA	ascorbic acid
As-PC	arsenic-phytochelatin
ATP	adenosine triphosphate
AUX	auxin
BABA	β-aminobutyric acid
BPH	brown plant hopper
CA	citric acid
CAT	catalase
CDF	common data format
CDPKs	calcium-dependent protein kinases
CE	capillary electrophoresis
ChEBI	Chemical Entities of Biological Interest
CI	chemical ionization
CK	creatine kinase
C-MS	gas chromatography-mass spectrometry
COR	cold regulated
CS	cold stress
CSV	comma separated values
Cys	cysteine
DEGs	differentially expressed genes
DHAR	dehydroascorbate reductase

DMDS	dimethyl disulphide
DMSP	dimethyl sulfoniopropionate
DS	drought stress
DW	dry weight
EI	electron ionization
EI	electron impact
ER	endoplasmic reticulum
ESI	electrospray ionization
ET	ethylene
ETC	electron transport chain
FPLC	fast protein liquid chromatography
GABA	γ-amino butyric acid
GAs	gibberellins
GB	glycine betaine
GC	gas chromatography
GCMS	gas chromatography–mass spectrometry
GMD	Golm Metabolome Database
GNPS	Global Natural Products Social Molecular Networking
GOPX	guaiacol peroxidase
GR	glutathione reductase
GRX	glutaredoxin
GSH	glutathione
GSLs	glucosinolates
GSSG	glutathione disulfide
GST	glutathione *S*-transferase
GWAS	genome-wide association studies
H_2O_2	hydrogen peroxide
HCN	hydrogen cyanides
HDF5	Hierarchical Data Format version 5
HETP	height equivalent to theoretical plates
HMs	heavy metals
HMs	herbal medicines
HPLC	high-performance liquid chromatography
HS	heat stress
HsfB1	heat stress factor B1
HSP	heat shock protein
HTPs	high-throughput platforms
IAA	indole-3-acetic acid
InChI	International Chemical Identifier
IPM	integrated pest management
JA	jasmonic acid
KEGG	Kyoto Encyclopedia of Genes and Genomes
LC	liquid chromatography
LCMS	liquid chromatography–mass spectrometry

MAB	metabolomics-assisted breeding
MAPK	mitogen-activated protein kinase
MDHAR	monodehydroascorbate reductase
Met	methionine
MG	methylglyoxal
mQTL	metabolite quantitative trait loci
MS	mass spectrometry
MSTs	mass spectral tags
MT	metallothionein
mtETC	mitochondrial electron transport chain
NADPH	nicotinamide adenine dinucleotide phosphate hydrogenoxidases
NCED3	9-cis-epoxycarotenoid dioxygenase 3
NetCDF	Network Common Data Form
NMR	nuclear magnetic resonance
NOS	nitric oxide synthase
NPs	nanoparticles
NRF2	nuclear factor-erythroid 2-related factor 2
OPs	organic pollutants
PA	pyrrolizidine alkaloids
PC	phytochelatin
PCA	principal component analysis
PCD	programmed cell death
PEPC	phosphoenolpyruvate carboxylase
PGPR	plant growth-promoting rhizobacteria
PMDB	Plant Metabolome Database
PMN	Plant Metabolic Network
PMR	plants metabolomics repository
POD	peroxidase
pQTL	protein quantitative trait locus
PRXs	peroxiredoxins
QC	quality control
RAM	root apical meristem
RBOH	respiratory burst oxidase homologues
RFOs	raffinose family oligosaccharides
RNS	reactive nitrogen species
ROS	reactive oxygen species
RRTF1	redox responsive transcription factor 1
RSA	root system architecture
RT	retention time
RT-PCR	reverse transcription polymerase chain reaction
RuBisCO	ribulose-1,5-bisphosphate carboxylase/oxygenase
RWC	relative water content
SA	salicylic acid
SAM	shoot apical meristem

SH	sulfydryl group
SMs	secondary metabolites
SNAT	alfalfa serotonin *N*-acetyltransferase
SNP	single nucleotide polymorphism
SOD	superoxide dismutase
SOD	sodium oxide dismutase
SOD	superoxide dismutase
SOPs	standard operating procedures
TAL	tyrosine ammonia-lyase
T-AOC	total antioxidant capacity
TCA	tricarboxylic acid
TFs	transcription factors
Thr	threonine
TRP	L-tryptophan
TRXs	thioredoxins
UPLC	ultraperformance liquid chromatography
UV	ultraviolet
VOCs	volatile organic compounds
XAS	X-ray absorption spectroscopy
XML	eXtensible Markup Language
XRF	X-ray fluorescence
ZEP	zeaxanthin epoxidase

Preface

Plants, being sessile organisms, are subjected to diverse stresses. Today, climate change has resulted in a significant increase in incidences of abiotic stresses, which has a strong impact on agriculture and food security. Thus, a comprehensive understanding of factors affecting plants' responses to these stresses remains crucial. Various approaches have been employed to comprehend the mechanisms of abiotic stresses and identification of traits. These traits remain crucial to improve stress adaptation and tolerance in plants.

One such approach is metabolomics, which deals with the functional characteristics of the metabolites. Though metabolomics has diverse applications in understanding complexity of biological life, it has emerged as one of the most preferred tools for dissecting plant responses to stresses.

In this book, diverse aspects of abiotic stresses have been discussed, where metabolomics has emerged as an important research perspective. These include the application of metabolomics in comprehending plant responses to metalloid stress, heavy metal stress, plant–insect interaction, plant–pathogen interaction, high- and low-temperature stress, and regulation of ROS production. In addition to this, a comprehensive account on metabolite fingerprinting, metabolomic databases, and analytical methods have also been included.

All the chapters have been contributed by experts in the relevant field and contain updated information on the topics. We believe that this book will provide important scientific material in the area of plant stress biology.

CHAPTER 1

Plant Metabolomic Responses to Metalloid Stress: A Special Focus on As and Se Tolerance

SWATI HAZRA[1], ANUPAM DAS[2], and ARKABANEE MUKHERJEE[3*]

[1]*School of Agricultural Sciences, Sharda University, Greater Noida, Uttar Pradesh, India*

[2]*Department of Soil Science and Agricultural Chemistry, Bihar Agricultural University, Bhagalpur, Bihar, India*

[3]*Indian Institute of Tropical Meteorology, Pune, Maharashtra, India*

*Corresponding authors

ABSTRACT

Elevated metalloid stress on agriculture is another facet of climate change. Metalloids like arsenic and selenium have created an alarming food safety issue due to their ubiquitous presence in the Earth's crust and relatively high interactions with plants. Increasing food demand and degraded cultivable lands, together with unpredictable climate conditions, are building tremendous pressure on the agricultural sector. High metalloid exposure in plant systems blocks essential functional groups or replaces active ions to immobilize important metabolic processes. The study of metabolomics using high-throughput screening helps in understanding the metabolite profile and has become the need of the hour to know the survival physiology of crops under such abiotic stresses. The same metabolite fingerprint can be replicated to generate more resistant or tolerant genotypes. Additionally, metalloid speciation studies can reveal the details of the origin, pathways, and fate of the metabolites present in the plant system. Both essential and nonessential metalloids contribute to plant stress responses and the formation of a wide array of metabolites in different crops. In this chapter, we aim to provide a comprehensive account of plant metabolomic responses to metalloid stress and their potential applications in the development of tolerant or resistant crop varieties.

1.1 INTRODUCTION

Crop behavioral prediction under multiple-stress situations has become difficult due to the interaction between abiotic conditions, metal/metalloid stress, and related biochemical

Exploring Plant Stress Tolerance: A Metabolomics Approach.
Shuvasish Choudhury, Debojyoti Moulick, and Mirza Hasanuzzaman (Eds.)

DOI: 10.1201/9781779641731-1

reactions within the plant system. Changes in the plant metabolic pathways to minimize the adverse effect of metalloid stress in the system are a common survival strategy for crops (Dusenge et al., 2019; Fernie et al., 2020). In the current scenario, abiotic stress, for example, salinity, water, and heat, are abundant. Metalloid stress is associated with abiotic stress complicates plant response. Primary metabolites play a majorly dominant role in plant growth, whereas secondary metabolites like phenolic compounds, terpenes, and nitrogen-containing compounds are responsible for not only growth but also act as a defense against pathogens. Another recent study reported that secondary metabolites like flavonoids, which regulate environmental responses, are also responsible (Erb and Kliebenstein, 2020). Identifying the metabolomic activities can target the stress-associated changes in the crop system and future breeding programs for selective stress-resistant crops (Abdelrahman et al., 2018). Reported stress-induced pathway from a combination of abiotic and two metals Co and Cu in a tolerant barley variety *Yann66* (Lwalaba et al., 2020). Amino acids are a regulating factor to mitigate metal stress in the crop by using the metal-binding capacity of proteins (Anjum et al., 2015; Yuan et al., 2020; Lwalaba et al., 2020). Publications reported various stress combinations, including heavy metal stress, showed different metabolites depending on the plant species. *Rhodiola rosea* seedlings with a combination of heat and heavy metal stress resulted in a rise in phenolic compounds as secondary metabolites (Zhao et al., 2016). Similarly, the primary metabolites pathway generates limonene as a secondary metabolite in case of salinity stress and heavy metal stress exposure in *Mentha spicata* (Chrysargyris et al., 2019).

In this chapter, we are discussing the metabolomic response of crops under metalloid stress with a special focus on arsenic and selenium.

1.2 ARSENIC TOXICITY AND CROP METABOLOMICS

Arsenic is ubiquitous in the Earth's crust in both organic and inorganic forms. Among these two forms of inorganic arsenic have been identified as Class I carcinogens by the International Agency for Research on Cancer (Menon et al., 2020). Arsenic is prevalent in the soil as arsenate (As^{V}) and arsenite (As^{III}). Among these two As^{V} is more abundant in the soil system. However, As^{III} prevails in paddy fields due to its reducing environment (Du et al., 2019; Farooq et al., 2016). Arsenic exposure leads to restricted organ development and physiological and metabolic disorders in crop systems. Overproduction of reactive oxygen species (ROS) is another side effect of the same. Production of vitamins, steroids, stress hormones, cyclic nucleotides, amino acids, and amines increases significantly under arsenic stress to mitigate the effect of arsenic toxicity (Ma et al., 2023). As^{V} showed affinity toward the P transporter in the plant system As^{V} can replace P from the system, which hinders plant growth and adenosine triphosphate (ATP) production (Sharma et al., 2020). Considering high water solubility, As^{III} enters through aquaporin channels in the plant system and binds with the protein –SH (sulfydryl group) group (Tripathi et al., 2012). As^{III} exposure leads to the deactivation of enzymes, membrane degradation, and cell death which is known as more toxic than As^{V} (Bhattacharya et al., 2021; Kumar et al., 2015). Organic arsenic species like monomethyl arsenic acid, dimethyl arsenic acid, and trimethyl arsenic oxide are less toxic and also enter through aquaporin channel (Lsi 1) in the crop system (Kalita et al., 2018).

Arsenic species transform at the cellular level from As^{V} to As^{III} by the arsenate reductase enzyme. The highest accumulation of arsenic occurs at the root of the plant after uptake, and it is translocated into the stem, leaf, flowers, and fruits through the xylem. Attachment of As^{III} to the thiol group reduced glutathione (GSH) or phytochelatin (PC), forming a complex called arsenic-phytochelatin (As-PC) (Kofronova et al., 2019). As-PC chelate sequestrates in the cell vacuoles by ABC transporters (ATP-binding cassette transporters) and accumulates in all the tissues. Conversion between As^{V} to As^{III} within the cell disturbs the electron transport chain of mitochondria, and chloroplast due to the production of ROS, followed by redox imbalance (Dwivedi et al., 2010). Primary and secondary metabolites analyzed due to arsenic stress are major antioxidants, thiol ligand compounds, PCs, and PC formation-related compounds (Table 1.1). Primary metabolites are responsible for growth and reproduction. Arsenic stress affects the formation of amino acids, nucleotides, sugars, and acyl glycerides (Taíz and Zeiger, 2006). Arsenic directly impacts the P bond in amino acids, which affects protein formation. Arsenic stress increases the concentration of certain amino acids like cysteine (Cys) in the crop. Increasing amino acid for GSH synthesis mitigates the heavy metal exposure effect in the plant system.

Secondary metabolites are nonessential for crop survival. Arsenic interacts with the –SH group in protein (Kofronova et al., 2019). Secondary metabolites like nonenzymatic antioxidants and some enzymes (Pavlík et al., 2010). Ascorbic acid acts as a scavenger for ROS under oxidative stress (Mishra et al., 2011). Pigments like chlorophyll (Chl) and carotenoid (Crt) also contributed to some extent in plants under As stress. The concentration of Chl and Crt increases under low concentration or a short period of As exposure. The same activity decreases with higher concentration and extends As exposure in plant systems. As stress causes free radicle to peroxidation of lipids of the chloroplast membrane (Gupta and Ahmad, 2014). Hence, under arsenic stress, Chl and Crt concentration decreases. Thiobarbituric acid-reactive substances are known to be a by-product of lipid peroxidation. As already mentioned, lipid peroxidation in the membrane is related to As stress (Mishra et al., 2011; Singh et al., 2006).

The effect of the HM stress in addition to other abiotic stress can cause much complex reaction to primary and secondary metabolites and compounds that contribute to its production. Under arsenic stress, amino acids in the primary metabolites and antioxidants and enzymes in the secondary metabolites are observed with significant changes in their activity.

1.3 SELENIUM: CHEMISTRY, FORMS, AVAILABILITY, AND ROLES IN SOIL–PLANT SYSTEMS

The nonmetal or metal-like (metalloid) element Selenium (Se) [origin "*Selene*" (Greek) = moon; atomic number: 34; and atomic weight: 78.96) shares many common chemical properties with sulfur (Hasanuzzaman et al., 2010). Selenium and sulfur are both included in Group 16 (previously, Group VI A) of the modern Periodic Table. The range of

TABLE 1.1 The Effect of Arsenic-Induced Stress on Secondary Metabolites and Antioxidants

Sl. No.	Plant	Arsenic Concentration Stress	Metabolite	Effect of As Stress on Metabolites	References
1	Chestnut mustard (*Brassica juncea* L.)	50–500 μM of As^{V} 25–250 μM of As^{III}	GR, CS, SAT, ECS	ECS, SAT, and CS activity increased, whereas GR decreased	Srivastava et al. (2009)
2.	Red clover (*Trifolium pratense* L.)	5, 10, and 50 mg As/kg soil	Chl, Crt	Chl production down from 9% to 12% and Crt to 14% with 50 mg As/kg soil	Mascher et al. (2002)
3.	Duckweed (*Lemna minor* L.)	1–64 μM of As^{III} ($NaAsO_2$) and As^{V}	MDA, Chl, Crt	Under 64 μM of As^{III} MDA content increased by 65% and the same for As^{V} is 49%, respectively The Chl content decreased by 57% and 42% at 64 μM of As^{III} and As^{V} Crt content decreased by 41% and 33% As^{III} and As^{V}, respectively	Duman et al. (2010)
4.	Rice (*Oryza sativa* L.)	50, 150, and 300 μM of As^{V}	Chl, SOD, APX, GR, GSH	SOD, APX, GSH, and GR concentrations increase proportionally with the rise in As concentration	Dwivedi et al. (2010), Tripathi et al. (2012)
5.	Spinach (*Spinacia oleracea* L. cv. Matador)	25–75 mg As/kg in soil exposure	GluK	GluK activity decreases 68% with higher As	Pavlík et al. (2010)

Notes: APX: ascorbate peroxidase, Chl: chlorophyll, Crt: carotenoids, CS: cysteine synthase, GSH: glutathione reduce, GR: glutathione reductase, GluK: glutamate kinase, ECS: glutamylcysteine synthetase, MDA: malondialdehyde, PCS: phytochelatin synthase, SAT: serine acetyltransferase, SOD: superoxide dismutase oxidation states of Se varies from −2 to +6. One of the unique properties of Se compounds is their photoconductivity. These compounds can absorb electromagnetic radiation (i.e., light), which increases their electrical conductivity, making them useful in the industry and medical fields (De Filippis, 2010).

Selenium is omnipresent. A variety of forms of Se, both inorganic and organic, can occur in the physical environment (atmosphere, soil, and water) as well as inside the cellular environment of the living beings (Hasanuzzaman et al., 2020). Selenides, the most abundant natural source of Se, are often found to exist in the Earth's crust along with sulfide minerals (El-Ramady et al., 2015). Other sources of Se minerals include copper and iron pyrites (Mehdi et al., 2013). A significant amount of Se is added to the terrestrial and aquatic ecosystems due to the mining and metal-processing activities (Etteieb et al., 2020). Biological organisms mostly are further responsible for the release of Se from these ecosystems to the atmosphere in the form of SeO_2 and SeO_3 (Winkel et al., 2015). Selenium is abundantly available in the soil as selenate (SeO_4^{2-}) and selenite (SeO_3^{2-}), which are taken up by the plants (Lanza and Dos Reis, 2021). However, Se in the soil is not uniformly distributed around the globe, resulting in the occurrence of seleniferous soils in the countries like USA, Switzerland, Finland, and Australia whereas some areas are deficient of the element as noted in sub-Saharan Africa, China, Spain, Italy, Sweden, and others (Hasanuzzaman et al., 2020). Inside the living cells or tissues, seven compounds are found in the form of Se-containing amino acids (Se-methionine [Met] and Se-Cys), selenoproteins (e.g., enzymes like GSH peroxidases, thioredoxin reductases, and deiodinases). All these play crucial roles in the growth and metabolism of living organisms, including antioxidant activities (Mehdi et al., 2013). Therefore, sufficient intake of Se has been proven necessary to minimize the risk of cancer (Kieliszek, 2019). The World Health Organization has recognized Se as one of the essential trace elements for humans along with other essential minerals like Fe and Zn (Hossain et al., 2021). However, the bioaccumulation potential of Se is a matter of concern. Very high Se accumulation can often become fatal for living organisms. Some forms of Se (catalytic selenite and selenocysteine) inside the cell can induce apoptosis and thus impart cytotoxic (prooxidative) effects (Stewart et al., 1999). Nevertheless, Se has been established as an essential micronutrient to several prokaryotes and eukaryotes, including bacteria, algae, fungi, animals, and human for their metabolism, growth, and development (El-Ramady et al., 2015). There is not much known about the essentiality of Se in plants. However, its importance in plant body has recently been the focus of several studies. Selenium is mostly associated with the antioxidant activities (Gupta and Gupta, 2017; Lanza and Dos Reis, 2021), and thus it often improves plants' responses to various types of abiotic stressors (Feng et al., 2013; Hasanuzzaman et al., 2020) like heavy metals (Mroczek-Zdyrska and Wójcik, 2012) and metalloids (Chauhan et al., 2017), salinity (Kamran et al., 2019), cold (Chongping et al., 2022), UV radiation (Mata-Ramírez et al., 2019), and drought (Sattar et al., 2019). Moreover, Se plays a significant role in the protection of plants against the biotic stressors, for example, insects, pests, and fungal pathogens as observed by Hanson et al. (2003). Nevertheless, the mode of Se action behind the plant stress responses is yet to be fully understood. Although the role of the element has been confirmed in the process of reactive oxygen species (ROS) scavenging by both enzymatic and nonenzymatic mediators (Lanza and Dos Reis, 2021) and thus in the amelioration of oxidative stress (Del Pino et al.,

2019). In addition, low levels of Se can enhance photosynthesis and stomatal conductance in plants (Jiang et al., 2015) leading to their better growth and development along with increased root activity (Hawrylak-Nowak et al., 2015) and carbohydrate metabolism (Hajiboland et al., 2015). It has been observed that application of Se through either foliar spray and/or soil incorporation improved wheat and canola yield as well as quality (D'Amato et al., 2018; Hemmati et al., 2019; Lara et al., 2019). Moreover, priming of paddy seeds with Se solution enhanced the accumulation of essential micronutrients, such as Mn, Zn, Fe, and Cu in grains (Moulick et al., 2023). Nevertheless, biofortification with Se antagonistically reduced the accumulation of Hg, As, and Cd in plant cells, which can be utilized to tackle the issue of harmful heavy metals contamination in foods (Zhou et al., 2020). Application of Se is also useful to increase the postharvest shelf-life of horticultural crops. For example, weight, firmness, and biochemical properties (total phenols, flavonoids, soluble solids, etc.) of guavas were retained for a longer period under storage at room temperature upon treatment with very low concentrations of Se along with chitosan and acetic acid (Choudhary et al., 2020). Although aeroponics was proven to be more effective in Se-biofortification of lettuce (Puccinelli et al., 2021), the yield, quality, and shelf-life of plants like chicory and lettuce were enhanced by the application of Se in hydroponics (floating) systems as demonstrated by Malorgio et al. (2009).

1.4 SE UPTAKE AND METABOLISM IN PLANTS

Uptake of Se from the environment can vary depending on the plant species, developmental phases, Se forms and concentrations, physiological conditions of the soil (pH and salinity), presence of other compounds, activity of membrane transporters present in root cells, and so on (Gupta and Gupta, 2017). Transport of selenite is similar to the mechanism of phosphate transport, whereas selenite shares transporters and channels with sulfate (Li et al., 2008; Zhang et al., 2003). Inside the plant cells, Se is stored mostly in the vacuoles (Mazej et al., 2008). The Se metabolism occurs in the leaf plastids via the sulfur assimilation pathway to form Se-Cys or Se-Met (Gupta and Gupta, 2017). In the first step of Se assimilation, inorganic Se is converted into selenite in the presence of ATP sulfurylase (APS) followed by APS reductase. Another enzyme sulfite reductase, then catalyzes the conversion of selenite into selenide. The latter is then incorporated into Se-Cys by the action of enzyme Cys synthase. The Se-Cys can then either be converted to elemental Se in the presence of enzyme Se-Cys lyase or can be methylated by selenocysteine methyltransferase or can be converted to Se-Met.

1.5 EXCESSIVE SELENIUM MAY CAUSE TOXICITY

Several cases of Se toxicity were reported in animals later in the form of alkali disease (Birringer et al., 2002). These animals mostly consumed high-Se containing plants (Whanger, 2002). Also, excess Se has been noted to cause human selenosis and has led to deformities and mortality in aquatic organisms (Etteieb et al., 2020).

Plants exhibit typical biphasic responses to Se, and there is a thin boundary lies in between the beneficial and harmful concentration ranges of Se. At low concentrations, Se promotes plant metabolism, growth, and development, whereas at higher than optimum concentrations, Se (≥24 μM) can disrupt photosynthesis; alter the levels of antioxidant enzymes, malondialdehyde, soluble sugars, and amino acids; and thus finally inhibits growth (Liu et al., 2024). Effects of Se toxicity are imparted by two mechanisms: (1) malformation of selenoproteins and (2) induction of oxidative stress (Gupta and Gupta, 2017). Misincorporation of Se-Cys and Se-Met, respectively, in the places of Cys and Met in the protein chains can lead to disruptions in protein structures and functions as a result of Se-toxicity. In the oxidative stress induction mechanism of Se-toxicity, excess Se can cause cellular damage through over-accumulation of lipid peroxides (Hawrylak-Nowak et al., 2015). Therefore, Se is considered to be a quasi-essential micronutrient (El-Ramady et al., 2016) for many plant species.

1.6 CONCLUSION

As stress added with abiotic stress needs to be further examined for detailed specifications. Apart from the primary metabolites, other secondary metabolites like glucosinolates, alkaloids, and phenolic compounds require more intense studies in situations of As stress. As for Se, recently the mode and mechanism of Se actions have come into the light. Because of the biphasic reactions of the plants to Se exposure, critical concentrations need to be determined for different crop plants to reap the benefits of Se-fortification. These critical concentrations may vary depending on the tolerance level of plants (hyperaccumulators, secondary accumulators, and nonaccumulators) to Se (Gupta and Gupta, 2017). Therefore, crop-wise, more definite research needs to be conducted to address this particular concern.

KEYWORDS

- **abiotic stress**
- **arsenic**
- **food safety**
- **metabolomics**
- **metalloids**

REFERENCES

Abdelrahman, M., Burritt, D., Tran, L. The use of metabolomic quantitative trait locus mapping and osmotic adjustment traits for the improvement of crop yields under environmental stresses. *Semin. Cell Dev. Biol.*, 2018, 83, 86–94.

Anjum, N. A., Hasanuzzaman, M., Hossain, M. A., et al. Jacks of metal/ metalloid chelation trade in plants—An overview. *Front. Plant Sci.*, 2015, 6, 192.

Bhattacharya, S., Sharma, P., Mitra, S., Mallick, I., Ghosh, A. Arsenic uptake and bioaccumulation in plants: A review on remediation and socio-economic perspective in Southeast Asia. *Environ. Nanotechnol. Monit. Manag.*, 2021, 15, 100430.

Birringer, M., Pilawa, S., Flohé, L. Trends in selenium biochemistry. *Nat. Prod. Rep.*, 2002, 19(6), 693–718. https://doi.org/10.1039/B205802M.

Chauhan, R., Awasthi, S., Tripathi, P., Mishra, S., Dwivedi, S., Niranjan, A., Mallick, S., Tripathi, P., Pande, V., Tripathi, R. D. Selenite modulates the level of phenolics and nutrient element to alleviate the toxicity of arsenite in rice (*Oryza sativa* L.). *Ecotoxicol. Environ. Saf.*, 2017, 138, 47–55. https://doi.org/10.1016/j.ecoenv.2016.11.015.

Chongping, H., Wenjie, H., Junlin, L. Selenium- and nano-selenium-mediated cold-stress tolerance in crop plants. In: Hossain, M. A., Ahammed, G. J., Kolbert, Z., El-Ramady, H., Islam, T., Schiavon, M. (Eds.), *Selenium and Nano-Selenium in Environmental Stress Management and Crop Quality Improvement* (pp. 173–190). Springer: Cham, 2022. https://doi.org/10.1007/978-3-031-07063-1_9.

Choudhary, P., Kumari, N., Jain, V. Influence of selenium and chitosan on physico-chemical properties of guava (*Psidium guajava*) under storage. *Indian J. Agric. Sci.*, 2020, 90(2), 351–355. https://doi.org/10.56093/ijas.v90i2.99021.

Chrysargyris, A., Papakyriakou, E., Petropoulos, S. A., Tzortzakis, N. The combined and single effect of salinity and copper stress on growth and quality of Mentha spicata plants. *Journal of Hazardous Materials*, 2019, 368, 584–593.

D'amato, R., Fontanella, M. C., Falcinelli, B., Beone, G. M., Bravi, E., Marconi, O., Benincasa, P., Businelli, D. Selenium biofortification in rice (*Oryza sativa* L.) sprouting: Effects on Se yield and nutritional traits with focus on phenolic acid profile. *J. Agric. Food Chem.*, 2018, 66(16), 4082–4090. https://doi.org/10.1021/acs.jafc.8b00127.

De Filippis, L. F. Biochemical and molecular aspects in phytoremediation of selenium. In: Ashraf, M., Ozturk, M., Ahmad, M. S. A. (Eds.), *Plant Adaptation and Phytoremediation* (pp. 193–226). Springer: Dordrecht, 2010. https://doi.org/10.1007/978-90-481-9370-7_10.

Del Pino, A. M., Guiducci, M., D'amato, R., Di Michele, A., Tosti, G., Datti, A., Palmerini, C. A. Selenium maintains cytosolic Ca^{2+} homeostasis and preserves germination rates of maize pollen under H_2O_2-induced oxidative stress. *Sci. Rep.*, 2019, 9(1), 1–9. https://doi.org/10.1038/s41598-019-49760-3.

Du, F., Liu, P., Wang, K., Yang, Z., Wang, L. Influence of different arsenic species on uptake, speciation and efflux of arsenic in hydroponic rice plants. *Ecotoxicol. Environ. Saf.*, 2019. 186. https://doi.org/10.1016/j.ecoenv.2019.109791.

Duman, F., Ozturk, F. Aydin, Z. Biological responses of duckweed (*Lemna minor* L.) exposed to the inorganic arsenic species As(III) and As(V): Effects of concentration and duration of exposure. *Ecotoxicology*, 2010, 19(5), 983–993. https://doi.org/10.1007/s10646-010-0480-5

Dusenge, M. E., Duarte, A. G., Way, D. A. Plant carbon metabolism and climate change: Elevated CO_2 and temperature impacts on photosynthesis, photorespiration and respiration. *New. Phytologist.*, 2019, 221, 32–49.

Dwivedi, S., Tripathi, P., Kumar, A., Dave, R., Mishra, S., Singh, R., Sharma, D., Rai, U. N., Chakrabarty, D., Trivedi, P. K., Adhikari, B., Bag, M. K., Dhankher, O. P., Tuli, R. Arsenate exposure affects amino acids, mineral nutrient status and antioxidants in rice (*Oryza sativa* L.) genotypes. *Environ. Sci. Technol.*, 2010, 44, 9542–9549.

El-Ramady, H., Abdalla, N., Alshaal, T., El-Henawy, A., Faizy, S. E. D., Shams, M. S., Shalaby, T., Bayoumi, Y., Elhawat, N., Shehata, S., Sztrik, A. Selenium and its role in higher plants. In: Lichtfouse, E., Schwarzbauer, J., Robert, D. (Eds.), *Pollutants in Buildings, Water and Living Organisms. Environmental Chemistry for a Sustainable World* (pp. 235–296). Springer: Cham, 2015. https://doi.org/10.1007/978-3-319-19276-5_6.

El-Ramady, H., Abdalla, N., Taha, H. S., Alshaal, T., El-Henawy, A., Faizy, S. E. D., Shams, M. S., Youssef, S. M., Shalaby, T., Bayoumi, Y., Elhawat, N. Selenium and nano-selenium in plant nutrition. *Environ. Chem. Lett.*, 2016, 14(1), 123–147. https://doi.org/10.1007/s10311-015-0535-1.

Erb, M., Kliebenstein, D. J. Plant secondary metabolites as defenses, regulators, and primary metabolites: The blurred functional trichotomy. *Plant Physiol.*, 2020, 184, 39–52.

Etteieb, S., Magdouli, S., Zolfaghari, M., Brar, S. Monitoring and analysis of selenium as an emerging contaminant in mining industry: A critical review. *Sci. Total Environ.*, 2020, 698, 134339. https://doi.org/10.1016/j.scitotenv.2019.134339.

Farooq, M. A., Islam, F., Ali, B., Najeeb, U., Mao, B., Gill, R. A., Yan, G., Siddique, K. H. M., Zhou, W. Arsenic toxicity in plants: Cellular and molecular mechanisms of its transport and metabolism. *Environ. Exp. Bot.*, 2016, 132, 42–52. https://doi.org/10.1016/j.envexpbot.2016.08.004.

Feng, R., Wei, C., Tu, S. The roles of selenium in protecting plants against abiotic stresses. *Environ. Exp. Bot.*, 2013, 87, 58–68. https://doi.org/10.1016/j.envexpbot.2012.09.002.

Fernie, A. R., Bachem, C. W. B., Helariutta, Y., et al. Synchronization of developmental, molecular and metabolic aspects of source–sink interactions. *Nat. Plants*, 2020, 6, 55–66.

Gupta, M., Ahmad, M. A. Arsenate induced differential response in rice genotypes. Ecotoxicol. *Environ. Saf.*, 2014, 107, 46–54. https://doi.org/10.1016/j.ecoenv.2014.04.030.

Gupta, M., Gupta, S. An overview of selenium uptake, metabolism, and toxicity in plants. *Front. Plant Sci.*, 2017, 7, 2074. https://doi.org/10.3389/fpls.2016.02074.

Hajiboland, R., Rahmat, S., Aliasgharzad, N., Hartikainen, H. Selenium-induced enhancement in carbohydrate metabolism in nodulated alfalfa (Medicago sativa L.) as related to the glutathione redox state. *Soil. Sci. Plant Nutr.*, 2015, 61(4), 676–687. https://doi.org/10.1080/00380768.2015.1032181.

Hanson, B., Garifullina, G. F., Lindblom, S. D., Wangeline, A., Ackley, A., Kramer, K., Norton, A. P., Lawrence, C. B., Pilon-Smits, E. A. Selenium accumulation protects Brassica juncea from invertebrate herbivory and fungal infection. *New Phytol.*, 2003, 159(2), 461–469. https://doi.org/10.1046/j.1469-8137.2003.00786.x.

Hasanuzzaman, M., Bhuyan, M. B., Raza, A., Hawrylak-Nowak, B., Matraszek-Gawron, R., Al Mahmud, J., Nahar, K., Fujita, M. Selenium in plants: Boon or bane? *Environ. Exp. Bot.*, 2020, 178, 104170. https://doi.org/10.1016/j.envexpbot.2020.104170.

Hasanuzzaman, M., Hossain, M. A., Fujita, M. Selenium in higher plants: Physiological role, antioxidant metabolism and abiotic stress tolerance. *J. Plant Sci.*, 2010, 5(4), 354–375. https://doi.org/10.3923/jps.2010.354.375.

Hawrylak-Nowak, B., Matraszek, R., Pogorzelec, M. The dual effects of two inorganic selenium forms on the growth, selected physiological parameters and macronutrients accumulation in cucumber plants. *Acta Physiol. Plant.*, 2015, 37(2), 1–13. https://doi.org/10.1007/s11738-015-1788-9.

Hemmati, M., Delkhosh, B., Rad, A. H. S., Mohammadi, G. N. Effect of the application of foliar selenium on canola cultivars as influenced by different irrigation regimes. *J. Agric. Sci.*, 2019, 25(3), 309–318. https://doi.org/10.15832/ankutbd.424899.

Hossain, A., Skalicky, M., Brestic, M., Maitra, S., Sarkar, S., Ahmad, Z., Vemuri, H., Garai, S., Mondal, M., Bhatt, R., Kumar, P. Selenium biofortification: Roles, mechanisms, responses and prospects. *Molecules*, 2021, 26(4), 881. https://doi.org/10.3390/molecules26040881.

Jiang, C., Zu, C., Shen, J., Shao, F., Li, T. Effects of selenium on the growth and photosynthetic characteristics of flue-cured tobacco (Nicotiana tabacum L.). *Acta Soc. Bot. Pol.*, 2015, 84, 1. https://doi.org/10.5586/asbp.2015.006.

Kalita, J., Pradhan, A. K., Shandilya, Z. M., Tanti, B. Arsenic stress responses and tolerance in rice: Physiological, cellular and molecular approaches. *Rice Sci.*, 2018, 25, 235–249. https://doi.org/10.1016/j.rsci.2018.06.007.

Kamran, M., Parveen, A., Ahmar, S., Malik, Z., Hussain, S., Chattha, M. S., Saleem, M. H., Adil, M., Heidari, P., Chen, J. T. An overview of hazardous impacts of soil salinity in crops, tolerance mechanisms, and amelioration through selenium supplementation. *Int. J. Mol. Sci.*, 2019, 21(1), 148. https://doi.org/10.3390/ijms21010148.

Kieliszek, M. Selenium–fascinating microelement, properties and sources in food. *Molecules*, 2019, 24(7), 1298. https://doi.org/10.3390/molecules24071298.

Kofronova, M., Hrdinová A., Maskova, P., Soudek, P., Tremlova, J., Pinkas, D., Lipavska, H. Strong antioxidant capacity of horseradish hairy root cultures under arsenic stress indicates the possible use of Armoracia rusticana plants for phytoremediation. *Ecotoxicol. Environ. Saf.*, 2019, 174, 295–304. https://doi.org/10.1016/j.ecoenv.2019.02.02.

Kumar, S., Dubey, R. S., Tripathi, R. D., Chakrabarty, D., Trivedi, K. P. Omics and biotechnology of arsenic stress and detoxification in plants: Current updates and prospective. *Environ. Int.*, 2015, 74, 221–230. https://doi.org/10.1016/j.envint.2014.10.019.

Lanza, M. G. D. B., Dos Reis, A. R. Roles of selenium in mineral plant nutrition: ROS scavenging responses against abiotic stresses. *Plant Physiol. Biochem.*, 2021, 164, 27–43. https://doi.org/10.1016/j.plaphy.2021.04.026.

Lara, T. S., de Lima Lessa, J. H., de Souza, K. R. D., Corguinha, A. P. B., Martins, F. A. D., Lopes, G., Guilherme, L. R. G. Selenium biofortification of wheat grain via foliar application and its effect on plant metabolism. *J. Food Compos. Anal.*, 2019, 81, 10–18. https://doi.org/10.1016/j.jfca.2019.05.002.

Li, H. F., McGrath, S. P., Zhao, F. J. Selenium uptake, translocation and speciation in wheat supplied with selenite or selenite. *New. Phytol.,* 2008, 178, 92–102. https://doi.org/10.1111/j.1469-8137.2007.02343.x.

Liu, C., Zhou, G., Qin, H., Guan, Y., Wang, T., Ni, W., Xie, H., Xing, Y., Tian, G., Lyu, M., Liu, J. Metabolomics combined with physiology and transcriptomics reveal key metabolic pathway responses in apple plants exposure to different selenium concentrations. *J. Hazard. Mater.*, 2024, 464, 132953. https://doi.org/10.1016/j.jhazmat.2023.132953.

Lwalaba, J. L. W., Zvobgo, G., Mwamba, T. M., et al. High accumulation of phenolics and amino acids confers tolerance to the combined stress of cobalt and copper in barley (*Hordeum vulagare*). *Plant Physiol. Biochem.,* 2020, 155, 927–937.

Ma, L., Zeng, J., Qi Zhang, R., Wang, L., Zhang, F., Zhao, X., Yuan, Y., Li, L. Integrated transcriptomic and metabolomic analysis the variation of rice cultivars response to arsenite stress. *Environ. Technol. Innov.*, 2023, 31, 103207.

Malorgio, F., Diaz, K. E., Ferrante, A., Mensuali-Sodi, A., Pezzarossa, B. Effects of selenium addition on minimally processed leafy vegetables grown in a floating system. *J. Sci. Food. Agric.,* 2009, 89(13), 2243–2251. https://doi.org/10.1002/jsfa.3714.

Mascher, R., Lippmann, B., Holzinger, S., Bergmann, H. Arsenate toxicity: Effects on oxidative stress response molecules and enzymes in red clover plants. *Plant. Sci.*, 2002, 163, 961–969.

Mata-Ramírez, D., Serna-Saldívar, S. O., Antunes-Ricardo, M. Enhancement of anti-inflammatory and antioxidant metabolites in soybean (Glycine max) calluses subjected to selenium or UV-light stresses. *Sci. Hortic.*, 2019, 257, 108669. https://doi.org/10.1016/j.scienta.2019.108669.

Mazej, D., Osvald, J., Stibilj, V. Selenium species in leaves of chicory, dandelion, lamb's lettuce and parsley. *Food Chem.,* 2008, 107, 75–83. https://doi.org/10.1016/j.foodchem.2007.07.036.

Mehdi, Y., Hornick, J. L., Istasse, L., Dufrasne, I. Selenium in the environment, metabolism and involvement in body functions. *Molecules*, 2013, 18(3), 3292–3311. https://doi.org/10.3390/molecules18033292.

Menon, M., Sarkar, B., Hufton, J., Reynolds, C., Reina, S. V., Young, S. Do arsenic levels in rice pose a health risk to the UK population. *Ecotoxicol. Environ. Saf.*, 2020, 197, 110601.

Mishra, S., Jha, A. B., Dubey, R. S. Arsenite treatment induces oxidative stress, upregulates antioxidant system, and causes phytochelatin synthesis in rice seedlings. *Protoplasma*, 2011, 248, 565–577. https://doi.org/10.1007/s00709-010-0210-0.

Moulick, D., Ghosh, D., Mandal, J., Bhowmick, S., Mondal, D., Choudhury, S., Santra, S. C., Vithanage, M., Biswas, J. K. A cumulative assessment of plant growth stages and selenium supplementation on arsenic and micronutrients accumulation in rice grains. *J. Clean. Prod.,* 2023, 386, 135764. https://doi.org/10.1016/j.jclepro.2022.135764.

Mroczek-Zdyrska, M., Wójcik, M. The influence of selenium on root growth and oxidative stress induced by lead in Vicia faba L. minor plants. *Biol. Trace Elem. Res.*, 2012, 147(1), 320–328. https://doi.org/10.1007/s12011-011-9292-6.

Pavlík, D., Staszkova, L., Neuberg, M., Kaliszova, R., Pavlı, M., Tlustos, P. The effect of arsenic contamination on amino acids metabolism in Spinacia oleracea L. *Ecotoxicol. Environ. Saf.*, 2010, 73, 1309–1313. https://doi.org/10.1016/j.ecoenv.2010.07.008.

Puccinelli, M., Malorgio, F., Incrocci, L., Rosellini, I., Pezzarossa, B. Effects of individual and simultaneous selenium and iodine biofortification of baby-leaf lettuce plants grown in two different hydroponic systems. *Horticulturae,* 2021, 7(12), 590. https://doi.org/10.3390/horticulturae7120590.

Sattar, A., Cheema, M. A., Sher, A., Ijaz, M., Ul-Allah, S., Nawaz, A., Abbas, T., Ali, Q. Physiological and biochemical attributes of bread wheat (*Triticum aestivum* L.) seedlings are influenced by foliar application of silicon and selenium under water deficit. *Acta Physiol. Plant.*, 2019, 41(8), 1–11. https://doi.org/10.1007/s11738-019-2938-2.

Sharma, S., Mehta, K., Narang, S., Kumar, A. Dose estimation from the exposure to radon, thoron and their progeny concentrations in the environs of Kangra area, Himachal Himalayas, India. *Groundw. Sustain. Dev.*, 2020, 11, 100409 https://doi.org/ 10.1016/j.gsd.2020.100409.

Singh, N., Ma, L. Q., Srivastava, M., Rathinasabapathi, B. Metabolic adaptations to arsenic-induced oxidative stress in Pteris vittata L and Pteris ensiformis L. *Plant Sci.*, 2006, 170, 274–282. https://doi.org/10.1016/j.plantsci.2005.08.013.

Srivastava, S., Srivastava, A. K., Suprasanna, P., D'Souza, S. F. Comparative biochemical and transcriptional profiling of two contrasting varieties of Brassica juncea L. in response to arsenic exposure reveals mechanisms of stress perception and tolerance. *J. Exp. Bot.*, 2009, 60, 3419–3431. https://doi.org/10.1093/jxb/crp181.

Stewart, M. S., Spallholz, J. E., Neldner, K. H., Pence, B. C. Selenium compounds have disparate abilities to impose oxidative stress and induce apoptosis. *Free Radic. Biol. Med.*, 1999, 26(1–2), 42–48. https://doi.org/10.1016/S0891-5849(98)00147-6.

Taíz, L., Zeiger, E. *Fisiología Vegetal, Tercer Edi.* Publicaciones de la Universitat de Jaume: Castelon de la Plana, Espanãed., 2006.

Tripathi, R. D., Tripathi, P., Dwivedi, S., Dubey, S., Chatterjee, S., Chakrabarty, D., Trivedi, P. K. Arsenomics: Omics of arsenic metabolism in plants. *Front. Physiol.*, 2012, 3, 1–14. https://doi.org/10.3389/fphys.2012.00275.

Whanger, P. D. Selenocompounds in plants and animals and their biological significance. *J. Am. Coll. Nutr.*, 2002, 21(3), 223–232. https://doi.org/10.1080/07315724.2002.10719214.

Winkel, L. H., Vriens, B., Jones, G. D., Schneider, L. S., Pilon-Smits, E., Bañuelos, G. S. Selenium cycling across soil-plant-atmosphere interfaces: A critical review. *Nutrients*, 2015, 7(6), 4199–4239. https://doi.org/10.3390/nu7064199.

Yuan, K., Wang, C., Zhang, C., Huang, Y., Wang, P., Liu, Z. Rice grains alleviate cadmium toxicity by expending glutamate and increasing manganese in the cadmium contaminated farmland. *Environ. Pollut.*, 2020, 262, 114236.

Zhang, Y., Pan, G., Chen, J., Hu, Q. Uptake and transport of selenite and selenite by soybean seedlings of two genotypes. *Plant. Soil.,* 2003, 253, 437–443. https://doi.org/10.1023/A:1024874529957.

Zhao, Y. H., Jia, X., Wang, W. K., Liu, T., Huang, S. P., Yang, M. Y. Growth under elevated air temperature alters secondary metabolites in Robinia pseudoacacia L. seedlings in Cd- and Pb-contaminated soils. *Sci. Total Environ.*, 2016, 565, 586–594.

Zhou, X., Yang, J., Kronzucker, H. J., Shi, W. Selenium biofortification and interaction with other elements in plants: A review. *Front. Plant Sci.*, 2020, 11, 586421. https://doi.org/10.3389/fpls.2020.586421.

Sharma, S., Mehta, [illegible] [illegible] [illegible] Dose estimation from the exposure of radium [illegible] and thorium [illegible] concentrations in the environs of [illegible] [illegible] [illegible]. [illegible] 2020, 11, 100[illegible]. https://doi.org/10.1016/[illegible].2020.100139.

Singh, [illegible] [illegible] Srivastava, A.K. [illegible]; [illegible]; [illegible]. Metabolic adaptations to arsenic-induced oxidative stress in *Pteris vittata* L. and *Pteris ensiformis* L. *Plant Sci.* 2006, 170, 274–282. https://doi.org/10.1016/j.plantsci.2005.08.013.

Srivastava, S.; Srivastava, A.K.; Suprasanna, P.; D'Souza, S.F. Comparative biochemical and transcriptional profiling of two contrasting varieties of *Brassica juncea* L. in response to arsenic exposure reveals mechanisms of stress perception and tolerance. *J. Exp. Bot.* 2009, 60, 3419–3431. https://doi.org/10.1093/jxb/erp181.

Stewart, [illegible] [illegible] [illegible] [illegible] [illegible] Identification of [illegible] [illegible] [illegible] [illegible] oxidative stress and [illegible] [illegible] [illegible] [illegible] 1999, [illegible]. https://doi.org/10.1[illegible].

[illegible] [illegible] [illegible] [illegible] [illegible] [illegible] [illegible] [illegible] [illegible] [illegible] [illegible] [illegible] [illegible] Espinosa, 2006.

[illegible] [illegible] [illegible] [illegible] [illegible] [illegible] [illegible] [illegible] [illegible] 2020, [illegible].

[illegible] [illegible] [illegible] [illegible] [illegible] [illegible] [illegible] [illegible] 2009, [illegible] https://doi.org/10.1[illegible].

[illegible] [illegible] [illegible] [illegible] [illegible] [illegible] [illegible] [illegible] [illegible] https://doi.org/[illegible].

[illegible] [illegible] [illegible] [illegible] [illegible] [illegible] [illegible] [illegible] 2019, 268, 113[illegible].

Zhang, [illegible] [illegible] [illegible] [illegible] [illegible] [illegible] [illegible] [illegible].

Zhao, [illegible] [illegible] [illegible] [illegible] [illegible] [illegible] [illegible] [illegible] [illegible] [illegible] 2010, [illegible].

Zhou, [illegible] [illegible] [illegible] [illegible] [illegible] [illegible] [illegible] [illegible] [illegible] [illegible] https://doi.org/10.[illegible].

CHAPTER 2

Metabolomics for Heavy Metal Stress Responses and Tolerance in Plants

HEND FOUAD[1,2*], RASHA FOUAD[1], PETER JACKSON[2], and ELSAYED A. OMER[1]

[1]*Medicinal and Aromatic Plants Research Department, National Research Centre, Giza, Egypt*

[2]*Institute for Sustainable Food, University of Sheffield, Western Bank, Sheffield, United Kingdom*

*Corresponding authors

ABSTRACT

Toxic heavy metals bioavailability and groundwater leaching have gradually increased during the past few years, mainly as an outcome of anthropogenic invasions. Heavy metals, particularly those classified as hazardous or cancer-causing nonessential ones, such as cadmium (Cd), lead (Pb), arsenic (As), mercury (Hg), and aluminum (Al) have emerged globally as the main soil, air, and water contaminants influencing plant growth, productivity, quality, and safety. A correlation exists between plant stress or toxicity and metal pollution of water and soil ecosystems. As a rule, plants are extremely sensitive to toxic metals and may experience a range of physiological symptoms from even mild heavy metal stress. Oxidative stress is brought on by heavy metals interfering with specific protein pathways and plant metabolism. Latest advances in different biological sciences like metabolomics, proteomics, and transcriptomics have helped within the metabolite characterization, stress-inducible proteins, and transcription factors associated with heavy metals tolerance that can be exploited for developing heavy metal-tolerant plants. This chapter covers metabolomics responses as a tolerance strategy of stressed plants that are subjected to heavy metals.

2.1 INTRODUCTION

Biotic and abiotic stress negatively impact a plant's advancement and growth, which reduces the plant's production. On the other hand, abiotic factors, including heavy metals (HMs), water, saline, heat, cold, pH of the soil, UV radiation, and ROS (reactive oxygen species)

Exploring Plant Stress Tolerance: A Metabolomics Approach.
Shuvasish Choudhury, Debojyoti Moulick, and Mirza Hasanuzzaman (Eds.)

DOI: 10.1201/9781779641731-2

are highly damaging and reduce crop output globally (Salam et al., 2023). Industrial waste and sewage disposal have long left HM deposits in soils (Ghori et al., 2019). Excessive HM build-up in plants' leaves, stems, and roots, as they are the essential producers in the food chain. This not only significantly impedes the growth as well as advancement of plants but also puts animals and people at risk by way of the food chain (Järup and Akesson, 2009; Anjum et al., 2016). Tissues of human or animal are mostly exposed to HMs from chains of tainted food, which puts them at risk for an assortment of illnesses from dermatitis to different kinds of cancer (McLaughlin et al., 1999).

Many micronutrients metals (Cd, Se, Co, Ag, Hg, Cr, Pb, Sb, and As) may not be important for plant growth or may have unidentified roles in plant systems, but some (Fe, Cu, Mg, Ni, Mn, and Zn) might be significant for plants to grow and develop normally (Rascio and Navari-Izzo, 2011). However, an excess of these elements can have negative consequences and directly affect plant development, metabolism, physiology, and senescence, numerous typical plant processes depend on several of these metals, which are essential micronutrients (Ghori et al., 2019).

Many different studies have reported that complicated intricate systems at the molecular, biochemical, physiological, cellular, tissue, and whole plant level are needed to balance the toxicity caused by HMs, which may lead to increased crop productivity (Singh et al., 2016). In addition to defending themselves against stress, plants are in charge of preserving the homeostasis of vital metals that they need to survive. These defensive mechanisms also concentrate on shielding plants from soil-borne HMs or giving them tolerance by detoxifying the metals. Additional mechanisms are unique and start working immediately after the appropriate stress is experienced (Ghori et al., 2019).

Nevertheless, it should be mentioned that a cell's regular defensive reactions may wear out if exposed to prolonged periods of trauma. Plants utilize various kinds of specialized detoxification mechanisms to remain alive during these periods. These mechanisms may involve transporting, chelating, sequestering, or detoxifying metals within the vacuole of plants. Upon activation of any of these systems by the plant, the production of stress-associated proteins and hormones, antioxidants, and signaling molecules, as heat-shock proteins, begins (Ghori et al., 2019). Plants' ability to balance harmful oxygen derivatives and their antioxidant defense mechanisms' potency, therefore, determines how tolerant they are to stress (Rastgoo and Alemzadeh, 2011).

Various adaptive strategies have evolved in plants over time to withstand harsh environmental stressors, and one crucial defensive mechanism is the metabolite accumulation (Jaitz et al., 2011; Fouad et al., 2023a). Plants contain over 200,000 types of metabolites. Amino acids, organic acids, and glycolysis metabolism were among the metabolic pathways altered by HM stress (Imtiaz et al., 2016). In toxicology, disease diagnosis, plant research, and other domains, metabolomics is a burgeoning technique that is employed for the qualitative and quantitative analysis of all metabolites with low molecular weight. It has grown in significance as a part of systems biology (Feng et al., 2021). The characterization of stress-inducible proteins, metabolites, and transcription factors related to the tolerance of HMs has been made easier by recent advancements in different biological science disciplines, including proteomics, transcriptomics, metabolomics, and so on. This information can be employed to generate crops that tolerate HMs (Singh et al., 2016).

Though not sufficient to completely resolve the disastrous issue associated with excess HMs, current developments in research clearly suggest that the potential harm that HMs could cause to plants and crops has been successfully mitigated by exogenous application (Feng et al., 2023a). It is crucial to investigate the mechanisms underlying crop damage and HMs resistance. The metabolomics (primary and secondary metabolites) function in preventing the toxicity of HMs is outlined in this chapter.

2.2 HEAVY METALS STRESS

Despite the fact that Earth's crust contains HMs at varying degrees naturally, an issue arises when these elements are released into excess and become toxic to living organisms (Miransari, 2011; Singh et al., 2016). Unlike other organic pollutants, these metals do not undergo biological degradation or conversion into innocuous compounds (Gisbert et al., 2003). Examples of naturally occurring HM sources are the weathering of rocks containing metal and lava flows. Industrial emissions, mining, smelting, and agricultural practices, like fertilizers with phosphate and pesticides, are examples of major human sources (Figure 2.1). Fossil fuel combustion also plays a role in the environmental release of HMs like cadmium (Cd) (Spiegel, 2002).

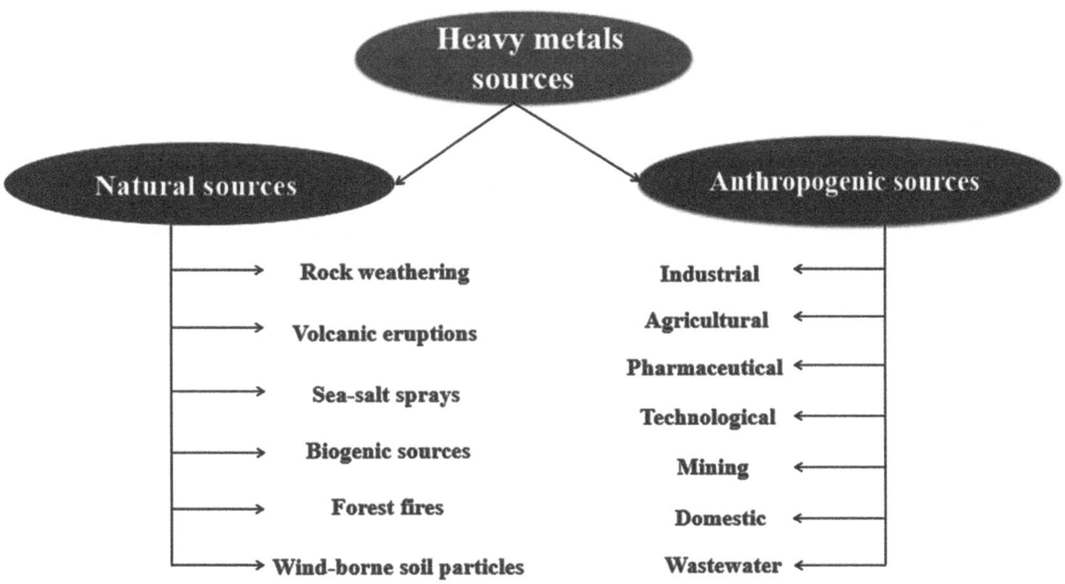

FIGURE 2.1 Different sources of HMs.

According to chemical definitions, metals are "elements, which have basic oxides, metallic luster, are malleable and ductile, conduct electricity" (Atkins and Jones, 1997). In environmental and biological studies, the terms "metal," "metalloid," "semimetal," "light metal," "heavy metal," "essential metal," "beneficial metal," "toxic metal," "available metal," "trace metal," and "micronutrient" are frequently used in reference to metals (Duffus, 2002). A metal which naturally has an atomic number greater than 20 and density more than 5 g/cm^3 (specific gravity greater than 5) is known as an HM (Ali and Khan, 2018). The term HM is frequently used as a collective term for metals and semimetals (metalloids) that have been linked to contamination and possible toxicity or ecotoxicity (Duffus, 2002).

HMs have two distinct roles in biological systems: essential and nonessential. Essential HMs are vital to life and may be needed in very small amounts within the body. There is no identified biological role for nonessential HMs in life organisms. Some examples of essential HMs are Fe, Zn, Cu, and Mn, while toxic and biologically nonessential HMs are Hg, Pb, and Cd (Türkmen et al., 2009; Jović et al., 2012; Ramírez, 2013; Rahim et al., 2016). The HMs that are trace elements or micronutrients for plants are Fe, Zn, Cu, Mn, Ni, Co, and Mo. They are very important for various biomolecules, including carbohydrates, chlorophyll, nucleic acids, growth chemicals, and secondary metabolites, as well as for growth and stress tolerance (Appenroth, 2010). The most dangerous metalloids and HMs that are relevant to the environment include As, Cr, Hg, Ni, Zn, Cu, Cd, and Pb (Barakat, 2011). The HMs Cr, Mn, Ni, Zn, Cu, Cd, and Pb are the most prevalent in the environment (Khan et al., 2011). Under the "12th 5-year plan for comprehensive prevention and control of HM pollution," China proposed in 2009 that the highest priority pollutants for control are four metals: Cd, Cr, Hg, Pb, and the metalloid As (Fu et al., 2017).

In the environment, HMs are considered as dangerous chemicals. Even at extremely low concentrations, unnecessary HMs are harmful to humans, animals, and plants. At high concentrations, even the main HMs have detrimental impacts on health (Mahboob et al., 2014). HMs have been discovered in vegetables that are irrigated with wastewater in greater concentration ranges than those irrigated with groundwater. Moreover, leafy vegetables have been found to contain higher concentrations of these metals than other vegetable varieties like tubers and bulbs (Mahmood and Malik, 2014).

Except for Cd, which does not appear to possess a specific transporter protein, the specific transporter proteins usually allow HMs to enter roots through the plasma membrane. From there, they travel via coplanar or plastid extracellular pathways to reach the xylem or phloem (Zhao et al., 2022). The toxicity of HMs in plants can occur in four different ways. These include (1) absorption competition at the root surface due to similarities with nutrient cations; Zn and P compete with Cd and As, respectively; (2) direct interaction between functional proteins' sulfhydryl group (–SH) and HMs, which causes structural and functional disruption and inactivation; (3) the removal of necessary cations from particular binding sites, which causes a function to collapse; and (4) ROS production, which causes ruin to the macromolecules (Sharma and Dietz, 2009; DalCorso et al., 2013).

Because HMs cause intense symptoms of toxicity in plants at higher concentrations, plant cells strictly regulate the absorption and use of these metals (Farias et al., 2013). HMs (nonessential and essential) usually cause similar detrimental effects on plants that can ultimately result in their death. These consequences comprise chlorosis, modified absorption of nutrients and water equilibrium, biomass reduction, photosynthesis and growth inhibition, and senescence (Singh et al., 2016).

Furthermore, enzyme functions and metabolic pathways can be hampered by excessive HM concentrations (Ali et al., 2013). A rise in HM concentrations leads to the production of cytotoxic substances like methylglyoxal that lead to oxidative stress and ROS or nonfree radical species like singlet oxygen and H_2O_2 (hydrogen peroxide) (Hossain et al., 2012; Sytar et al., 2013). According to Rascio and Navari-Izzo (2011), HMs stress resulted in ion leakage, redox imbalance, and damage to DNA and membranes, which results in programmed cell death (Rascio and Navari-Izzo, 2011).

2.3 THE RESPONSE OF PLANTS TO HEAVY METAL STRESS

It is interesting to note that plants are bio-accumulators naturally, which means that they concentrate various HMs that they extract from the water and soil, whether or not they are necessary for their healthy growth (Ozturk et al., 2017). Different plant species have different rates of accumulation and tolerance to HMs; some develop toxicity more quickly than others. According to Ozturk et al. (2008), plants under HMs toxicity exhibit visible symptoms like chlorosis, browning of the roots, growth reduction, decline, and death. Many significant proteins and enzymes are inactivated or denatured by HMs stress, which also obstructs the processes that replace lost metal ions in biomolecules. Basic plant metabolic processes like photosynthesis, respiration, and homeostasis are changed as a consequence of this reaction, which compromises the membrane's integrity (Hossain et al., 2012).

Additionally, it promotes the ROS synthesis, including hydroxyl radical (OH), superoxide radical (O^{2-}), and H_2O_2. Methylglyoxa (cytotoxic compound), during HMs stress, is increased. Oxidative stress is caused by these extremely reactive species and results in lipid peroxidation, particularly of cell membranes, which causes them to leak, damage to biomolecules, and cleavage of DNA strands (Hossain et al., 2010; Ahmad et al., 2012).

Plants react to the toxicity of HM in the same ways as they do to other external stress stimuli. These mechanisms include (1) sensing the external stressor, (2) signal transduction and signal transmission into the cell, and (3) initiating appropriate responses to counterbalance the negative effects of stressors by modifying the physiological, biochemical, and molecular status of the cell. Measuring sensing and changes in signal transmission in plants exposed to HM stress is challenging at the whole plant level. On the other hand, keeping an eye out for early reactions like oxidative stress, changes in transcriptome and proteome, or metabolite accumulation, may be helpful (Singh et al., 2016).

2.3.1 RESTRICTING THE ABSORPTION OF HEAVY METALS BY ROOTS

Typically, plants possess multiple defense mechanisms to withstand distinct metal stresses that operate at various locations throughout the plant. By isolating and excluding certain metal ions in chambers and preventing their exposure to other cellular organelles, these mechanisms work against specific metal ions. Adaptive mechanisms, in contrast to constitutive mechanisms, only become active when stress is experienced. Limiting the metal ions' absorption by roots is a plant's first line of defense when it comes to stress. The majority of metals get in roots through either generic or specific energy-driven ion channels or pumps. Plants restrict HMs uptake by altering the rhizosphere's pH, which causes HMs to precipitate. The different root exudates that concentrate these metal ions to apoplast and prevent them from permeating into the cell are the reason of the pH variations.

Metal ions are actively transported from symplastic to apoplastic space by carrier protein and transporting HMs. Numerous studies demonstrate that plants that secrete root exudates, such as oxalate and malate, limited the Cd absorption from the soil (Di Toppi and Gabbrielli, 1999; Pinto et al., 2008). It has been found that resistant plant varieties have evolved a more effective secretion system and generate significant amounts of particular root exudates, such as citric acid, histidine, nitric oxide (NO), malic acid (Huang et al., 1996), oxygen and oxidants (Chen et al., 1980), and so on (Kramer et al., 1996 and Hakeem et al., 2017). Root exudates often bind or attach metal ions to cell walls to prevent their migration, especially over long distances (Ghori et al., 2019).

2.3.2 METALS' DISTRIBUTION TO DIFFERENT PLANT ORGANS

Comparing the tolerant *Vigna unguiculata* variety to the sensitive variety, which displayed brown spots in the areas where manganese was locally accumulated, Mn is uniformly distributed throughout the leaves of the tolerant variety (Horst, 1983). In the event that metal ions have the ability to penetrate plant cells, the plant will activate additional mechanisms of defense to deal with stress. When metal ions enter a cell, they are often kept apart from other cellular organelles in a distinct compartment within the cytosol. One of the main locations for metal storage inside plant cells, these compartments are frequently the central vacuole. These metal ions are pumped inside the vacuole by two proton pumps called vacuolar proton-ATPase (V-ATPase) and vacuolar proton pyrophosphatase (V-Ppase). Peptides and other molecules that bind to the metals to sequester and detoxify them appear to be present in the vacuole.

Metal-binding ligands in the cytosol chelate metals upon entry as well. According to Gupta et al. (2013), plant cells produce two different kinds of metal-binding ligands: phytochelatins and metallo-thioneins. Many metals stimulate the phytochelatins synthesis, but Cd appears to be the most potent. Contrarily, metalochellatins are low molecular weight polypeptides with residues rich in cysteine that bind to metals through their thiol functional group. There are different isoforms of metallochelatins that can combine with different metal ions to form complexes.

Additionally, phytate which is primarily present in seeds may act as a complexing agent for HMs within cells. Furthermore, the concentration of HMs is used to regulate the phytate concentration. Most plants typically limit the metal ions' absorption in order to detoxify them via their roots or sequester them in internal cell compartments; however, hyperaccumulators (a unique class of plants) effectively translocate HMs from roots to shoots via xylem. These plants also tend to have a higher HM tolerance. HMs are effectively transported through xylem by the presence of HM transporters, nicotinamines, and organic chelators in the cell membrane of shoot (Hossain et al., 2012). These plants also contain enzymes that can change metals and incorporate them into different oxidative states or organic compounds (Yang et al., 2005).

2.4 SIGNAL TRANSDUCTION PATHWAY IN PLANTS DUE TO THE STRESS OF HEAVY METALS

Plants have defense mechanisms that are triggered in times of stress to ensure their survival. A sophisticated transmission of signal networks that is determined by various signaling molecules and the stress-related proteins synthesis and activation is immediately triggered upon detection of HM stress. The signal transmission pathway usually activates the transcription factors that cause several HM stress-responsive genes' transcription. It is widely known that plants have four primary common signal transmission pathways that are able to correspond with one another: signaling based on phosphorylation/dephosphorylation of proteins by kinase cascades, plant hormone signaling, reactive species signaling, and calcium-dependent signaling (Tiwari and Lata, 2018).

2.4.1 ROS AND OXIDATIVE STRESS

The generation of ROS is one of the most common and well-studied plant responses to HM stress (Jalmi et al., 2018). Reactive oxygen species, also known as superoxide radicals, hydroxyl radicals, and H_2O_2, are highly reactive intermediate products that are released under the stress of HMs. These free radicals, also known as ROS, start a chain reaction that breaks down proteins and lipids and causes significant damage to DNA and cell membranes. ROS have been regarded as dangerous species for a long time, but they are not just dangerous. In actuality, they serve as markers for specific stages of the plant's development and are crucial to plant defense. As stated by Yuan et al. (2013), ROS are also produced by the cell under normal circumstances and are important regulators of many physiological processes, as germination, senescence, and stomatal closure.

ROS's small size and short diffusion distance enable them to function as signaling molecules and control the process of gene expression. Recently, it has been discovered that the temporal–spatial coordination of ROS and other signals causes stress responses in plants. According to recent research, ROS and other signals coordinate temporally and spatially to cause stress responses in plants. Recently, it has been discovered that the temporal–spatial coordination of ROS and other signals generated by chemicals, hormones, and stress-dependent chemicals is what causes stress responses in plants (Baxter et al., 2014). Due to these species' adaptability, the plant maintains tight control over them using enzymes like superoxide dismutase, catalase, glutathione reductase, peroxidase, and ascorbate peroxidase as well as antioxidants like ascorbic acid, tocopherol, and glutathione (Sytar et al., 2013). Since H_2O_2 is the only reactive oxygen species that can pass through the plasma membrane, it is believed to play a direct role in cell-to-cell communication (Pitzschke et al., 2006).

2.4.2 CALCIUM-DEPENDENT SIGNALING

Based on multiple studies, calcium (Ca^{2+}) functions as a comprehensive secondary messenger in plant normalcy and in response to different stress types (Sanders et al., 2002).

Various stresses cause changes in the cytosolic free Ca^{2+} level, which sets off intricate reactions and pathways of signal transmission (Rudd and Franklin-Tong, 2001). The highly sensitive calcium-sensing proteins sense this transitory rise in cytosolic concentration and translate it into a response (biological reaction).

2.4.3 THE CALCIUM–CALMODULIN PATHWAY

When plants are exposed to HM stress, calcium flux into the cell is increased and calcium channel activity is altered. One more messenger that affects the absorption of HM, transport, and metabolism is calcium. It does this by stimulating the protein calmodulin. Numerous metals have been shown to target calmodulin in their cells, as evidenced by studies on the Ca–calmodulin pathway in Cd, Ni, and Pb stress. As per Ouyang and Vogel (1998), Pb has the capacity to bind to each of calmodulin's four calcium-binding sites simultaneously, thereby stimulating its activity to in excess of 90%.

Similar to this, the ions Mg, K, Na, and Zn either interact with calmodulin via auxiliary sites or bind to specific calcium-binding sites on it with low affinity, resulting in different effects on calmodulin functions. According to Snedden and Fromm (2001), a majority of metals function as calcium analogs, causing calmodulin to trigger signal transduction. The calcium concentration has also been linked to NO, which is said to play a role in raising the cytosolic Ca^{2+} concentration as a result of abiotic stressors like HMs. Increased calcium levels have a synergistic effect in the control of abiotic stress because they alter the nitric acid concentration and induce biochemical and physiological alterations in the cell (Khan et al., 2014).

2.4.4 PLANT HORMONE SIGNALING

Two approaches have been applied to enhance stress tolerance in plants and yield through the action of plant hormones. These tactics include genetically modifying the endogenous contents of plants and applying plant hormones exogenously. According to Vriet et al. (2012), both strategies have shown promise in raising crop and improving stress tolerance in a range of crop species. The high cost of synthetic brassinosteroids (BRs) and the unpredictability of the results have discouraged the use of exogenous BRs in agriculture and horticulture, despite the fact that both gibberellins (GAs) and SA (salicylic acid) are readily available and cost-effective for their exogenous application in crop fields under stress (Khripach et al., 2000; Gomes, 2011). In this scenario, genetic engineering-induced modulation of endogenous BRs levels has been demonstrated as an effective strategy for improving crop yield in both favorable and unfavorable growth conditions (Divi and Krishna, 2009). Here, we have given a summary of current advancements in the application of GA, BRs, and SA to increase the tolerance for HMs and achieve high yield with desired agronomic traits.

2.4.5 PHOSPHORYLATION/DEPHOSPHORYLATION OF PROTEINS BY KINASE CASCADES

Distinct antioxidant enzymes are expressed due to distinct transcriptional factors being activated through various intermediate molecules' HM stress signals (Kaur et al., 2021). Among these, mitogen-activated protein kinases (MAPKs) catalyze protein phosphorylation, a crucial signal transduction pathway in plants. Among the most important and widely conserved signaling molecules in eukaryotes are MAPKs. They work in tandem with the sensor/receptor to facilitate normal cellular responses necessary for the organism's advancement and growth (Sharma et al., 2020). Based on sequence similarity and the distinctive TXY activation motif, MAPKs are easily recognized. They facilitate the transmission of signals produced by ligand–receptor interactions with downstream substrates (Komis et al., 2018).

Cultivated plants respond under HM stress by activating the MAPK cascade. Jonak et al. (2002) reported that this cascade is important for the signal transmission pathway activation involved in hormone synthesis. Three kinases—MAPKK (MAPK kinase), MAPKKK (MAPK kinase kinase), and MAPK—that are activated by the phosphorylation process participate in this cascade. Substrates in cells, including transcription factors in the nucleus, are phosphorylated by the phosphorylated MAPK cascade that has finally formed. Consequently, downstream information transduction is regulated by MAPK.

According to Jonak et al. (2004), *Medicago sativa* exhibits four isoforms of MAPK that become activated as a result of Cu or Cd stress. At the end of the day, all of these pathways result in the regulation of transcription factors, which then drive the expression of genes that activate metal transporters, chelating compound biosynthesis, and other defensive compounds. The stress response is triggered by the MAPK cascade, which targets different effector proteins or transcriptional factors and phosphorylates and dephosphorylates signals sensed by cell membrane surface receptors. ROS, NO, and plant hormones can all trigger the MAPK cascade by differentially expressed genes, the activation of the antioxidant system and synergistic cross-talk between various signal molecules, which in turn regulates how plants respond to HMs. Transcriptional factors, which are found downstream of MAPK are significant elements in controlling plant reactions to HMs and improving plant tolerance and accumulation of HMs. Consequently, comprehending how HMs induce the MAPK cascade pathway associated with genes expression and subsequently phosphorylate those transcriptional factors could facilitate the creation of a regulatory network that enhances our comprehension of HM accumulation and tolerance.

2.5 METABOLOMICS FOR HEAVY METAL STRESS

According to Arbona et al. (2013) and Singh et al. (2016), metabolomics is the study of identifying and measuring all low-molecular-weight metabolites that organisms need for growth, and some metabolites have been linked to HM stress tolerance techniques. First of all, metabolites are bioindicators of biochemical and physiological characteristics of plants, revealing metabolic pathways, gene expression, and signaling. Plant growth, development, and reproduction are intrinsic processes that are correlated with both primary and secondary metabolites (Figure 2.2) (Rastgoo and Alemzadeh, 2011).

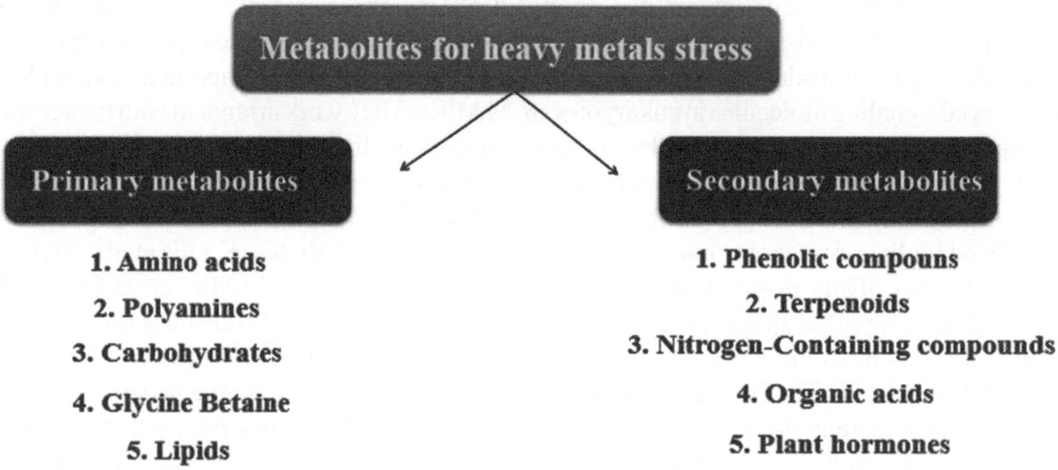

FIGURE 2.2 Primary and secondary metabolites for heavy metals stress.

In addition to causing oxygen stress, HMs obstruct certain protein functions and plant metabolism. Through a wide range of intricate biological processes, including chelation, antioxidant mechanisms, and regionalization of HM ions, plants fight excess HMs in the environment. Plant metabolomics is the science and technology used to study plant metabolic networks and associated gene functions, as well as to qualitatively and quantitatively analyze the metabolic components of plants both before and after they are exposed to external disturbances or stimuli (Rochfort, 2005; Ruan and da Silva, 2011). When an organism is under adverse stress, subtle changes in its metabolism can be sensitively reflected through metabolomics. Breeding extremely resistant plants and developing phytoremediation technologies benefit from research on HM injury and the mechanisms of resistance in plants.

2.5.1 PRIMARY METABOLOMICS

Primary metabolites are substances that have a connection to vital physiological processes in living things. They are, therefore, typically present in all plant species and play a direct role in development, growth, and reproduction (Wink, 2010). To verify their presence in harsh environmental conditions, plants developed several mechanisms of adaptation to withstand abiotic factors, such as variations in metabolism in different directions (Ashapkin et al., 2020). Many plant metabolites may help to mitigate the negative effects of severe stress (Teklić et al., 2021). GB (glycine betaine) and dimethylsulfoniopropionate are two examples of these metabolites; other examples include sugars like fructan, trehalose, and sucrose; amino acids like proline and ectoine; and certain metabolites of polyols like mannitol and sorbitol (Dikilitas et al., 2020; Kebert and Vuksanović, 2022).

2.5.1.1 AMINO ACIDS

Amino acids are thought to be a building block for proteins and other organic molecules, as nucleic acids, which play an important function in how a plant reacts to various stressors. Additionally, amino acids may be important for signaling and molecular control (Heinemann and Hildebrandt, 2021). Numerous studies revealed that abiotic stressors can affect the amino acids that plants store (Hildebrandt, 2018; Khan et al., 2021; Fouad et al., 2022). Furthermore, when plants are exposed to these kinds of stressors, proline and other amino acids accumulate (Fouad et al., 2023b).

The function played by stored amino acids in plants differs after its role as an osmolyte to modify ion passage and reduce the opening of stomata (Khan et al., 2020). Additionally, gene expression, multiple enzymes synthesis and activation, and the redox state of homeostasis can all be disrupted by amino acids (Rodziewicz et al., 2014). Since ectoine and proline build-up serve as osmo-protectants to protect plants from damaging effects and demanding environmental stresses, they are thought to be the most widespread osmolytes (Dikilitas et al., 2020; Verbruggen and Hermans, 2008). Proline is thought to carry out numerous other crucial tasks associated with plant resistance as well as osmolytes, such as redox balance, ROS scavenging, molecular chaperon, buffer cytosolic pH, and a stabilizer of protein structure (Meena et al., 2019). Over a period of years, it was observed that plants' enlarged proline levels were a stress-responsive characteristic in response to abiotic stimuli. Because proline accumulates as osmolytes and can be applied to various crops, this relationship has a significant impact on stress tolerance (Hanif et al., 2021; Rashedy et al., 2022).

According to reports, proline can build up when exposed to HMs (Yusuf et al., 2012). Siripornadulsil et al. (2002) conducted a study on microalgae and found that proline levels rose as a result of Cd stress and increasing proline levels work by reducing the process by which free radicals are created rather than sequestering electrons. It was investigated in mung beans how exogenous proline-induced tolerance against Cd stress promotes the glutathione synthesis and enzymes that metabolize glutathione (Hossain et al., 2011).

It has been found that histidine (an additional crucial amino acid) is crucial when exposed to HMs. When *Alyssum lesbiacum* (a Ni hyperaccumulator) is under Ni stress, Krämer et al. (1996) found elevated histidine levels in the xylem sap. Parallel uptake of Ni and Histidine was also noted in *Brassica juncea* by Kerkeb and Krämer (2003). Sharma and Dietz (2006) posit that alterations in the histidine composition play a crucial role in metal stress resilience.

Of all the low molecular weight tripeptide thiols present in plants, glutathione is the most prevalent. The tripeptide has a gamma peptide bond forming between cysteine and the carboxyl group of the glutamate side chain. Normal peptide linkage binds the carboxyl group of the cysteine residue to glycine. In addition to offering protection against ROS and numerous other cellular processes, it is essential for the sequestration of HMs (Freeman et al., 2004; Mullineaux and Rausch, 2005). The primary way that glutathione responds to HM stress is by regulating ROS, which is accomplished by regulating H_2O_2 levels. Glutathione oxidizes when H_2O_2 degrades, and the proportion of reduced to oxidized glutathione is frequently used as a gauge for redox balance and the ROS generation in cells. During oxidative stress, the reduced form of glutathione has a direct impact on the breakdown of free radical species (Foyer and Noctor, 2005; Shao et al., 2008). It also plays a role in shielding the cell from reactive metal exposure and the detoxifying effects of HMs.

2.5.1.2 POLYAMINES

All living things contain polyamines, which are organic cations with a low molecular weight. The most prevalent polyamines in plants are putrescine, spermidine, and spermine. They are categorized as second messengers and plant growth regulators, and they are necessary for many stages of plant development and growth. By scavenging ROS, they also actively participate in the mechanism that provides tolerance against stress (both abiotic and biotic) (Hossain et al., 2012).

The spermidine build-up resulting from the increased regulation of *Cucurbita ficifolia*'s spermidine synthase enhanced several stress-related reactions in a recombinant Arabidopsis plant (Capell et al., 2004; Khan et al., 2020). It has also been demonstrated that polyamines are involved in protecting membranes from harm in stressful conditions (Zhao et al., 2021), regulating nucleic acid production and enzymes' activity (Vickers, 2017). Furthermore, various results showed that polyamines are crucial in cases of oxidative harm since they either directly or indirectly control the antioxidant mechanism and reduce ROS generation, thereby mitigating the balance state of ROS. Additionally, some authors proposed that polyamines may serve in plants as a cellular signal during the response to stress (Gill and Tuteja, 2010).

2.5.1.3 CARBOHYDRATES

The primary building blocks that give the biomass of plants power and assistance are the carbohydrates created during photosynthesis (Saxena et al., 2009). Numerous investigations have demonstrated that nonliving agents cause various species of plants to accumulate nonstructural saccharides like lactose and sucrose, simple sugars, or polyhydric compounds like phenols and alcohols (Ruppert et al., 2012). In particular, osmotic stress resistance and carbohydrate accretion are strongly correlated (Khan et al., 2020). Soluble carbohydrates are a major source of energy and carbon that are used by the cells in the plant metabolism. Since the build-up of carbohydrates is connected to photosynthesis, various stressors may affect the concentration of these soluble carbohydrates (Cao et al., 2018).

By upregulating growth-related genes and down-regulating stress-related genes, soluble sugars (as hexoses and sucrose) can increase stress tolerance (Rosa et al., 2009). However, the contents of some carbohydrates, like maltose, glucose, fructose, and raffinose, increase in reaction to environmental stressors. The findings of Sperdouli and Moustakas (2012) demonstrated an increase of enhanced soluble carbohydrates within dry environmental stress, maintaining a strong antioxidant safeguard in the leaves of *Arabidopsis thaliana*.

Under stress, the carbohydrates' hydrolysis usually results in an increase in the amount of sugars, which requires the use of hydrolytic enzymes (Kaplan and Guy, 2004). Furthermore, soluble carbohydrates, including polymers of fructose molecules (fructans), oligosaccharides (raffinose and stachyose), and disaccharides (sucrose and trehalose), along with their associated metabolic enzymes, are vitally compatible osmolytes linked to the scavenging of unstable molecules (ROS) throughout their assortment in the tissues of plant (Keunen et al., 2013).

In order to give the plant tolerance, HM build-up is believed to occur primarily in the area between the root cell's plasma membrane and cell wall. Additionally, the cell wall produces callose, which prevents HMs from penetrating the cell's interior. When exposed to metal stress, plant cell walls change in different ways that help the plant survive. Compounds with functional groups (–COOH, –SH, –OH), including proteins, amino acids, carbohydrates, and phenolics are widely distributed in the cell wall. These groups bind divalent and trivalent metal cations. A critical element of cell walls that is needed for storing metal ions is pectin. All cells contain pectin, which consists of four polysaccharide domains, one of which is primarily in charge of ensnaring the metal ions (Krzesłowska, 2011). According to Renard and Jarvis (1999), the charge density of pectin, the ionic interaction's strength between pectin and cation, and the solution's ionic strength all affect the interaction and strength of the pectin and metal ion complex.

Various metals are deposited in changed conditions; for example, zinc was deposited as a silicate within *Minuaria verna* cell wall (Neumann and zur Nieden, 2001). According to Kopittke et al. (2008), Pb was also observed to be deposited in cell walls as a nontoxic form of chloropyromorphite. Additionally, it is observed in different plants that when pectin deposition rises and cell wall thickening takes place, the cell wall functions as a barrier when exposed to HMs. Thus, HMs that are extracted from the protoplast may also be sequestered by the thicker cell wall, rendering them immobile.

2.5.1.4 GLYCINE BETAINE

As an osmoprotectant, stabilizer of macromolecules, protector against photosynthesis, and detoxifier of reactive oxygen radicals, GB is an important compound (Raza et al., 2007; Chen and Murata, 2008). Numerous studies proved their significance to improve the tolerance of plants to diverse abiotic factors. Studies reveal that some plants, such as spinach, barley, wheat, and maize can be recognized by the quantity of GB they generate and store in their chloroplasts. Nevertheless, some plant species, like tobacco and rice, are not able to obtain significant amounts of GB under stress (Giri, 2011).

According to earlier research, GB can lessen oxidative stress and promote root growth. Furthermore, chickpea plants' tolerance to Cr stress is enhanced by the external application of GB (Singh et al., 2022). Therefore, it is important to take into account extra safeguards for GB, such as defense against ROS and stress from HMs, as these could raise the tolerance level (Emamverdian et al., 2015).

2.5.1.5 LIPIDS

Biological membranes are primarily composed of lipids, especially the plasma membrane, which serves as a wall separating the cell from its environment (Hou et al., 2016). Based on their chemical structure and unique hydrophobic and hydrophilic components, lipids are categorized into eight major types: fatty acids, glycerides, phosphoglycerides, sphingolipids, steroids, isoprenoids, glycolipids, and polyketides (Fahy et al., 2005). Lipid-mediated signaling is the plant's reaction to various stressors, including abiotic and biotic elements like temperature, drought, HMs, salinity, and pathogen invasion. Plants use the plasma membrane, which is typically the origin of lipid signals, to sense these stimuli and translate them into subsequently biochemical metabolism.

Additionally, it has been demonstrated that plants' fatty acids are lipids that react to stress. In *Arabidopsis*, oleic acids reduce tolerance by altering NO-related proteins, that consequently controls NO (Mandal et al., 2012). The most potent form of vitamin E, α-tocopherol, is produced in the plastids of higher plants. By quenching 1 O_2 in the chloroplast, it is discovered to have a role in scavenging lipid peroxides and ROS (Munne-Bosch, 2005), preventing cell membrane damage under stress. Numerous studies have demonstrated variations in α-tocopherol levels under HM stress (Collin et al., 2008; Yusuf et al., 2010; Kumar et al., 2012; Lushchak and Semchuk, 2012). An upregulation of genes linked to α-tocopherol's biosynthesis was proposed by Collin et al. (2008), which found that *Arabidopsis* treated with Cd had greater levels of the plant's vitamin.

2.5.2 SECONDARY METABOLOMICS

Secondary metabolites improve plant survival under abiotic and biotic stressors but have less of an impact on the fundamental plant processes than primary metabolites (Rastgoo and Alemzadeh, 2011). In cells and tissues of plants, secondary metabolites yield various compounds that are crucial for many biochemical and biophysical processes. Nevertheless, they lack the shared physiological processes that are known to occur in plants, like respiration, photosynthesis, solute translocation, nutrient acclimatization, and differentiation (Divekar et al., 2022). Typically, they are produced under specific circumstances with a specific goal in mind, like improved resistance to abiotic stresses and defense against pathogen infection (Debona et al., 2017).

Furthermore, the designated plant species that produced these natural products, and the growth period, environments, and degree of adjustment all influence these products' concentration (Ochoa-Villarreal et al., 2016). Current studies aim to determine the important functions that secondary metabolites perform in plants, as antioxidants and indicators (Pagare et al., 2015). Phenolics, terpenes, and compounds comprising S and N are the three main secondary metabolite groups (Grace, 2005; Yazici, 2022).

2.5.2.1 PHENOLICS

Phenolics are known to be the most prevalent and important secondary metabolites in plants, transitioning from simpler aromatic rings to more complex ones like lignin. They are also known to have a major physiological impact on the plant's ability to adapt and increase resistance to unsuitable environmental circumstances throughout its life cycle (Waterman, 2007; Kabera et al., 2014). Phenolic compounds are produced from plants in both optimum and suboptimal environments, and they are very important to many developmental mechanisms, including cell division, hormone balancing, photosynthetic processes, reproduction, and nutrient mineralization (Sharma et al., 2019). These substances, which include flavonoids, anthocyanins, tannins, lignins, isoflavonoids, and coumarins, are classified as secondary metabolites (Saltveit, 2017).

Furthermore, the phenylpropanoid pathway is in charge of producing all of these chemical compounds in plants, with phenylalanine serving as the primary building block that helps plants effectively fend off environmental stressors (Dong and Lin, 2021). Abiotic and biotic variables control the phenylpropanoid pathway, which causes various phenolics' build-up (Valifard et al., 2015; Sharma et al., 2019). As a result, groups of phenolic compounds found in plant materials are thought to be a crucial way for plants to survive in hostile environments. In order to increase evolutionary aptness, react to these stresses and help eliminate ROS, catalyze oxygenated reactions with the formation of metabolic structures, and obstruct the actions of oxidative enzymes (Amarowicz and Weidner, 2009; Nada et al., 2022). Furthermore, it is thought that phenolic cumulation act as a crucial defense mechanism and dependable characteristic in times of stress, promoting the free radicals production and other oxidative species in plants (Cheynier et al., 2013).

Metals are chelated by the antioxidant qualities of phenolics. They bind to metals, especially iron and copper, thanks to the hydroxyl group they contain. When lipid alkoxyl radicals generated during stress are trapped, phenolic compounds prevent lipid peroxidation. A class of phenolics known as flavanols is capable of binding proteins and forming hydrogen bonds with phospholipids to form an anchor. They have the ability to gather close to the lipid membrane's extracellular and cytosolic sides as a result. They also maintain the bound membrane's safety in this way (Michalak, 2006).

Phenolics are linked to gene expression reforms in plants (André et al., 2009), besides their protective function due to their antioxidant qualities (Li et al., 2021). Additionally, Parida et al. (2007) and Rodziewicz et al. (2014) report that they have a function in maintaining osmotic potential in cells under stress. Furthermore, phenols influence where inorganic and organic soil nutrients come from, how they are transferred, and how plants and microbes respond to nutrient shortages. This means that phenols could be utilized to detect nutrient disorders before symptoms become apparent (Naikoo et al., 2019). Numerous investigations revealed that the phenolics production, such as flavonoids and phenolic acids (i.e., iso-flavonoids, flavonolignan silymarin, and anthocyanins), was mediated by environmental stressors (Nakabayashi et al., 2014; Elateeq et al., 2020; Li et al., 2021). According to Diáz et al. (2001), when *capsicum annum* leaves were subjected to copper sulfate, phenols accumulated in the leaves. It has been noted that phenols play a direct role in chelating iron ions and thereby inhibiting Fenton's reaction, a significant source of ROS production. According to Lavid et al. (2001), tea plants with high tannin content can withstand great concentrations of manganese and can avoid toxicity by directly chelating the metal. In contrast, the sensitive genotypes showed a larger decline in flavonoids; this suggests that flavonoid content was essential for maintaining higher levels of antioxidant activity during stressful situations (Djoukeng et al., 2008; Rodziewicz et al., 2014). Comparably, *Phaseolus vulgaris* accumulates both soluble and insoluble phenolics when exposed to Cd (Winkel-Shirley, 2002).

2.5.2.2 TERPENOIDS

To adapt to various kinds of stress within an ecosystem, plants have developed a sophisticated system of resistance that is dependent on the quick perception and secondary metabolites stimulation (Andersen et al., 2018). Terpenoids, which are made by plants from isoprene units (C_5H_8), comprise a large and structurally varied class of lipophilic secondary metabolites (Toffolatti et al., 2021). Terpenoids, like gibberellic acid (diterpenoids) and abscisic acid (ABA) (sesquiterpenoids), function as phytohormones in opposition to biotic and abiotic stressors and are important in physiological processes. Through altering the characteristics of the membrane, the phytohormone ABA stimulates defense mechanisms and abiotic stress (Kumar et al., 2022; Zhao et al., 2023).

Furthermore, terpenoids exhibit antibiotic and antioxidant activity that preserves lipid membranes and raise resistance to environmental stress (Al Musayeib et al., 2022). As a result of biotic and abiotic stimuli, plants produce terpenoids as a component of their defense mechanism, which also functions as a phytoalexin. For instance, *Oryza sativa* has been demonstrated to contain numerous diterpene phytoalexins (Gu et al., 2019). Similarly, sesquiterpenoid phytoalexins accumulated under stress in cotton, including heliocides, gossypol, and hemigossypolone (Yang et al., 2013). Furthermore, by stabilizing the *B. juncea* cell membranes, Yusuf et al. (2010) observed that the elevated concentration of the antioxidant-rich soluble alcohol tocopherol has a vital function in reducing stress. Additionally, it was discovered that among the essential secondary metabolites linked to stress resistance in soybean plants is their saponin content (Wu et al., 2008).

2.5.2.3 NITROGEN-CONTAINING SECONDARY METABOLITES

Glycosides, alkaloids, glucosinolates, cyanogenic and nonprotein amino acids are examples of the complex secondary metabolism defense mechanisms that plants have developed to fend off stressors (Miyagawa, 2009; Divekar et al., 2022). Substrates involving nitrogen, which were once thought to be undesirable by plants, are now recognized for their resistance to various stressors (Cao et al., 2018).

Alkaloids are a diverse secondary metabolites group found in phytochemicals that arise from abiotic stress and contain a single or multiple nitrogen atoms. It has been found that more alkaloid contents and derivatives are produced in situations of abiotic stress. For instance, stressed poppies produce more alkaloids (Davodnia et al., 2017). Stress also affected the amount of alkaloids in cultivars of lupins (*Lupinus termis*) (Taha et al., 2020).

Moreover, cyanogenic glycosides and glucosinolates are nitrogen- and sulfur-containing secondary metabolites that are produced from amino acids and glucose. Likewise, Rodziewicz et al. (2014) showed that each of these organic substances contributes significantly to the defense against various abiotic and biotic factors. Mewis et al. (2012)'s research revealed that *A. thaliana* tends to produce more aliphatic glucosinolate compounds when stressed. Furthermore, under stressful conditions, an elevated glucosinolate level was noted in *B. juncea* during the vegetative stage.

2.5.2.4 ORGANIC ACIDS

Organic acids such as citrate, malate, and oxalate help sequester ions in vacuoles and carry metals through xylem (Rauser, 1999). Citrate, which is made from citrate synthase, has been shown to have a strong affinity for Ni, Fe, and Cd, but its main purpose is to chelate Fe (Cataldo et al., 1988). Zinc has been reported to be chelated by malate, which primarily chelates cytosolic ions (Mathys, 1977).

2.5.2.5 PLANT HORMONES

Phytohormones like ethylene, JA (jasmonic acid), SA, BRs, and ABA can increase a plant's ability to withstand HMs. Plant hormones have the potential to act as signaling molecules by inducing resistance system activation and acclimation reactions within the defense system (Shafeeq-Ur-Rahman et al., 2023).

In addition to increasing antioxidant activity (Vardhini, 2016) and inhibiting lipid degradation brought on by excessive ROS production during stress, BRs are thought to be possible chelators of HM stress in plants. Additionally, they can stop protein reduction, sugars, and chlorophyll and encourage phytochelatins synthesis (Noriega et al., 2012; Wang et al., 2020).

Another significant phytohormone linked to the stress defense response is JA, which works by triggering ascorbate mechanisms or glutathione antioxidant to inhibit lipid peroxidase activity (Herrera-Vásquez et al., 2015).

Additionally, reports have indicated that SA concentration rises under metal stress. It has been observed that pretreating plants with SA causes glutathione to accumulate and increases their toxicity resistance to nickel. As stated by Hossain et al. (2012), SA's primary functions in cells of plant are to regulate ROS, antioxidant concentrations and to induce the way various genes are expressed.

Plants require ABA as a phytohormone, as well as being exposed to HMs that raise the endogenous levels of ABA in plants (Fediuc et al., 2005; Zhao et al., 2023).

2.6 PLANT RESISTANCE TO HEAVY METAL STRESS

To counteract stress caused by HMs, plants have a lot of physiological and molecular defenses, including intricate biochemical and genomic systems. Certain mechanisms are constitutive in nature, as they are a component of the homeostatic process. Certain metals are toxic, and other mechanisms are exclusive and only become active when they are encountered. According to Krzesłowska (2011), all reactions could be broadly categorized as avoidance or tolerant types. Many defense mechanisms have been developed by plants over time to withstand the negative consequences of stress by HMs. By keeping out or prohibiting entry, they can prevent metal stress before it even arises (Viehweger, 2014).

Plants impede the metals' entry by complexing them with organic acids secreted from roots and immobilizing them through mycorrhizal association, according to Dalvi and Bhalerao (2013).

Plants employ various tolerance mechanisms to eliminate metals from their systems if they happen to get into them. Metal binding to cell walls, phytochelatin biosynthesis, metallothionein synthesis, and metal sequestration into vacuoles are some of these strategies (Patra et al., 2004; John et al., 2009).

Plants produce osmolytes like proline and activate antioxidant defense mechanisms along with metal sequestration mechanisms (Manara, 2012). Thus, to counteract the toxicity of HMs, plants adopt various strategies. However, the complex molecular mechanisms linked to these strategies are still not fully understood (Parihar et al., 2019).

2.7 TOLERANCE OF HEAVY METALS BY DETOXIFICATION

When HMs enter the cytosol, plants either transport it outside the cell or confine it to the vacuole; this process, known as sequestration/detoxification, leads to detoxification (Kanoun-Boulé et al., 2009). Therefore, it is prohibited for HMs to interfere with essential metabolic pathways. Plants can thrive in metal-contaminated environments thanks to this process, which has no harmful effects (Pollard et al., 2021). Numerous investigations demonstrate that a cell's vacuole contains high values of zinc and copper. As an example, Ernst et al. (1992) stated that *Alyssum serpytllifolium* (a nickel hyperaccumulator) was able to accumulate up to 72% of Ni within the vacuole.

The primary purpose of exudates from roots, which are primarily composed of sugars, exoenzymes, organic acids, and amino acids, is to mobilize and detoxify nutrients that are restrictive and recalcitrant, like HMs (Jaitz et al., 2011; van Dam and Bouwmeester, 2016). Hyperaccumulator plants secrete large amounts of organic acids with low molecular weight (like citric, malic, and oxalic acids) (Xie et al., 2013; Sun et al., 2020), which can help repair contaminated soil, activate insoluble HM compounds in the soil, and promote absorption Furthermore, by secreting REs, certain plants—which are not hyperaccumulators—can prevent HMs to enter the root (Bali et al., 2020).

2.8 ALLEVIATING HEAVY METAL STRESS IN PLANTS BY EXOGENOUS APPLICATION

Unfortunately, plants, like other living things, have a limited ability to control HMs through self-regulation before an excess of HMs overcomes their protective resistance (Feng et al., 2023b). To protect plants against abiotic stresses, some researchers have developed novel techniques, such as applying exogenous substances. The term "exogenous substance" here refers to any material applied externally that influences a plant's tolerance to HMs; this does not include substances that directly interact with soil-borne HMs or any vital nutrients that a plant needs to grow and develop.

Application of exogenous substances has the following effects on reducing stress of HMs in plants: (1) increases antioxidant system capacity, (2) stimulates osmoregulatory substance synthesis, (3) improves photochemical system, (4) diverts HM building up and migration, (5) controls endogenous hormone secretion, (6) modifies gene expressions, and (7) engages in microbe-involved regulations (Feng et al., 2023a).

These exogenous materials involve minerals (like Ca, S, Se, and Si), polyamines, plant hormones (like SA, auxin, gibberellic acid, indole-3-butyric acid, and indole acetic acid precursor), amino acids (e.g. proline), organic aids, and microorganisms that were tested and investigated to lessen the adverse effects of the stress that HMs impose on plants (Figure 2.3) (Tang et al., 2005; Popova et al., 2009; Chen et al., 2020; Wang et al., 2022; Sharma et al., 2023).

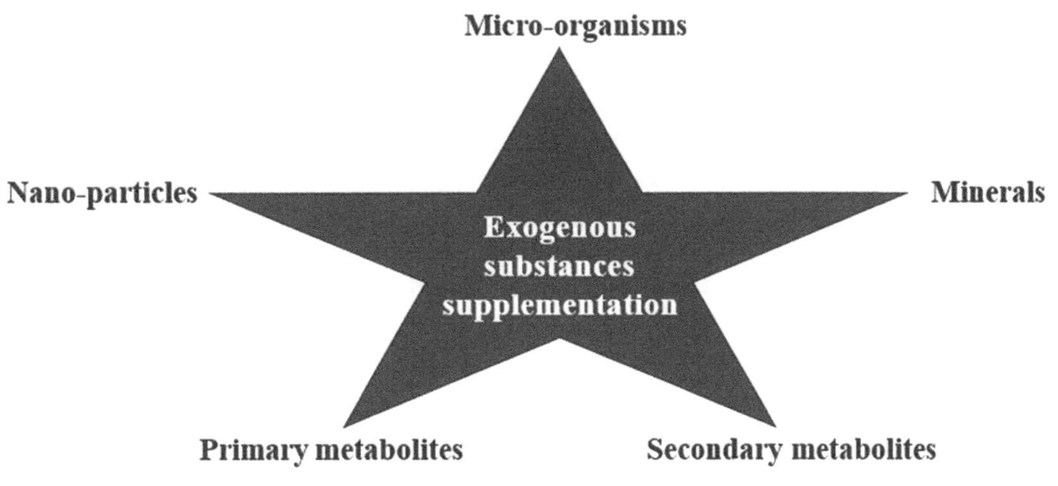

FIGURE 2.3 Exogenous substances supplementation.

Two special attributes of exogenous NO and calcium are their aptitude to reduce HMs in wheat (Liu et al., 2015) and mustard (Ahmad et al., 2015) plant attributed to Cd. When ascorbic acid was applied exogenously, the levels of Pb, Cd, and Hg in tissues of plant were decreased (Kováćik et al., 2017), the mineral nutrients (Mg, K, Ca, and P) absorption was improved (Atta-Ullah et al., 2016) and wheat plants' yield was increased (Moghadam, 2016). Exogenous glutathione supplementation lessens the impacts of copper and zinc on plant photosynthesis (Son et al., 2014; Cao et al., 2015). JA application under HM stress can raise the levels of organic acids, penetrants, metal-chelating compounds, and secondary metabolites (such as total phenols, polyphenols, polyamines, flavonoids, and anthocyanins) (Bali et al., 2019; Dai et al., 2020).

According to Rajab et al. (2020), sulfur metabolic engineering can help rapeseed (*Brassica napus*) to improve its resistance to Cd stress by raising its levels of glutathione and cysteine. Additionally, α-tocopherols have been demonstrated to shield cell membranes by scavenging ROS and preventing lipid peroxidation (genotypes with higher tocopherol content were better able to withstand Cd stress [Ali et al., 2020]).

Moreover, using microorganisms like *Pseudomonas aeruginosa* and *Bacillus xiameni*, a mycorrhizal fungus, has been proven to help reduce HM levels in plants (Oves et al., 2017).

Using nanoparticles (NPs) in agriculture—such as nanopesticides and nanofertilizers (Fraceto et al., 2016)—is becoming more and more popular because of the quick advancement of nanotechnology. There are several environmental problems which need to be further considered due to the rapidly expanding industrial applications of metal oxides and metal NPs, the most common of which are cerium dioxide, titanium dioxide, silver, copper, aluminum, nickel, iron, zinc oxide, and copper oxide. Because of their unique physicochemical characteristics—such as their small size and surface effects—NPs are readily absorbed by living things. Metal oxide NPs, such as TiO_2 NPs, CeO_2 NPs, ZnO NPs, and Fe_3O_4 NPs, have been demonstrated in numerous studies to be able to effectively prevent crops from absorbing and accumulating HMs. TiO_2 NPs are widely used nanomaterials that have the capability for penetrating soil through various means such as sewage and landfill waste. This can impact the biotoxicity of coexisting contaminants, including Cd, as mentioned previously (Lian et al., 2020; Spanò et al., 2019). According to Singh and Lee's research, soybean plants' Cd uptake was enhanced by nano-TiO_2, which also lessened Cd stress (as evidenced by decreased proline and lipid peroxidation) (2016). TiO_2 NPs had no impact on Cd toxicity of radish seeds, according to Manesh et al. (2018). ZnO and Fe NPs, respectively, enhanced the content of Zn and Fe, and reduced Cd cumulation in wheat, according to Rizwan et al. (2019).

2.9 CONCLUSION

Toxicity of plants with HMs is a serious problem that not only impacts the physiological performance of the plant, but also results in a significant loss in yield because of many damages that the plant suffers from. These metals also move into the food chain, affecting consumers of these contaminated plants. Different plants respond differently to different metals, and different growing conditions affect their responses. There are certain balances in plant cells that can be disturbed by HMs, tilt the balance in favor of one, stimulate reactive oxygen species, and interact directly with working cellular molecules and organelles. Toxic HM ions also substitute basic cations, and HM stress frequently causes these ions to bind to active groups of cofactors, leading to various degenerative symptoms. This clarifies the clear similarity between the apparent symptoms observed between plants stressed by HMs and those suffering from a scarcity of essential nutrients.

Either resistant or sensitive plant species to HM stress activate a basic defense system when they perceive the threat from HMs. Recent developments and the use of technologies in the molecular and biological fields have played a major role in understanding some of the intricate cellular and molecular defense mechanisms plants employ against metal stress. The activation of transcription factors that trigger the transcription of different HM stress-responsive genes is typically the result of the signal transduction pathway. It is now understood that plants have four primary common signal transduction pathways that can interact with one another: phytohormone signaling, species-reactive signaling, calcium-dependent signaling, and signaling based on protein phosphorylation/dephosphorylation by kinase cascades. Because they can lessen or even completely stop the uptake of metal ions, plants are able to lessen the toxic effects of HMs. This increases the complexity of the metals in the extracellular space, chelates them in the cytoplasm, and may even sequester them in the vacuole. For this reason, the plants start defense mechanisms, which include antioxidant enzyme expression and activation as well as mechanisms to stop or fix secondary oxidative stress-related defects.

It is crucial to comprehend the function of primary and secondary metabolites in the mechanisms of stress resistance in order to enhance crop species development and stress tolerance. Less is known, though, about how these metabolites protect plants from environmental stressors, particularly abiotic ones. It is necessary to advance "omics" tools, bioinformatics, and improved data assimilation from different molecular levels (Salam et al., 2023).

Thus, a great deal more research needs to be done to remove HMs from our ecosystem in order to support sustainable agriculture and clean environments. Some of these strategies include phyto-detoxicating contaminated landscapes, recovering HMs from plants or crops that are toxic, breeding more tolerant cultivars to produce crops that are both highly productive and resistant to HMs, and looking for ways to combine multiple exogenous substances to reduce the effects of HMs in our future research (Feng et al., 2023a).

Future research should identify the important genes influencing HMs accumulation. Understanding the molecular interactions between signal transduction and signal cascades in plants exposed to heavy metals is crucial.

KEYWORDS

- **metabolomics**
- **heavy metals**
- **abiotic stress**
- **plant responses**
- **plant tolerance**

REFERENCES

Ahmad, P., Ozturk, M., Gucel, S. Oxidative damage and antioxidants induced by heavy metal stress in two cultivars of mustard (*Brassica juncea* L.) plants. *Fresenius Environ. Bull.*, 2012, 21(10), 2953–2961.

Ahmad, P., Sarwat, M., Bhat, N. A., Wani,, M. R., Kazi, A. G., Tran, L. Ph. Alleviation of cadmium toxicity in *Brassica juncea* L. (Czern. & Coss.) by calcium application involves various physiological and biochemical strategies. *PLoS One*, 2015, 10(1), e0114571.

Al Musayeib, N. M., Musarat, A., Maqsood, F. Terpenes: A source of novel antimicrobials, applications and recent advances. In: *Eco-Friendly Biobased Products Used in Microbial Diseases* (pp. 247–270). CRC Press: Boca Raton, FL, USA, 2022.

Ali, B., Wang, B., Ali, S., Ghani, M. A., Hayat, M. T., Yang, C., et al. 5-Aminolevulinic acid ameliorates the growth, photosynthetic gas exchange capacity, and ultra structural changes under cadmium stress in *Brassica napus* L. *J. Plant Growth Regul.*, 2013, 32, 604–614.

Ali, E., Hassan, Z., Irfan, M., Hussain, S., Haseeb-ur-Rehman, Shah, J. M., et al. Indigenous tocopherol improves tolerance of oilseed rape to cadmium stress. *Front. Plant Sci.*, 2020, 11, 547133.

Ali, H. and Khan, E. What are heavy metals? Long-standing controversy over the scientific use of the term' heavy metals'—Proposal of a comprehensive definition. *Toxicol. Environ. Chem.*, 2018, 100(1), 6–19.

Amarowicz, R., Weidner, S. Phenolic compounds and properties of antioxidants in grapevine roots (*Vitis vinifera* L.) under drought stress followed by recovery. *Acta Soc. Bot. Pol.*, 2009, 78, 389–405.

Andersen, E. J., Ali, S., Byamukama, E., Yen, Y., Nepal, M. P. Disease resistance mechanisms in plants. *Genes*, 2018, 9, 339.

André, C. M., Schafleitner, R., Legay, S., Lefèvre, I., Aliaga, C. A. A., Nomberto, G., et al. Gene expression changes related to the production of phenolic compounds in potato tubers grown under drought stress. *Phytochemistry*, 2009, 70, 1107–1116.

Anjum, S. A., Tanveer, M., Hussain, Ehsan-Ullah, E., Wang, L., Khan, I., et al. Morpho physiological growth and yield responses of two contrasting maize cultivars to cadmium exposure, *Clean –Soil, Air, Water*, 2016, 44 (1), 29–36.

Appenroth, K.-J. What are "heavy metals" in plant sciences? *Acta Physiologiae Plantarum*, 2010, 32(4), 615–619.

Arbona, V., Manzi, M., Ollas, C. D., Gonez-Cadenao, A. Metabolites as a tool to investigate abiotic stress tolerance in plants. *Int. J. Mol. Sci.*, 2013, 14, 4885–4911.

Ashapkin, V. V., Kutueva, L. I., Aleksandrushkina, N. I., Vanyushin, B. F. Epigenetic mechanisms of plant adaptation to biotic and abiotic stresses. *Int. J. Mol. Sci.*, 2020, 21, 7457.

Atkins, P., Jones, L. *Chemistry: Molecules, Matter and Change*, 3rd edn., W. H. Freeman: New York, NY, USA, 1997.

Atta-Ullah, H., Javed, F., Wahid, A., Sadia, B. Alleviating effect of exogenous application of ascorbic acid on growth and mineral nutrients in cadmium stressed barley (*Hordeum vulgare*) seedlings. *Int. J. Agric. Biol.*, 2016, 18, 73–79.

Bali, A. S., Sidhu, G. P. S., Kumar, V. Root exudates ameliorate cadmium tolerance in plants: A review. *Environ. Chem. Lett.*, 2020, 18, 1243–1275.

Bali, S., Jamwal, V. L., Kohli, S. K., Kaur, P., Tejpal, R., Bhalla, V., et al. Jasmonic acid application triggers detoxification of lead (Pb) toxicity in tomato through the modifications of secondary metabolites and gene expression. *Chemosphere*, 2019, 235, 734–748.

Barakat, M. A. New trends in removing heavy metals from industrial wastewater. *Arab. J. Chem.*, 2011, 4(4), 361–377.

Baxter, A., Mittler, R., Suzuki, N. ROS as key players in plant stress signalling. *J Exp Bot.*, 2014, 65(5), 1229–1240.

Cao, F., Cai, Y., Liu, L., Zhang, M., He, M., Zhang, G., Wu, F. Differences in photosynthesis, yield and grain cadmium accumulation as affected by exogenous cadmium and glutathione in the two rice genotypes. *Plant Growth Regul.*, 2015, 75, 715–723.

Cao, X., Zhu, C., Zhong, C., Hussain, S., Zhu, L., Wu, L., Jin, Q. Mixed-nitrogen nutrition-mediated enhancement of drought tolerance of rice seedlings associated with photosynthesis, hormone balance and carbohydrate partitioning. *Plant Growth Regul.*, 2018, 84, 451–465.

Capell, T., Bassie, L., Christou, P. Modulation of the polyamine biosynthetic pathway in transgenic rice confers tolerance to drought stress. *Proc. Natl. Acad. Sci. USA*, 2004, 101, 9909–9914.

Cataldo, D. A., McFadden, K. M., Garland, T. R., Wildung, R. E. Organic constituents and complexation of nickel (II), iron (III), cadmium (II), and plutonium (IV) in soy bean xylem exudates. *Plant Physiol.*, 1988, 86, 734–739.

Chen, C. C., Dixon, J. B., Turner, F. T. Iron coatings on rice roots: Morphology and models of development. *Soil Sci Soc Am J.*, 1980, 44, 1113–1119.

Chen, H. C., Zhang, S. L., Wu, K. J., Li, R., He, X. R., He, D. N. et al. The effects of exogenous organic acids on the growth, photosynthesis and cellular ultrastructure of *Salix variegata* Franch. under Cd stress. *Ecotoxicol. Environ. Saf.*, 2020, 187, 109790.

Chen, T. H., Murata, N. Glycinebetaine: An effective protectant against abiotic stress in plants. *Trends Plant Sci.*, 2008, 13, 499–505.

Cheynier, V., Comte, G., Davies, K. M., Lattanzio, V., Martens, S. Plant phenolics: Recent advances on their biosynthesis, genetics, and ecophysiology. *Plant Physiol. Biochem.*, 2013, 72, 1–20.

Collin, V. C., Eymery, F., Genty, B., Rey, P., Havaux, M. Vitamin E is essential for the tolerance of *Arabidopsis thaliana* to metal induced oxidative stress. *Plant Cell Environ.*, 2008, 31, 244–257.

Dai, H., Wei, S., Pogrzeba, M., Rusinowski, S., Krzyżak, J., Jia, G. Exogenous jasmonic acid decreased Cu accumulation by alfalfa and improved its photosynthetic pigments and antioxidant system. *Ecotoxicol. Environ. Saf.*, 2020, 190, 110176.

DalCorso, G., Manara, A., Furini, A. An overview of heavy metal challenge in plants: From roots to shoots. *Metallomics*, 2013, 5, 1117–1132.

Dalvi, A. A., Bhalerao, S. A. Response of plants towards heavy metal toxicity: An overview of avoidance, tolerance and uptake mechanism. *Ann. Plant Sci.*, 2013, 2, 362–368.

Davodnia, B., Ahmahdi, J., Fabriki Ourang, S. Evaluation of drought and salinity stresses on morphological and biochemical characteristics in four species of Papaver. *Eco-Phytochem. J. Med. Plants*, 2017, 5, 24–36.

Debona, D., Rodrigues, F. A., Datnoff, L. E. Silicon's role in abiotic and biotic plant stresses. Annu. Rev. *Phytopathol.*, 2017, 55, 85–107.

Di Toppi S. L., Gabbrielli, R. Response to cadmium in higher plants. *Environ Exp Bot.*, 1999, 41(2), 105–130.

Diáz, J., Bernal, A., PoMar, F., Merino, F. Induction of shikimate dehydrogenase and peroxidase in pepper (*Capsicum annum* L.) seedlings in response to copper stress and its relation to lignification. *Plant Sci.*, 2001, 161, 179.

Dikilitas, M., Simsek, E., Roychoudhury, A. Role of proline and glycine betaine in overcoming abiotic stresses. In: Roychoudhury, A., Tripathi, D.K. (Eds.), *Protective Chemical Agents in the Amelioration of Plant Abiotic Stress: Biochemical and Molecular Perspectives* (pp. 1–23.), Wiley: Hoboken, NJ, USA, 2020.

Divekar, P. A., Narayana, S., Divekar, B. A., Kumar, R., Gadratagi, B. G., Ray, A., et al. Plant secondary metabolites as defense tools against herbivores for sustainable crop protection. *Int. J. Mol. Sci.*, 2022, 23, 2690.

Divi, U. K., Krishna, P. Brassinosteroid: A biotechnological target for enhancing crop yield and stress tolerance. *New Biotechnol.*, 2009, 26, 131–136.

Djoukeng, J. D., Arbona, V., Argamasilla, R., Gomez-Cadenas, A. Flavonoid profiling in leaves of citrus genotypes under different environmental situations. *J. Agric. Food Chem.*, 2008, 56, 11087–11097.

Dong, N. Q., Lin, H. X. Contribution of phenylpropanoid metabolism to plant development and plant–environment interactions. *J. Integr. Plant Biol.*, 2021, 63, 180–209.

Duffus, J. H. "Heavy metals" a meaningless term? (IUPAC Technical Report)," *Pure Appl. Chem.*, 2002, 74(5), 793–807.

Elateeq, A. A., Sun, Y., Nxumalo, W., Gabr, A. M. Biotechnological production of silymarin in *Silybum marianum* L.: A review. *Biocatal. Agric. Biotechnol.*, 2020, 29, 101775.

Emamverdian, A., Ding, Y., Mokhberdoran, F., Xie, Y. Heavy metal stress and some mechanisms of plant defense response. *Sci. World J.*, 2015, 2015, 756120.

Ernst, W., Verkleij, J., Schat, H. Metal tolerance in plants. *Acta Botanica Neerlandica*, 1992, 41, 229–248.

Fahy, E., Subramaniam, S., Brown, H. A., Glass, C. K., Merrill, A. H., Murphy, R. C., et al. A comprehensive classification system for lipids1. *J. Lipid Res.*, 2005, 46, 839–861.

Farias J. G., Antes F. L. G., Nunes P. A. A., Nunes S. T., Schaich G., Rossato L. V., et al. Effects of excess copper in vineyard soils on the mineral nutrition of potato genotypes. *Food Energy Security*, 2013, 2, 49–69.

Fediuc, E., Lips, S. H., Erdei, L. O-Acetylserine (thiol) lyase activity in *Phragmites* and *Typha* plants under cadmium and NaCl stress conditions and the involvement of ABA in the stress response. *J. Plant Physiol.*, 2005, 162, 865–872.

Feng, D., Gao, Q., Liu, J., Tang, J., Hua, Z., Sun, X. Categories of exogenous substances and their effect on alleviation of plant salt stress. *Eur. J. Agron.*, 2023b, 142, 126656.

Feng, D., Wang, R., Sun, X., Liu, L., Liu, P., Tang, J., Zhang, C., Liu, H. Heavy metal stress in plants: Ways to alleviate with exogenous substances. *Sci. Total Environ.*, 2023a, 897, 165397.

Feng, Z., Ji, Sh., Ping, J., Cui, D. Recent advances in metabolomics for studying heavy metal stress in plants. *Trends in Analytical Chemistry*, 2021, 143, 116402.

Fouad, H., Elsayed, S. I. M., Fouad, R., Hendawy, S. F., Omer, E. A. Influence of exogenous tryptophan application on production and carotenoids of *Calendula officinalis* under drip irrigation treatments. *Int. J. Health Sci.*, 2022, 6(S6), 10840–10852.

Fouad, R., Fouad, H., El-Desoky, A. H., Omer, E. A. Secondary metabolism and its role in enhancing drought stress tolerance. In: Hasanuzzaman, M. (Ed.), *Climate-Resilient Agriculture* (Vol 2, pp. 603–640). Springer: Cham, 2023a.

Fouad, R., Fouad, H., Elsayed, Sh. I. M., Hendawy, S. F., Omer, E. A. Response of growth, productivity, and active constituents of *Hyssopus officinalis* to irrigation and salicylic acid foliar application. *Egyptian Pharm. J.*, 2023b, 22(2), 311–323.

Foyer, C. H., Noctor, G. Redox homeostasis and antioxidant signaling: A metabolic interface between stress perception and physiological responses. *Plant Cell*, 2005, 17, 1866–1875.

Fraceto, L. F., Grillo, R., de Medeiros, G. A., Scognamiglio, V., Rea, G., Bartolucci, C. Nanotechnology in agriculture: Which innovation potential does it have? *Front. Environ. Sci.*, 2016, 4, 20.

Freeman, J. L., Persans, M. W., Nieman, K., Albrecht, C., Peer, W., Pickering, I. J., Salt, D. E. Increased glutathione biosynthesis plays a role in nickel tolerance in *Thlaspi* nickel hyperaccumulators. *Plant Cell*, 2004, 16, 2176–2191.

Fu, Z., Guo, W., Dang, Z., Hu, Q., Wu, F., Feng, Ch., et al. Refocusing on nonpriority toxic metals in the aquatic environment in China. *Environ. Sci. Technol.*, 2017, 51(6), 3117–3118.

Ghori, N.-H., Ghori, T., Hayat, M. Q., Imadi, S. R., Gul, A., Altay, V., Ozturk, M. Heavy metal stress and responses in plants. *Int. J. Environ. Sci. Technol.*, 2019, 16, 1807–1828.

Gill, S. S., Tuteja, N. Polyamines and abiotic stress tolerance in plants. Plant Signal. Behav., 2010, 5, 26–33.

Giri, J. Glycinebetaine and abiotic stress tolerance in plants. *Plant Signal. Behav.*, 2011, 6, 1746–1751.

Gisbert, C., Ros, R., Deharo, A., Walker, D. J., Pilarbernal, M., Serrano, R., Navarro-Avino, J. A plant genetically modified that accumulates Pb is especially promising for phytoremediation. *Biochem. Biophys. Res. Commun.*, 2003, 303, 440–445.

Gomes, M. M. A. Physiological effects related to brassinosteroid application in plants. In: Hayat, S., Ahmad, A. (Eds.), *Brassinosteroids: A Class of Plant Hormone* (pp.193–242). Springer: Dordrecht, 2011.

Grace, S. C. Phenolics as antioxidants. *Antioxid. React. Oxyg. Species Plants*, 2005, 141, 168.

Gu, C.-Z., Xia, X.-M., Lv, J., Tan, J.-W., Baerson, S. R., Pan, Z.-q., Song, Y.-Y., Zeng, R.-S. Diterpenoids with herbicidal and antifungal activities from hulls of rice (*Oryza sativa*). *Fitoterapia*, 2019, 136, 104183.

Gupta, D. K., Vandenhove, H., Inouhe, M. Role of phytochelatins in heavy metal stress and detoxification mechanisms in plants. In: Gupta, D., Corpas, F., Palma, J. (Eds.), *Heavy Metal Stress in Plants* (pp. 73–94). Springer: Berlin, 2013.

Hakeem, K. R., Sabir, M., Ozturk, M., Akhtar, M. S., Ibrahim, F. H. Nitrate and nitrogen oxides: Sources, health effects and their remediation. In: Gunther, F. A., de Voogt, P. (Eds.), *Reviews of Environmental Contamination and Toxicology* (242, pp. 183–217), Springer: Cham, 2017.

Hanif, S., Saleem, M. F., Sarwar, M., Irshad, M., Shakoor, A., Wahid, M. A., Khan, H. Z. Biochemically triggered heat and drought stress tolerance in rice by proline application. *J. Plant Growth Regul.*, 2021, 40, 305–312.

Heinemann, B., Hildebrandt, T. M. The role of amino acid metabolism in signaling and metabolic adaptation to stress-induced energy deficiency in plants. *J. Exp. Bot.*, 2021, 72, 4634–4645.

Herrera-Vásquez, A., Salinas, P., Holuigue,, L. Salicylic acid and reactive oxygen species interplay in the transcriptional control of defense genes expression. *Front Plant Sci.*, 2015, 6, 171.

Hildebrandt, T. M. Synthesis versus degradation: Directions of amino acid metabolism during *Arabidopsis* abiotic stress response. *Plant Mol. Biol.*, 2018, 98, 121–135.
Horst, W. J. Factors responsible for genotypic manganese tolerance in cowpea (*Vigna unguiculata*). *Plant Soil*, 1983, 72(2–3), 213–218.
Hossain, M. A., Hasanuzzaman, M., Fujita, M. Coordinate induction of antioxidant defense and glyoxalase system by exogenous proline and glycine betaine is correlated with salt tolerance in mung bean. *Front. Agric. China*, 2011, 5, 1–14.
Hossain, M. A., Piyatida, P., daSilva, J. A. T., Fujita, M. Molecular mechanism of heavy metal toxicity and tolerance in plants: Central role of glutathione in detoxification of reactive oxygen species and methylglyoxal and in heavy metal chelation. *J. Bot.*, 2012, 2012, 872–875.
Hossain, M. A., Hasanuzzaman, M., Fujita, M. Up-regulation of antioxidant and glyoxalase systems by exogenous glycinebetaine and proline in mung bean confer tolerance to cadmium stress. *Physiol. Mol. Biol. Plants*, 2010, 16(3), 259–272.
Hou, Q., Ufer, G., Bartels, D. Lipid signalling in plant responses to abiotic stress. *Plant Cell Environ.*, 2016, 39, 1029–1048.
Huang, J. W., Pellet, D. M., Papernik, L. A., Kochian, L. V. Aluminum interactions with voltage dependent calcium transport in plasma membrane vesicles isolated from roots of aluminum-sensitive and -resistant wheat cultivars. *Plant Physiol.*, 1996, 110(2), 561–569.
Imtiaz, M., Rizwan, M. S., Mushtaq, M. A., Ashraf, M., Shahzad, S. M., Yousaf, B., et al. Silicon occurrence, uptake, transport and mechanisms of heavy metals, minerals and salinity enhanced tolerance in plants with future prospects: A review, *J. Environ. Manag.*, 2016, 183, 521–529.
Jaitz, L., Mueller, B., Koellensperger, G., Huber, D., Oburger, E., Puschenreiter, M., Hann, S. LC-MS analysis of low molecular weight organic acids derived from root exudation. *Anal. Bioanal. Chem.*, 2011, 400, 2587–2596.
Jalmi, S. K., Bhagat, P. K., Verma, D., Noryang, S., Tayyeba, S., Singh, K., et al. Traversing the links between heavy metal stress and plant signaling. *Front Plant Sci.*, 2018, 9, 12.
Järup, L., Akesson, A. Current status of cadmium as an environmental health problem. *Toxicol. Appl. Pharmacol.*, 2009, 238(3), 201–218.
John, R., Ahmad, P., Gadgil, K., Sharma, S. Heavy metal toxicity: Effect on plant growth, biochemical parameters and metal accumulation by *Brassica juncea* L. *Int. J. Plant*, 2009, 3, 65–76.
Jonak, C., Nakagami, H., Hirt, H. Heavy metal stress. Activation of distinct mitogen-activated protein kinase pathways by copper and cadmium. *Plant Physiol.*, 2004, 136, 3276–3283.
Jonak, C., Okrész, L., Bögre, L., Hirt, H. Complexity, cross talk and integration of plant MAP kinase signalling. *Curr. Opin. Plant Biol.*, 2002, 5, 415–424.
Jović, M., Onjia, A., Stanković, S. Toxic metal health risk by mussel consumption. *Environ. Chem. Lett.*, 2012, 10(1), 69–77.
Kabera, J. N., Semana, E., Mussa, A. R., He, X. Plant secondary metabolites: Biosynthesis, classification, function and pharmacological properties. *J. Pharm. Pharm.*, 2014, 2, 377–392.
Kanoun-Boulé, M., Vicente, J. A., Nabais, C., Prasad, M. N. V., Freitas, H. Ecophysiological tolerance of duck weeds exposed to copper. *Aquat. Toxicol.*, 2009, 91, 1–9.
Kaplan, F., Guy, C. L. β-Amylase induction and the protective role of maltose during temperature shock. *Plant Physiol.*, 2004, 135, 1674–1684.
Kaur, R., Das, S., Bansal, S., Singh, G., Sardar, S., Dhar, H., Ram, H. Heavy metal stress in rice: Uptake, transport, signaling, and tolerance mechanisms. *Physiol. Plant*, 2021, 173, 430–448.
Kebert, M., Vuksanović, V. Species-level differences in osmoprotectants and antioxidants contribute to stress tolerance of *Quercus robur* L., and *Q. cerris* L. seedlings under water deficit and high temperatures. *Plants*, 2022, 11, 1744.
Kerkeb, L., Krämer, U. The role of free histidine in xylem loading of nickel in *Alyssum lesbiacum* and *Brassica juncea*. *PlantPhysiol.*, 2003, 131, 716–724.
Keunen, E., Peshev, D., Vangronsveld, J., Van Den Ende, W., Cuypers, A. Plant sugars are crucial players in the oxidative challenge during abiotic stress: Extending the traditional concept. *Plant Cell Environ.*, 2013, 36, 1242–1255.

Khan, A., Ali, S., Murad, W., Hayat, K., Siraj, S., Jawad, M., et al. Phytochemical and pharmacological uses of medicinal plants to treat cancer: A case study from Khyber Pakhtunkhwa, North Pakistan. *J. Ethnopharmacol.*, 2021, 281, 114437.

Khan, M. N., Mohammad, F., Mobin, M., Saqib, M. A. Tolerance of plants to abiotic stress: A role of nitric oxide and calcium. In: Khan, M., Mobin, M., Mohammad, F., Corpas, F. (Eds.), *Nitric Oxide in Plants: Metabolism and Role in Stress Physiology* (pp. 225–242). Springer: Cham, 2014.

Khan, N., Ali, S., Zandi, P., Mehmood, A., Ullah, S., Ikram, M., et al. Role of sugars, amino acids and organic acids in improving plant abiotic stress tolerance. *Pak. J. Bot.*, 2020, 52, 355–363.

Khan, T., Muhammad, S., Khan, B., Khan, H. Investigating the levels of selected heavy metals in surface water of Shah Alam River (a tributary of River Kabul, Khyber Pakhtunkhwa). *J. Himalayan Earth Sci.*, 2011, 44(2), 71–79.

Khripach, V., Zhabinskii, V., DeGroot, A. Twenty years of brassinosteroids: Steroidal plant hormones warrant better crops for the XXI century. *Ann. Bot.*, 2000, 86, 441–447.

Komis, G., Šamajová, O., Ovečka, M., Šamaj, J. Cell and developmental biology of plant mitogen-activated protein kinases. *Annu. Rev. Plant Biol.*, 2018, 69, 237–265.

Kopittke, P. M., Asher, C. J., Blamey, F. P., Auchterlonie, G. J., Guo, Y. N., Menzies, N. W. Localization and chemical speciation of Pb in roots of signal grass (*Brachiaria decumbens*) and Rhodes grass (*Chloris gayana*). *Environ. Sci. Technol.*, 2008, 42, 4595–4599.

Kováćik, J., Rotková, G., Bujdoś, M., Babula, P., Peterková, V., Matúś, P. Ascorbic acid protects *Coccomyxa subellipsoidea* against metal toxicity through modulation of ROS/NO balance and metal uptake. *J. Hazard Mater.*, 2017, 339, 200–207.

Krämer, U., Cotter-Howells, J. D., Charnock, J. M., Baker, A. J. M., Smith, J. A. C. Free histidine as a metal chelator in plants that accumulate nickel. *Nature,* 1996, 379(6566), 635–638.

Krzesłowska, M. The cell wall in plant cell response to trace metals: Polysaccharide remodeling and its role in defense strategy. *Acta Physiol. Plant*, 2011, 33(1), 35–51.

Kumar, A., Prasad, M. N. V., Sytar, O. Lead toxicity, defense strategies and associated in dicative biomarkers in *Talinum triangulare* grown hydroponically. *Chemosphere*, 2012, 89, 1056–1165.

Kumar, S., Shah, S. H., Vimala, Y., Jatav, H. S., Ahmad, P., Chen, Y., Siddique, K. H. Abscisic acid: Metabolism, transport, crosstalk with other plant growth regulators, and its role in heavy metal stress mitigation. *Front. Plant Sci.*, 2022, 13, 972856.

Lavid, N., Schwartz, A., YarDen, O., Tel-Or, E. The involvement of poly phenols and peroxidase activities in heavy metal accumulation by epidermal glands of water lily (*Nymphaeceaea*). *Planta,* 2001, 212, 323.

Li, B., Fan, R., Sun, G., Sun, T., Fan, Y., Bai, S., et al. Flavonoids improve drought tolerance of maize seedlings by regulating the homeostasis of reactive oxygen species. *Plant Soil*, 2021, 461, 389–405.

Lian, J., Zhao, L., Wu, J., Xiong, H., Bao, Y., Zeb, A., Tang, J., Liu, W. Foliar spray of TiO_2 nanoparticles prevails over root application in reducing Cd accumulation and mitigating Cd-induced phytotoxicity in maize (*Zea mays* L.). *Chemosphere*, 2020, 239, 124794.

Liu, Sh., Yang, R. Pan, Y., Ma, J., Pan, M., Zhao, Y., et al. Nitric oxide contributes to minerals absorption, proton pumps and hormone equilibrium under cadmium excess in *Trifolium repens* L. plants. *Ecotoxicol. Environ. Saf.*, 2015, 119, 35–46.

Lushchak, V. I., Semchuk, N. M. Tocopherol biosynthesis: Chemistry, regulation and effects of environmental factors. *Acta Physiol. Plant.*, 2012, 34, 1607–1628.

Mahboob, S., Alkkahem Al-Balwai, H. F., Al-Misned, F., Al-Ghanim, K. A., Ahmad, Z. A study on the accumulation of nine heavy metals in some important fish species from a natural reservoir in Riyadh, Saudi Arabia, *Toxicol. Environ. Chem.*, 2014, 96(5), 783–798.

Mahmood, A., Malik, R. N. Human health risk assessment of heavy metals via consumption of contaminated vegetables collected from different irrigation sources in Lahore, Pakistan. *Arab. J. Chem.*, 2014, 7(1), 91–99.

Manara, A. Plant responses to heavy metal toxicity. In: Furini, A. (Ed.), *Plants and Heavy Metals, Springer Briefs in Molecular Science* (pp. 27–53). Springer: Dordrecht, 2012.

Mandal, M. K., Chandra-Shekara, A., Jeong, R.-D., Yu, K., Zhu, S., Chanda, B., et al. Oleic acid-dependent modulation of NITRIC OXIDE ASSOCIATED1 protein levels regulates nitric oxide-mediated defense signalling in *Arabidopsis*. *Plant Cell*, 2012, 24, 1654–1674.

Manesh, R. R., Grassi, G., Bergami, E., Marques-Santos, L. F., Faleri, C., Liberatori, G., Corsi, I. Co-exposure to titanium dioxide nanoparticles does not affect cadmium toxicity in radish seeds (*Raphanus sativus*). *Ecotoxicol. Environ. Saf.*, 2018, 148, 359–366.
Mathys, W. The role of malate, oxalate, and mustard oil glucosides in the evolution of zinc resistance in herbage plants. *Physiol. Plant.*, 1977, 40, 130–136.
McLaughlin, M. J., Parker, D. R., Clark, J. M. Meta land micronutrients-food safety issues. *Field Crops Res.*, 1999, 60, 143–163.
Meena, M., Divyanshu, K., Kumar, S., Swapnil, P., Zehra, A., Shukla, V., et al. Regulation of L-proline biosynthesis, signal transduction, transport, accumulation and its vital role in plants during variable environmental conditions. *Heliyon*, 2019, 5, e02952.
Mewis, I., Khan, M. A., Glawischnig, E., Schreiner, M., Ulrichs, C. Water stress and aphid feeding differentially influence metabolite composition in *Arabidopsis thaliana* (L.). *PLoS ONE*, 2012, 7, e48661.
Michalak, A. Phenolic compounds and their antioxidant activity in plants growing under heavy metal stress. *Pol. J. Environ. Stud.*, 2006, 15(4), 523–530.
Miransari, M. Hyper accumulators, arbuscular mycorrhizal fungi and stress of heavy metal. *Biotechnol. Adv.*, 2011, 29, 645–653.
Miyagawa, H. Studies on nitrogen-containing secondary metabolites playing a defensive role in plants. *J. Pestic. Sci.*, 2009, 34, 110–112.
Moghadam, H. R. T. Application of super absorbent polymer and ascorbic acid to mitigate deleterious effects of cadmium in wheat. *Pesqui. Agropecu_aria Trop.*, 2016, 46, 9–18.
Mullineaux, P., Rausch, T. Glutathione, photosynthesis and the redox regulation of stress-responsive gene expression. *Photosynth Res.*, 2005, 86, 459–474.
Munne-Bosch, S. The role of a9-tocopherol in plant stress tolerance. *J. Plant Physiol.*, 2005, 162, 743–748.
Nada, R. S., Ashmawi, A. E., Mady, E., Randhir, T. O., Elateeq, A. A. Effect of organic manure and plant growth promoting microbes on yield, quality and essential oil constituents of fennel bulb (*Foeniculum vulgare* Mill.). *J. Ecol. Eng.*, 2022, 23, 149–164.
Naikoo, M. I., Dar, M. I., Raghib, F., Jaleel, H., Ahmad, B., Raina, A., et al. Role and regulation of plants phenolics in abiotic stress tolerance: An overview. *Plant Signal. Mol.*, 2019, 2, 157–168.
Nakabayashi, R., Yonekura-Sakakibara, K., Urano, K., Suzuki, M., Yamada, Y., Nishizawa, T., et al. Enhancement of oxidative and drought tolerance in Arabidopsis by overaccumulation of antioxidant flavonoids. *Plant J.*, 2014, 77, 367–379.
Neumann, D., Zur Nieden, U. Silicon and heavy metal tolerance in higher plants. *Phytochemistry*, 2001, 56, 685–692.
Noriega, G., Cruz, D. S., Batlle, A., Tomaro, M., Balestrasse, K. Heme oxygenase is involved in the protection exerted by jasmonic acid against cadmium stress in soybean roots. *J. Plant Growth Regul.*, 2012, 31, 79–89.
Ochoa-Villarreal, M., Howat, S., Hong, S., Jang, M. O., Jin, Y.-W., Lee, E.-K., Loake, G. J. Plant cell culture strategies for the production of natural products. *BMB Rep.*, 2016, 49, 149.
Ouyang, H., Vogel, H. J. Metal ion binding to calmodulin: NMR and fluorescence studies. *Biometals*, 1998, 11(3), 213–222.
Oves, M., Khan, M. S., Qari, H. A. *Ensifer adhaerens* for heavy metal bioaccumulation, biosorption, and phosphate solubilization under metal stress condition. *Taiwan Inst. Chem. Eng.*, 2017, 80, 540–552.
Ozturk, M., Altay, V., Karahan, F. Studies on trace elements in *Glycyrrhiza* taxa distributed in Hatay-Turkey. *Int. J. Plant Environ.*, 2017, 3(2), 1–7.
Ozturk, M., Yucel, E., Gucel, S., Sakcali, S., Aksoy, A. Plants as biomonitors of trace elements pollution in soil. In: Prasad, M. N. V. (Ed.), *Trace Elements: Environmental Contamination, Nutritional Benefits and Health Implications* (pp. 723–744). Wiley: New York, 2008.
Pagare, S., Bhatia, M., Tripathi, N., Pagare, S., Bansal, Y. Secondary metabolites of plants and their role: Overview. *Curr. Trends Biotechnol. Pharm.*, 2015, 9, 293–304.
Parida, A. K., Dagaonkar, V. S., Phalak, M. S., Umalkar, G., Aurangabadkar, L. P. Alterations in photosynthetic pigments, protein and osmotic components in cotton genotypes subjected to short-term drought stress followed by recovery. *Plant Biotechnol. Rep.*, 2007, 1, 37–48.

Parihar, P., Singh, S., Singh, R., Rajasheker, G., Rathnagiri, P., Srivastava, R. K., et al. An integrated transcriptomic, proteomic, and metabolomic approach to unravel the molecular mechanisms of metal stress tolerance in plants. In: Srivastava, S. et al. (Eds.), *Plant-Metal Interactions* (pp. 1–28). Springer Nature: Switzerland AG, 2019.

Patra, M., Bhowmik, N., Bandopadhyay, B., Sharma, A. Comparison of mercury, lead and arsenic with respect to genotoxic effects on plant systems and the development of genetic tolerance. *Environ. Exp. Bot.*, 2004, 52, 199–223.

Pinto, A. P., Simoes, I., Mota, A. M. Cadmium impact on root exudates of sorghum and maize plants: A speciation study. *J. Plant Nutr.*, 2008, 31(10), 1746–1755.

Pitzschke, A., Forzani, C., Hirt, H. Reactive oxygen species signalling in plants. *Antioxid. Redox. Signal.*, 2006, 8, 1757–1764.

Pollard, A. J., McCartha, G. L., Quintela-Sabarís, C., Flynn, Th. A., Sobczyk, M. K., Smith, J. A. C. Intraspecific variation in nickel tolerance and hyperaccumulation among serpentine and limestone populations of *Odontarrhena serpyllifolia* (Brassicaceae: Alysseae) from the Iberian Peninsula. *Plants,* 2021, 10, 800.

Popova L. P., Maslenkova L. T., Yordanova R. Y., Ivanova A. P., Krantev A. P., Szalai G., et al. Exogenous treatment with salicylic acid attenuates cadmium toxicity in pea seedlings. *Plant Physiol. Biochem.*, 2009, 47, 224–231.

Rahim, M., Ullah, I., Khan, A., Haris, M. R. H. M. Health risk from heavy metals via consumption of food crops in the vicinity of district Shangla. *J. Chem. Soc. Pak.*, 2016, 38(1), 177–185.

Rajab, H., Khan, M. S., Wirtz, M., Malagoli, M., Qahar, F., Hell, R. Sulfur metabolic engineering enhances cadmium stress tolerance and root to shoot iron translocation in *Brassica napus* L. *Plant Physiol. Biochem.*, 2020 152, 32–43.

Ramírez, R. The gastropodOsilinus atrataas a bioindicator of Cd, Cu, Pb and Zn contamination in the coastal waters of the Canary Islands. *Chem. Ecol.*, 2013, 29(3), 208–220.

Rascio, N., Navari-Izzo, F. Heavy metal hyperaccumulating plants: How and why do they do it? And what makes them so interesting? *Plant Sci.*, 2011, 180(2), 169–181.

Rashedy, A. A., Abd-ElNafea, M. H., Khedr, E. H. Co-application of proline or calcium and humic acid enhances productivity of salt stressed pomegranate by improving nutritional status and osmoregulation mechanisms. *Sci. Rep.*, 2022, 12, 1–10.

Rastgoo, L., Alemzadeh, A. Biochemical responses of Gouan (*Aeluropus littoralis*) to heavy metals stress. *Australian J. Crop Sci.*, 2011, 5(4), 375–383.

Rauser, W. E. Structure and function of metal chelators produced by plants. The case for organic acids, amino acids, phytin and metallothioneins. *Cell Biochem. Biophys.*, 1999, 31, 19–48.

Raza, S. H., Athar, H. R., Ashraf, M., Hameed, A. Glycinebetaine-induced modulation of antioxidant enzymes activities and ion accumulation in two wheat cultivars differing in salt tolerance. *Environ. Exp. Bot.*, 2007, 60, 368–376.

Renard, C. M. G. C., Jarvis, M. C. Acetylation and methylation of homogalacturonans. 2: Effect on ion-binding properties and conformations. *Carbohydr. Polym.*, 1999, 39, 209–216.

Rizwan, M., Ali, S., Ali, B., Adrees, M., Arshad, M., Hussain, A., et al. Zinc and iron oxide nanoparticles improved the plant growth and reduced the oxidative stress and cadmium concentration in wheat. *Chemosphere*, 2019, 214, 269–277.

Rochfort, S. Metabolomics reviewed: A new "omics" platform technology for systems biology and implications for natural products research. *J. Nat. Prod.*, 2005, 68, 1813–1820.

Rodziewicz, P., Swarcewicz, B., Chmielewska, K., Wojakowska, A., Stobiecki, M. Influence of abiotic stresses on plant proteome and metabolome changes. *Acta Physiol. Plant.*, 2014, 36, 1–19.

Rosa, M., Prado, C., Podazza, G., Interdonato, R., González, J. A., Hilal, M., Prado, F. E. Soluble sugars: Metabolism, sensing and abiotic stress: A complex network in the life of plants. *Plant Signal. Behav.*, 2009, 4, 388–393.

Ruan, C., da Silva, T. J. A. Metabolomics: Creating new potentials for unraveling the mechanisms in response to salt and drought stress and for the biotechnological improvement of xero-halophytes. *Crit. Rev. Biotechnol.*, 2011, 31, 153–169.

Rudd, J. J., Franklin-Tong, V. E. Unravelling response-specificity in Ca^{2+} signalling pathways in plant cells. *New Phytol.*, 2001, 151, 7–33.

Ruppert, A. M., Weinberg, K., Palkovits, R. Hydrogenolysis goes bio: From carbohydrates and sugar alcohols to platform chemicals. *Angew. Chem. Int. Ed.*, 2012, 51, 2564–2601.

Salam, U., Shakir-Ullah, Tang, Z., Elateeq, A. A., Khan, Y., Khan, J., Khan, A., Ali, S. Plant metabolomics: An overview of the role of primary and secondary metabolites against different environmental stress factors. *Life.*, 2023, 13, 706–730.

Saltveit, M. E. Synthesis and metabolism of phenolic compounds. *Fruit Veget. Phytochem. Chem. Hum. Health*, 2017, 2, 115.

Sanders, D., Pelloux, J., Brownlee, C., Harper, J. F. Calcium at the crossroads of signaling. *Plant Cell*, 2002, 14, S401–S417.

Saxena, R., Adhikari, D., Goyal, H. Biomass-based energy fuel through biochemical routes: A review. *Renew. Sustain. Energy Rev.*, 2009, 13, 167–178.

Shafeeq-Ur-Rahman, Sh. U., Li, Y., Hussain, S., Hussain, B., Khan, W., Riaz, L., et al. Role of phytohormones in heavy metal tolerance in plants: A review. *Ecol. Indic.*, 2023, 146, 109844.

Shao, H. B., Chu, L. Y., Lu, Z. H., Kang, C. M. Primary antioxidant free radical scavenging and redox signaling pathways in higher plant cells. *Int. J. Biol. Sci.*, 2008, 4, 8–14.

Sharma, A., Shahzad, B., Rehman, A., Bhardwaj, R., Landi, M., Zheng, B. Response of phenylpropanoid pathway and the role of polyphenols in plants under abiotic stress. *Molecules*, 2019, 24, 2452.

Sharma, B., Walia, A., Sharma, S., Soni, U. 18 - Exogenous application of nanomaterials as biostimulants for heavy metal stress tolerance. In: Gill, S. S., Tuteja, N., Khan, N. A., Gill, R. (Eds.), *Biostimulants and Protective Biochemical Agents, Biostimulants in Alleviation of Metal Toxicity in Plants* (pp. 423–448). Academic Press, 2023.

Sharma, D., Verma, N., Pandey, C., Verma, D., Bhagat, P. K., Noryang, S., et al. MAP kinase as regulators for stress responses in plants: An overview. In: *Protein Kinases and Stress Signaling in Plants*. John Wiley & Sons Ltd.: Hoboken, NJ, USA, 2020.

Sharma, S. S., Dietz, K. J. The significance of amino acids and amino acid-derived molecules in plant responses and adaptation to heavy metal stress. *J. Exp. Bot.*, 2006, 57, 711–726.

Sharma, S. S., Dietz, K. J. The relationship between metal toxicity and cellular redox imbalance. *Trends Plant Sci.*, 2009, 14, 43–50.

Singh, D., Singh, C. K., Singh, D., Sarkar, S. K., Prasad, S. K., Sharma, N. L., Singh, I. Glycine betaine modulates chromium (VI)-induced morpho-physiological and biochemical responses to mitigate chromium toxicity in chickpea (*Cicer arietinum* L.) cultivars. *Sci. Rep.*, 2022, 12, 8005.

Singh, J., Lee, B.-K. Influence of nano-TiO_2 particles on the bioaccumulation of Cd in soybean plants (*Glycine max*): A possible mechanism for the removal of Cd from the contaminated soil. *J. Environ. Manag.*, 2016, 170, 88e96.

Singh, S., Parihar, P., Singh, R., Singh, V. P., Prasad, Sh. M. Heavy metal tolerance in plants: Role of transcriptomics, proteomics, metabolomics, and ionomics. *Front. Plant Sci.*, 2016, 6, 1143–1179.

Siripornadulsil, S., Traina, S., Verma, D. P. S., Sayre, R. T. Molecular mechanisms of proline-mediated tolerance to toxic heavy metal in transgenic microalgae. *Plant Cell*, 2002, 14, 2837–2847.

Snedden, W. A., Fromm, H. Calmodulin as a versatile calcium signal transducer in plants. *New Phytol.*, 2001, 151, 35–66.

Son, J. A., Narayanankutty, D P., Roh, K. S. Influence of exogenous application of glutathione on rubisco and rubisco activase in heavy metal-stressed tobacco plant grown in vitro. *Saudi J. Biol. Sci.*, 2014, 21, 89–97.

Spanò, S., Bottega, C., Sorce, G., Bartoli, M., Castiglione, R. TiO_2 nanoparticles may alleviate cadmium toxicity in co-treatment experiments on the model hydrophyte *Azolla filiculoides*. *Environ. Sci. Pollut. Res.*, 2019, 26, 29872–29882.

Sperdouli, I., Moustakas, M. Interaction of proline, sugars, and anthocyanins during photosynthetic acclimation of Arabidopsis thaliana to drought stress. *J. Plant Physiol.*, 2012, 169, 577–585.

Spiegel, H. Trace element accumulation in selected bioindicators exposed to emissions along the industrial facilities of Danube Lowland. *Turk. J. Chem.*, 2002, 26(6), 815–823.

Sun, L., Cao, X., Tan, C., Deng, Y., Cai, R., Peng, X., Bai, J. Analysis of the effect of cadmium stress on root exudates of *Sedum plumbizincicola* based on metabolomics. *Ecotoxicol. Environ. Saf.*, 2020, 205, 111152.

Sytar, O., Kumar, A., Latowski, D., Kuczynska, P., Strzałka, K., Prasad, M. N. V. Heavy metal-induced oxidative damage, defense reactions, and detoxification mechanisms in plants. *Acta Physiol. Plant*, 2013, 35(4), 985–999.

Taha, R. S., Seleiman, M. F., Alhammad, B. A., Alkahtani, J., Alwahibi, M. S., Mahdi, A. H. Activated Yeast extract enhances growth, anatomical structure, and productivity of *Lupinus termis* L. plants under actual salinity conditions. *Agronomy*, 2020, 11, 74.

Tang W., Charles T. M., Newton R. J. Overexpression of the pepper transcription factor CaPF1 in transgenic *Virginia pine* (*Pinus Virginiana* Mill.) confers multiple stress tolerance and enhances organ growth. *Plant Mol. Biol.*, 2005, 59, 603–617.

Teklić, T., Paradiković, N., Špoljarević, M., Zeljković, S., Loncarić, Z., Lisjak, M. Linking abiotic stress, plant metabolites, biostimulants and functional food. *Ann. Appl. Biol.*, 2021, 178, 169–191.

Tiwari, S., Lata, C. Heavy metal stress, signaling, and tolerance due to plant-associated microbes: An overview. *Front. Plant Sci.*, 2018, 9, 452.

Toffolatti, S. L., Maddalena, G., Passera, A., Casati, P., Bianco, P. A., Quaglino, F. Role of terpenes in plant defense to biotic stress. In: *Biocontrol Agents and Secondary Metabolites* (pp. 401–417). Elsevier: Amsterdam, the Netherlands, 2021.

Türkmen, M., Türkmen, A., Tepe, Y., Töre, Y., Ates, A. Determination of metals in fish species from Aegean and Mediterranean seas, *Food Chem.*, 2009, 113(1), 233–237.

Valifard, M., Mohsenzadeh, S., Niazi, A., Moghadam, A. Phenylalanine ammonia lyase isolation and functional analysis of phenylpropanoid pathway under salinity stress in '*Salvia*' species. *Aust. J. Crop Sci.*, 2015, 9, 656–665.

van Dam, N. M., Bouwmeester, H. J., Metabolomics in the rhizosphere: Tapping into belowground chemical communication. *Trends Plant Sci.*, 2016, 21, 256–265.

Vardhini, B. V. Brassinosteroids are potential ameliorators of heavy metal stresses in plants. In: Ahmad, P. (Ed.), *Plant Metal Interaction: Emerging Remediation Techniques* (pp. 209–237). Elsevier, 2016.

Verbruggen, N., Hermans, C. Proline accumulation in plants: A review. *Amino Acids*, 2008, 35, 753–759.

Vickers, N. J. Animal communication: When I'm calling you, will you answer too? *Curr. Biol.*, 2017, 27, R713–R715.

Viehweger, K. How plants cope with heavy metals. *Bot Stud.*, 2014, 55, 1–12.

Vriet, C., Russinova, E., Reuzeau, C. Boosting crop yields with plant steroids. *Plant Cell*, 2012, 24, 842–857.

Wang, J., Song, L., Gong, X., Xu, J., Li, M. Functions of jasmonic acid in plant regulation and response to abiotic stress. *Int. J. Mol. Sci.*, 2020, 21, 1446.

Wang, Y., Tan, P., Chang, L., Yue, Z., Zeng, Ch., Li, M., et al. Exogenous proline mitigates toxic effects of cadmium via the decrease of cadmium accumulation and reestablishment of redox homeostasis in *Brassica juncea*. *BMC Plant Biol.*, 2022, 22(182), 1–19.

Waterman, P. G. Roles for secondary metabolites in plants. In: *Ciba Foundation Symposium 171-Secondary Metabolites: Their Function and Evolution: Secondary Metabolites: Their Function and Evolution.* Ciba Foundation Symposium 171; Wiley Online Library: Hoboken, NJ, USA, 2007.

Wink, M. Introduction: Biochemistry, physiology and ecological functions of secondary metabolites. In: *Annual Plant Reviews Volume 40: Biochemistry of Plant Secondary Metabolism* (pp. 1–19). Wiley: Hoboken, NJ, USA, 2010.

Winkel-Shirley, B. Biosynthesis of flavonoids and effects of stress. *Curr. Opin. Plant Biol.*, 2002, 5, 218.

Wu, W., Zhang, Q., Zhu, Y., Lam, H.-M., Cai, Z., Guo, D. Comparative metabolic profiling reveals secondary metabolites correlated with soybean salt tolerance. *J. Agric. Food Chem.*, 2008, 56, 11132–11138.

Xie, X., Weiss, D. J., Weng, B., Liu, J., Lu, H., Yan, C. The short-term effect of cadmium on low molecular weight organic acid and amino acid exudation from mangrove (*Kandelia obovata* (S., L.) Yong) roots. *Environ. Sci. Pollut. Res.*, 2013, 20, 997–1008.

Yang, C.-Q., Wu, X.-M., Ruan, J.-X., Hu, W.-L., Mao, Y.-B., Chen, X.-Y., Wang, L.-J. Isolation and characterization of terpene synthases in cotton (*Gossypium hirsutum*). *Phytochemistry*, 2013, 96, 46–56.

Yang, X. E., Jin, X. F., Feng, Y., Islam, E. Molecular mechanisms and genetic basis of heavy metal tolerance/hyperaccumulation in plants. *J. Integr. Plant. Biol.*, 2005, 47(9), 1025–1035.

Yazici, L. Influence of different sowing times on yield and biochemical characteristics of different opium poppy (*Papaver somniferum* L.) genotypes. *J. King Saud Univ.-Sci.*, 2022, 34, 102337.

Yuan, H. M., Liu, W. C., Jin, Y., Lu, Y. T. Role of ROS and auxin in plant response to metal-mediated stress. *Plant Signal Behav.*, 2013, 8(7), e24671.

Yusuf, M. A., Kumar, D., Rajwanshi, R., Strasser, R. J., Tsimilli-Michael, M., Govindjee, et al. Over expression of c-tocopherol methyl transferase gene in transgenic *Brassica juncea* plants alleviates abiotic stress: Physiological and chlorophyll a fluorescence measurements. *Biochim. Biophys. Acta,* 2010, 1797, 1428–1438.

Yusuf, M., Fariduddin, Q., Ahmad, A. 24-Epibrassinolide modulates growth, nodulation, antioxidant system, and osmolyte in tolerant and sensitive varieties of *Vigna radiata* under different levels of nickel: A shot gun approach. *Plant Physiol. Biochem.,* 2012, 57, 143–153.

Zhao, H., Lan, X., Yu, F., Li, Z., Yang, J., Du L. Comprehensive assessment of heavy metals in soil-crop system based on PMF and evolutionary game theory. *Sci. Total Environ.*, 2022, 849, 157549.

Zhao, J., Wang, X., Pan, X., Jiang, Q., Xi, Z. Exogenous putrescine alleviates drought stress by altering reactive oxygen species scavenging and biosynthesis of polyamines in the seedlings of Cabernet Sauvignon. *Front. Plant Sci.*, 2021, 12, 767992.

Zhao, Y., Wang, J., Huang, W., Zhang, D., Wu, J., Li, B., et al. Abscisic-acid-regulated responses to alleviate cadmium toxicity in plants. *Plants (Basel)*, 2023, 12(5), 1023.

CHAPTER 3

Concept of Metabolites Fingerprinting

RUPSHIKHA NATH[1], SIBASHISH KITYANIA[1], RAJAT NATH[2], DEEPA NATH[3], and ANUPAM DAS TALUKDAR[1*]

[1]*Department of Life Science and Bioinformatics, Assam University, Silchar, Cachar, Assam, India*

[2]*Department of Biotechnology and Microbiology, School of Natural Sciences, Techno India University Tripura (TIUT), Agartala, Tripura, India*

[3]*Department of Botany, Guru Charan University, Silchar, Cachar, Assam, India*

*Corresponding authors

ABSTRACT

Metabolite fingerprinting, a subfield of metabolomics, has become an effective tool for characterizing the dynamic metabolic profiles of biological systems. Metabolomics, the investigation of metabolites of very small molecules, plays a pivotal role in understanding cellular metabolism and its alterations in different physiological and pathological circumstances. Metabolite fingerprinting involves the qualitative and quantitative analysis of these metabolites to generate unique metabolic profiles, akin to fingerprints, which reflect the underlying biochemical processes within a biological system. The use of advanced analytical techniques, such as mass spectrometry, liquid chromatography, and nuclear magnetic resonance spectroscopy, has significantly enhanced the scope and accuracy of metabolite fingerprinting. By comparing metabolic profiles between healthy and diseased individuals or different disease stages, unique metabolite signatures associated with specific conditions can be uncovered. Metabolic signatures offer insights into disease pathogenesis, helping to find possible drug targets and making it easier to create cutting-edge treatment approaches. Metabolite fingerprinting faces challenges in standardization, data processing, interpretation, and combining methods from systems biology and multi-omics. Metabolite fingerprinting has the potential to revolutionize the identification of illness biomarkers, deepen our understanding of biological mechanisms, and enhance human health.

3.1 INTRODUCTION

Metabolomics, an essential part of systems biology, studies metabolites' quantification and their interactions within a biological system. It provides a top–down approach by assessing

Exploring Plant Stress Tolerance: A Metabolomics Approach.
Shuvasish Choudhury, Debojyoti Moulick, and Mirza Hasanuzzaman (Eds.)

DOI: 10.1201/9781779641731-3

the end result of cellular processes affected by various stimuli (Nicholson and Lindon, 2008). Unlike genomics or proteomics, metabolomics offers a comprehensive view of phenotype changes influenced by factors such as the microbiome, diet, drugs, and more (Kosmides et al., 2013). An early screening technique called "metabolite fingerprinting" classifies data according to their intrinsic biological properties and separates them from irrelevant background noise without identifying specific metabolites. This method unveils unique metabolic fingerprints specific to an organism's state at a given time. Its integration with other omics fields enhances understanding, compensating for limitations in solely relying on genomics or proteomics, especially in complex diseases such as cancer or autoimmune conditions (Kosmides et al., 2013). Metabolomics relies on specific biofluids or tissues to capture an organism's metabolic fingerprint. Additionally, metabolomics simplifies analyses, offers reproducibility, and permits ethical studies on laboratory animals, enabling insights transferable to humans, all while leveraging easily collectible biofluids such as urine and blood for temporal evaluations (Nicholson et al., 2012). Urine and plasma, frequently used due to easy collection and noninvasiveness, provide dynamic reflections of metabolic changes. Although plasma offers a snapshot, urine presents a time-averaged representation (Maher et al., 2007). Techniques like nuclear magnetic resonance (NMR) spectroscopy, despite lower sensitivity, offer advantages such as chemical structure insights and nondestructive sampling (Kell, 2004; Kosmides et al., 2013; Lenz and Wilson, 2007). Techniques used in mass spectrometry (MS) that including gas chromatography-mass spectrometry (GC-MS) and liquid chromatography-mass spectrometry (LC-MS) provide higher sensitivity but demand more complex sample preparation (Lenz and Wilson, 2007). GC-MS stands out for its resolution (O'Hagan et al., 2007), whereas LC-MS benefits from minimal sample prep (Lu et al., 2008). Additionally, magic-angle spinning NMR spectroscopy proves useful for tissue-specific metabolic phenotyping, offering insights into metabolite compartmentalization and the cellular environment (Lenz and Wilson, 2007; Rooney et al., 2003). These techniques vary in sensitivity, resolution, and applicability across biofluids and tissues, each contributing uniquely to metabolomic data generation. Metabolomics faces the challenge of handling vast data volumes from various collection methods, with ^{1}H-NMR spectroscopy providing insights into up to 100 metabolites in urine and 30 in plasma and tissues, whereas MS-coupled techniques can yield data on over 1000 metabolites (Griffin, 2006). Computational analysis, a crucial aspect, involves techniques such as metabolite identification through NMR spectroscopy (Cui et al., 2008), preprocessing for MS datasets, and multivariate analysis methods. Principal component analysis (PCA) (Nicholson et al., 1999) aids in visualizing metabolic fingerprints and potential biomarkers (Kosmides et al., 2013). Class distinction is improved by the partial least squares method with discriminant analysis (PLS-DA) (Mao et al., 2009; Nicholson and Lindon, 2008). Geometric trajectory analysis studies temporal responses (Keun et al., 2004), and entropy-based models assess uncertainty (Veselkov et al., 2010). Genetic algorithms improve model accuracy and biomarker selection (Cavill et al., 2009). Moreover, metabolomics reveals how an organism's metabolism responds to internal and external factors such as diet, drugs, and lifestyle, crucial in understanding health and diseases like cardiovascular disease, including hypertension (Huang et al., 2020; Nicholson et al., 2012; Onuh and Qiu, 2021). Because of their variety, the millions of small molecules they analyze within cells, tissues, and fluids are a challenge (Tzoulaki et al., 2018). Multiple techniques, such as NMR, LC-MS, GC-MS, and others, are combined to capture these metabolites comprehensively (Nikolic et al., 2014; Theodoridis et al., 2008). Statistical tools such as PCA and PLS-DA help identify distinct metabolic profiles (Onuh and Aluko, 2019). To ensure consistency and comparability across studies, researchers employ specific methods, set thresholds, and use databases such as METLIN, HMDB, NIST, and KEGG for biomarker identification and pathway analysis (Onuh and Aliani, 2020; Onuh and Qiu, 2021; Rinschen and Ivanisevic, 2019). Various software tools, such as MeltDB (Neuweger et al., 2008), Metabo Analyst (Xia et al., 2009), Metabolite Set Enrichment Analysis (Xia and Wishart, 2010), and MassTrix (Suhre and Schmitt-Kopplin, 2008), assist in data preprocessing, metabolite identification, and pathway analysis, contributing to a comprehensive understanding of metabolic changes (Kosmides et al., 2013). Harmonizing techniques and careful data processing remain essential, reflecting the complexity and precision required in metabolomics research.

3.2 METHODS USED IN METABOLITE FINGERPRINTING

3.2.1 ADVANCED ANALYTICAL TECHNIQUE

3.2.1.1 NMR

NMR detects atoms with nonzero magnetic moments, such as ^{1}H, ^{13}C, ^{14}N, ^{15}N, and ^{31}P, which enables the measurement and identification of metabolites in biological samples. It offers detailed information on molecule identity via frequency, intensity, and relaxation properties. NMR's versatility lies in nondestructive analysis from various biological samples, tailored detection schemes, and isotopic labeling enabling metabolic pathway exploration (Krishnan et al., 2004; Ratcliffe and Shachar-Hill, 2001). High-throughput metabolomics demands sensitivity and broad detection; hence, ^{1}H-NMR, sensitive to most compounds, is commonly used (Exarchou et al., 2003; Hinse et al., 2003). Metabolite fingerprinting through ^{1}H-NMR spectroscopy offers a transformative approach in characterizing plant samples. This method, despite being a recent application in plant analysis, holds immense promise in identifying diverse metabolites efficiently (Krishnan et al., 2004). Metabolite fingerprinting via NMR involves multivariate analysis to compare spectra sets from biological samples, prioritizing similarities or differences over specific metabolite identification (Defernez and Colquhoun, 2003; Lindon et al., 2001). The strength of ^{1}H-NMR lies in its minimal sample preparation requirement (Defernez and Colquhoun, 2003; Ott et al., 2003), its capacity to examine a variety of substances, nondestructiveness, and flexibility in analyzing samples without prior knowledge of their composition. By using PCA and pattern recognition, it bypasses the signal assignment complexities and identifies critical spectral components relevant to sample classification. Applied in various plant studies, fingerprinting discerns differences in metabolite composition due to growth conditions (Ratcliffe and Shachar-Hill, 2001), stressors, or genetic modifications, showcasing the technique's utility in revealing metabolic variations even under controlled growth (Sumner et al., 2003). Notably, these analyses unearth profound biological insights, delineating metabolic responses to stressors such as cadmium exposure (Bailey et al., 2003;

Ott et al., 2003) or herbicide action and illuminating nuances in wild-type versus transgenic plants (Charlton et al., 2004), exemplifying the potential of NMR fingerprinting beyond mere sample classification (Krishnan et al., 2004).

3.2.1.2 CHROMATOGRAPHIC TECHNIQUE COUPLED WITH MS

To get further insight into metabolic fingerprinting via the application of GC-MS and LC-MS techniques, we have taken two examples of human and plant studies from some recent research work. Here is a brief overview described below.

3.2.1.2.1 In Human

Prostate cancer (CaP) presents a significant health concern due to late detection (Gerlinger et al., 2015), limited diagnostic accuracy of prostate-specific antigen (Kilpeläinen et al., 2011), and invasive biopsy methods. Its pathogenesis involves multifactorial influences, including age, genetics, lifestyle, and hormonal effects (Jerónimo et al., 2011; Key, 2014). Understanding these processes is crucial. This study employs urine metabolic fingerprinting, considering its noninvasiveness and stability for potential biomarker discovery (Struck-Lewicka et al., 2015). Metabolomic approaches have been pivotal in determining markers in various biological samples (Szymańska et al., 2010). Although NMR offers robust metabolite qualification, MS surpasses it in sensitivity (Struck-Lewicka et al., 2015). In Struck-Lewicka et al study (Struck-Lewicka et al., 2015), Concurrently used methods include GC with triple quadrupole MS and also using LC combined with time-of-flight MS, allowing for a more comprehensive urinary metabolite investigation to identify potential CaP markers, a pioneering endeavor in this domain (Struck-Lewicka et al., 2015).

Urine samples were obtained for the study from 32 individuals with CaP and 32 healthy participants who were matched for age and body mass index. These samples underwent distinct preparation processes for LC-MS and GC-MS analysis. For LC-MS, after thawing, the urine was diluted and filtered before being transferred to high-performance liquid chromatography (HPLC) vials (Struck-Lewicka et al., 2015). Conversely, for GC-MS, the urine underwent a treatment involving urease addition, methanol inclusion, vortexing, centrifugation, and silylation, eventually evaporating to dryness before GC-MS analysis (Webb-Robertson et al., 2014). Quality control (QC) samples ensured system stability and were analyzed alongside the urine samples. LC-MS analysis used HPLC coupled with TOF/MS, whereas GC-MS used GC-MS-TQ8030. Both analyses were conducted in a randomized sequence, and QC samples were assessed regularly to ensure stability (Struck-Lewicka et al., 2015). The outcomes showed metabolic fingerprints that distinguished between CaP patients and healthy people based on variations in metabolites.

Many processes and analyses were performed on the data from the LC-MS and GC-MS investigations. LC-ESI-TOF/MS data underwent MFE extraction for ion chromatograms, followed by alignment and filtration based on quality assurance criteria. GC-MS underwent deconvolution and identification steps using Automated Mass Spectral Deconvolution and Identification System with NIST spectra library, followed by alignment and data filtration (Dunn et al., 2011; Struck-Lewicka et al., 2015). The PCA models exhibited good clustering of QC samples without outliers for both LC-ESI-TOF/MS and GC-MS. Univariate analysis identified a significant number of variables for LC-MS in positive and negative ionization modes, as well as for GC-MS. PLS-DA models, based on the important variables' normalized intensities, revealed R2 and Q2 values indicating the quality of the model. Metabolites identification involved database searching and comparison, with details of potential identities, measured mass, retention time, formula, and regulation between healthy volunteers and patients with CaP (Struck-Lewicka et al., 2015). This study delves into the metabolic alterations distinguishing CaP from healthy conditions. Amino acids such as glycine, alanine, serine, and threonine crucial in gluconeogenesis, were notably reduced in cancer patients' urine, as seen in their interconnected metabolic pathways (Kami et al., 2013; McDunn et al., 2013; Wu et al., 2011). Additional metabolites related to amino acid, TCA cycle, and energy metabolism showed distinctive patterns in cancer patients, pointing to disruptions in cellular energy production (Hirayama et al., 2009). These findings hint at potential new markers for diagnosing and understanding CaP, though further validation in larger populations is essential to confirm their diagnostic significance and independence from external factors (Struck-Lewicka et al., 2015).

3.2.1.2.2 In Plants

Legumes, known for their health benefits, contain various bioactive compounds (Ha et al., 2014). The investigation concentrated on the comprehensive analysis of legumes' bioactive compounds using LC-MS (Llorach et al., 2019). The research aimed to compare the composition of chickpeas, lentils, and white beans through untargeted metabolomics to discern discriminating compounds (Llorach et al., 2019). The metabolite fingerprinting process involved obtaining various chemicals and reagents, followed by the preparation of samples (Madrid-Gambin et al., 2018), which were then washed, ground, homogenized, and freeze-dried before storage and subsequent extraction (Llorach et al., 2019). The extraction procedure involved sonication and centrifugation, yielding concentrated supernatants for analysis (Abu-Reidah et al., 2014; Konar et al., 2012). The analysis was conducted through LC-ESI-LTQ-Orbitrap MS, for which an HPLC system was used with specific gradient elution conditions (Llorach et al., 2013, 2019). This was followed by data processing and statistical analysis using MAIT and Metabolanalyst, where the metabolomic fingerprints of different legumes were compared (Fernandez-Albert et al., 2014a, 2014b). The compounds were identified and characterized by MSn Orbitrap experiments, applying computational-assisted compound identification along with LC-MS pattern analysis, along with comparison to various databases and in-house collections (Gómez-Canela et al., 2014; Sumner et al., 2007). The study used LC-ESI-LTQ-Orbitrap-MS to identify and characterize 43 compounds across three legume types through statistical analyses (Sumner et al., 2007). These compounds belonged to six phytochemical classes (Marinangeli et al., 2017). Visualization methods such as PCA and HCA showcased distinct clustering between legume types, emphasizing biomarkers unique to each class and shared patterns across samples (Llorach et al., 2019). The study detailed differences in polyphenols, prenol lipids, nucleosides, α-galactosides, fatty acyls, and organic compounds among beans, chickpeas, and lentils, pinpointing specific compounds contributing to differentiation (Llorach et al., 2019). This comprehensive comparison aids in understanding the distinct metabolomic profiles of these legumes, showcasing their potential utility in food QC and authentication.

3.2.3 STATISTICAL TOOL USED FOR ANALYSIS

In the study of metabolite fingerprinting, several analytical tools are used, where unsupervised machine learning involves data-driven group separations without prior group information. The number and identity of significant NMR buckets vary based on normalization strategies, impacting statistical analysis reproducibility (Zacharias et al., 2018). Supervised machine learning methods (PLS-DA; Barker and Rayens, 2003), orthogonal partial least squares-discriminant analysis (OPLS-DA) (Trygg and Wold, 2002), elastic net regression (Zou and Hastie, 2005), and least absolute shrinkage and selection operator (Tibshirani, 1996) aid in classifying samples into known phenotypic classes, with PLS-DA and OPLS-DA extensively employed in metabolomic data (Hochrein et al., 2012). High-dimensional data challenges in supervised learning led to overfitting issues, demanding robust validation methods such as bootstrapping or cross-validation (CV). Preprocessing, especially data normalization, profoundly affects classification performance, with normalization influencing feature selection and classification outcomes (Gromski et al., 2015; Zacharias et al., 2017). PCA is prominent for visualization, whereas independent component analysis and various clustering methods aid in analyzing metabolic data (Klein et al., 2011; Scholz et al., 2004). Hypothesis testing, pivotal in identifying biomarkers, uses t tests or analysis of variance, with corrections for multiple hypothesis testing crucial for robust results (Zacharias et al., 2018). Although various analytical tools are used, of which most extensively used tools PCA and PLS-DA are described below.

3.2.4 PCA

PCA is a fundamental tool for multivariate data analysis (Wold, 1968; Wold et al., 1987a, 1987b). PCA, a vital method in data analysis across sciences, excels with large, highly correlated, multivariable datasets, aiming to represent data in a lower-dimensional space while preserving information. It enables data visualization, reduces storage, eliminates collinearity, and reduces noise. Operating by projecting data onto new dimensions (principal components), PCA organizes these components based on their ability to capture variance, where the first axis represents the most significant variation. Despite being a linear method, PCA's simplicity makes it a preferred choice for preprocessing data before complex statistical or machine learning methods. Its applications span diverse fields, including neuroscience, where it aids in integrating diverse data sources for comprehensive disease models, especially now that personalized medicine and big data are prevalent (Kherif and Latypova, 2020). PCA finds broad applications across various data matrices, typically used as an initial step in multivariate analysis for data exploration and outlier identification (Albano et al., 1978). It is particularly effective for obtaining an overview of data structures via score plots and for reducing dimensionality, extracting latent variables, and resolving spectra of multicomponent mixtures. Pretreatment of data involves transformations, scaling, and handling missing values, ensuring proper scaling, especially concerning variance scaling for equal influence. PCA also extends to two-block regression (PLS analysis) and multivariate image analysis, allowing for broader application potentials in diverse fields, emphasizing the theoretical background and diverse application possibilities for these extensions (Labrín and Urdinez, 2020; Wold et al., 1987b).

3.2.5 PLS-DA

PLS-DA is a widely used tool in chemometrics and omics data analyses (Eriksson et al., 2004; Gottfries et al., 1995). It operates with class labels, identifying it as a version under the supervision of PCA. It is also used for feature selection and classification (Christin et al., 2013; Nguyen and Rocke, 2002). However, its growing popularity brings forth the risks of misuse and overfitting, necessitating CV for accurate application (Ruiz-Perez et al., 2020; Westerhuis et al., 2008). Understanding PLS-DA's performance across diverse datasets remains limited, prompting experiments to illuminate its strengths, weaknesses, and suitability across various data distributions. Both PCA and PLS-DA aim to transform data into lower-dimensional spaces, with PCA prioritizing variance preservation, whereas PLS-DA emphasizes covariance between data and labels (Kjeldahl and Bro, 2010; Ruiz-Perez et al., 2020). PLS-DA has superiority in clustered data by selecting the correct hyperplanes, even with fewer samples or low cluster separations. PLS-DA's blind application without considering underlying data models risks identifying "noise" as significant, emphasizing the need for cautious use and reliance on CV for effectiveness verification (Ruiz-Perez et al., 2020). The overall diagrammatic sketch of the process involved in metabolite fingerprinting is shown in Figure 3.1.

3.3 APPLICATION OF METABOLITE FINGERPRINTING

3.3.1 TOXICOLOGY ANALYSIS

The consortium assessed xenobiotic toxicity using metabonomics, developing an expert system to predict toxicity organ sites based on spectra. Reference data on normal human urine metabotypes aided in studying toxin effects, storage variations, and biological variation sources such as age, sex, and diet (Lindon et al., 2000; Shockcor and Holmes, 2002; Zuppi et al., 1997).

3.3.2 PHARMACOMETABONOMICS

It explores predicting individual toxicity and detecting drug-induced liver injuries before symptoms (Winnike et al., 2010). Studies focus on discovering biomarkers indicating drug toxicity, especially for liver and kidney, aiding drug development and testing (Beckwith-Hall et al., 1998; Clayton et al., 2006).

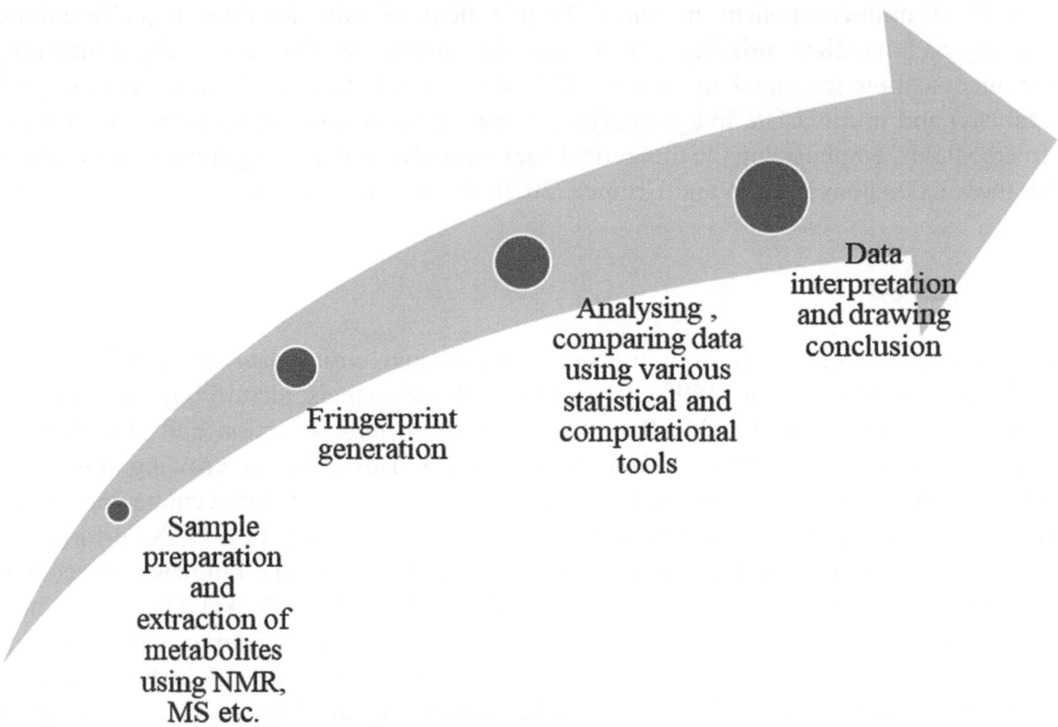

FIGURE 3.1 Diagrammatic sketch of the process involved in metabolite fingerprinting.

3.3.3 PATTERN OF PATHOLOGY

Metabolomics is used in disease models to detect and track diseases, distinguishing illnesses such as systemic inflammatory response syndrome and multiple organ dysfunction syndrome (Aller et al., 2010; Mao et al., 2009). It explores cancer diagnosis potential (Chan et al., 2009; Fan et al., 2005), reveals metabolic changes in obesity (Calvani et al., 2010), hypertension, and coronary heart disease, and uncovers the interplay between gut microbiota and diseases (Brindle et al., 2002; Holmes et al., 2008).

3.3.4 OMICS STUDY

Integrating genomics, proteomics, and metabolomics broadens our understanding of biological systems. Metabolic phenotypes are linked to cytokine levels during infections, and integrative omics-metabolic analysis offers a constraint-based approach to predict metabolic fluxes (Shimizu et al., 2006; Yizhak et al., 2010). Studies combine gene expression and metabolomic data to visualize affected pathways (Griffin et al., 2004).

The integration of various omics fields, especially the growing depth of metabolomic data, holds promise for a comprehensive understanding of biological systems and diseases.

3.4 SUCCESS STORIES OF METABOLITE FINGERPRINTING

Some recent works using metabolite fingerprinting have suggested the following results in different fields.

3.4.1 IN HUMAN HYPERTENSION

Traditional screening methods like blood pressure measurements are reliable but often detect symptoms too late (Ameta et al., 2017). Metabolomics delves deeper, unveiling the underlying molecular mechanisms behind hypertension's onset and progression (Bai et al., 2018; Onuh and Qiu, 2021). It is a powerful tool, identifying potential biomarkers and shedding light on intricate pathways linked to hypertension, including glucose, amino acid, inflammation, steroid hormone biosynthesis, fatty acid metabolism, and oxidative stress (Onuh and Qiu, 2021). For instance, glucose metabolism disturbances correlate with increased blood pressure, showcasing a potential avenue for intervention (Boller et al., 2012; Deng et al., 2021). Similarly, amino acid imbalances, especially methionine and tyrosine, are implicated in hypertension's pathogenesis (Zhao et al., 2018). Fatty acid metabolism abnormalities and their association with oxidative stress contribute significantly, with specific metabolites acting as potential biomarkers, particularly phosphatidylcholines' role, highlights its impact on hypertension (Ke et al., 2018), whereas steroid hormone biosynthesis disturbances tie into sodium absorption regulation, influencing blood pressure (Goodwin and Geller, 2012; Zhao et al., 2018). These pathways intertwine, suggesting that interventions targeting multiple pathways simultaneously could offer promising avenues for treating hypertension (Onuh and Qiu, 2021). Additionally, insights from metabolomics open doors for developing novel therapeutic approaches, including AMP-activated protein kinase modulators, SGLT2 inhibitors, and other pharmacological compounds showing potential in hypertension management (Quagliariello et al., 2021; Steinberg and Carling, 2019). Overall, metabolomics' intricate analysis illuminates the complexity of hypertension's molecular landscape, presenting opportunities for more targeted, effective treatments.

3.4.2 IN HERBAL MEDICINE

In the realm of herbal medicines (HMs), there is a rich history of traditional use globally, especially in areas with limited access to modern medications (Indrayanto, 2018; Riswanto et al., 2022). However, ensuring the authenticity and quality of these products poses challenges due to the complex mixture of herbs and the variability in their chemical compositions affected by factors such as plant species, growth conditions, and harvesting techniques (Li et al., 2020). To address this, analytical methods focusing on metabolite fingerprinting have emerged, aiming to authenticate these medicines without necessarily identifying all their compounds (Castro-Puyana et al., 2017). Techniques such as DNA fingerprinting, spectroscopy, and chromatography, coupled with chemometrics, play vital roles in this authentication process (Kumar et al., 2014; Purwakusumah et al., 2014). The increased use of herbal products has raised concerns about adulteration and abuse, prompting the need for authentication methods to detect mislabeling, substitution, or addition of contaminants. These practices have been identified in various markets worldwide, stressing the need for strong authentication procedures to guarantee the efficacy and security of herbal remedies (Gad et al., 2013; Jordan et al., 2010; Riswanto et al., 2022).

The authentication of HMs is crucial for ensuring their quality, and various methods are employed for this purpose. The European Medicines Agency, World Health Organization, and US Food and Drug Administration emphasize the importance of authenticity identification as a key assay for quality assurance (Wang and Yu, 2015). Classical methods involve single-component analysis using targeted compounds, such as curcuminoids for Curcuma species identification (Lestari et al., 2017). However, these methods require extensive sample preparation (Esslinger et al., 2014). Metabolite fingerprinting, an alternative approach, uses raw analytical data to discriminate between genuine and tampered herbal specimens (Kharyuk et al., 2018). Liquid chromatography–mass spectrometry tandem mass spectrometry (LC–MS/MS) is employed for bioactive component analysis (Mansur et al., 2016). NMR spectroscopy-based metabolite fingerprinting is used for herbal authentication, revealing differences in metabolite levels. The integration of chemometrics with NMR aids in distinguishing herbal components like *Curcuma longa* (Nurani et al., 2021; Venditti et al., 2016). Various herbs, including *Andrographis paniculata* (Kharyuk et al., 2018), *Rosa rugosa* (Nicoletti, 2011), and *Saffron*, are authenticated using HPLC, ultra-performance liquid chromatography, and LC-MS. Additionally, ^{1}H-NMR spectroscopy is used for metabolite fingerprinting, revealing the levels of metabolites such as curcuminoids (Nurani et al., 2021), pirocrocin (Petrakis et al., 2015), and rosmarinic acid (Venditti et al., 2017). This approach proves effective in discerning herbal authenticity and detecting adulteration. Moreover, Chemometrics plays a crucial role in handling complex chemical data derived from chromatographic and spectroscopic techniques used in HM authentication (Rohman and Putri, 2019). Techniques like PCA aid in data interpretation and dimensionality reduction (Jolliffe and Cadima, 2016), whereas similarity analysis and HCA are used for similarity and dissimilarity assessments (Chuchote and Somwong, 2019). Supervised methods like linear discriminant analysis (Berrueta et al., 2007; Luo and Shao, 2013), PLS-DA (Razmovski-Naumovski et al., 2010), and least-squares–support vector machine (Samui and Kothari, 2011) classify HMs based on pattern properties, maximizing classification distance or handling small sample-size issues. These techniques play a pivotal role in identifying and authenticating herbal products using metabolite fingerprinting. Further developments in understanding metabolite systems and pathways are suggested to enhance the reliability of metabolite fingerprinting in HM studies (Kruger et al., 2008).

3.4.3 IN FOOD CHEMISTRY

The food industry constantly introduces new products to meet diverse nutritional needs, leading to concerns about the authenticity of functional foods (Brigante et al., 2022; Spink and Moyer, 2011). Metabolomics, focusing on small molecules, emerges as a method for food authentication, offering targeted and untargeted approaches. Although targeted profiling quantifies specific metabolites, untargeted fingerprinting observes overall metabolite patterns (Dettmer et al., 2007; Jacobs et al., 2021; Mozzi et al., 2013). Chemometrics aids in handling the data generated, employing techniques such as PCA and OPLS-DA for classification (Brereton, 2009). Despite their popularity for their high nutritional content, chia, flax, and sesame seeds may contain adulterants. These seeds are sought after for their health benefits, prompting their widespread inclusion in various foodstuffs (Brigante et al., 2020; Paddon-Jones et al., 2015). The comparative study of flax, sesame seeds, and chia revealed distinct metabolic profiles and allowed for robust discrimination between these seeds. Using advanced analytical techniques such as OPLS-DA and unsupervised PCA, the study achieved high accuracy and sensitivity in distinguishing these seeds based on their metabolomic profiles (Brigante et al., 2022; Bylesjö et al., 2006; Wiklund et al., 2008). These methods enabled the detection of distinctive metabolites, such as sesamolinol diglucoside-rhamnoside (MOAZZAMI et al., 2006) and sesaminol diglucoside, among others, although some compounds remained unidentified due to limitations in spectral data or databases (Brigante et al., 2022). The analysis highlighted the potency of polyphenols as potential markers for authenticity, echoing findings from previous research (Brigante et al., 2022; Moing et al., 2020; Teegarden et al., 2019). The research emphasizes the importance of considering sets of markers rather than individual compounds for evaluating food authenticity and the challenges inherent in identifying comprehensive metabolites in untargeted metabolomics studies. Overall, the research provided valuable insights into the distinct metabolic fingerprints of these seeds, aiding in their differentiation and potential authentication (Brigante et al., 2022).

3.5 LIMITATIONS

Metabolomics has made strides in understanding diseases, but applying it, for example, to clinical hypertension remains challenging due to human complexity and species differences from animal models (Ameta et al., 2017). Clinical studies using blood for biomarker discovery are limited, and ethnic, diet, and gender variations affect metabolite profiles (Mels et al., 2019). Small sample sizes in studies hinder capturing comprehensive metabolite profiles (Ke et al., 2018). Overcoming these hurdles requires larger clinical studies, improved instrumentation, and better data analysis. Lowering technology costs and collective efforts for continuous development are crucial (Savoia et al., 2017). Integrating metabolomics with other omics can enhance disease understanding and enable real-time monitoring for predictive biomarkers, aiding personalized medicine (Cheng et al., 2017). These advancements can streamline healthcare costs and improve patient outcomes.

3.6 CONCLUSION

The metabolite fingerprinting approach, comparing samples to highlight specific features, emerges as a powerful tool to discern biomarkers for environmental, developmental, or genetic alterations. Metabolite fingerprinting, although promising, faces challenges in biological variation in various aspects. High-throughput methods like ^{1}H-NMR, although economical, lack sensitivity for low-concentration metabolites and struggle to pinpoint specific pathways. Unknown metabolites pose hurdles, yet their identification improves with larger databases and techniques like LC-MS. Managing complex data remains a challenge, though evolving software aids in analysis. Despite these, metabolomics fuels personalized healthcare, aiding drug development and treatment prediction. Integration with genomics and proteomics enhances biomedical insights, enabling accurate disease modeling and treatment evaluation. More sensitive methods and sophisticated tools are needed for robust biomarkers, facilitating diagnosis across various health conditions, including critical illnesses.

KEYWORDS

- **metabolomics**
- **advanced analytical technique**
- **biomarkers**
- **personalized medicine**
- **profiling**
- **drug discovery**

REFERENCES

Abu-Reidah, I. M., del Mar Contreras, M., Arráez-Román, D., Fernández-Gutiérrez, A., Segura-Carretero, A. UHPLC-ESI-QTOF-MS-based metabolic profiling of Vicia faba L. (Fabaceae) seeds as a key strategy for characterization in foodomics. *Electrophoresis*, 2014, 35, 1571–1581.

Albano, C., Dunn III, W., Edlund, U., Johansson, E., Nordén, B., Sjöström, M., Wold, S. Four levels of pattern recognition. *Anal. Chim. Acta*, 1978, 103, 429–443.

Aller, M. A., Arias, J. I., Alonso-Poza, A., Arias, J. A review of metabolic staging in severely injured patients. *Scand. J. Trauma. Resusc. Emerg. Med.*, 2010, 18, 27, https://doi.org/10.1186/1757-7241-18-27.

Ameta, K., Gupta, A., Kumar, S., Sethi, R., Kumar, D., Mahdi, A. A. Essential hypertension: A filtered serum based metabolomics study. *Sci. Rep.*, 2017, 7, 2153, https://doi.org/10.1038/s41598-017-02289-9.

Bai, Q., Peng, B., Wu, X., Cao, Y., Sun, X., Hong, M., Na, R., Liu, B., Li, Q., Li, Z., Fang, W., Zhu, N., Zong, C., Yu, Q. Metabolomic study for essential hypertension patients based on dried blood spot mass spectrometry approach. *IUBMB Life*, 2018, 70, 777–785, https://doi.org/10.1002/iub.1885.

Bailey, N. J., Oven, M., Holmes, E., Nicholson, J. K., Zenk, M. H. Metabolomic analysis of the consequences of cadmium exposure in Silene cucubalus cell cultures via 1H NMR spectroscopy and chemometrics. *Phytochemistry*, 2003, 62, 851–858.

Barker, M., Rayens, W. Partial least squares for discrimination. *J. Chemom.: J. Chemom. Soc.*, 2003, 17, 166–173.

Beckwith-Hall, B. M., Nicholson, J. K., Nicholls, A. W., Foxall, P. J., Lindon, J. C., Connor, S. C., Abdi, M., Connelly, J., Holmes, E. Nuclear magnetic resonance spectroscopic and principal components analysis investigations into biochemical effects of three model hepatotoxins. *Chem. Res. Toxicol.*, 1998,11, 260–72, https://doi.org/10.1021/tx9700679.

Berrueta, L. A., Alonso-Salces, R. M., Héberger, K. Supervised pattern recognition in food analysis. *J. Chromatogr. A*, 2007, 1158, 196–214.

Boller, S., Joblin, B. A., Xu, L., Item, F., Trüb, T., Boschetti, N., Spinas, G. A., Niessen, M. From signal transduction to signal interpretation: An alternative model for the molecular function of insulin receptor substrates. *Arch. Physiol. Biochem.*, 2012, 118, 148–55, https://doi.org/10.3109/13813455.2012.671333.

Brereton, R. G., *Chemometrics for Pattern Recognition*. John Wiley & Sons, 2009.

Brigante, F. I., Podio, N. S., Wunderlin, D. A., Baroni, M. V. Comparative metabolite fingerprinting of chia, flax and sesame seeds using LC-MS untargeted metabolomics. *Food Chem.*, 2022, 371, 131355, https://doi.org/10.1016/j.foodchem.2021.131355.

Brigante, F. I., Lucini Mas, A., Pigni, N. B., Wunderlin, D. A., Baroni, M. V. Targeted metabolomics to assess the authenticity of bakery products containing chia, sesame and flax seeds. *Food Chem.*, 2020, 312, 126059, https://doi.org/10.1016/j.foodchem.2019.126059.

Brindle, J. T., Antti, H., Holmes, E., Tranter, G., Nicholson, J. K., Bethell, H. W., Clarke, S., Schofield, P. M., McKilligin, E., Mosedale, D. E., Grainger, D. J. Rapid and noninvasive diagnosis of the presence and severity of coronary heart disease using 1H-NMR-based metabonomics, *Nat. Med.*, 2002, 8, 1439–44, https://doi.org/10.1038/nm1202-802.

Bylesjö, M., Rantalainen, M., Cloarec, O., Nicholson, J. K., Holmes, E., Trygg, J. OPLS discriminant analysis: Combining the strengths of PLS-DA and SIMCA classification. *J. Chemom.*, 2006, 20, 341–351, https://doi.org/10.1002/cem.1006.

Calvani, R., Miccheli, A., Capuani, G., Tomassini Miccheli, A., Puccetti, C., Delfini, M., Iaconelli, A., Nanni, G., Mingrone, G. Gut microbiome-derived metabolites characterize a peculiar obese urinary metabotype. *Int. J. Obes. (Lond)*, 2010, 34, 1095–8, https://doi.org/10.1038/ijo.2010.44.

Castro-Puyana, M., Pérez-Míguez, R., Montero, L., Herrero, M. Reprint of: Application of mass spectrometry-based metabolomics approaches for food safety, quality and traceability. *TrAC Trends Anal. Chem.*, 2017, 96, 62–78.

Cavill, R., Keun, H. C., Holmes, E., Lindon, J. C., Nicholson, J. K., Ebbels, T. M. Genetic algorithms for simultaneous variable and sample selection in metabonomics. *Bioinformatics*, 2009, 25, 112–8, https://doi.org/10.1093/bioinformatics/btn586.

Chan, E. C., Koh, P. K., Mal, M., Cheah, P. Y., Eu, K. W., Backshall, A., Cavill, R., Nicholson, J. K., Keun, H. C. Metabolic profiling of human colorectal cancer using high-resolution magic angle spinning nuclear magnetic resonance (HR-MAS NMR) spectroscopy and gas chromatography mass spectrometry (GC/MS). *J. Proteome Res.*, 2009, 8, 352–61, https://doi.org/10.1021/pr8006232.

Charlton, A., Allnutt, T., Holmes, S., Chisholm, J., Bean, S., Ellis, N., Mullineaux, P., Oehlschlager, S. NMR profiling of transgenic peas. *Plant Biotechnol. J.*, 2004, 2, 27–35.

Cheng, S., Shah, S. H., Corwin, E. J., Fiehn, O., Fitzgerald, R. L., Gerszten, R. E., Illig, T., Rhee, E. P., Srinivas, P. R., Wang, T. J., Jain, M. Potential impact and study considerations of metabolomics in cardiovascular health and disease: A scientific statement from the American Heart Association. *Circ. Cardiovasc. Genet.*, 2017, 10, https://doi.org/10.1161/hcg.0000000000000032.

Christin, C., Hoefsloot, H. C., Smilde, A. K., Hoekman, B., Suits, F., Bischoff, R., Horvatovich, P. A critical assessment of feature selection methods for biomarker discovery in clinical proteomics. *Mol. Cell. Proteom.*, 2013, 12, 263–276.

Chuchote, C., Somwong, P. Similarity analysis of the chromatographic fingerprints of Thai herbal Ya-Ha-Rak remedy using HPLC. *Interprof. J. Health Sci.*, 2019, 17, 55–63.

Clayton, T. A., Lindon, J. C., Cloarec, O., Antti, H., Charuel, C., Hanton, G., Provost, J. P., Le Net, J. L., Baker, D., Walley, R. J., Everett, J. R., Nicholson, J. K. Pharmaco-metabonomic phenotyping and personalized drug treatment. *Nature*, 2006, 440, 1073–7, https://doi.org/10.1038/nature04648.

Cui, Q., Lewis, I. A., Hegeman, A. D., Anderson, M. E., Li, J., Schulte, C. F., Westler, W. M., Eghbalnia, H. R., Sussman, M. R., Markley, J. L. Metabolite identification via the madison metabolomics consortium database. *Nat. Biotechnol.*, 2008, 26, 162–4, https://doi.org/10.1038/nbt0208-162.

Defernez, M., Colquhoun, I. J. Factors affecting the robustness of metabolite fingerprinting using 1H NMR spectra. *Phytochemistry*, 2003, 62, 1009–1017.

Deng, Y., Huang, C., Su, J., Pan, C. W., Ke, C. Identification of biomarkers for essential hypertension based on metabolomics. *Nutr. Metab. Cardiovasc Dis.*, 2021, 31, 382–395, https://doi.org/10.1016/j.numecd.2020.11.023.

Dettmer, K., Aronov, P. A., Hammock, B. D. Mass spectrometry-based metabolomics. *Mass Spectrom. Rev.*, 2007, 26, 51–78, https://doi.org/10.1002/mas.20108.

Dunn, W. B., Broadhurst, D., Begley, P., Zelena, E., Francis-McIntyre, S., Anderson, N., Brown, M., Knowles, J. D., Halsall, A., Haselden, J. N. Procedures for large-scale metabolic profiling of serum and plasma using gas chromatography and liquid chromatography coupled to mass spectrometry. *Nat. Protoc.*, 2011, 6, 1060–1083.

Eriksson, L., Antti, H., Gottfries, J., Holmes, E., Johansson, E., Lindgren, F., Long, I., Lundstedt, T., Trygg, J., Wold, S. Using chemometrics for navigating in the large data sets of genomics, proteomics, and metabonomics (gpm). *Anal. Bioanal. Chem.*, 2004, 380, 419–429.

Esslinger, S., Riedl, J., Fauhl-Hassek, C. Potential and limitations of non-targeted fingerprinting for authentication of food in official control. *Food Res. Int.*, 2014, 60, 189–204.

Exarchou, V., Godejohann, M., van Beek, T. A., Gerothanassis, I. P., Vervoort, J. LC-UV-solid-phase extraction-NMR-MS combined with a cryogenic flow probe and its application to the identification of compounds present in Greek oregano. *Anal. Chem.*, 2003, 75, 6288–6294.

Fan, X., Bai, J., Shen, P. Diagnosis of breast cancer using HPLC metabonomics fingerprints coupled with computational methods. *Conf. Proc. IEEE Eng. Med. Biol. Soc.*, 2005, 2005, 6081–4, https://doi.org/10.1109/iembs.2005.1615880.

Fernandez-Albert, F., Llorach, R., Andres-Lacueva, C., Perera-Lluna, A., Peak aggregation as an innovative strategy for improving the predictive power of LC-MS metabolomic profiles. *Anal. Chem.*, 2014a, 86, 2320–2325.

Fernández-Albert, F., Llorach, R., Andrés-Lacueva, C., Perera, A., An R package to analyse LC/MS metabolomic data: MAIT (Metabolite Automatic Identification Toolkit). *Bioinformatics*, 2014b, 30, 1937–1939.

Gad, H. A., El-Ahmady, S. H., Abou-Shoer, M. I., Al-Azizi, M. M. Application of chemometrics in authentication of herbal medicines: A review. *Phytochem. Anal.*, 2013, 24, 1–24.

Gerlinger, M., Catto, J. W., Orntoft, T. F., Real, F. X., Zwarthoff, E. C., Swanton, C. Intratumour heterogeneity in urologic cancers: From molecular evidence to clinical implications. *Eur. Urol.*, 2015, 67, 729–737.

Gómez-Canela, C., Ventura, F., Caixach, J., Lacorte, S. Occurrence of cytostatic compounds in hospital effluents and wastewaters, determined by liquid chromatography coupled to high-resolution mass spectrometry. *Anal. Bioanal. Chem.*, 2014, 406, 3801–3814.

Goodwin, J. E., Geller, D. S. Glucocorticoid-induced hypertension. *Pediatr. Nephrol.*, 2012, 27, 1059–66, https://doi.org/10.1007/s00467-011-1928-4.

Gottfries, J., Blennow, K., Wallin, A., Gottfries, C. Diagnosis of dementias using partial least squares discriminant analysis. *Dement. Geriatr. Cogn. Disord.*, 1995, 6, 83–88.

Griffin, J. L. The Cinderella story of metabolic profiling: Does metabolomics get to go to the functional genomics ball? *Philos. Trans. R. Soc. Lond. B Biol. Sci.*, 2006, 361, 147–61, https://doi.org/10.1098/rstb.2005.1734.

Griffin, J. L., Bonney, S. A., Mann, C., Hebbachi, A. M., Gibbons, G. F., Nicholson, J. K., Shoulders, C. C., Scott, J. An integrated reverse functional genomic and metabolic approach to understanding orotic acid-induced fatty liver. *Physiol. Genomics.*, 2004, 17, 140–9, https://doi.org/10.1152/physiolgenomics.00158.2003.

Gromski, P. S., Xu, Y., Hollywood, K. A., Turner, M. L., Goodacre, R. The influence of scaling metabolomics data on model classification accuracy. *Metabolomics*, 2015, 11, 684–695.

Ha, T. J., Lee, B. W., Park, K. H., Jeong, S. H., Kim, H.-T., Ko, J.-M., Baek, I.-Y., Lee, J. H. Rapid characterisation and comparison of saponin profiles in the seeds of Korean Leguminous species using ultra performance liquid chromatography with photodiode array detector and electrospray ionisation/mass spectrometry (UPLC–PDA–ESI/MS) analysis. *Food Chem.*, 2014, 146, 270–277.

Hinse, C., Richter, C., Provenzani, A., Stöckigt, J. In vivo monitoring of alkaloid metabolism in hybrid plant cell cultures by 2D cryo-NMR without labelling. *Bioorg. Med. Chem.*, 2003, 11, 3913–3919.

Hirayama, A., Kami, K., Sugimoto, M., Sugawara, M., Toki, N., Onozuka, H., Kinoshita, T., Saito, N., Ochiai, A., Tomita, M. Quantitative metabolome profiling of colon and stomach cancer microenvironment by capillary electrophoresis time-of-flight mass spectrometry. *Cancer Res.*, 2009, 69, 4918–4925.

Hochrein, J., Klein, M. S., Zacharias, H. U., Li, J., Wijffels, G., Schirra, H. J., Spang, R., Oefner, P. J., Gronwald, W. Performance evaluation of algorithms for the classification of metabolic 1H NMR fingerprints. *J. Proteome Res.*, 2012, 11, 6242–6251.

Holmes, E., Loo, R. L., Stamler, J., Bictash, M., Yap, I. K., Chan, Q., Ebbels, T., De Iorio, M., Brown, I. J., Veselkov, K. A., Daviglus, M. L., Kesteloot, H., Ueshima, H., Zhao, L., Nicholson, J. K., Elliott, P. Human metabolic phenotype diversity and its association with diet and blood pressure. *Nature*, 2008, 453, 396–400, https://doi.org/10.1038/nature06882.

Huang, Y., Zheng, H., Tan, K., Sun, X., Ye, J., Zhang, Y. Circulating metabolomics profiling reveals novel pathways associated with cognitive decline in patients with hypertension. *Ther. Adv. Neurol. Disord.*, 2020, 13, 1756286420947973, https://doi.org/10.1177/1756286420947973.

Indrayanto, G. Recent development of quality control methods for herbal derived drug preparations. *Nat. Prod. Commun.*, 2018, 13, 1934578X1801301208.

Jacobs, D. M., van den Berg, M. A., Hall, R. D. Towards superior plant-based foods using metabolomics. *Curr. Opin. Biotechnol.*, 2021, 70, 23–28, https://doi.org/10.1016/j.copbio.2020.08.010.

Jerónimo, C., Bastian, P. J., Bjartell, A., Carbone, G. M., Catto, J. W., Clark, S. J., Henrique, R., Nelson, W. G., Shariat, S. F. Epigenetics in prostate cancer: Biologic and clinical relevance. *Eur. Urol.*, 2011, 60, 753–766.

Jolliffe, I. T., Cadima, J. Principal component analysis: A review and recent developments. *Philos. Trans. R. Soc. A: Math. Phys. Eng. Sci.*, 2016, 374, 20150202.

Jordan, S. A., Cunningham, D. G., Marles, R. J. Assessment of herbal medicinal products: Challenges, and opportunities to increase the knowledge base for safety assessment. *Toxicol. Appl. Pharmacol.*, 2010, 243, 198–216.

Kami, K., Fujimori, T., Sato, H., Sato, M., Yamamoto, H., Ohashi, Y., Sugiyama, N., Ishihama, Y., Onozuka, H., Ochiai, A. Metabolomic profiling of lung and prostate tumor tissues by capillary electrophoresis time-of-flight mass spectrometry. *Metabolomics*, 2013, 9, 444–453.

Ke, C., Zhu, X., Zhang, Y., Shen, Y. Metabolomic characterization of hypertension and dyslipidemia. *Metabolomics*, 2018, 14, 117, https://doi.org/10.1007/s11306-018-1408-y.

Kell, D. B. Metabolomics and systems biology: Making sense of the soup. *Curr. Opin. Microbiol.*, 2004, 7, 296–307, https://doi.org/10.1016/j.mib.2004.04.012.

Keun, H. C., Ebbels, T. M., Bollard, M. E., Beckonert, O., Antti, H., Holmes, E., Lindon, J. C., Nicholson, J. K. Geometric trajectory analysis of metabolic responses to toxicity can define treatment specific profiles. *Chem. Res. Toxicol.*, 2004, 17, 579–87, https://doi.org/10.1021/tx034212w.

Key, T. J. Nutrition, hormones and prostate cancer risk: Results from the European prospective investigation into cancer and nutrition. In: *Prostate Cancer Prevention* (pp. 39–46). Springer, 2014.

Kharyuk, P., Nazarenko, D., Oseledets, I., Rodin, I., Shpigun, O., Tsitsilin, A., Lavrentyev, M. Employing fingerprinting of medicinal plants by means of LC-MS and machine learning for species identification task. *Sci. Rep.*, 2018, 8, 17053.

Kherif, F., Latypova, A. Principal component analysis. In: *Machine Learning* (pp. 209–225). Elsevier, 2020.

Kilpeläinen, T. P., Tammela, T. L., Roobol, M., Hugosson, J., Ciatto, S., Nelen, V., Moss, S., Määttänen, L., Auvinen, A. False-positive screening results in the European randomized study of screening for prostate cancer. *Eur. J. Cancer*, 2011, 47, 2698–2705.

Kjeldahl, K., Bro, R. Some common misunderstandings in chemometrics. *J. Chemom.*, 2010, 24, 558–564.

Klein, M. S., Dorn, C., Saugspier, M., Hellerbrand, C., Oefner, P. J., Gronwald, W. Discrimination of steatosis and NASH in mice using nuclear magnetic resonance spectroscopy. *Metabolomics*, 2011, 7, 237–246.

Konar, N., Poyrazoğlu, E. S., Demir, K., Artik, N. Determination of conjugated and free isoflavones in some legumes by LC–MS/MS. *J. Food Composit. Anal.*, 2012, 25, 173–178.

Kosmides, A. K., Kamisoglu, K., Calvano, S. E., Corbett, S. A., Androulakis, I. Metabolomic fingerprinting: Challenges and opportunities. *Crit. Rev. Biomed. Engineer.*, 2013, 41.

Krishnan, P., Kruger, N. J., Ratcliffe, R. G. Metabolite fingerprinting and profiling in plants using NMR. *J. Exp. Bot.*, 2004, 56, 255–265, https://doi.org/10.1093/jxb/eri010.

Kruger, N. J., Troncoso-Ponce, M. A., Ratcliffe, R. G. 1H NMR metabolite fingerprinting and metabolomic analysis of perchloric acid extracts from plant tissues. *Nat. Protoc.*, 2008, 3, 1001–1012.

Kumar, S., Ketkar, P., Nayak, S., Roy, S. Application of DNA fingerprinting tools for authentication of ayurvedic herbal medicines—A review. *J. Sci. Innov. Res. JSIR*, 2014, 3, 606–612.

Labrín, C., Urdinez, F. Principal component analysis. In: *R for Political Data Science* (pp. 375–393). Chapman and Hall/CRC, 2020.

Lenz, E. M., Wilson, I. D. Analytical strategies in metabonomics. *J. Proteome Res.*, 2007, 6, 443–58, https://doi.org/10.1021/pr0605217.

Lestari, H., Martono, S., Wulandari, R., Rohman, A. Simultaneous analysis of Curcumin and demethoxycurcumin in Curcuma xanthorriza using FTIR spectroscopy and chemometrics. *Int. Food Res. J.*, 2017, 24.

Li, Y., Shen, Y., Yao, C.-l., Guo, D.-A. Quality assessment of herbal medicines based on chemical fingerprints combined with chemometrics approach: A review. *J. Pharm. Biomed. Anal.*, 2020, 185, 113215.

Lindon, J., Holmes, E., Nicholson, J. Pattern recognition methods and applications in biomedical magnetic resonance. *Progress in Nuclear Magnetic Resonance Spectroscopy*, 2001, 39, 1–40.

Lindon, J. C., Nicholson, J. K., Holmes, E., Everett, J. R. Metabonomics: Metabolic processes studied by NMR spectroscopy of biofluids. *Concept. Magn. Res*, 2000, 12, 289–320.

Llorach, R., Urpi-Sarda, M., Tulipani, S., Garcia-Aloy, M., Monagas, M., Andres-Lacueva, C. Metabolomic fingerprint in patients at high risk of cardiovascular disease by cocoa intervention. *Mol. Nutrit. Food Res.*, 2013, 57, 962–973.

Llorach, R., Favari, C., Alonso, D., Garcia-Aloy, M., Andres-Lacueva, C., Urpi-Sarda, M. Comparative metabolite fingerprinting of legumes using LC-MS-based untargeted metabolomics. *Food Res. Int.*, 2019, 126, 108666.

Lu, X., Zhao, X., Bai, C., Zhao, C., Lu, G., Xu, G. LC-MS-based metabonomics analysis. *J. Chromatogr. B Analyt. Technol. Biomed. Life Sci.*, 2008, 866, 64–76, https://doi.org/10.1016/j.jchromb.2007.10.022.

Luo, D. H., Shao, Y. W. Classification of Chinese herbal medicine based on improved LDA algorithm using machine olfaction. *Appl. Mech. Mater.*, 2013, 239, 1532–1536.

Madrid-Gambin, F., Brunius, C., Garcia-Aloy, M., Estruel-Amades, S., Landberg, R., Andres-Lacueva, C. Untargeted 1H NMR-based metabolomics analysis of urine and serum profiles after consumption of lentils, chickpeas, and beans: An extended meal study to discover dietary biomarkers of pulses. *Agric. Food Chem.*, 2018, 66, 6997–7005.

Maher, A. D., Zirah, S. F., Holmes, E., Nicholson, J. K. Experimental and analytical variation in human urine in 1H NMR spectroscopy-based metabolic phenotyping studies. *Anal. Chem.*, 2007, 79, 5204–11, https://doi.org/10.1021/ac070212f.

Mansur, S., Abdulla, R., Ayupbec, A., Aisa, H. A. Chemical fingerprint analysis and quantitative analysis of Rosa rugosa by UPLC-DAD. *Molecules*, 2016, 21, 1754.

Mao, H., Wang, H., Wang, B., Liu, X., Gao, H., Xu, M., Zhao, H., Deng, X., Lin, D. Systemic metabolic changes of traumatic critically ill patients revealed by an NMR-based metabonomic approach. *J. Proteome Res.*, 2009, 8, 5423–30, https://doi.org/10.1021/pr900576y.

Marinangeli, C. P., Curran, J., Barr, S. I., Slavin, J., Puri, S., Swaminathan, S., Tapsell, L., Patterson, C. A. Enhancing nutrition with pulses: Defining a recommended serving size for adults. *Nutrit. Rev.*, 2017, 75, 990–1006.

McDunn, J. E., Li, Z., Adam, K. P., Neri, B. P., Wolfert, R. L., Milburn, M. V., Lotan, Y., Wheeler, T. M. Metabolomic signatures of aggressive prostate cancer. *Prostate*, 2013, 73, 1547–1560.

Mels, C. M., Delles, C., Louw, R., Schutte, A. E. Central systolic pressure and a nonessential amino acid metabolomics profile: The African prospective study on the early detection and identification of cardiovascular disease and hypertension. *J Hypertens*, 2019, 37, 1157–1166, https://doi.org/10.1097/hjh.0000000000002040.

Moazzami, A. A., Andersson, R. E., Kamal-Eldin, A. Characterization and analysis of sesamolinol diglucoside in sesame seeds. *Biosci. Biotechnol. Biochem.*, 2006, 70, 1478–1481, https://doi.org/10.1271/bbb.60013.

Moing, A., Allwood, J. W., Aharoni, A., Baker, J., Beale, M. H., Ben-Dor, S., Biais, B., Brigante, F., Burger, Y., Deborde, C. Comparative metabolomics and molecular phylogenetics of melon (Cucumis melo, Cucurbitaceae) biodiversity. *Metabolites*, 2020, 10, 121.

Mozzi, F., Ortiz, M. E., Bleckwedel, J., De Vuyst, L., Pescuma, M. Metabolomics as a tool for the comprehensive understanding of fermented and functional foods with lactic acid bacteria. *Food Res. Int.*, 2013, 54, 1152–1161, https://doi.org/10.1016/j.foodres.2012.11.010.

Neuweger, H., Albaum, S. P., Dondrup, M., Persicke, M., Watt, T., Niehaus, K., Stoye, J., Goesmann, A. MeltDB: A software platform for the analysis and integration of metabolomics experiment data. *Bioinformatics*, 2008, 24, 2726–32, https://doi.org/10.1093/bioinformatics/btn452.

Nguyen, D. V., Rocke, D. M. *Classification of Acute Leukemia based on DNA Microarray Gene Expressions using Partial Least Squares*. Springer, 2002.

Nicholson, J. K., Lindon, J. C. Systems biology: Metabonomics. *Nature*, 2008, 455, 1054–6, https://doi.org/10.1038/4551054a.

Nicholson, J. K., Lindon, J. C., Holmes, E. 'Metabonomics': Understanding the metabolic responses of living systems to pathophysiological stimuli via multivariate statistical analysis of biological NMR spectroscopic data. *Xenobiotica*, 1999, 29, 1181–9, https://doi.org/10.1080/004982599238047.

Nicholson, J. K., Holmes, E., Kinross, J. M., Darzi, A. W., Takats, Z., Lindon, J. C. Metabolic phenotyping in clinical and surgical environments. *Nature*, 2012, 491, 384–92, https://doi.org/10.1038/nature11708.

Nicoletti, M. HPTLC fingerprint: A modern approach for the analytical determination of botanicals. *Rev. Bras. Farmacogn.*, 2011, 21, 818–823.

Nikolic, S. B., Sharman, J. E., Adams, M. J., Edwards, L. M. Metabolomics in hypertension. *J. Hypertens*, 2014, 32, 1159–69, https://doi.org/10.1097/hjh.0000000000000168.

Nurani, L. H., Rohman, A., Windarsih, A., Guntarti, A., Riswanto, F. D. O., Lukitaningsih, E., Fadzillah, N. A., Rafi, M. Metabolite fingerprinting using 1H-NMR spectroscopy and chemometrics for classification of three curcuma species from different origins. *Molecules*, 2021, 26, 7626.

O'Hagan, S., Dunn, W. B., Knowles, J. D., Broadhurst, D., Williams, R., Ashworth, J. J., Cameron, M., Kell, D. B. Closed-loop, multiobjective optimization of two-dimensional gas chromatography/mass spectrometry for serum metabolomics. *Anal. Chem.*, 2007, 79, 464–76, https://doi.org/10.1021/ac061443+.

Onuh, J. O., Aluko, R. E. Metabolomics as a tool to study the mechanism of action of bioactive protein hydrolysates and peptides: A review of current literature. *Trends Food Sci. Technol.*, 2019, 91, 625–633.

Onuh, J. O., Aliani, M. Metabolomics profiling in hypertension and blood pressure regulation: A review. *Clin. Hypertens*, 2020, 26, 23, https://doi.org/10.1186/s40885-020-00157-9.

Onuh, J. O., Qiu, H. Metabolic profiling and metabolites fingerprints in human hypertension: Discovery and potential. *Metabolites*, 2021, 11, 687.

Ott, K.-H., Araníbar, N., Singh, B., Stockton, G. W. Metabonomics classifies pathways affected by bioactive compounds. Artificial neural network classification of NMR spectra of plant extracts. *Phytochemistry*, 2003, 62, 971–985.

Paddon-Jones, D., Campbell, W. W., Jacques, P. F., Kritchevsky, S. B., Moore, L. L., Rodriguez, N. R., van Loon, L. J. C. Protein and healthy aging234. *Am. J. Clin. Nutrit.*, 2015, 101, 1339S–1345S, https://doi.org/10.3945/ajcn.114.084061.

Petrakis, E. A., Cagliani, L. R., Polissiou, M. G., Consonni, R. Evaluation of saffron (Crocus sativus L.) adulteration with plant adulterants by 1H NMR metabolite fingerprinting. *Food Chem.*, 2015, 173, 890–896.

Purwakusumah, E., Rafi, M., Safitri, U., Nurcholis, W., Adzkiya, M. Identification and authentication of Jahe Merah using combination of FTIR spectrocopy and chemometrics. *J. Agritech.*, 2014, 34, 82–87.

Quagliariello, V., De Laurentiis, M., Rea, D., Barbieri, A., Monti, M. G., Carbone, A., Paccone, A., Altucci, L., Conte, M., Canale, M. L., Botti, G., Maurea, N. The SGLT-2 inhibitor empagliflozin improves myocardial strain, reduces cardiac fibrosis and pro-inflammatory cytokines in non-diabetic mice treated with doxorubicin. *Cardiovasc Diabetol.*, 2021, 20, 150, https://doi.org/10.1186/s12933-021-01346-y.

Ratcliffe, R. G., Shachar-Hill, Y. Probing plant metabolism with NMR. *Ann. Rev. Plant Biol.*, 2001, 52, 499–526.

Razmovski-Naumovski, V., Tongkao-on, W., Kimble, B., Qiao, V. L., Beilun, L., Li, K. M., Roufogalis, B., Depo, Y., Meicun, Y., Li, G. Q. Multiple chromatographic and chemometric methods for quality standardisation of Chinese herbal medicines. *World Sci. Technol.*, 2010, 12, 99–106.

Rinschen, M. M., Ivanisevic, J. Identification of bioactive metabolites using activity metabolomics. *Nat. Rev. Mol. Cell Biol.*, 2019, 20, 353–367, https://doi.org/10.1038/s41580-019-0108-4.

Riswanto, F. D. O., Windarsih, A., Lukitaningsih, E., Rafi, M., Fadzilah, N. A., Rohman, A. Metabolite fingerprinting based on 1H-NMR spectroscopy and liquid chromatography for the authentication of herbal products. *Molecules*, 2022, 27, 1198.

Rohman, A., Putri, A. R. The chemometrics techniques in combination with instrumental analytical methods applied in Halal authentication analysis. *Indones. J. Chem.*, 2019, 19, 262–272.

Rooney, O. M., Troke, J., Nicholson, J. K., Griffin, J. L. High-resolution diffusion and relaxation-edited magic angle spinning 1H NMR spectroscopy of intact liver tissue. *Magn. Reson. Med.*, 2003, 50, 925–30, https://doi.org/10.1002/mrm.10620.

Ruiz-Perez, D., Guan, H., Madhivanan, P., Mathee, K., Narasimhan, G. So you think you can PLS-DA? *BMC Bioinform.*, 2020, 21, 1–10.

Samui, P., Kothari, D. Utilization of a least square support vector machine (LSSVM) for slope stability analysis. *Sci. Iranica*, 2011, 18, 53–58.

Savoia, C., Volpe, M., Grassi, G., Borghi, C., Agabiti Rosei, E., Touyz, R. M. Personalized medicine-a modern approach for the diagnosis and management of hypertension. *Clin. Sci. (Lond)*, 2017. 131, 2671–2685, https://doi.org/10.1042/cs20160407.

Scholz, M., Gatzek, S., Sterling, A., Fiehn, O., Selbig, J. Metabolite fingerprinting: Detecting biological features by independent component analysis. *Bioinformatics*, 2004, 20, 2447–2454.

Shimizu, K., Ogura, H., Goto, M., Asahara, T., Nomoto, K., Morotomi, M., Yoshiya, K., Matsushima, A., Sumi, Y., Kuwagata, Y., Tanaka, H., Shimazu, T., Sugimoto, H. Altered gut flora and environment in patients with severe SIRS. *J. Trauma*, 2006, 60, 126–33, https://doi.org/10.1097/01.ta.0000197374.99755.fe.

Shockcor, J. P., Holmes, E. Metabonomic applications in toxicity screening and disease diagnosis. *Curr. Top. Med. Chem.*, 2002, 2, 35–51, https://doi.org/10.2174/1568026023394498.

Spink, J., Moyer, D. C. Defining the public health threat of food fraud. *J. Food Sci.*, 2011, 76, R157–R163, https://doi.org/10.1111/j.1750-3841.2011.02417.x.

Steinberg, G. R., Carling, D. AMP-activated protein kinase: The current landscape for drug development. *Nat. Rev. Drug Discov.*, 2019, 18, 527–551, https://doi.org/10.1038/s41573-019-0019-2.

Struck-Lewicka, W., Kordalewska, M., Bujak, R., Mpanga, A. Y., Markuszewski, M., Jacyna, J., Matuszewski, M., Kaliszan, R., Markuszewski, M. J. Urine metabolic fingerprinting using LC–MS and GC–MS reveals metabolite changes in prostate cancer: A pilot study. *J. Pharm. Biomed. Anal.*, 2015, 111, 351–361.

Suhre, K., Schmitt-Kopplin, P. MassTRIX: Mass translator into pathways. *Nucleic Acids Res.*, 2008, 36, W481–4, https://doi.org/10.1093/nar/gkn194.

Sumner, L. W., Mendes, P., Dixon, R. A. Plant metabolomics: Large-scale phytochemistry in the functional genomics era. *Phytochemistry*, 2003, 62, 817–836.

Sumner, L. W., Amberg, A., Barrett, D., Beale, M. H., Beger, R., Daykin, C. A., Fan, T. W.-M., Fiehn, O., Goodacre, R., Griffin, J. L. Proposed minimum reporting standards for chemical analysis: Chemical analysis working group (CAWG) metabolomics standards initiative (MSI). *Metabolomics*, 2007, 3, 211–221.

Szymańska, E., Markuszewski, M. J., Markuszewski, M., Kaliszan, R. Altered levels of nucleoside metabolite profiles in urogenital tract cancer measured by capillary electrophoresis. *J. Pharm. Biomed. Anal.*, 2010, 53, 1305–1312.

Teegarden, M. D., Schwartz, S. J., Cooperstone, J. L. Profiling the impact of thermal processing on black raspberry phytochemicals using untargeted metabolomics. *Food Chem.*, 2019, 274, 782–788, https://doi.org/10.1016/j.foodchem.2018.09.053.

Theodoridis, G., Gika, H. G., Wilson, I. D. LC-MS-based methodology for global metabolite profiling in metabonomics/metabolomics. *TrAC Trends Anal. Chem.*, 2008, 27, 251–260.

Tibshirani, R. Regression shrinkage and selection via the lasso. *J. R. Statist. Soc. B: Statist. Methodol.*, 1996, 58, 267–288.

Trygg, J., Wold, S. Orthogonal projections to latent structures (O-PLS). *J. Chemom.: J. Chemom. Soc.*, 2002, 16, 119–128.

Tzoulaki, I., Iliou, A., Mikros, E., Elliott, P. An overview of metabolic phenotyping in blood pressure research. *Curr. Hypertens Rep.*, 2018, 20, 78, https://doi.org/10.1007/s11906-018-0877-8.

Venditti, A., Frezza, C., Celona, D., Sciubba, F., Foddai, S., Delfini, M., Serafini, M., Bianco, A. Phytochemical comparison with quantitative analysis between two flower phenotypes of Mentha aquatica L.: Pink-violet and white. *AIMS Mol. Sci.*, 2017, 4, 288–300.

Venditti, A., Frezza, C., Maggi, F., Lupidi, G., Bramucci, M., Quassinti, L., Giuliani, C., Cianfaglione, K., Papa, F., Serafini, M. Phytochemistry, micromorphology and bioactivities of Ajuga chamaepitys (L.) Schreb. (Lamiaceae, Ajugoideae): Two new harpagide derivatives and an unusual iridoid glycosides pattern. *Fitoterapia*, 2016, 113, 35–43.

Veselkov, K. A., Pahomov, V. I., Lindon, J. C., Volynkin, V. S., Crockford, D., Osipenko, G. S., Davies, D. B., Barton, R. H., Bang, J. W., Holmes, E., Nicholson, J. K. A metabolic entropy approach for measurements of systemic metabolic disruptions in patho-physiological States. *J. Proteome Res.*, 2010, 9, 3537–3544, https://doi.org/10.1021/pr1000576.

Wang, P., Yu, Z. Species authentication and geographical origin discrimination of herbal medicines by near infrared spectroscopy: A review. *J. Pharm. Anal.*, 2015, 5, 277–284, https://doi.org/10.1016/j.jpha.2015.04.001.

Webb-Robertson, B.-J., Kim, Y.-M., Zink, E. M., Hallaian, K. A., Zhang, Q., Madupu, R., Waters, K. M., Metz, T. O. A statistical analysis of the effects of urease pre-treatment on the measurement of the urinary metabolome by gas chromatography–mass spectrometry. *Metabolomics*, 2014, 10, 897–908.

Westerhuis, J. A., Hoefsloot, H. C., Smit, S., Vis, D. J., Smilde, A. K., van Velzen, E. J., van Duijnhoven, J. P., van Dorsten, F. A. Assessment of PLSDA cross validation. *Metabolomics*, 2008, 4, 81–89.

Wiklund, S., Johansson, E., Sjöström, L., Mellerowicz, E. J., Edlund, U., Shockcor, J. P., Gottfries, J., Moritz, T., Trygg, J. Visualization of GC/TOF-MS-based metabolomics data for identification of biochemically interesting compounds using OPLS class models. *Anal. Chem.*, 2008, 80, 115–122, https://doi.org/10.1021/ac0713510.

Winnike, J. H., Li, Z., Wright, F. A., Macdonald, J. M., O'Connell, T. M., Watkins, P. B. Use of pharmaco-metabonomics for early prediction of acetaminophen-induced hepatotoxicity in humans. *Clin. Pharmacol. Ther.*, 2010, 88, 45–51, https://doi.org/10.1038/clpt.2009.240.

Wold, H. O. A. Nonlinear estimation by iterative least square procedures, 1968.

Wold, S., Esbensen, K., Geladi, P. Principal component analysis. *Chemom. Intell. Lab. Syst.*, 1987, 2, 37–52.

Wold, S., Geladi, P., Esbensen, K., Öhman, J. Multi-way principal components-and PLS-analysis. *J. Chemom.*, 1987, 1, 41–56.

Wu, H., Liu, T., Ma, C., Xue, R., Deng, C., Zeng, H., Shen, X. GC/MS-based metabolomic approach to validate the role of urinary sarcosine and target biomarkers for human prostate cancer by microwave-assisted derivatization. *Anal. Bioanal. Chem.*, 2011, 401, 635–646.

Xia, J., Wishart, D. S. MSEA: A web-based tool to identify biologically meaningful patterns in quantitative metabolomic data. *Nucleic Acids Res.*, 2010, 38, W71–7, https://doi.org/10.1093/nar/gkq329.

Xia, J., Psychogios, N., Young, N., Wishart, D. S. MetaboAnalyst: A web server for metabolomic data analysis and interpretation. *Nucleic Acids Res.*, 2009, 37, W652–60, https://doi.org/10.1093/nar/gkp356.

Yizhak, K., Benyamini, T., Liebermeister, W., Ruppin, E., Shlomi, T. Integrating quantitative proteomics and metabolomics with a genome-scale metabolic network model. *Bioinformatics*, 2010, 26, i255–60, https://doi.org/10.1093/bioinformatics/btq183.

Zacharias, H. U., Altenbuchinger, M., Gronwald, W. Statistical analysis of NMR metabolic fingerprints: Established methods and recent advances. *Metabolites*, 2018, 8, 47.

Zacharias, H. U., Rehberg, T., Mehrl, S., Richtmann, D., Wettig, T., Oefner, P. J., Spang, R., Gronwald, W., Altenbuchinger, M. Scale-invariant biomarker discovery in urine and plasma metabolite fingerprints. *J. Proteome Res.*, 2017, 16, 3596–3605.

Zhao, H., Liu, Y., Li, Z., Song, Y., Cai, X., Liu, Y., Zhang, T., Yang, L., Li, L., Gao, S., Li, Y., Yu, C. Identification of essential hypertension biomarkers in human urine by non-targeted metabolomics based on UPLC-Q-TOF/MS. *Clin. Chim. Acta.*, 2018, 486, 192–198, https://doi.org/10.1016/j.cca.2018.08.006.

Zou, H., Hastie, T. Regularization and variable selection via the elastic net. *J. R. Stat. Soc. B: Stat. Methodol.*, 2005, 67, 301–320.

Zuppi, C., Messana, I., Forni, F., Rossi, C., Pennacchietti, L., Ferrari, F., Giardina, B. 1H NMR spectra of normal urines: Reference ranges of the major metabolites. *Clin. Chim. Acta.*, 1997, 265, 85–97, https://doi.org/10.1016/s0009-8981 (97)00110-1.

CHAPTER 4

Biochemical Dynamics in Plant–Insect Interaction: A Metabolomics Approach

SUKANYA SONOWAL[1], KABYASHREE BURAGOHAIN[1], ANKITA GOGOI[1], NIKHA BORAH[1], RATUL NATH[1*], BHABEN CHOWARDHARA[2], and AKBAR HOSSAIN[3]

[1]*Department of Life Sciences, Microbiology Laboratory, Dibrugarh University, Dibrugarh, Assam, India*

[2]*Department of Botany, Arunachal University of Studies, Namsai, Arunachal Pradesh, India*

[3]*Division of Soil Science, Bangladesh Wheat and Maize Research Institute (BWMRI), Nashipur, Dinajpur, Bangladesh*

*Corresponding authors

ABSTRACT

Metabolomics has emerged as a potent tool for deciphering the intricate biochemical dynamics within plant–insect interactions. This advancing field employs an untargeted approach to analyze metabolite fluxes in both plants and insects during their encounters, offering valuable insights into functional genomics and systems biology aspects of these relationships. Positioned downstream from transcriptomics and proteomics, metabolomics stands out as an advanced postgenomic analysis, capable of revealing subtle changes in gene expression or post-translational modifications that influence enzyme activity. Its applicability spans various plant–insect interactions, even in organisms with less-characterized metabolomes or genomes. Success in this field hinges on the availability of specific metabolome databases. Metabolomics is increasingly pivotal in ecology and environmental sciences, providing deeper insights into plant resistance mechanisms and insect herbivore adaptations. This powerful technique offers a holistic perspective on biochemical responses in plants and insects, contributing to sustainable strategies for pest management and crop protection. This chapter explores the multifaceted role of metabolomics, emphasizing its applications via the vast array of compounds synthesized by them and the significance of metabolomics databases in understanding plant–insect interactions.

Exploring Plant Stress Tolerance: A Metabolomics Approach.
Shuvasish Choudhury, Debojyoti Moulick, and Mirza Hasanuzzaman (Eds.)

DOI: 10.1201/9781779641731-4

4.1 INTRODUCTION

About 18–20% of agricultural crop losses are directly related to insect damage, and between 30% and 40% of crop yield reductions are thought to result from a combination of insect damage and the indirect effects these insects cause by spreading microbes (Weintraub and Beanland, 2006; Sharma et al., 2017). The range of hosts for insects primarily relies on adaptations that evolved during their coevolution with plants. This has led to some insects being specialized, meaning they are limited to specific plant families or species (known as specialists, either oligophages or monophages). In contrast, others have a wide range of host plants, referred to as generalists or polyphages (Ali and Agrawal, 2012). Insects have a significant impact on some of the most economically valuable food crops, as indicated by their presence in the list of the top 10 crops with the highest agricultural gross production value, as reported by FAO in 2016. These crops encompass rice, corn, wheat, soybeans, tomatoes, potatoes, sugarcane, a variety of vegetables (such as bamboo shoots, beets, celery, parsley, etc.), grapes, and apples. Floriculture and forestry play significant roles in the economy and face challenges related to diverse insect populations. In floriculture, the cultivation primarily occurs in greenhouses, which create ideal conditions for pests like thrips and aphids. Conversely, forestry tackles issues caused by larger, leaf-chomping insects such as grasshoppers and beetles (Coolen et al., 2022).

Many insects that feed on plants rely on microbes for their well-being. The presence of these symbiotic microorganisms is crucial for the insects, as their absence can significantly harm the insects' health and overall survival, and in some cases, make their existence impossible (Douglas, 2017; Singh et al., 2019). Furthermore, it has been demonstrated that microorganisms linked to insects play a role in influencing the relationships between insects and the plants they inhabit (Frago et al., 2012). Because about 80% of the roughly 900 recognized plant viruses rely on insect vectors for transmission.

Metabolomics is a firmly established scientific discipline focused on investigating naturally occurring, small organic molecules in cells, tissues, or biofluids (Bundy et al., 2009). Multiple factors make plant–insect interactions a compelling field for utilizing and advancing metabolomic methods (Maag et al., 2015). Metabolomics employs analytical chemistry techniques to assess a broad spectrum of small molecules resulting from metabolic processes (typically less than 1500 Da) to gain insights into biological systems. This encompasses understanding how biochemical pathways react to treatments or vary across species, as well as identifying new metabolites that can act as biomarkers for specific biological conditions. Metabolomics is particularly valuable for exploratory research where the precise molecular targets related to a scientific inquiry may be unclear. For instance, when faced with two different plant pathogens causing distinct symptoms in a plant, the molecular underpinnings of this discrepancy may be unknown. In this context, metabolomics presents a potent opportunity to uncover unconventional indicators of biological traits, valuable in the realm of biosecurity (Ross et al., 2023). It has been widely used to study the relationships between plants and insects, offering a potent tool for comprehending biological interfaces. Because metabolomics is nontargeted, it can now uncover bioactive compounds and disentangle metabolic networks more quickly and thoroughly than in the past. Nevertheless, there are still certain restrictions associated with this method, and a metabolomic analysis requires the making of several choices and trade-offs. The chapter provides only a concise overview of viral transmission and its impact on interactions between insects and plants. To gain a deeper insight into the complex relationships between insects and the plants they feed on, this chapter delves into the significance of microorganisms associated with insects in influencing how insects interact with their host plants.

4.2 SYNTHESIS OF METABOLITES USED IN PLANT DEFENSE MECHANISM

Plants deploy an array of metabolites in their defense mechanisms against various stresses including pests. These include phenolics like flavonoids and phenolic acids, terpenoids like terpenes and isoprenoids, alkaloids such as nicotine and caffeine, glycosides like cyanogenic glycosides, and glucosinolates (GSLs). Specific biochemical pathways govern the synthesis of these metabolites. The phenylpropanoid pathway produces phenolics, starting with the conversion of phenylalanine to cinnamic acid (Kumar et al., 2020). Terpenoids are synthesized through both the mevalonate and nonmevalonate pathways, occurring in different cellular compartments. Alkaloids are derived from amino acids, like tryptophan and tyrosine, while GSLs are formed from amino acids such as methionine. These mechanisms contribute to the plant's ability to fend off pathogens, herbivores, and other environmental stresses.

Plants possess inherent defense mechanisms against pests, employing both natural chemicals (metabolites) and self-protective proteins. These defense strategies are typically triggered when the plant detects an insect threat. This process supports specialized interactions between plants and insects, as mentioned by Karban et al. (1997), ultimately resulting in the development of novel defense tactics crucial for successful pest control.

4.2.1 PHENOLICS

Phenolics are molecules characterized by having one or more hydroxylated benzene-ring structures. In a divergence from the main metabolic pathway, L-phenylalanine undergoes a transformation into cinnamic acid through the pivotal enzyme L-phenylalanine ammonia-lyase. This marks the entry point into the phenylpropanoid pathway, which ultimately results in the creation of a wide array of phenolic compounds, such as flavonoids, anthocyanins, benzoic acids, and polyphenols like lignins (Xue Bin and Chang Jun, 2015).

4.2.1.1 FLAVONOIDS

Flavonoids are compounds found widely in plants and are considered secondary metabolites. They are derived from 2-phenyl-benzyl-γ-pyrone and are present in various plant species across the plant kingdom (Mierziak et al., 2014). They are produced through the phenylpropanoid pathway and play crucial roles in various physiological and biochemical processes in plants, acting both as antioxidants and antimicrobials. However, because of the intricate regulation of pathways and diverse transformations, focusing on individual genes has not been highly effective in increasing the concentration of specific molecules. Misra et al. (2010) found that the *Arabidopsis* transcription factor AtMYB12 in tobacco plants resulted in heightened expression of genes associated with the phenylpropanoid pathway, which led to a significant increase in the levels of flavonols. By analyzing the overall gene activity and specific metabolites in the leaves of these transgenic tobacco plants, they found that AtMYB12 influenced several pathways, creating a greater availability of resources for the phenylpropanoid pathway, especially in the production of flavonols. Moreover, these genetically modified tobacco plants exhibited resistance against insect pests like *Spodoptera litura* and *Helicoverpa armigera*, primarily due to the increased accumulation of a specific flavonol called rutin. When we deliberately suppressed the production of flavonols using artificial microRNA, the insect resistance in the AtMYB12-expressing tobacco plants was reversed. This study proposes that a strategic utilization of AtMYB12 could serve as a promising strategy for creating genetically modified plants with enhanced resistance to insect pests, all while maintaining safety considerations.

Flavonoids can be found in different forms, such as aglycones, sugar-bound compounds, and polymers like tannins. Even though the presence of anthocyanin is boosted by the plant's reaction to herbivores through the hormone jasmonic acid (JA) (Shan et al., 2009), flavonols and anthocyanins themselves appear not to have direct insecticidal properties. Apart from that, they play a role in initiating the creation of tannins, which are substances known for their ability to deter insects and exert toxic effects. When these compounds undergo oxygenation within the insect's digestive system, they trigger the production of reactive oxygen species (Barbehenn and Peter Constabel, 2011). The observed impact on aphid reproduction could not be attributed to the heightened buildup of anthocyanins at the location where aphids were feeding on Sorghum (Costa-Arbulú et al., 2001). Phenolic compounds become harmful when consumed by herbivorous insects. When the caterpillars of the oriental leafworm *S. litura* feed on pepper plants (*Capsicum annum*), it triggers the production of phenolic substances like rutin, vanillic acid, and synaptic acid, which subsequently impact both larval growth and the behavior of the adult moth (Movva and Pathipati, 2017). Nonetheless, it seems that the insect gut commonly employs prophenoloxidases for the effective detoxification of a wide range of phenolic compounds (Wu et al., 2015).

4.2.1.2 PHENOLIC ACIDS

Phenolic acids include compounds like salicylic acid (SA), which is involved in signaling during stress responses. Usha Rani and Pratyusha (2013) explored how cotton plants respond to herbivory by *S. litura F.* The research focused on specific plant metabolites as defense mechanisms. The findings revealed increased levels of total phenols and proteins, along with reduced carbohydrates and amino acids after insect feeding. Antioxidative enzyme activity was also assessed, showing higher levels of peroxidase, catalase, and superoxide dismutase, and lower levels of polyphenol oxidase and phenylalanine ammonia-lyase.

High-performance liquid chromatography identified significant elevations in gallic acid, catechin, and caffeic acid levels in infested cotton plants. Feeding experiments with these phenolic compounds activated detoxifying enzymes in the insect gut, indicating their toxicity. This research sheds light on the chemical changes occurring in response to *S. litura* herbivory in cotton plants, revealing their potential yet overlooked defensive roles, while specifically stating the role of specific phenolic acids as antifeedants and detoxification enzyme inducers.

4.2.1.3 TANNINS

Tannins are polyphenolic secondary metabolites with a molecular weight ranging from 300 to 3000 Da. Their functional characteristic lies in their capacity to cause protein precipitation (Peter Constabel et al., 2014). They are especially prone to oxidize in insects with high pH guts, forming semiquinone radicals and quinones, as well as other reactive oxygen species. Tannin toxicity in insects is thought to result from the production of high levels of reactive oxygen species (Barbehenn and Peter Constabel, 2011).

Using tannic acid as a standard, Yuan et al. (2020) investigated its impact on nutritional aspects and the activity of specific enzymes in fourth-instar larvae of *Hyphantria cunea* (Drury) when fed a diet containing tannic acid under different conditions. The findings reveal that tannic acid significantly influences the larvae's digestive capacity and their ability to utilize food. At concentrations below 2.0%, tannic acid promotes feeding and the utilization of less desirable food sources, while higher concentrations (>2.5%) have inhibitory effects. Additionally, tannic acid demonstrates a notable concentration- and time-dependent effect on the activity of detoxification enzymes and acetylcholinesterase in the larvae. These results offer valuable insights into the potential mechanisms employed by *H. cunea* larvae for detoxification against tannic acid in their host plants (Yuan et al., 2020).

4.2.1.4 LIGNANS

Lignans are a group of plant-derived secondary compounds formed by the oxidative merging of two phenylpropanoid components. Despite having a basic structure composed of only two phenylpropane (C6–C3) units, lignans exhibit a wide range of structural variations (Saleem et al., 2005). Qie et al (2022) found that the effectiveness of Haedoxan A (HA), a potential insecticidal lignan derived from *Phryma leptostachya*, against *Aedes aegypti* mosquitoes. Through laboratory selection, a HA-resistant strain was developed, revealing that enhanced metabolic enzyme activity, specifically cytochrome P450-dependent monooxygenases, was the main mechanism of resistance. Importantly, HA showed low fitness cost and resistance risk, making it a promising biopesticide for mosquito control. Furthermore, it exhibited minimal cross-resistance compared to other commonly used insecticides. This suggests that HA derived from *P. leptostachya* holds the potential as an effective and safe tool for mosquito management (Qie et al., 2022).

4.2.1.5 LIGNINS

Lignins are polymers made up of phenylpropanoid units originating from three types of cinnamyl alcohols, namely, *p*-coumaryl, coniferyl, and sinapyl alcohols (Hatfield and Vermerris, 2001). The *Cm*MYB19 gene, which codes for the MYB (the v-myb avian myeloblastosis viral oncogene homolog) transcription factor, was isolated from *Chrysanthemum*. It belongs to the R2R3–MYB subfamily encodes a 200-amino acid protein and was found to be located in the nucleus. *Cm*MYB19 expression was induced by aphid infestation, with higher levels in leaves and stems compared to roots. Overexpressing *Cm*MYB19 led to a reduction in aphid multiplication due to increased lignin accumulation. This suggests a potential role of *Cm*MYB19 in plant defense against aphids through lignin synthesis. *Chrysanthemum*, a valuable ornamental species, faces challenges from the aphid *Macrosiphoniella sanborni*, which hinders its growth and yield. *R2R3–MYB* genes play diverse roles in biological processes, and in association with this, Wang et al. (2017) described *Cm*MYB19's association with lignin synthesis and defense against aphid feeding, highlighting its potential significance in chrysanthemum–aphid interactions.

4.2.2 TERPENES

Terpenes consist of basic hydrocarbons formed from combinations of DMAPP (dimethylallyl pyrophosphate) and ID (isoprenyl diphosphate/pyrophosphate). Terpenoids, also known as isoprenoids, are derived from terpenes but include additional structural modifications and an oxygen component. Despite this distinction, the terms terpenes and terpenoids are often used interchangeably. Terpenoids are categorized based on the number of carbon atoms they contain: hemiterpenoids (C5), monoterpenoids (C10), homoterpenoids (C11, 16), sesquiterpenoids (C15), diterpenoids (C20), sesterpenoids (C25), triterpenoids (C30), tetraterpenoids (C40), and polyterpenoids (C>40, which are higher order terpenoids) (Boncan et al., 2020).

Terpenoids play a role in plant–insect interactions, benefiting plants and deterring herbivores. These interactions are thought to be facilitated by the release of volatile organic compounds (VOCs), which mainly comprise terpenoids. The specific blend and proportion of VOC components influence their characteristics, and herbivores are responsive to particular combinations. The emitted VOCs may have toxic, repellent, or attractive properties (Boncan et al., 2020). Isoprene-derived terpenoids are specialized metabolites that have been shown to have insecticidal or anti-insect activity the most frequently, followed by alkaloids and phenolic compounds (Boulogne et al., 2012). Significantly less discussed but equally important are cyanogenic glycosides and acylsugars (ASs).

Terpenoids have the capability to initiate plant defense responses against pests. β-ocimene, when overexpressed in tobacco, led to a significant release of methyl salicylate and cis-3-hexen-1-ol in tomato (Cascone et al., 2015). β-Ocimene is a type of monoterpenoid with the chemical formula $C_{10}H_{16}$. It comes in two forms, *cis*- and *trans*-β-ocimene (also known as (*Z*)- and (*E*)-β-ocimene), distinguished by the arrangement of the central double bond. The trans form is more prevalent and is released in higher quantities in floral scents compared to the cis form. This compound is produced through the methyl-erythritol-phosphate pathway from the precursor molecules isopentenyl pyrophosphate and DMAPP (Farré-Armengol et al., 2017). These compounds are believed to hinder the development and reproduction of aphids, thereby enhancing the tomato defense mechanisms against them (Cascone et al., 2015). Cinnamon and clove essential oils, along with five terpenoids (eugenol, caryophyllene oxide, α-pinene, α-humulene, and α-phellandrene), were identified as toxic and repellent to *Sitophilus granaries*, a significant grain pest. Among these, eugenol exhibited the highest contact toxicity (Plata-Rueda et al., 2018). A study observed reduced damage by *Paropsisterna tigrina* adults in *Melaleuca alternifolia* when exposed to high concentrations of terpinolene (Bustos-Segura et al., 2015). Additionally, volatiles containing these terpenoids, found in both intact and infested rice plants, demonstrated the ability to either attract or repel the rice pest *Orseolia oryzivora*, depending on their concentrations (Ogah et al., 2017).

Plants use attracting predators or parasitoids as an effective defense against herbivores. *Aphidiuservi*, a versatile predator, is enticed by specific compounds, enhancing its parasitoid abilities. 1,8-cineole serves as both insecticide and parasitoid attractant. Damage inflicted by larvae corresponds to VOC blend upregulation. Iso-pinocarveol may recruit parasitoids. Exposure to β-ocimene from neighboring tobacco plants induces tomato to release compounds, attracting aphid parasitoids (Cascone et al., 2015). Through stacking, subsequent cyclization, or chimeric modification by terpene synthases, a multitude of terpenoid compounds are generated (Hemmerlin et al., 2012). Terpenoids are frequently cited as bioactive compounds when it comes to insects (Aharoni et al., 2005). Volatile allelochemicals have the potential to function as deterrence agents, serving to hinder or reduce the interaction between plants and insects, thus mitigating the spread of (viral) diseases. For instance, the compound sesquiterpene 7-epizingiberene found in *Solanum habrochaites* can discourage the silverleaf whitefly *Bemisia tabaci*, which is a carrier of begomoviruses (Kortbeek et al., 2019). The incorporation of the biosynthetic pathway into a particular tomato variety led to the synthesis of 7-epizingiberene, causing the plant to become less appealing to whiteflies and seemingly harmful to spider mites, *Tetranychus urticae* (Bleeker et al., 2012). One other extensively characterized sesquiterpene with repellent properties is (E)-β-caryophyllene. When the β-caryophyllene synthase is overexpressed in *Arabidopsis,* it results in heightened emission of (*E*)-β-caryophyllene from the flowers, which in turn deters the Asian citrus psyllid *Diaphorina citri.* This psyllid is a carrier of the pathogen responsible for citrus greening disease (Kortbeek et al., 2019).

While some terpenoids play a role in defending against invading herbivores, certain terpenoids serve as chemical attractants to herbivores. This characteristic can be utilized in pest management by deliberately drawing herbivores into traps. Introducing genetically modified plants that produce higher levels of these attractants can effectively lure herbivores into entrapments, preventing their escape. This approach holds significant promise for mitigating the agricultural and economic impacts of these unwanted insect pests. Naidoo et al. (2018) mentioned that an important link was established between increased γ-terpinene and α-pinene production in *Eucalyptus grandis* and its heightened susceptibility to infestation by the pest *Leptocybe invasa*. This suggests the potential value of these terpenes as attractants for *L. invasa*, or as decoys/traps when expressed in nonhost plant species.

4.2.3 ALKALOIDS

Alkaloids are nitrogen-containing compounds with relatively low molecular weight, characterized by the presence of a nitrogen-containing heterocyclic ring, which imparts an alkaline quality to them. They are recognized for their various pharmacological effects on vertebrates. These metabolites can be categorized into distinct classes based on their precursor molecules, such as indole alkaloids, which originate from tryptophan. In total, there are over 20 different classes of alkaloids, including pyrrolidine alkaloids, tropane alkaloids, piperidine alkaloids, pyridine alkaloids, quinolizidine alkaloids, and indole alkaloids, among others (Yang and Stöckigt, 2010).

Alkaloids, which constitute a diverse group of bioactive natural compounds, are widely distributed and can be found in ~20% of all vascular plants, with over 15,000 different types identified. They are particularly prevalent in plant families such as Leguminosae (legumes), Liliaceae (lilies), Solanaceae (nightshade plants), and Amaryllidaceae (Amaryllis). Alkaloids, known for their metabolic effects in mammals, include well-known substances like caffeine, nicotine, morphine, strychnine, and cocaine. They likely evolved as a defense mechanism against insect herbivores (Howe and Jander, 2008).

True alkaloids are synthesized from amino acids in the plant roots and are stored above ground. They possess an alkaline nature and feature nitrogen within a heterocyclic ring structure, as seen in compounds like nicotine and atropine. The ring structure encompasses various types, including pyridines, pyrroles, indoles, pyrrolidines, isoquinolines, and piperidines. Pseudoalkaloids, like caffeine and solanidine, also exhibit an alkaline nature but are not derived from amino acids. Protoalkaloids, such as mescaline, are alkaline and originate from amino acids, but their nitrogen is not part of a heterocycle (Fattorusso and Taglialatela-Scafati, 2007).

Alkaloids derived from quinolizidine, such as cytisine and sparteine, serve as effective deterrents against several herbivores (Petterson et al., 1991). For instance, *Solanum demissum* (nightshade potato) containing the alkaloid demissine exhibits resistance against *Leptinotarsa decemlineata* (Colorado beetle) and *Empoasca fabae* (potato leafhopper). Conversely, *S. tuberosum* contains sterol-derived pseudoalkaloid solanine, which the beetles can detoxify, even though its structure is similar to demissine (Harborne, 2014). Pyrrolizidine alkaloids (PAs) are derived from ornithine or arginine and are naturally present in numerous plants in nontoxic N-oxide forms. However, once they encounter the often-alkaline digestive tracts of certain insect herbivores, they undergo rapid reduction, forming toxic, uncharged, hydrophobic tertiary alkaloids. These can easily traverse membranes (Hartmann, 1999). For example, in *Festuca arundinacea* (tall fescue) colonized by the fungal endophyte *Acremonium coenophialum*, the consumption by two aphids was reduced due to the presence of fungal PAs (Johnson, 2011). Furthermore, PAs demonstrated significant antifeedant properties and proved extremely toxic to the aphid *Rhopalosiphum padi* and the milkweed bug *Oncopeltus fasciatus*. However, some herbivores like *Utetheisea ornatrix* (ornate moth) can detoxify PAs, storing them in their bodies for defense against their own predators, such as the lacewing *Ceraeochrysa cubana* (Eisner et al., 2000).

Approximately 20% of plant species manufacture alkaloids, with around 12,000 different structures having been recognized, encompassing famous substances like caffeine, nicotine, cocaine, and morphine (Ziegler and Facchini, 2008). Alkaloids are organic compounds with a cyclic-ring structure that include a nitrogen base and are not easily evaporated. They are usually produced from nitrogen-containing molecules found in the primary metabolism of plants, such as phenylalanine, arginine, purines, and putrescine (Luca et al., 2000; Ashihara et al., 2008).

Nicotine is perhaps the most renowned alkaloid, and it has been demonstrated to play a vital role in safeguarding wild tobacco species against a diverse array of indigenous herbivorous insects, such as *Diabrotica undecimpunctata*, *Spodoptera exigua*, and *Trimerotropis* spp. When the production of nicotine in *Nicotiana attenuate* was suppressed, it resulted in a threefold rise in the damage caused by various insects that fed on the plant (Steppuhn et al., 2004; Steppuhn and Baldwin, 2007). Nevertheless, some insects thrive on cultivated tobacco, such as *Helicoverpa zea* (Gog et al., 2014) and the tobacco hornworm caterpillar, *Manduca sexta*, which not only can endure nicotine but also employs the metabolite as a means to repel its natural predator (Kumar et al., 2014). Tomatines, which are the primary alkaloids found in tomatoes, are glycoalkaloids of the spyrosolane type, comprising α-tomatine and dehydrotomatine (Kortbeek et al., 2019). While numerous studies have explored the insect-killing characteristics of tomatines, their findings have primarily relied on experiments using synthetic diets, and there is currently no proof to support the idea that tomatines naturally exhibit insecticidal properties (Kortbeek et al., 2019).

4.2.4 GLYCOSIDES

4.2.4.1 CYANOGENIC GLYCOSIDES

These can release toxic cyanide upon breakdown, acting as a defense against herbivores. The stable, comparatively nontoxic, sugar-linked cyanide groups that are stored inside the vacuole and result from the breakdown of amino acids are called cyanogenic glucosides (CNglc) (Gleadow and Møller, 2014). However, they assemble with β-glycosidases upon cell disruption, that is, by herbivory, which cleaves off the sugar moiety to create unstable cyanohydrins. Toxic hydrogen cyanides (HCN) are released through either spontaneous or α-hydroxynitrilase-mediated subsequent hydrolysis (Gleadowand Møller, 2014). While ethylene (ET) biosynthesis results in the minute amounts of HCN that all plants produce, between 5% and 11% of all plant species are thought to actively produce CNglc as a defense mechanism (Gleadow and Møller, 2014). Similar to this, *Brassica* species use myrosinases to hydrolyse GSLs, a process referred to as the "mustard-oil bomb" to produce isothiocyanates (ICN). CNglc plays a less significant role in defense as toxins and is more of a deterrent to insect feeding (Kortbeek et al., 2019). Phloem-feeding insects are among the insects that can stop tissue rupture and prevent the release of these metabolites (Gleadow and Woodrow, 2002).

CNglcs are present in a wide range of plant species, spanning over 2600 species from more than 550 genera and 150 families. This diverse group includes vascular plants across various classes, encompassing angiosperms, both monocotyledons and dicotyledons, as well as gymnosperms and pteridophytes (Nilsen and Orcutt, 1996; Møller, 2010; Rosenthal and Berenbaum, 2012). Derived from amino acids, CNglcs originate from either aromatic or branched-chain amino acids. For example, in *Sorghum bicolor* (sorghum), dhurrin is derived from tyrosine, while in *Lotus japonicus* (lotus) and *Manihot esculenta* (cassava), linamarin and lotaustralin originate from valine and isoleucine. Additionally, in plants belonging to the Rosaceae family (including apples, plums, cherries, peaches, and strawberries), amygdalin and prunasin are derived from phenylalanine (Sánchez-Pérez et al., 2008).

Within intact plant tissues, CNglcs are stored in vacuoles. When plant tissues are disrupted, such as through feeding, β-glucosidases located in either plastids or the apoplast catalyze their hydrolysis. This leads to the formation of a sugar and a cyanohydrin, which subsequently decomposes into the toxic HCN and a ketone or aldehyde. The conversion can also be facilitated by α-hydroxynitrile lyase (Vetter, 2000). HCN is well-known for its toxicity, as it inhibits the enzyme cytochrome c oxidase in the mitochondrial respiratory pathway. The lethal dose of cyanide for vertebrates, if administered in a single dose, ranges from 1.3 to 5.5 mg/kg (Davis and Nahrstedt, 1985). For instance, consumption of 1 kg of white clover could potentially reach this lethal dose (Zagrobelny et al., 2009). CNglcs have also been proposed to serve as nitrogen storage compounds or osmoprotectants (Forslund and Jonsson, 1997; Busk and Møller, 2002). In the case of *M. esculenta* tubers, the presence of CNglcs increases resistance against the generalist *Cyrtomenus bergi* (cassava burrower bug) (Bellotti and Arias, 1992). It is worth noting that the production of CNglcs can be costly for plants, resulting in reduced growth and development (Kakes, 1989). Additionally, the release of HCN may lead to damage to the plant itself, inhibiting the production of phytoalexins used in defense against microorganisms (Lieberi, 1986).

4.2.4.2 GSLS

GSLs are compounds containing sulfur and nitrogen, extensively present in plant families like Brassicaceae and order like Capparales. They are glucosides derived from amino acids, with over 120 distinct structures identified (Hopkins et al., 2009). GSLs are categorized into four groups based on their amino acid precursors: aliphatic GSLs (50%) originating from methionine, indole GSLs (10%) synthesized from tryptophan, aromatic GSLs (10%) derived from phenylalanine or tyrosine, and structures formed from various amino acids (30%), or with unknown biosynthetic origins. Additional diversity arises from processes like chain elongation, oxidation, or hydroxylation of the side chain (Hopkins et al., 2009). GSLs are more abundant in roots compared to shoots. In shoots, indol-3-ylglucosinolate is the most dominant, while its methoxy derivatives and aromatic 2-phenylethyl GSL prevail in roots. This tissue-specific distribution is attributed to differences in volatility, stability in soil, and membrane permeability (van Dam et al., 2009). In roots, GSL levels are largely constitutive, whereas in shoots, they are inducible, likely due to varying selection pressures above and below ground (Karban et al., 1999).

Similar to CNglcs, GSLs are stored in the vacuole, safeguarded from thioglucosidases called myrosinases. When tissues are disrupted, myrosinases come into contact with GSLs and catalyze their hydrolysis, resulting in the formation of toxic breakdown products like ICN (R–N=C=S), nitriles, and thiocyanates (R–S–C≡N), collectively known as "mustard oils." These compounds contribute to the distinctive flavors of various vegetables such as Brassicaceae (mustards, cabbages, and radishes) and serve as both herbivore toxins and feeding deterrents (Bennett and Wallsgrove, 1994). For example, GSLs deter *Deroceras reticulatum* (field slug) from feeding on emerging cotyledons of *Brassica napus* (oilseed rape) (Glen et al., 1990). Additionally, certain herbivores like the amphipod *Gammarus pseudolimnaeus*, the physid snail *Physella* sp., and limnephilid caddisflies *Hesperophylax designates* and *Limnephilus* sp. show a preference for senescent *Nasturtium officinale* (watercress) over fresh watercress due to the significantly lower GSL content (Newman et al., 1992). Moreover, the flea beetle *Phyllotreta cruciferae* displays a preference for older cotyledons of *Sinapis alba* (white mustard) due to lower levels of the GSL sinalbin (Bodnaryk, 1991).

Just like CNglcs, GSLs serve as deterrents for generalist herbivores and attractants for specialists. *Psylliodes chrysocephala* (cabbage stem flea beetle) exclusively feeds on plants containing GSLs (Bartlet and Williams, 1991). *Ceutorhynchus assimilis* (cabbage seed weevil) utilize antennal receptors to detect ICN derived from GSL breakdown, aiding in the localization and discrimination of host plants (Blight et al., 1989; Moyes and Raybould, 2001).

Some insects redirect the hydrolysis typically catalyzed by myrosinase toward the formation of less toxic nitriles, effectively inhibiting the production of toxic ICN. The specialist *Pieris rapae* (cabbage white butterfly) employs the nitrile specifier protein to excrete nitriles in their frass (Wittstock et al., 2004) while *Plutella xylostella* (diamondback moth) uses a GSL sulfatase enzyme to convert ICN and nitriles into desulfoglucosinolates (Ratzka et al., 2002). The example of homozygous lines of *Brassica juncea* (mustard greens) with varying myrosinase activity and GSL profiles illustrates how plants utilize the GSL system for defense against both generalist and specialist herbivores. While the generalist *Spodoptera eridania* prefers lines with low GSL concentrations, the specialist *P. xylostella* shows a preference for lines with low myrosinase activity (Li et al., 2000).

The metabolic diversity in toxin production within individual plants can also provide defense against herbivores with different feeding strategies or resistance mechanisms. For instance, the breakdown of the nearly 40 different GSLs found in *Arabidopsis thaliana* yields over 100 breakdown products (Howe and Jander, 2008). Indole GSLs, which break down in the absence of myrosinase, offer better defense against *Myzuspersicae* (green peach aphid) compared to the more stable aliphatic GSLs (Kim and Jander, 2007).

4.2.4.3 ASS

During the early 1960s, researchers found a harmful substance in the wild tobacco plant known as *Nicotiana gossei*, which was later identified as compounds referred to as AS (Kortbeek et al., 2019). AS are sugar moieties esterified at a particular point within the ring structure to a few C2–C12 aliphatic acyl-chains. A wide variety of insects are severely impacted by AS, which are abundantly synthesized within the Solanaceae family and can collect up to 20% of the leaf dry weight when secreted to the leaf surface. Here are some examples from wild tomato species, *B. tabaci*, *Tuta absoluta*, *T. urticae*, and many thrips species are all affected by different AS in *Solanum galapagense*, *Solanum cheesmaniae*, *Solanum pimpinellifolium*, and *Solanum pennellii* (Kortbeek et al., 2019). The varying AS compositions among and within plant species raise concerns regarding the specificity of plant–herbivore interactions as well as the pressure of selection. The length, quantity, and backbone of the acyl chains all have an impact on the death rate of *Myzuspersicae* aphids and *B. tabaci*, according to bioassays conducted with synthetic AS (Puterka et al., 2003). According to bioassays conducted using fractionated AS extracts from wild tomatoes, the mortality of *B. tabaci* and thrips *Frankliniella occidentalis* was shown to be significantly different between fractions with acylglucoses with C5 acyl chains and fractions with acylsucroses with a mixture of C5/C10/C12 acyl chains (Figure 4.1) (Leckie et al., 2016).

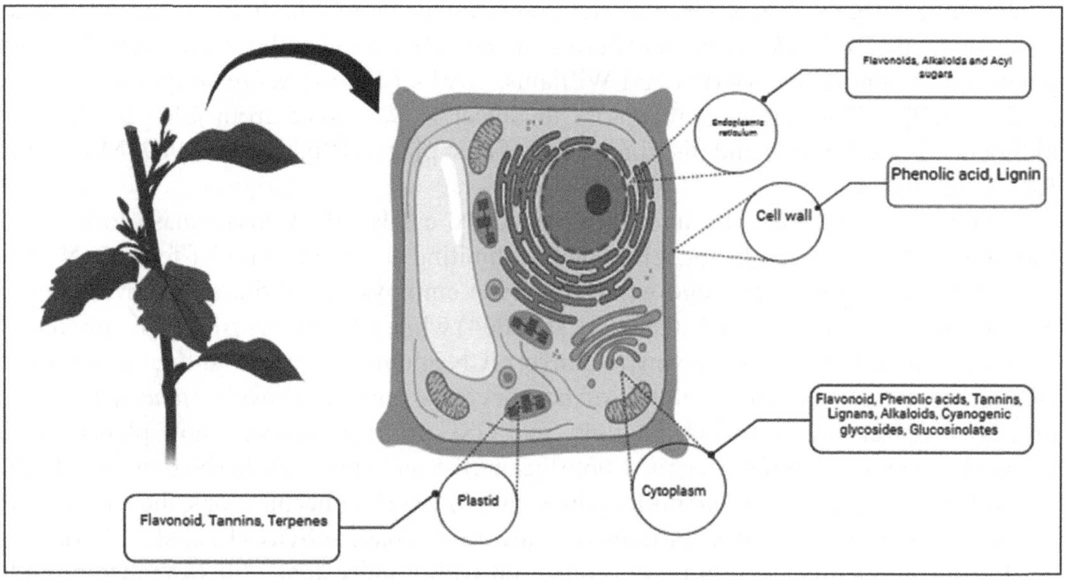

FIGURE 4.1 Primary sites of synthesis of various classes of compounds

Source: Created with BioRender.com.

4.3 APPLICATIONS OF METABOLOMICS IN STUDYING PLANT–INSECT INTERACTIONS

Metabolomic analyses offer a promising avenue for biological sciences researchers to uncover connections between an organism's metabolome and various aspects of its physical state. Although relatively underutilized in insect studies, the application of metabolomics in this field began in 1990 with an examination of how parasitism affects moth larvae's metabolism. Since then, researchers have extended metabolomics to investigate diverse facets of insect biology, including behavior, infection, responses to temperature stress, sedation, and symbiotic relationships with bacteria. In these studies, a variety of established experimental techniques have been employed. However, there is room for enhancement in several areas, such as refining sampling protocols, minimizing variations in sampling and equipment, employing sample extracts, improving statistical analyses, and ensuring metabolite confirmation and identification. Overall, it is evident that metabolomics enables the identification of associations between an organism's observable traits and its underlying cellular metabolic processes, surpassing the capabilities of previous, more focused approaches. The distinctive combination of comprehensive, untargeted analyses with precise quantitative measurements presents a powerful tool for advancing future entomological investigations (Snart et al., 2015).

Several studies have indirectly employed a metabolomic approach to gain insights into plant–insect interactions (Allwood et al., 2008; Jansen et al., 2009; Misra et al., 2010; Leiss et al., 2011).One particular investigation focused on how different developmental stages of the beet armyworm, *Spodoptera exigua* (Hübner), affected *Brassica rapa* L. leaves. Using a combination of 1D and 2D ^{1}H NMR, Widarto et al. (2006) were able to discern significant changes in the metabolome of the leaves after feeding damage commenced. Through spectral analysis, employing principal component analysis, they observed an increase in the levels of glucose, ferulic acid, and gluconapin in response to second-instar *S. exigua* feeding, while fourth-instar feeding led to an increase in alanine and sinapoyl malate. In contrast, larvae of the moth *P. xylostella* L., which is a more specialized herbivore, caused an increase in gluconapin, glucose, feruloyl malate, sinapoyl malate, and threonine. This study once again highlighted the advantages of employing two-dimensional NMR, as it helped mitigate assignment issues associated with overlapping spectral signals (Snart et al., 2015). Following are a few applications of metabolomics that aid in plant–insect interactions.

4.3.1 IDENTIFICATION OF DEFENSE COMPOUNDS AND ANTIFUNGAL METABOLITES

In a metabolomics investigation focused on necrotrophic fungal pathogens, Buxdorf et al. (2013) demonstrated that the fungus *Alternaria brassicicola*, specific to the Brassicaceae family, exhibits greater resilience to specific GSL hydrolysis byproducts produced by *A. thaliana* in comparison to the broad–host range pathogen *Botrytis cinerea*. Another metabolomics-based study corroborated that canola, also a member of the *Brassica* genus, employs GSLs in its root tissue as a response to the biotrophic fungal pathogen *Plasmodiophora brassicae*, which leads to the development of clubroot (Pedras et al., 2008). Furthermore, in addition to the detection of well-known defense compounds, this metabolomics approach unveiled novel anti-fungal metabolites like 4-methoxycyclobrassinin and dehydrocyclobrassinin, previously unreported in plants, and could be categorized as newly identified phytoalexins (Pedras et al., 2008).

Rice blast, caused by the fungus *Magnaporthe oryzae*, is a major threat to rice production, leading to annual losses of over 30% (Nalley et al., 2016). This pathogen exhibits a hemibiotrophic lifestyle, infecting rice at any growth stage and causing damage to various tissues, including leaves, stems, nodes, and panicles (Wilson and Talbot, 2009). To invade rice cells, it forms an appressorium, crucial for intracellular colonization, a process supported by sphingolipids revealed through metabolomic analysis (Park et al., 2009; Liu et al., 2019).

The infection process is tissue-specific, indicating a specialized penetration ability. *M. oryzae* develops haustoria for feeding, representing the biotrophic phase of the fungus (Marcel et al., 2010). In compatible interactions, hyphal multiplication leads to characteristic necrotrophic lesions. Resistant rice genotypes respond with serotonin production, triggering a hypersensitive response and cell death to hinder colonization in the early biotrophic phase (Hayashi et al., 2016). Rice employs a multifaceted pattern-triggered immunity response, activating mitogen-activated protein (MAP) kinase (MAPK) cascades, producing antimicrobial compounds, stimulating hormonal signaling, and inducing callose deposition to reinforce cell walls (Nasir et al., 2018). Microarray studies validate this interaction, showing the upregulation of genes associated with signaling (MAPK) and regulators linked to blast resistance (Wei et al., 2013). ET plays a pivotal role in rice's defense against blast, as ET-deficient rice lines fail to produce phytoalexins, crucial secondary metabolites for defense (Helliwell et al., 2016). Additionally, other hormones like methyl jasmonate and SA are implicated in the upregulation of the disease-related gene *OsERF83* upon *M. oryzae* inoculation (Tezuka et al., 2019).

4.3.2 PLANT PRIMING

Plants can bolster their resistance by heightening the reactivity of their immune system when exposed to specific danger signals from the environment. This phenomenon, termed "priming" or "preconditioning," entails inducing a state in which a plant is pre-exposed to a triggering agent, resulting in heightened resistance to subsequent stresses. Essentially, a "primed" plant reacts more promptly and effectively to subsequent stress compared to a nonprimed one facing the same conditions (Conrath et al., 2002, 2015; Balmer et al., 2015; Martinez-Medina et al., 2016). Priming agents serve as modifiers of the plant's response, potentially leading to a more robust, rapid, early, or sensitive defense response (Hilker et al., 2016).

Experimental studies have provided evidence that priming in *A. thaliana*, induced by β-aminobutyric acid (BABA), leads to changes in tricarboxylic acid fluxes involving malate, oxoglutarate, and fumarate. It also enhances phenylpropanoid biosynthesis and the octadecanoic pathway (Pastor et al., 2013). Comparative metabolomic analyses of chemically-induced priming with BABA in *Arabidopsis* plants have revealed significant impacts on sugar metabolism, cell-wall remodeling, and levels of shikimic acid derivatives. This reprogramming of metabolism results in specific alterations in amino acid profiles and the accumulation of certain compounds like camalexin, indole-acetic acid, and indole-3-carboxaldehyde (Gamir et al., 2014; Pastor et al., 2014). Additionally, when Arabidopsis plants are treated with bacterial lipopolysaccharides as a priming agent, untargeted metabolomics analyses highlight the importance of tryptophan-derived indolic metabolites, including camalexin, indoleglucosinolates, indole-3-carboxylic acid, and indole-acetic acid (Finnegan et al., 2016).

In cultured tobacco cells, exposure to various chemical and pathogen-derived priming agents leads to distinct metabolic changes, particularly in early phenylpropanoid pathway intermediates and products. Activation of these pathways is a crucial aspect of priming events, as well as modifications in secondary metabolism pathways, involving the conjugation of hydroxycinnamic acid derivatives to quinic acid, tyramine, polyamines, or glucose (Mhlongo et al., 2016). On the other hand, analysis of resistant progeny has shown that transgenerational priming is associated with elevated levels of primary metabolites like amino acids and sugars (Luna et al., 2012; Mandal et al., 2012; Hilker et al., 2016; Pastor et al., 2014). However, the specific pathways affected largely depend on the nature of the pathogen. Biotrophic stimuli primarily impact primary metabolism and involve SA signaling, while insects and necrotrophic fungi trigger secondary metabolism via JA/ET-dependent pathways (Gamir et al., 2014).

Recent research has demonstrated that the symbiotic relationship between *Microbacterium* sp. 3J1 and pepper plants protect against drought through metabolic reprogramming. This involves the production of osmoprotectants and antioxidants, along with changes in carbon and nitrogen metabolism, resulting in increased levels of sugars, amino acids, phenolics, and lignin precursors (Vílchez et al., 2018). Another recent research, focusing on the interaction between poplar roots and the ectomycorrhizal fungus *Laccaria bicolor*, revealed systemic adjustments in defense mechanisms in leaves. This includes transcriptional and metabolic reprogramming, leading to heightened levels of chitinases, nitrogen-bearing compounds, volatiles, and reduced levels of phenolics. This mycorrhiza-primed state influences interactions between plants and insects, providing protection to the plant (Kaling et al., 2018).

4.3.3 IDENTIFYING LEADS FOR BREEDING EFFORTS EMPHASIZING NATURAL PEST RESISTANCE

Metabolomics, in conjunction with bioassays and genetic analyses, offers a swift approach to pinpoint specific candidates for breeding programs prioritizing natural pest resistance. Conducting extensive screening of existing accessions and wild relatives on a large scale using bioassays is a potent method to uncover new sources of chemical resistance (Visschers et al., 2019). By focusing on subsets with varying levels of resistance, metabolomic analyses can reveal distinct compounds whose concentrations differ between susceptible and resistant accessions. For example, a metabolomics approach based on LC–MS was employed to identify dimer acyclic diterpene glycosides (capsianosides) as factors conferring resistance against thrips in Capsicum. Furthermore, a GC–MS-based metabolomics approach was utilized to determine which wax components were associated with thrips resistance (Macel et al., 2020).

4.4 PLANT METABOLOMICS RESEARCH: BREEDING FOR RESISTANCE TO PESTS

Insect herbivores often pose a threat to crops even with farmers' best efforts at phytosanitary control. An estimated 15% of agricultural productivity is lost each year to damage by insect herbivores (Mitchell et al., 2016). Since the 1950s, synthetic insecticides have been employed to successfully lower pest pressure and raise crop output. Many effective pesticides, including neonicotinoids, have been banned as a result of contemporary concerns about the nontarget effects of synthetic pesticides, including those on pollinating insects (Allema et al., 2017). Furthermore, several herbivore species have evolved a resistance to pesticides. This has increased the demand for herbivore control strategies that are more sustainable. These tactics correlate with the plants' increased capacity for self-defense. Thus, plant breeders are faced with the difficult task of striking a balance between producing innovative, pest- and pathogen-resistant cultivars that are suitable for the food market and tolerant of changing climatic circumstances. Combined with genetic analysis and bioassays, metabolomics can swiftly provide particular leads for breeding efforts when natural resistance to pests is critical.

Using bioassays for large-scale panel screening of current accessions and wild relatives is an effective way to find (new) sources of chemical resistance (Visschers et al., 2019). Metabolite analysis can reveal individual molecules whose concentrations change between susceptible and resistant accessions by focusing on subsets with differing levels of resistance. For instance, capsianosides, or dimer acyclic diterpene glycosides, were identified as thrips-resistance components in capsicum using an LC–MS-based metabolomics approach (Macel et al., 2020). Additionally, the wax components associated with thrips resistance were examined using a GC–MS-based metabolomics technique (Macel et al., 2020). Plant breeders must connect these resistance features to genetic markers. Genome-wide association studies (GWAS) mapping and quantitative trait loci (QTL) analysis are the typical methods (Omena-Garcia et al., 2019). Natural accessions are genotyped by sequencing using GWAS.

The metabolomics analysis performed on these identical accessions can be integrated with QTL maps to provide insights into the specific genomic areas linked to certain metabolites, or mQTL. By comparing an individual's metabolome with their gene sequence, resistance factors have been found (Omena-Garcia et al., 2019). Combining metabolomics with both mapping techniques can help uncover accessions linked to certain metabolic characteristics and resistance. The responsible regulatory or biosynthetic genes can only be discovered by additional genetic fine mapping (Omena-Garcia et al., 2019; Wahyuni et al., 2014).

Furthermore, these methods could be coupled with specific genome editing instruments, like CRISPR–Cas9, which accurately target particular DNA sequences (Ran et al., 2013). Targeted mutations can offer proof-of-principle on the function of particular genes and metabolites when combined with bioassays. To swiftly produce naturally insect-resistant accessions, high-throughput metagenomic analysis can be used to discover required compounds in the breeding stock. This can then be combined with other speed-breeding techniques (Watson et al., 2018). One problem, though, is that the nutritional content and palatability of crops and fodder are frequently significantly influenced by plant defenses. Plant defense can be poisonous or useful to health, depending on the chemistry of the constituent compounds. As a result, most crops have been chosen to be more palatable by lowering their toxicity and bitterness. The lower levels of the GSL progoitrin in oil seed rape are a well-known example. Progoitrin impairs thyroid metabolism in mammals, leading to health problems (Björkman et al., 2011). However, poor progoitrin accessions become more vulnerable to herbivores as a trade-off (Baaij et al., 2018). When paired with bioassays, metabolic studies may assist in maximizing the levels of chemical defenses in crops without sacrificing the produce's nutritional content or marketability (Table 4.1).

TABLE 4.1 Few Metabolomics Databases

Sl. No.	Metabolomics Databases	Description	Accessible Websites	References
1	PhenoPlant	It contains information on metabolite profiles of various plant species, including those involved in interactions with insects. The emphasis on PhenoPlant lies in its user-friendly interface, automated procedures, and valuable insights into plant stress discrimination based on CF parameters	http://www.phenoplant.org	Rousseau et al. (2015)
2	MassBank	MassBank is a publicly accessible repository dedicated to metabolomics and lipidomics research, hosting a vast collection of mass spectra derived from small compounds. As of February 1, 2020, it encompasses 19,963 MS spectra and 36,505 MS/MS spectra	https://mona.fiehnlab.ucdavis.edu/	Liu et al. (2021)
3	KEGG (Kyoto Encyclopedia of Genes and Genomes)	The KEGG database is widely recognized as a prominent resource for metabolomics information, offering valuable data on metabolic pathways and interaction networks. It can be used to explore the metabolic interactions between plants and insects	https://www.genome.jp/kegg/	Kanehisa et al. (2012), Liu et al. (2021)
4	The Arabidopsis Information Resource (AraCyc)	Its goal is to comprehensively represent Arabidopsis metabolism through a user-friendly web-based interface. Currently, it encompasses over 170 pathways, offering detailed information on compounds, intermediates, cofactors, reactions, genes, proteins, and their subcellular locations. This resource utilizes Pathway Tools software, enabling users to gain a comprehensive overview of all pathways, right down to the specific chemical structures of the compounds	http://www.arabidopsis.org	Mueller et al. (2003)
5	METLIN	METLIN is a metabolite database that contains information on a wide range of metabolites. It can be a valuable resource for identifying and studying metabolites involved in plant–insect interactions. The METLIN database played a crucial role in a study focused on *Centella asiatica* (CA) extract, a plant with potential medicinal properties. This database, used in conjunction with liquid chromatography-mass spectrometry, enabled the identification of over 3200 compounds in the extract	https://metlin.scripps.edu/	Ho et al. (2017)
6	Pep2Path	It can be useful for understanding the biochemical pathways involved in plant defense against insects. It serves as a genome mining tool dedicated to the exploration of gene clusters with similarity to those involved in the biosynthesis of both ribosomal and nonribosomal peptides. It specifically focuses on predicting these peptides through tandem MS fragmentation analysis, providing a detailed examination of gene clusters associated with the biosynthetic pathways. This tool is designed to enhance our understanding of the molecular mechanisms underlying the production of peptides by systematically analyzing genomic data	http://pep2path.sourceforge.net/	Medema et al. (2014)

4.5 PROSPECTS FOR THE FUTURE: WHICH NEW INSTRUMENTS AND METHODS WILL HAVE AN EFFECT?

When it comes to identifying metabolic signals for which the process of compound identification is laborious and inadequate, metabolomics is a useful technique (Allwood et al., 2021). Though metabolomic data processing has greatly improved due to bioinformatics advancements (Tsugawa et al., 2015; Domingo-Almenara et al., 2019), hundreds of signals are still unexplained (da Silva et al., 2015). Therefore, with the creation of better annotation methods and more extensive plant metabolite libraries, metabolomics may have a much bigger influence. Nowadays, metabolomics relies heavily on machine learning (LC–MS is frequently used (Domingo-Almenara et al., 2018). It forecasts a wide range of molecular processes and attributes, including chromatographic retention duration, MS/MS deconvolution, adduct species, and MS fragmentation (Tsugawa et al., 2015; Domingo-Almenara et al., 2019; Gomes et al., 2021). Nevertheless, it is important to remember that there are a large number of closely related isomeric compounds found in plants, as well as their derivatives that are fused with amino acids or decorated with different isomeric sugar moieties. These compounds produce nearly identical MS/MS and MS^n fragmentation patterns and co-elute even during lengthy (60+ min) ultra-high-performance liquid chromatography separations. As a result, research and development efforts should be directed toward creating higher resolution separation techniques and technologies that can be used in conjunction with 2D NMR and MS.

A significant step forward in creating retention time (RT)-MS/MS and MS^n libraries that are transferrable between various LC–MS systems and laboratories would be the development of 2D LC separations and time-series retention index standards that are applicable to LC, as *n*-alkanes and fatty acid methyl esters are applied in GC. Overcoming these obstacles and improving our metabolite annotation skills would be especially helpful for the identification of viral and antibiotic compounds in plant-pathogen studies. In addition to offering intriguing antimicrobial targets, a recent study (Rubio et al., 2021) reveals the metabolic context of plant–virus interactions in a natural habitat and its predictive relationship to susceptibility. Similarly, predictive metabolomics based on machine learning shows great promise in identifying and validating metabolic predictors in biotic conditions (Luna et al., 2020). Robotized experiments, for example, will enable future bioinformatics advancements in metabolomics to decrease human intervention and boost data analysis throughput (Allwood et al., 2021; Luna et al., 2020).

Although it lacks the ability to distinguish structural isomers that exhibit identical MS/MS and MS^n fragmentation patterns, MSi is a prominent tool among the new tools that have been developed for metabolome analysis (Walker, 2021) and offers promising perspectives for spatial dissection of metabolic responses to infection. However, isomeric molecules may be resolved without the need for prior chromatographic separation by using ion mobility separations prior to HRMS detection. The sulfur-containing GSLs in Arabidopsis that influence fungal and pest responses (Shroff et al., 2008) and the barley defense metabolite serotonin and other metabolites of the melanin pathway against *M. oryzae* are two recent notable examples of MSi (Kalmar et al., 2020). In addition to MSi, other methods that enable the separation of cells and organelles—such as sorting cells and organelles, flow cytometry, and microfluidics—will significantly enhance the potential effects of metabolomics on plant stress research in the future.

Plant–pathogen interactions can also be studied with tremendous potential thanks to genome-scale metabolic models. Protein–protein interactions, transcriptomics, and structural and comparative genomics are all implied by this computational systems biology viewpoint (Pinzón et al., 2009). A typical experimental design of phytopathological metabolomics emphasizes the need for high-resolution temporal sampling and high-resolution metabolome data, which can be readily obtained with today's Orbitrap and Q-TOF technologies, in order to calculate metabolite coefficients for genome-scale metabolic flux predictions. In this regard, the metabolic fluxes during infection have been simulated using genome-scale metabolic models of Phytophthora, potato, and tomato leaves (Rodenburg et al., 2019; Botero et al., 2018). Phytophthora's consumption of nutrients beyond the infection cycle phases was revealed, along with interesting molecular elements including the pathogen's suppression of photosynthesis. It should be mentioned that by using high-resolution data from the tomato infection metabolome, stage-specific profiles of the combined metabolism of the pathogen and host could be improved (Rodenburg et al., 2019).

To reiterate the final point, differentiating between the metabolomes of plants and microbes is a major problem for metabolomics in plant pathology (Allwood et al., 2010). Primary chemicals in particular are frequently found in a wide variety of plant, microbial, and animal taxa. They are also frequently transferred as metabolic responses or signals between pathogens and hosts. Plants and bacteria are co-incubated in pathological investigations, but the pathogens are rarely eliminated prior to metabolite extraction, resulting in the mixing of metabolomes from various origins. Given that other omics, such as transcriptomics and proteomics, make use of species-specific datasets, this certainly seems insignificant. Nevertheless, because of the cross-contamination of host and nonhost metabolites in metabolomics, it is frequently challenging to determine with precision where the metabolic changes were found.

While it is possible to search data for nonplant metabolites, such as arachidonic acid in relation to fungal infections, it is not reasonable to conclude that their lack indicates a lack of fungal metabolites. This urgent issue might be resolved by labeling microorganisms with stable isotopes. An incubation approach using ^{13}C- and ^{15}N-labeled (98%) Pst DC3000 with *Arabidopsis* cells was used in a recent metabolomics analysis of the early stage of plant cell–microbe interactions. This was followed by a washing phase to remove the excess bacterial cells (Pang et al., 2018). Targeted LC–MS analysis of "light" plant metabolites revealed an infection-induced reprogramming of plant signaling and main metabolic pathways, which include amino, organic, and carbohydrate acids. As an alternative, uniformly tagged plants with ^{13}C and ^{15}N labels can be produced, providing a precise tool for metabolomic investigation in the presence of biotic stress (Ćeranić et al., 2020). The development of additional analytical methods, such as NMR and high-throughput GC– and LC–HRMS, which enable the distinction between native and tagged metabolites, corresponds with the increased interest in isotopically labeled biological material. Elegant investigations of this kind might be conducted less infrequently to shed light on the cause of alterations in the metabolomes of hosts and pathogens.

In conclusion, future developments that will significantly boost the effectiveness of metabolomics in the investigation of plant immunity will encompass the following: (1) improving the ability to annotate and resolve plant and pathogen metabolomes chromatographically, (2) developing new methods and instruments for subcellular metabolome analysis, and (3) developing and putting into practice modeling to forecast and/or compute the dynamics of metabolic variations.

4.6 METABOLOMICS PROVIDES A POWERFUL TOOL FOR UNRAVELLING THE INTRICACIES OF PLANT–INSECT INTERACTIONS

In agricultural ecosystems, the impact of insect damage on crop losses is substantial, accounting for ~18–20% of agricultural crop losses directly attributed to insect-related factors. Additionally, the combined effects of insect damage and the indirect consequences arising from the spread of microbes by these insects contribute to yield reductions ranging from 30% to 40% (Sharma et al., 2017; Weintraub and Beanland, 2006). The coevolution of insects with plants has given rise to specialized feeders, known as oligophages or monophages, limited to specific plant families or species, and generalists or polyphages, which can exploit a wide range of host plants (Ali and Agrawal, 2012). Notably, insects exert a significant influence on economically vital food crops, such as rice, corn, wheat, soybeans, tomatoes, potatoes, sugarcane, various vegetables, grapes, and apples, as well as impacting floriculture and forestry.

The intricate relationships between insects and plants are further complicated by the symbiotic microorganisms crucial for the well-being and survival of many plant-feeding insects (Douglas, 2017; Singh et al., 2019). The absence of these microorganisms can have detrimental effects on insect health, potentially making their existence impossible. Moreover, microorganisms associated with insects play a role in shaping the interactions between insects and their host plants (Frago et al., 2012). Given that ~80% of recognized plant viruses rely on insect vectors for transmission, understanding viral transmission becomes pivotal in deciphering the complex web of interactions between insects and plants (Hohn, 2007; Roossinck, 2012).

Metabolomics, as a scientific discipline, provides a powerful tool for unraveling the intricacies of plant–insect interactions. By investigating small organic molecules resulting from metabolic processes, metabolomics offers insights into biochemical pathways, variations across species, and the identification of potential biomarkers for specific biological conditions (Bundy et al., 2009; Maag et al., 2015). In the context of plant–insect interactions, metabolomics allows for the rapid and thorough dissection of bioactive compounds and metabolic networks. Despite its advantages, metabolomics does come with certain limitations and requires careful consideration of choices and trade-offs in the analytical process (Ross et al., 2023). The application of metabolomic methods in this field presents a promising avenue for advancing our understanding of the complex interplay between insects, plants, and microorganisms in agricultural and ecological systems.

4.7 CONCLUSIONS

Over the past decade, the application of metabolomics in the investigation of plant–insect interactions has emerged as a pivotal and valuable method, offering profound insights into the intricate biological connections within ecological systems. The untargeted nature of metabolomics has revolutionized the efficiency and comprehensiveness with which bioactive substances and metabolic networks are uncovered, marking a significant advancement from traditional approaches. Despite its successes, it is imperative to recognize the existing constraints associated with metabolomics. Researchers navigating the complexities of this method are compelled to make nuanced decisions and trade-offs throughout the analytical process. This acknowledgement underscores the evolving nature of metabolomic research and the ongoing need for methodological refinement.

Looking ahead, there is a promising trajectory for metabolomics in the realm of plant–insect interactions. Anticipated developments include enhanced accessibility for researchers, leading to a reduction in the time spent on methodological intricacies. This optimistic outlook hinges on the establishment of standardized analytical methods. While the notion of standardization may appear divergent from the customary practices of analytical chemists who often prefer developing unique methods, it holds the potential to yield substantial benefits. The establishment of standardized analytical methods is poised to streamline processes, allowing researchers to focus more intently on the biological nuances of plant–insect interactions rather than methodological intricacies. Importantly, this standardization facilitates cross-comparisons of results among diverse research teams, fostering a more cohesive understanding of metabolomic outcomes. Moreover, it instills a heightened level of confidence in the identification of metabolites, thereby strengthening the robustness of findings in this field.

In conclusion, the future trajectory of metabolomics in plant–insect interaction research promises not only increased accessibility but also a more standardized and collaborative approach. As we move toward a future of streamlined processes, researchers can anticipate a landscape where the focus shifts from methodological challenges to a deeper exploration of the complex relationships between plants, insects, and the associated microorganisms. This evolution holds the potential to unlock new dimensions of understanding in the dynamic and intricate world of ecological interactions.

KEYWORDS

- **metabolome**
- **plant-insect interaction**
- **metabolites**
- **metabolomics**

REFERENCES

Aharoni, A., Jongsma, M. A., Bouwmeester, H. J. Volatile science? Metabolic engineering of terpenoids in plants. *Trends Plant Sci.*, 2005, 10(12), 594–602. https://doi.org/10.1016/j.tplants.2005.10.005

Ali, J. G., Agrawal, A. A. Specialist versus generalist insect herbivores and plant defense. *Trends Plant Sci.*, 2012, 17(5), 293–302. https://doi.org/10.1016/j.tplants.2012.02.006

Allema, B., Hoogendoorn, M., van Beek, J., Leendertse, P. *Neonicotinoids in European Agriculture: Main Applications, Main Crops and Scope for Alternatives*. CLM, 2017.

Allwood, J. W., Clarke, A., Goodacre, R., Mur, L. A. Dual metabolomics: A novel approach to understanding plant–pathogen interactions. *Phytochemistry*, 2010, 71(5–6), 590–597.

Allwood, J. W., Ellis, D. I., Goodacre, R. Metabolomic technologies and their application to the study of plants and plant–host interactions. *Physiol. Plant.*, 2008, 132(2), 117–135. https://doi.org/10.1111/j.1399-3054.2007.01001.x

Allwood, J. W., Williams, A., Uthe, H., van Dam, N. M., Mur, L. A., Grant, M. R., Pétriacq, P. Unravelling plant responses to stress—The importance of targeted and untargeted metabolomics. *Metabolites*, 2021, 11(8), 558.

Ashihara, H., Sano, H., Crozier, A. Caffeine and related purine alkaloids: Biosynthesis, catabolism, function and genetic engineering. *Phytochemistry*, 2008, 69(4), 841–856. https://doi.org/10.1016/j.phytochem.2007.10.029

Baaij, B. M., Kim, H. K., Grosser, K., Worrich, A., de Jong, T. J. Slug herbivory on hybrids of the crop Brassica napus and its wild relative B. rapa. *Basic Appl. Ecol.*, 2018, 31, 52–60. https://doi.org/10.1016/j.baae.2018.06.001

Balmer, A., Pastor, V., Gamir, J., Flors, V., Mauch-Mani, B. The "prime-ome": Towards a holistic approach to priming. *Trends Plant Sci.*, 2015, 20(7), 443–452. https://doi.org/10.1016/j.tplants.2015.04.002

Barbehenn, R. V., Peter Constabel, C. Tannins in plant–herbivore interactions. *Phytochemistry*, 2011, 72(13), 1551–1565. https://doi.org/10.1016/j.phytochem.2011.01.040

Bartlet, E., Williams, I. H. Factors restricting the feeding of the cabbage stem flea beetle (*Psylliodes chrysocephala*). *Entomol. Exp. Appl.*, 1991, 60(3), 3.

Bellotti, A. C., Arias, B. The possible role of HCN on the biology and feeding behaviours of the cassava burrowing bug (Cyrtomenus Bergi Froeschner). In: Roca, W. M., Thro, A. M. (Eds.), *Proceedings on the 1st International Scientific Meeting of the Cassava Biotechnology Network* (pp. 406–409), Cali, Columbia, 25–28 August 1992; Cali: Columbia.

Bennett, R. N., Wallsgrove, R. M. Secondary metabolites in plant defence mechanisms. *New Phytol.*, 1994, 127(4), 617–633. https://doi.org/10.1111/j.1469-8137.1994.tb02968.x

Björkman, M., Klingen, I., Birch, A. N. E., Bones, A. M., Bruce, T. J. A., Johansen, T. J., Meadow, R., Mølmann, J., Seljåsen, R., Smart, L. E., Stewart, D. Phytochemicals of Brassicaceae in plant protection and human health—Influences of climate, environment and agronomic practice. *Phytochemistry*, 2011, 72(7), 538–556. https://doi.org/10.1016/j.phytochem.2011.01.014

Bleeker, P. M., Mirabella, R., Diergaarde, P. J., VanDoorn, A., Tissier, A., Kant, M. R., Prins, M., de Vos, M., Haring, M. A., Schuurink, R. C. Improved herbivore resistance in cultivated tomato with the sesquiterpene biosynthetic pathway from a wild relative. *Proc. Natl. Acad. Sci.*, 2012, 109(49), 20124–20129. https://doi.org/10.1073/pnas.1208756109

Blight, M. M., Pickett, J. A., Wadhams, L. J., Woodcock, C. M. Antennal responses of Ceutorhynchusassimilis and Psylliodes chrysocephala to volatiles from oilseed rape. *Aspects Appl. Biol.*, 1989, 23, 329–334.

Bodnaryk, R. P. Developmental profile of sinalbin (p-hydroxybenzyl glucosinolate) in mustard seedlings, Sinapis alba L., and its relationship to insect resistance. *J. Chem. Ecol.*, 1991, 17(8), 1543–1556. https://doi.org/10.1007/BF00984687

Boncan, D. A. T., Tsang, S. S. K., Li, C., Lee, I. H. T., Lam, H.-M., Chan, T.-F., Hui, J. H. L. Terpenes and terpenoids in plants: Interactions with environment and insects. *Int. J. Mol. Sci.*, 2020, 21(19), 19. https://doi.org/10.3390/ijms21197382

Botero, K., Restrepo, S., Pinzón, A. A genome-scale metabolic model of potato late blight suggests a photosynthesis suppression mechanism. *BMC Genom. [Electronic Resource]*, 2018, 19, 31–44.

Boulogne, I., Petit, P., Ozier-Lafontaine, H., Desfontaines, L., Loranger-Merciris, G. Insecticidal and antifungal chemicals produced by plants: A review. *Environ. Chem. Lett.*, 2012, 10(4), 325–347. https://doi.org/10.1007/s10311-012-0359-1

Bundy, J. G., Davey, M. P., Viant, M. R. Environmental metabolomics: A critical review and future perspectives. *Metabolomics*, 2009, 5(1), 3–21. https://doi.org/10.1007/s11306-008-0152-0

Busk, P. K., Møller, B. L. Dhurrin synthesis in sorghum is regulated at the transcriptional level and induced by nitrogen fertilization in older plants. *Plant Physiol.*, 2002, 129(3), 1222–1231. https://doi.org/10.1104/pp.000687

Bustos-Segura, C., Külheim, C., Foley, W. Effects of terpene chemotypes of melaleuca alternifolia on two specialist leaf beetles and susceptibility to myrtle rust. *J. Chem. Ecol.*, 2015, 41(10), 937–947. https://doi.org/10.1007/s10886-015-0628-0

Buxdorf, K., Yaffe, H., Barda, O., Levy, M. The effects of glucosinolates and their breakdown products on necrotrophic fungi. *PLoS One*, 2013, 8(8), e70771. https://doi.org/10.1371/journal.pone.0070771

Cascone, P., Iodice, L., Maffei, M. E., Bossi, S., Arimura, G.-I., Guerrieri, E. Tobacco overexpressing β-ocimene induces direct and indirect responses against aphids in receiver tomato plants. *J. Plant Physiol.*, 2015, 173, 28–32. https://doi.org/10.1016/j.jplph.2014.08.011

Ćeranić, A., Doppler, M., Büschl, C., Parich, A., Xu, K., Koutnik, A., Bürstmayr, H., Lemmens, M. and Schuhmacher, R. Preparation of uniformly labelled 13 C-and 15 N-plants using customised growth chamber rs. *Plant Methods*, 2020, 16, 1–15.

Conrath, U., Beckers, G. J. M., Langenbach, C. J. G., Jaskiewicz, M. R. Priming for enhanced defense. *Annu. Rev. Phytopathol.*, 2015, 53, 97–119. https://doi.org/10.1146/annurev-phyto-080614-120132

Conrath, U., Pieterse, C. M. J., Mauch-Mani, B. Priming in plant–pathogen interactions. *Trends Plant Sci.*, 2002, 7(5), 210–216. https://doi.org/10.1016/S1360-1385(02)02244-6

Coolen, S., Rogowska-van der Molen, M., Welte, C. U. The secret life of insect-associated microbes and how they shape insect–plant interactions. *FEMS Microbiol. Ecol.*, 2022, 98(9), fiac083. https://doi.org/10.1093/femsec/fiac083

Costa-Arbulú, C., Gianoli, E., Gonzáles, W. L., Niemeyer, H. M. Feeding by the aphid sipha flava produces a reddish spot on leaves of sorghum halepense: An induced defense. *J. Chem. Ecol.*, 2001, 27(2), 273–283. https://doi.org/10.1023/A:1005676321251

da Silva, R. R., Dorrestein, P. C., Quinn, R. A. Illuminating the dark matter in metabolomics. *Proc. Natl. Acad. Sci.*, 2015, 112(41), 12549–12550.

Davis, R. H., Nahrstedt, A. Cyanogenesis in insects. In: Kerkut, G. A., Gilbert, L. I. (Eds.), *Comprehensive Insect Physiology, Biochemistry Pharmacology* (pp. 635–654). Pergamon Press: Oxford, UK, 1985.

Domingo-Almenara, X., Guijas, C., Billings, E., Montenegro-Burke, J. R., Uritboonthai, W., Aisporna, A. E., Chen, E., Benton, H. P. and Siuzdak, G. The METLIN small molecule dataset for machine learning-based retention time prediction. *Nat. Commun.*, 2019, 10(1), 5811.

Domingo-Almenara, X., Montenegro-Burke, J. R., Ivanisevic, J., Thomas, A., Sidibé, J., Teav, T., Guijas, C., Aisporna, A. E., Rinehart, D., Hoang, L. and Nordström, A. XCMS-MRM and METLIN-MRM: A cloud library and public resource for targeted analysis of small molecules. *Nat. Methods*, 2018, 15(9), 681–684.

Douglas, A. E. The B vitamin nutrition of insects: The contributions of diet, microbiome and horizontally acquired genes. *Curr. Opin. Insect Sci.*, 2017, 23, 65–69. https://doi.org/10.1016/j.cois.2017.07.012

Eisner, T., Eisner, M., Rossini, C., Iyengar, V. K., Roach, B. L., Benedikt, E., Meinwald, J. Chemical defense against predation in an insect egg. *Proc. Natl. Acad. Sci.*, 2000, 97(4), 1634–1639. https://doi.org/10.1073/pnas.030532797

Farré-Armengol, G., Filella, I., Llusià, J., Peñuelas, J. β-Ocimene, a key floral and foliar volatile involved in multiple interactions between plants and other organisms. *Molecules*, 2017, 22(7), 7. https://doi.org/10.3390/molecules22071148

Fattorusso, E., Taglialatela-Scafati, O. *Modern Alkaloids: Structure, Isolation, Synthesis, and Biology*. John Wiley & Sons, 2007.

Finnegan, T., Steenkamp, P. A., Piater, L. A., Dubery, I. A. The Lipopolysaccharide-induced metabolome signature in arabidopsis thaliana reveals dynamic reprogramming of phytoalexin and phytoanticipin pathways. *PLoS One*, 2016, 11(9), e0163572. https://doi.org/10.1371/journal.pone.0163572

Forslund, K., Jonsson, L. Cyanogenic glycosides and their metabolic enzymes in barley, in relation to nitrogen levels. *Physiol. Plant.*, 1997, 101(2), 367–372. https://doi.org/10.1111/j.1399-3054.1997.tb01010.x

Frago, E., Dicke, M., Godfray, H. C. J. Insect symbionts as hidden players in insect–plant interactions. *Trends Ecol. Evol.*, 2012, 27(12), 705–711. https://doi.org/10.1016/j.tree.2012.08.013

Gamir, J., Pastor, V., Kaever, A., Cerezo, M., Flors, V. Targeting novel chemical and constitutive primed metabolites against *Plectosphaerella cucumerina*. *Plant J.*, 2014, 78(2), 227–240. https://doi.org/10.1111/tpj.12465

Gamir, J., Sánchez-Bel, P., Flors, V. Molecular and physiological stages of priming: How plants prepare for environmental challenges. *Plant Cell Rep.*, 2014, 33(12), 1935–1949. https://doi.org/10.1007/s00299-014-1665-9

Gleadow, R. M., Møller, B. L. Cyanogenic glycosides: Synthesis, physiology, and phenotypic plasticity. *Annu. Rev. Plant Biol.*, 2014, 65(1), 155–185. https://doi.org/10.1146/annurev-arplant-050213-040027

Gleadow, R. M., Woodrow, I. E. Mini-review: Constraints on effectiveness of cyanogenic glycosides in herbivore defense. *J. Chem. Ecol.*, 2002, 28(7), 1301–1313. https://doi.org/10.1023/A:1016298100201

Glen, D. M., Jones, H., Fieldsend, J. K. Damage to oilseed rape seedlings by the field slug *Deroceras reticulatum* in relation to glucosinolate concentration. *Ann. Appl. Biol.*, 1990, 117(1), 197–207. https://doi.org/10.1111/j.1744-7348.1990.tb04207.x

Gog, L., Vogel, H., Hum-Musser, S. M., Tuter, J., Musser, R. O. Larval helicoverpazea transcriptional, growth and behavioral responses to nicotine and nicotiana tabacum. *Insects*, 2014, 5(3), 3. https://doi.org/10.3390/insects5030668

Gomes, P., Quirós-Guerrero, L., Silva, C., Pamplona, S., Boutin, J. A., Eberlin, M., Wolfender, J. L., Silva, M. Feature-based molecular network-guided dereplication of natural bioactive products from leaves of *Stryphnodendron pulcherrimum* (Willd.) Hochr. *Metabolites*, 2021, 11(5), 281.

Harborne, J. B. *Introduction to Ecological Biochemistry*. Academic Press, 2014.

Hartmann, T. Chemical ecology of pyrrolizidine alkaloids. *Planta*, 1999, 207(4), 483–495.

Hatfield, R., Vermerris, W. Lignin formation in plants. the dilemma of linkage specificity. *Plant Physiol.*, 2001, 126(4), 1351–1357. https://doi.org/10.1104/pp.126.4.1351

Hayashi, K., Fujita, Y., Ashizawa, T., Suzuki, F., Nagamura, Y., Hayano-Saito, Y. Serotonin attenuates biotic stress and leads to lesion browning caused by a hypersensitive response to *Magnaporthe oryzae* penetration in rice. *Plant J.*, 2016, 85(1), 46–56. https://doi.org/10.1111/tpj.13083

Helliwell, E. E., Wang, Q., Yang, Y. Ethylene biosynthesis and signaling is required for rice immune response and basal resistance against magnaportheoryzae infection. *Mol. Plant-Microbe Interactions®*, 2016, 29(11), 831–843. https://doi.org/10.1094/MPMI-06-16-0121-R

Hemmerlin, A., Harwood, J. L., Bach, T. J. A raison d'être for two distinct pathways in the early steps of plant isoprenoid biosynthesis? *Prog. Lipid Res.*, 2012, 51(2), 95–148. https://doi.org/10.1016/j.plipres.2011.12.001

Hilker, M., Schwachtje, J., Baier, M., Balazadeh, S., Bäurle, I., Geiselhardt, S., Hincha, D. K., Kunze, R., Mueller-Roeber, B., Rillig, M. C., Rolff, J., Romeis, T., Schmülling, T., Steppuhn, A., van Dongen, J., Whitcomb, S. J., Wurst, S., Zuther, E., Kopka, J. Priming and memory of stress responses in organisms lacking a nervous system. *Biol. Rev. Camb. Philos. Soc.*, 2016, 91(4), 1118–1133. https://doi.org/10.1111/brv.12215

Ho, Y.-N., Shu, L.-J., Yang, Y.-L. Imaging mass spectrometry for metabolites: Technical progress, multimodal imaging, and biological interactions. *WIREs Syst. Biol. Med.*, 2017, 9(5), e1387. https://doi.org/10.1002/wsbm.1387

Hohn, T. Plant virus transmission from the insect point of view. *Proc. Natl. Acad. Sci.*, 2007, 104(46), 17905–17906. https://doi.org/10.1073/pnas.0709178104

Hopkins, R. J., van Dam, N. M., van Loon, J. J. A. Role of Glucosinolates in Insect-Plant Relationships and Multitrophic Interactions. *Annu. Rev. Entomol.*, 2009, 54(1), 57–83. https://doi.org/10.1146/annurev.ento.54.110807.090623

Howe, G. A., Jander, G. Plant immunity to insect herbivores. *Annu. Rev. Plant Biol.*, 2008, 59(1), 41–66. https://doi.org/10.1146/annurev.arplant.59.032607.092825

Jansen, J. J., Allwood, J. W., Marsden-Edwards, E., van der Putten, W. H., Goodacre, R., van Dam, N. M. Metabolomic analysis of the interaction between plants and herbivores. *Metabolomics*, 2009, 5(1), 150–161. https://doi.org/10.1007/s11306-008-0124-4, https://app.biorender.com/illustrations/654c7945539d47266ec6b50

Johnson, M. T. J. Evolutionary ecology of plant defences against herbivores. *Funct. Ecol.*, 2011, 25(2), 305–311. https://doi.org/10.1111/j.1365-2435.2011.01838.x

Kakes, P. An analysis of the costs and benefits of the cyanogenic system in Trifolium repens L. *Theor. Appl. Genet.*, 1989, 77(1), 111–118. https://doi.org/10.1007/BF00292324

Kaling, M., Schmidt, A., Moritz, F., Rosenkranz, M., Witting, M., Kasper, K., Janz, D., Schmitt-Kopplin, P., Schnitzler, J.-P., Polle, A. Mycorrhiza-triggered transcriptomic and metabolomic networks impinge on herbivore fitness. *Plant Physiol.*, 2018, 176(4), 2639–2656. https://doi.org/10.1104/pp.17.01810

Kalmar, J. G., Oh, Y., Dean, R. A., Muddiman, D. C. Investigating host-pathogen meta-metabolic interactions of *Magnaporthe oryzae* infected barley using infrared matrix-assisted laser desorption electrospray ionization mass spectrometry. *Anal. Bioanal.Chem.*, 2020, 412, 139–147.

Kanehisa, M., Goto, S., Sato, Y., Furumichi, M., Tanabe, M. KEGG for integration and interpretation of large-scale molecular data sets. *Nucleic Acids Res.*, 2012, 40(Database issue), D109–D114. https://doi.org/10.1093/nar/gkr988

Karban, R., Agrawal, A. A., Mangel, M. The benefits of induced defenses against herbivores. *Ecology*, 1997, 78(5), 1351–1355. https://doi.org/10.1890/0012-9658(1997)078[1351:TBOIDA]2.0.CO;2

Karban, R., Agrawal, A. A., Thaler, J. S., Adler, L. S., Karban, R., Agrawal, A. A., Thaler, J. S., Adler, L. S., Karban, R., Agrawal, A. A., Thaler, J. S., Adler, L. S. Induced plant responses and information content about risk of herbivory. *Trends Ecol. Evol.*, 1999, 14(11), 443–447. https://doi.org/10.1016/S0169-5347(99)01678-X

Kim, J. H., Jander, G. Myzuspersicae (green peach aphid) feeding on Arabidopsis induces the formation of a deterrent indole glucosinolate. *Plant J.*, 2007, 49(6), 1008–1019. https://doi.org/10.1111/j.1365-313X.2006.03019.x

Kortbeek, R. W. J., van der Gragt, M., Bleeker, P. M. Endogenous plant metabolites against insects. *Eur. J. Plant Pathol.*, 2019, 154(1), 67–90. https://doi.org/10.1007/s10658-018-1540-6

Kumar, P., Rathi, P., Schöttner, M., Baldwin, I. T., Pandit, S. Differences in nicotine metabolism of two nicotiana attenuata herbivores render them differentially susceptible to a common native predator. *PLoS One*, 2014, 9(4), e95982. https://doi.org/10.1371/journal.pone.0095982

Kumar, S., Abedin, Md. M., Singh, A. K., Das, S. Role of phenolic compounds in plant-defensive mechanisms. In: Lone, R., Shuab, R., Kamili, A. N. (Eds.), *Plant Phenolics in Sustainable Agriculture* (Vol. 1, pp. 517–532). Springer, 2020. https://doi.org/10.1007/978-981-15-4890-1_22

Leckie, B. M., D'Ambrosio, D. A., Chappell, T. M., Halitschke, R., Jong, D. M. D., Kessler, A., Kennedy, G. G., Mutschler, M. A. Differential and synergistic functionality of acylsugars in suppressing oviposition by insect herbivores. *PLoS One*, 2016, 11(4), e0153345. https://doi.org/10.1371/journal.pone.0153345

Leiss, K. A., Choi, Y. H., Verpoorte, R., Klinkhamer, P. G. L. An overview of NMR-based metabolomics to identify secondary plant compounds involved in host plant resistance. *Phytochem. Rev.*, 2011, 10(2), 205–216. https://doi.org/10.1007/s11101-010-9175-z

Li, Q., Eigenbrode, S. D., Stringam, G. R., Thiagarajah, M. R. Feeding and growth of plutellaxylostella and spodoptera eridania on brassica juncea with varying glucosinolate concentrations and myrosinase activities. *J. Chem. Ecol.*, 2000, 26(10), 2401–2419. https://doi.org/10.1023/A:1005535129399

Lieberi, R. Cyanogenesis of Hevea brasiliensis during infection with Microcyclusulei. *J. Phytopathol.*, 1986, 115, 134–146.

Liu, R., Bao, Z.-X., Zhao, P.-J., Li, G.-H. Advances in the study of metabolomics and metabolites in some species interactions. *Molecules*, 2021, 26(11), 11. https://doi.org/10.3390/molecules26113311

Liu, X. H., Liang, S., Wei, Y. Y., Zhu, X. M., Li, L., Liu, P. P. Metabolomics analysis identifies sphingolipids as key signaling moieties in appressorium morphogenesis and function in *Magnaporthe oryzae*. *mBio*, 2019, 10, e01467–19.

Luca, V. D., Pierre, B. S., Luca, V. D., Pierre, B. S. The cell and developmental biology of alkaloid biosynthesis. *Trends Plant Sci.*, 2000, 5(4), 168–173. https://doi.org/10.1016/S1360-1385(00)01575-2

Luna, E., Bruce, T. J. A., Roberts, M. R., Flors, V., Ton, J. Next-generation systemic acquired resistance. *Plant Physiol.*, 2012, 158(2), 844–853. https://doi.org/10.1104/pp.111.187468

Luna, E., Flandin, A., Cassan, C., Prigent, S., Chevanne, C., Kadiri, C. F., Gibon, Y. and Pétriacq, P. Metabolomics to exploit the primed immune system of tomato fruit. *Metabolites*, 2020, 10(3), 96.

Maag, D., Erb, M., Glauser, G. Metabolomics in plant–herbivore interactions: Challenges and applications. *Entomol. Exp. Appl.*, 2015, 157(1), 18–29. https://doi.org/10.1111/eea.12336

Macel, M., Visschers, I. G. S., Peters, J. L., van Dam, N. M., de Graaf, R. M. High concentrations of very long chain leaf wax alkanes of thrips susceptible pepper accessions (Capsicum spp). *J. Chem. Ecol.*, 2020, 46(11), 1082–1089. https://doi.org/10.1007/s10886-020-01226-x

Mandal, R., Kathiria, P., Psychogios, N., Bouatra, S., Krishnamurthy, R., Wishart, D., Kovalchuk, I. Progeny of tobacco mosaic virus-infected Nicotiana tabacum plants exhibit trans-generational changes in metabolic profiles. *Biocatal. Agric. Biotechnol.*, 2012, 1(2), 115–123. https://doi.org/10.1016/j.bcab.2012.01.004

Marcel, S., Sawers, R., Oakeley, E., Angliker, H., Paszkowski, U. Tissue-adapted invasion strategies of the rice blast fungus *Magnaporthe oryzae*. *Plant Cell*, 2010, 22(9), 3177–3187. https://doi.org/10.1105/tpc.110.078048

Martinez-Medina, A., Flors, V., Heil, M., Mauch-Mani, B., Pieterse, C. M. J., Pozo, M. J., Ton, J., van Dam, N. M., Conrath, U. Recognizing plant defense priming. *Trends Plant Sci.*, 2016, 21(10), 818–822. https://doi.org/10.1016/j.tplants.2016.07.009

Medema, M. H., Paalvast, Y., Nguyen, D. D., Melnik, A., Dorrestein, P. C., Takano, E., Breitling, R. Pep2Path: Automated mass spectrometry-guided genome mining of peptidic natural products. *PLoS Comput. Biol.*, 2014, 10(9), e1003822. https://doi.org/10.1371/journal.pcbi.1003822

Mhlongo, M. I., Steenkamp, P. A., Piater, L. A., Madala, N. E., Dubery, I. A. Profiling of altered metabolomic states in nicotiana tabacum cells induced by priming agents. *Front. Plant Sci.*, 2016, 7, 1527. https://doi.org/10.3389/fpls.2016.01527

Mierziak, J., Kostyn, K., Kulma, A. Flavonoids as important molecules of plant interactions with the environment. *Molecules*, 2014, 19(10), 10. https://doi.org/10.3390/molecules191016240

Misra, P., Pandey, A., Tiwari, M., Chandrashekar, K., Sidhu, O. P., Asif, M. H., Chakrabarty, D., Singh, P. K., Trivedi, P. K., Nath, P., Tuli, R. Modulation of transcriptome and metabolome of tobacco by arabidopsis transcription factor, AtMYB12, leads to insect resistance. *Plant Physiol.*, 2010, 152(4), 2258–2268. https://doi.org/10.1104/pp.109.150979

Mitchell, C., Brennan, R. M., Graham, J., Karley, A. J. Plant defense against herbivorous pests: Exploiting resistance and tolerance traits for sustainable crop protection. *Front. Plant Sci.*, 2016, 7, 1132.

Møller, B. L. Functional diversifications of cyanogenic glucosides. *Curr. Opin. Plant Biol.*, 2010, 13(3), 337–346. https://doi.org/10.1016/j.pbi.2010.01.009

Movva, V., Pathipati, U. R. Feeding-induced phenol production in Capsicum annuum L. influences Spodoptera litura F. larval growth and physiology. *Arch. Insect Biochem. Physiol.*, 2017, 95(1), e21387. https://doi.org/10.1002/arch.21387

Moyes, C. L., Raybould, A. F. The role of spatial scale and intraspecific variation in secondary chemistry in host–plant location by *Ceutorhynchus assimilis* (Coleoptera: Curculionidae). *Proc. Roy. Soc. London. Ser. B: Biol. Sci.*, 2001, 268(1476), 1567–1573. https://doi.org/10.1098/rspb.2001.1685

Mueller, L. A., Zhang, P., Rhee, S. Y. AraCyc: A biochemical pathway database for arabidopsis. *Plant Physiol.*, 2003, 132(2), 453–460. https://doi.org/10.1104/pp.102.017236

Naidoo, S., Christie, N., Acosta, J. J., Mphahlele, M. M., Payn, K. G., Myburg, A. A., Külheim, C. Terpenes associated with resistance against the gall wasp, Leptocybeinvasa, in Eucalyptus grandis. *Plant, Cell Environ.*, 2018, 41(8), 1840–1851. https://doi.org/10.1111/pce.13323

Nalley, L., Tsiboe, F., Durand-Morat, A., Shew, A., Thoma, G. Economic and environmental impact of rice blast pathogen (*Magnaporthe oryzae*) alleviation in the United States. *PLoS One*, 2016, 11(12), e0167295. https://doi.org/10.1371/journal.pone.0167295

Nasir, F., Tian, L., Chang, C., Li, X., Gao, Y., Tran, L.-S. P., Tian, C. Current understanding of pattern-triggered immunity and hormone-mediated defense in rice (Oryza sativa) in response to *Magnaporthe oryzae* infection. *Semin. Cell Dev. Biol.*, 2018, 83, 95–105. https://doi.org/10.1016/j.semcdb.2017.10.020

Newman, R. M., Hanscom, Z., Kerfoot, W. C. The watercress glucosinolate-myrosinase system: A feeding deterrent to caddisflies, snails and amphipods. *Oecologia*, 1992, 92(1), 1–7. https://doi.org/10.1007/BF00317255

Nilsen, E. T., Orcutt, D. M. Physiology of plants under stress. Abiotic factors. Physiology of Plants under Stress. *Abiotic Factors*, 1996. https://www.cabdirect.org/cabdirect/abstract/19970701860

Ogah, E. O., Smart, L. E., Woodcock, C. M., Caulfield, J. C., Birkett, M. A., Pickett, J. A., Nwilene, F. E., Bruce, T. J. Electrophysiological and behavioral responses of female African rice gall midge, Orseoliaoryzivora Harris and Gagné, to host plant volatiles. *J. Chem. Ecol.*, 2017, 43(1), 13–16. https://doi.org/10.1007/s10886-016-0788-6

Omena-Garcia, R. P., de Ávila Silva, L., Vital, C. E., Araújo, W. L., Ramos, H. J. O., Nunes-Nesi, A. Identification of metabolite traits from the current metabolomic approaches. *Theor. Exp. Plant Physiol.*, 2019, 31(1), 1–19. https://doi.org/10.1007/s40626-019-00147-9

Pang, Q., Zhang, T., Wang, Y., Kong, W., Guan, Q., Yan, X., Chen, S. Metabolomics of early stage plant cell–microbe interaction using stable isotope labeling. *Front. Plant Sci.*, 2018, 9, 760.

Park, J.-Y., Jin, J., Lee, Y.-W., Kang, S., Lee, Y.-H. Rice blast fungus (*Magnaporthe oryzae*) infects arabidopsis via a mechanism distinct from that required for the infection of rice. *Plant Physiol.*, 2009, 149(1), 474–486. https://doi.org/10.1104/pp.108.129536

Pastor, V., Balmer, A., Gamir, J., Flors, V., Mauch-Mani, B. Preparing to fight back: Generation and storage of priming compounds. *Front. Plant Sci.*, 2014, 5, 295. https://doi.org/10.3389/fpls.2014.00295

Pastor, V., Luna, E., Ton, J., Cerezo, M., García-Agustín, P., Flors, V. Fine tuning of reactive oxygen species homeostasis regulates primed immune responses in Arabidopsis. *Mole. Plant-Microbe Interact.: MPMI*, 2013, 26(11), 1334–1344. https://doi.org/10.1094/MPMI-04-13-0117-R

Pedras, M. S. C., Zheng, Q.-A., Strelkov, S. Metabolic changes in roots of the oilseed canola infected with the biotroph *Plasmodiophora brassicae*: Phytoalexins and phytoanticipins. *J. Agric. Food Chem.*, 2008, 56(21), 9949–9961. https://doi.org/10.1021/jf802192f

Peter Constabel, C., Yoshida, K., Walker, V. Diverse ecological roles of plant tannins: Plant defense and beyond. In: *Recent Advances in Polyphenol Research* (pp. 115–142). John Wiley & Sons, Ltd, 2014. https://doi.org/10.1002/9781118329634.ch5

Petterson, D. S., Harris, D. J., Allen, D. G. Alkaloids. In: D'Mello, J. P. F., Duffus, C. M., Duffus, J. H. (Eds.), *Toxic Substances in Crop Plants* (pp. 148–179). The Royal Society of Chemistry: Cambridge, UK, 1991.

Pinzón, A., Barreto, E., Bernal, A., Achenie, L., Gonzalez Barrios, A. F., Isea, R., Restrepo, S. Computational models in plant-pathogen interactions: The case of Phytophthora infestans. *Theor. Biol. Med. Model.*, 2009, 6(1), 1–11.

Plata-Rueda, A., Campos, J. M., da Silva Rolim, G., Martínez, L. C., Dos Santos, M. H., Fernandes, F. L., Serrão, J. E., Zanuncio, J. C. Terpenoid constituents of cinnamon and clove essential oils cause toxic effects and behavior repellency response on granary weevil, Sitophilus granarius. *Ecotoxicol. Environ. Saf.*, 2018, 156, 263–270. https://doi.org/10.1016/j.ecoenv.2018.03.033

Puterka, G. J., Farone, W., Palmer, T., Barrington, A. Structure-function relationships affecting the insecticidal and miticidal activity of sugar esters. *J. Econ. Entomol.*, 2003, 96(3), 636–644. https://doi.org/10.1093/jee/96.3.636

Qie, X., Sun, A., Hao, H., Lv, B., Wu, W., Hu, Z. A potential lignan botanical insecticide from *Phryma leptostachya* against Aedes aegypti: Laboratory selection, metabolic mechanism, and resistance risk assessment. *J. Pest Sci.*, 2022, 95(1), 397–408. https://doi.org/10.1007/s10340-021-01389-w

Ran, F. A., Hsu, P. D., Wright, J., Agarwala, V., Scott, D. A., Zhang, F. Genome engineering using the CRISPR-Cas9 system. *Nat. Protoc.*, 2013, 8(11), 11. https://doi.org/10.1038/nprot.2013.143

Ran, F. A., Hsu, P. D., Wright, J., Agarwala, V., Scott, D. A., Zhang, F. Genome engineering using the CRISPR-Cas9 system. *Nat. Protoc.*, 2013, 8(11), 2281–2308.

Ratzka, A., Vogel, H., Kliebenstein, D. J., Mitchell-Olds, T., Kroymann, J. Disarming the mustard oil bomb. *Proc. Natl. Acad. Sci.*, 2002, 99(17), 11223–11228. https://doi.org/10.1073/pnas.172112899

Rodenburg, S. Y., Seidl, M. F., Judelson, H. S., Vu, A. L., Govers, F., de Ridder, D. Metabolic model of the Phytophthora infestans-tomato interaction reveals metabolic switches during host colonization. *mBio*, 2019, 10(4), 10–1128.

Roossinck, M. J. Plant virus metagenomics: Biodiversity and ecology. *Ann. Rev. Genet.*, 2012, 46(1), 359–369. https://doi.org/10.1146/annurev-genet-110711-155600

Rosenthal, G. A., Berenbaum, M. R. *Herbivores: Their Interactions with Secondary Plant Metabolites: The Chemical Participants*. Academic Press, 2012.

Ross, A. B., Muller, H., Subbaraj, A., Homewood, I., Mas, F., Hardwick, S., Stringer, L., Vereijssen, J., Visnovsky, S., Najar-Rodriguez, A., Armstrong, K. F. Metabolomics for plant health biosecurity diagnostics and response. *Sustainability*, 2023, 15(5), 5. https://doi.org/10.3390/su15054654

Rousseau, C., Hunault, G., Gaillard, S., Bourbeillon, J., Montiel, G., Simier, P., Campion, C., Jacques, M.-A., Belin, E., Boureau, T. Phenoplant: A web resource for the exploration of large chlorophyll fluorescence image datasets. *Plant Methods*, 2015, 11(1), 24. https://doi.org/10.1186/s13007-015-0068-4

Rubio, B., Fernandez, O., Cosson, P., Berton, T., Caballero, M., Lion, R., Roux, F., Bergelson, J., Gibon, Y. and Schurdi-Levraud, V. Metabolic profile discriminates and predicts Arabidopsis susceptibility to virus under field conditions. *Metabolites*, 2021, 11(4), 230.

Saleem, M., Kim, H. J., Ali, M. S., Lee, Y. S. An update on bioactive plant lignans. *Nat. Prod. Rep.*, 2005, 22(6), 696–716. https://doi.org/10.1039/B514045P

Sánchez-Pérez, R., Jørgensen, K., Olsen, C. E., Dicenta, F., Møller, B. L. Bitterness in almonds. *Plant Physiol.*, 2008, 146(3), 1040–1052. https://doi.org/10.1104/pp.107.112979

Shan, X., Zhang, Y., Peng, W., Wang, Z., Xie, D. Molecular mechanism for jasmonate-induction of anthocyanin accumulation in Arabidopsis. *J. Exp. Bot.*, 2009, 60(13), 3849–3860. https://doi.org/10.1093/jxb/erp223

Sharma, S., Kooner, R., Arora, R. Insect pests and crop losses. In: Arora, R., Sandhu, S. (Eds.), *Breeding Insect Resistant Crops for Sustainable Agriculture* (pp. 45–66). Springer, 2017. https://doi.org/10.1007/978-981-10-6056-4_2

Shroff, R., Vergara, F., Muck, A., Svatoš, A., Gershenzon, J. Nonuniform distribution of glucosinolates in Arabidopsis thaliana leaves has important consequences for plant defense. *Proc. Natl. Acad. Sci.*, 2008, 105(16), 6196–6201.

Singh, S., Arya, S. K., Kaur, G., Saxena, G., Verma, P. C. Role of endosymbionts in nutritional uptake of sap sucking insects. In: Singh, S. P., Upadhyay, S. K., Pandey, A., Kumar, S. (Eds.), *Molecular Approaches in Plant Biology and Environmental Challenges* (pp. 487–499). Springer, 2019. https://doi.org/10.1007/978-981-15-0690-1_22

Snart, C. J. P., Hardy, I. C. W., Barrett, D. A. Entometabolomics: Applications of modern analytical techniques to insect studies. *Entomol. Exp. Appl.*, 2015, 155(1), 1–17. https://doi.org/10.1111/eea.12281

Steppuhn, A., Baldwin, I. T. Resistance management in a native plant: Nicotine prevents herbivores from compensating for plant protease inhibitors. *Ecol. Lett.*, 2007, 10(6), 499–511. https://doi.org/10.1111/j.1461-0248.2007.01045.x

Steppuhn, A., Gase, K., Krock, B., Halitschke, R., Baldwin, I. T. Nicotine's defensive function in nature. *PLoS Biol.*, 2004, 2(8), e217. https://doi.org/10.1371/journal.pbio.0020217

Tezuka, D., Kawamata, A., Kato, H., Saburi, W., Mori, H., Imai, R. The rice ethylene response factor OsERF83 positively regulates disease resistance to *Magnaporthe oryzae*. *Plant Physiol. Biochem.: PPB*, 2019, 135, 263–271. https://doi.org/10.1016/j.plaphy.2018.12.017

Tsugawa, H., Cajka, T., Kind, T., Ma, Y., Higgins, B., Ikeda, K., Kanazawa, M., VanderGheynst, J., Fiehn, O. and Arita, M. MS-DIAL: Data-independent MS/MS deconvolution for comprehensive metabolome analysis. *Nat. Methods*, 2015, 12(6), 523–526.

Usha Rani, P., Pratyusha, S. Defensive role of Gossypium hirsutum L. anti-oxidative enzymes and phenolic acids in response to Spodoptera litura F. feeding. *J. Asia Pac. Entomol.*, 2013, 16(2), 131–136. https://doi.org/10.1016/j.aspen.2013.01.001

van Dam, N. M., Tytgat, T. O. G., Kirkegaard, J. A. Root and shoot glucosinolates: A comparison of their diversity, function and interactions in natural and managed ecosystems. *Phytochem. Rev.*, 2009, 8(1), 171–186. https://doi.org/10.1007/s11101-008-9101-9

Vetter, J. Plant cyanogenic glycosides. *Toxicon*, 2000, 38(1), 11–36. https://doi.org/10.1016/S0041-0101(99)00128-2

Vílchez, J. I., Niehaus, K., Dowling, D. N., González-López, J., Manzanera, M. Protection of pepper plants from drought by microbacterium sp. 3J1 by modulation of the Plant's glutamine and α-ketoglutarate content: A comparative metabolomics approach. *Front. Microbiol.*, 2018, 9, 284. https://doi.org/10.3389/fmicb.2018.00284

Visschers, I. G. S., Peters, J. L., van de Vondervoort, J. A. H., Hoogveld, R. H. M., van Dam, N. M. Thrips resistance screening is coming of age: Leaf position and ontogeny are important determinants of leaf-based resistance in pepper. *Front. Plant Sci.*, 2019, 10, 510. https://doi.org/10.3389/fpls.2019.00510

Visschers, I. G., Peters, J. L., van de Vondervoort, J. A., Hoogveld, R. H., van Dam, N. M. Thrips resistance screening is coming of age: Leaf position and ontogeny are important determinants of leaf-based resistance in pepper. *Front. Plant Sci.*, 2019, 10, 510.

Wahyuni, Y., Stahl-Hermes, V., Ballester, A.-R., de Vos, R. C. H., Voorrips, R. E., Maharijaya, A., Molthoff, J., Zamora, M. V., Sudarmonowati, E., Arisi, A. C. M., Bino, R. J., Bovy, A. G. Genetic mapping of semi-polar metabolites in pepper fruits (Capsicum sp.): Towards unravelling the molecular regulation of flavonoid quantitative trait loci. *Mol. Breed.*, 2014, 33(3), 503–518. https://doi.org/10.1007/s11032-013-9967-0

Wahyuni, Y., Stahl-Hermes, V., Ballester, A. R., de Vos, R. C., Voorrips, R. E., Maharijaya, A., Molthoff, J., Zamora, M. V., Sudarmonowati, E., Arisi, A. C. M. and Bino, R. J. Genetic mapping of semi-polar metabolites in pepper fruits (Capsicum sp.): Towards unravelling the molecular regulation of flavonoid quantitative trait loci. *Mol. Breed.*, 2014, 33, 503–518.

Walker, H. J. Metabolite imaging by mass spectrometry: A new discovery tool. In: *Advances in Botanical Research* (Vol. 98, pp. 271–296). Academic Press, 2021.

Wang, Y., Sheng, L., Zhang, H., Du, X., An, C., Xia, X., Chen, F., Jiang, J., Chen, S. CmMYB19 over-expression improves aphid tolerance in chrysanthemum by promoting lignin synthesis. *Int. J. Mol. Sci.*, 2017, 18(3), 3. https://doi.org/10.3390/ijms18030619

Watson, A., Ghosh, S., Williams, M. J., Cuddy, W. S., Simmonds, J., Rey, M.-D., Asyraf Md Hatta, M., Hinchliffe, A., Steed, A., Reynolds, D., Adamski, N. M., Breakspear, A., Korolev, A., Rayner, T., Dixon, L. E., Riaz, A., Martin, W., Ryan, M., Edwards, D., Batley, J., Hickey, L. T. Speed breeding is a powerful tool to accelerate crop research and breeding. *Nat. Plants*, 2018, 4(1), 23–29. https://doi.org/10.1038/s41477-017-0083-8

Wei, T., Ou, B., Li, J., Zhao, Y., Guo, D., Zhu, Y., Chen, Z., Gu, H., Li, C., Qin, G., Qu, L.-J. Transcriptional profiling of rice early response to *Magnaporthe oryzae* identified OsWRKYs as important regulators in rice blast resistance. *PLoS One*, 2013, 8(3), e59720. https://doi.org/10.1371/journal.pone.0059720

Weintraub, P. G., Beanland, L. Insect vectors of phytoplasmas. *Annu. Rev. Entomol.*, 2006, 51(1), 91–111. https://doi.org/10.1146/annurev.ento.51.110104.151039

Widarto, H. T., Van Der Meijden, E., Lefeber, A. W. M., Erkelens, C., Kim, H. K., Choi, Y. H., Verpoorte, R. Metabolomic differentiation of Brassica rapa following herbivory by different Insect instars using two-dimensional nuclear magnetic resonance spectroscopy. *J. Chem. Ecol.*, 2006, 32(11), 2417–2428. https://doi.org/10.1007/s10886-006-9152-6

Wilson, R. A., Talbot, N. J. Under pressure: Investigating the biology of plant infection by *Magnaporthe oryzae*. Nat. *Rev. Microbiol.*, 2009, 7(3), 3. https://doi.org/10.1038/nrmicro2032

Wittstock, U., Agerbirk, N., Stauber, E. J., Olsen, C. E., Hippler, M., Mitchell-Olds, T., Gershenzon, J., Vogel, H. Successful herbivore attack due to metabolic diversion of a plant chemical defense. *Proc. Natl. Acad. Sci.*, 2004, 101(14), 4859–4864. https://doi.org/10.1073/pnas.0308007101

Wu, K., Zhang, J., Zhang, Q., Zhu, S., Shao, Q., Clark, K. D., Liu, Y., Ling, E. Plant phenolics are detoxified by prophenoloxidase in the insect gut. *Sci. Rep.*, 2015, 5(1), 1. https://doi.org/10.1038/srep16823

Xue Bin, Z., ChangJun, L. Multifaceted regulations of gateway enzyme phenylalanine ammonia-lyase in the biosynthesis of phenylpropanoids. *Mol. Plant*, 2015, 8(1), 17–27.

Yang, L., Stöckigt, J. Trends for diverse production strategies of plant medicinal alkaloids. *Nat. Prod. Rep.*, 2010, 27(10), 1469–1479. https://doi.org/10.1039/C005378C

Yuan, Y., Li, L., Zhao, J., Chen, M. Effect of tannic acid on nutrition and activities of detoxification enzymes and acetylcholinesterase of the fall webworm (Lepidoptera: Arctiidae). *J. Insect. Sci.*, 2020, 20(1), 8. https://doi.org/10.1093/jisesa/ieaa001

Zagrobelny, M., Dreon, A. L., Gomiero, T., Marcazzan, G. L., Glaring, M. A., MøLler, B. L., Paoletti, M. G. Toxic moths: Source of a truly safe delicacy. *J. Ethnobiol.*, 2009, 29(1), 64–76. https://doi.org/10.2993/0278-0771-29.1.64

Ziegler, J., Facchini, P. J. Alkaloid biosynthesis: Metabolism and trafficking. *Annu. Rev. Plant Biol.*, 2008, 59(1), 735–769. https://doi.org/10.1146/annurev.arplant.59.032607.092730

CHAPTER 5

Plant Metabolome: Database, Dataset Processing Methods, and Application

ASHISH KUMAR* and MD. ZISHAN ANSARI

Department of Biotechnology, Guru Ghasidas Vishwavidyalaya (A Central University), Bilaspur, Chhattisgarh, India

*Corresponding authors

ABSTRACT

Metabolomics is the systematic study of unique chemical fingerprints of specific cellular processes. In metabolomics, a metabolite is defined as a molecule less than 1.5 kDa in size. Based on their metabolic roles, plant metabolomics generally classify metabolites as "primary" or "secondary." Plant Metabolome Database (PMDB) is a database of secondary metabolites of plants in the three-dimensional structures available in biological data banks and databases. Tools specific to a metabolomics plant form or technology, that is, liquid chromatography-mass spectrometry (LC-MS), gas chromatography-mass spectrometry (GC-MS), and nuclear magnetic resonance (NMR), are available. PMDB finds applications in and is not limited to plant breeding, crop improvement, drug discovery, functional genomics, plant–environment interaction, bioinformatics, and metabolic engineering. Spectral data is obtained from the plant sample using, not limited to, NMR, LC-MS, or GC-MS and followed by spectral processing to get a feature matrix. Data analysis is performed to identify the candidate biomarkers. Metabolites are identified through spectral databases. Finally, dataset features of the identified metabolites, including peak positions, correlation patterns, and relative intensities, are obtained. Metabolomics dataset processing involves data preprocessing, feature extraction, statistical analysis, identification, pathway analysis, and validation. Software tools such as TidyMS, PMAT, MetaboAnalyst, MetFID, MetaCyc, and Metabolite Standards can be employed for metabolomics dataset processing.

5.1 INTRODUCTION

Metabolomics is one of the latest branches of science that is concerned with the study of the metabolome. The term "metabolome" refers to the complete set of small metabolites, usually secondary, that are produced as an intermediate or end-product of a

Exploring Plant Stress Tolerance: A Metabolomics Approach.
Shuvasish Choudhury, Debojyoti Moulick, and Mirza Hasanuzzaman (Eds.)

DOI: 10.1201/9781779641731-5

metabolic pathway. Metabolomics is one of the newest additions in the list "omics" sciences that also include genomics, proteomics, and transcriptomics. Metabolomics combines the latest analytical tools and techniques with the knowledge of bioinformatics to make the process of disease diagnosis easier and predict prognosis (Adamski, 2020).

Plants, as complex living organisms, consist of interconnected metabolic pathways that determine their genotypic and phenotypic characteristics. The metabolic profiling (Hou et al., 2023) of a complete set of plant metabolites gives insights into the biochemical pathways that support normal growth, stress response mechanisms, and energy production. Metabolomics enables the study of stress-induced metabolic changes in plants and their repercussions on normal growth and developmental processes. Keeping a record of the formation and movement of metabolites within the plant body helps understand and explore nutrient recycling, storage, and allocation.

Data on a plant metabolite is collected from the sample using different various analytical techniques, including liquid chromatography-mass spectroscopy (Rai et al., 2023) (LC-MS), gas chromatography-mass spectroscopy (Pokajewicz, 2023) (GC-MS), and nuclear magnetic resonance (NMR) spectroscopy (Deborde et al., 2017). The raw data is transformed into standardized data formats, units, and metadata. The metabolites are then identified using spectral libraries and computational tools. The annotated metabolites are assigned scientifically accurate names, chemical formulas, and pathway associations. Quality control checks are performed to eliminate errors and inconsistencies in the datasets. An interactive web interface is designed to search, retrieve, and visualize metabolomics data. Cross-references are established to link the plant metabolome database with other "omics" databases, including genomics (Naithani and Van de Peer, 2016) and proteomics databases (Figure 5.1).

Plant metabolome databases can evolve into robust platforms and lead the drive for scientific discoveries. A combination of technological, collaborative, and quality control strategies is essential to overcome the underlying challenges of resource data invariability and resource limitation. Standardized protocols, metadata, annotations, and data types will ensure the integrity of metabolomic data. Data quality control measures are necessary to identify and correct errors and inconsistencies within the datasets. The database needs to be updated regularly in collaboration with researchers, institutions, and domain experts. Data security measures are implemented to protect sensitive data from theft while ensuring data accessibility. By the combination of these approaches, the plant metabolome database (Ma and Qi, 2021) continues to serve as a powerful resource for research- and data-driven discoveries. As metabolomics continues to advance, its importance in unraveling the biochemical mysteries in plants (Rozin, 2023) and the horizon of applications in diverse fields will continue to expand.

5.2 PREPROCESSING OF RAW DATA

5.2.1 DATA ACQUISITION TECHNIQUES

The plant metabolome data is acquired from the sample using bioanalytical techniques, including LC-MS, GC-MS, and NMR spectroscopy. Some of the most commonly used techniques are described below.

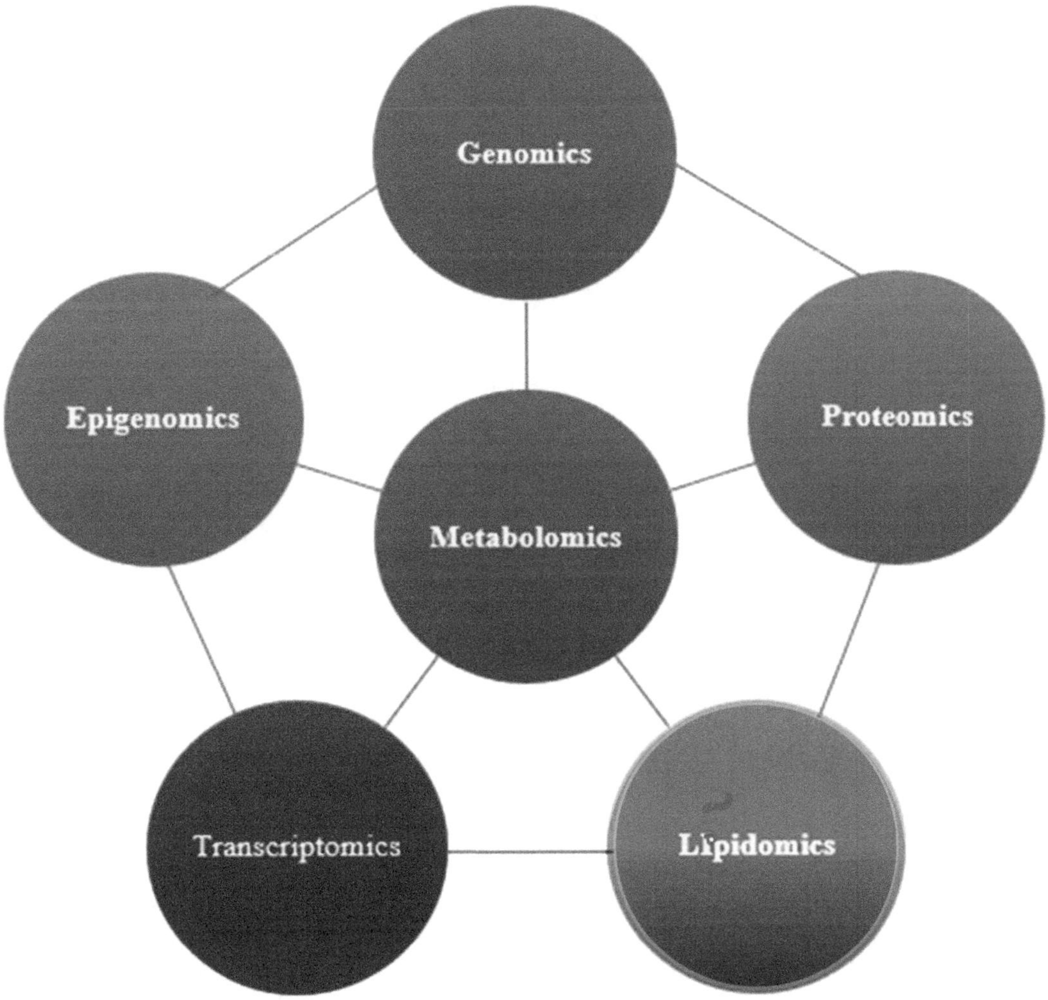

FIGURE 5.1 Integrated "omics" sciences.

5.2.1.1 LIQUID CHROMATOGRAPHY-MASS SPECTROSCOPY

LC separates the individual components of a mixture based on their distinct chemical properties. A column is filled with a stationary phase, and a liquid solvent (mobile phase (Souihi et al., 2023)) is allowed to flow through the column. The different components of the mixture interact differently with the stationary phase and finally flow out from the column at different durations. For MS (Dias et al., 2023), the molecules under investigation are first ionized. The ionized particles are accelerated, detected, and finally separated based on their mass and charge in an electric or magnetic field. The mass-to-charge ratio (m/z) of the ions is measured. The mass spectrum (Zhang et al., 2021), thus, obtained provides detailed information regarding the molecular composition and structural features of the sample. LC in combination with MS (Figure 5.2) aids in the identification (Cabañas-García et al., 2021), separation (Ozaki et al., 2022), and structural analysis of plant metabolites.

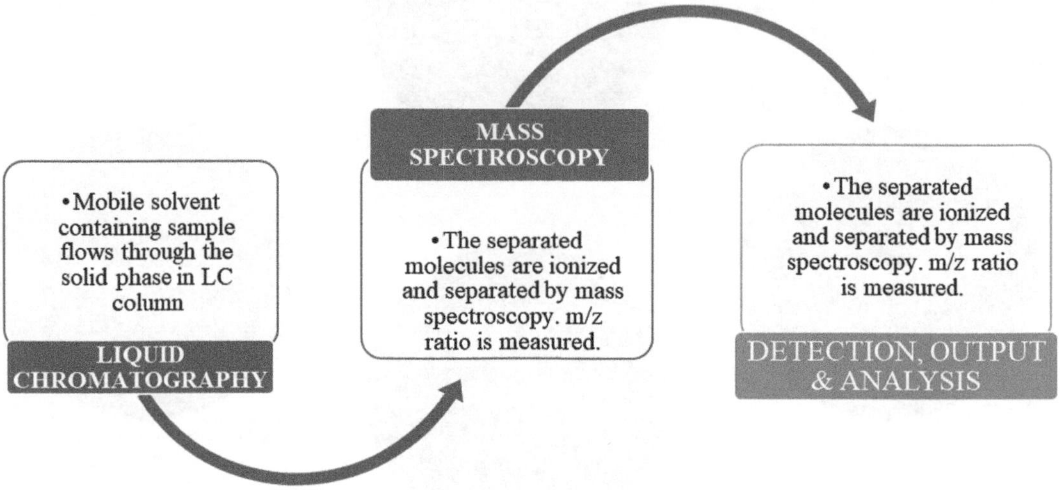

FIGURE 5.2 Schematic diagram of LC-MS showing different stages.

5.2.1.2 GAS CHROMATOGRAPHY-MASS SPECTROSCOPY

In GC, the sample is vaporized and introduced into the GC (Zheng et al., 2022) column. Different components of the sample have different volatilities (Zhang et al., 2023) and affinities for the stationary phase in the GC column. The more volatile components move faster through the column compared to the less volatile components. Consequently, the different components of the mixture are separated as they move out through the chromatography column. GC-MS is widely employed in pharmaceutical industries and research labs to quantify and selectively detect plant metabolites based on volatility.

5.2.1.3 NUCLEAR MAGNETIC RESONANCE SPECTROSCOPY

NMR spectroscopy provides structural details at the atomic or nuclear level of plant metabolites. The desired metabolites are extracted from plant tissue using appropriate solvents such as D_2O. The sample is loaded into NMR tube (Waschina and Seeger, 2022) and sealed properly. The tube is then placed in an NMR spectrometer (De Souza Braga Neto et al., 2023), and spectral data (Hariharan et al., 2023) of the metabolite is acquired, as shown in Figure 5.3.

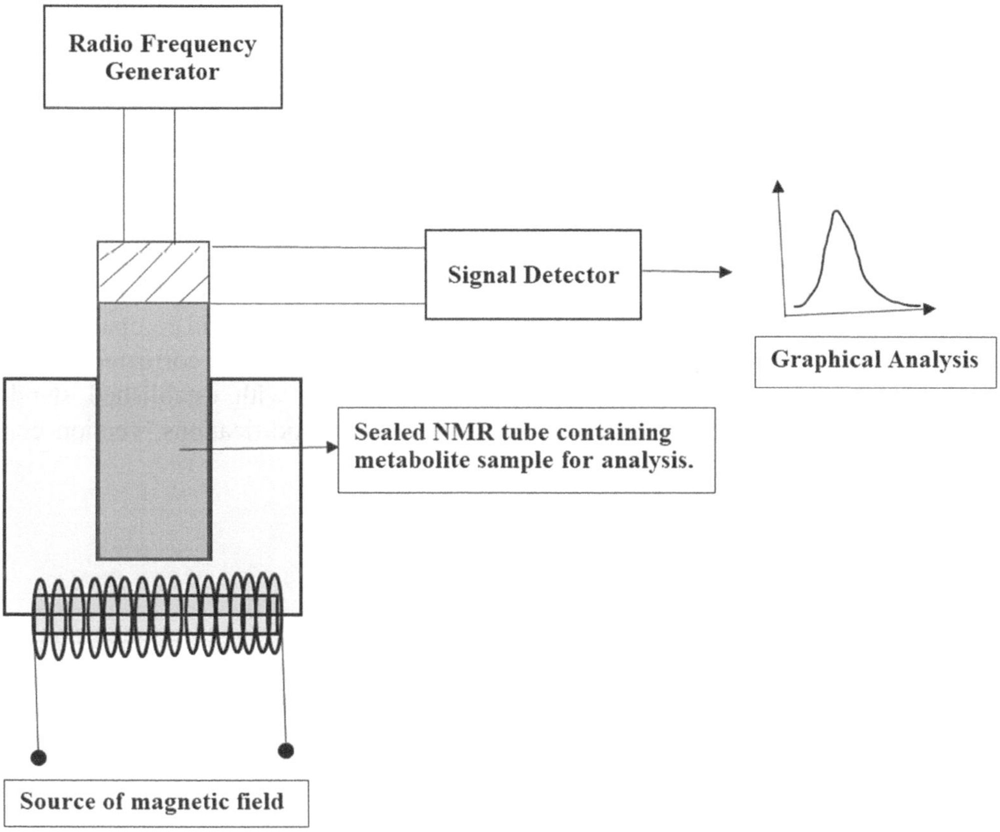

FIGURE 5.3 Schematic representation of NMR spectroscopy.

5.2.2 FORMATS FOR RAW DATA STORAGE (Table 5.1)

TABLE 5.1 Overview of Commonly Used Data Formats for the Storage of Raw Plant Metabolome Datasets

S. No.	Data Format	Description
5.2.2.1	*NetCDF* (Liu et al., 2016)	Scientific data can be created, stored, and accessed using the Network Common Data Form (NetCDF), a group of software libraries. It gives comprehensive details about the data it holds. From a bigger dataset, the desired small data subsets may be extracted. To save data effectively, it is usually combined with Analytical Data Interchange for Mass Spectroscopy (ANDI-MS).
5.2.2.2	CDF	The Common Data Format (CDF) is a software that allows efficient storage of multidimensional datasets. It is appropriate for the storage of mass spectral data.
5.2.2.3	mzXML (Wilhelm et al., 2012)	MS-based proteomics data, including mass spectra, retention times, and other information, can be stored and accessed using the innovative open data format known as mzXML, which is based on eXtensible Markup Language (XML).
5.2.2.4	CSV	MS data is stored in tabular form using the Comma Separated Values (CSV) format. A data point is represented by each row, while other properties, including charge, intensities, m/z values, and retention times, are represented by each column.
5.2.2.5	HDF5 (Chen et al., 2016)	The Hierarchical Data Format version 5 (HDF5) is an open-source format for heterogeneous scientific data storage in a structured manner.

5.2.3 DATA CLEANING

The raw plant metabolome data that has been stored may occasionally be irrelevant or inconsistent (Qi et al., 2020). The trustworthiness of the dataset for the plant metabolome is improved by effective data cleaning. To protect the integrity of the recorded data, data cleaning is an important step in processing the plant metabolome information. To find the discrepancies, the raw data that has been stored is examined. The missing values are either completely excluded or substituted with estimated values. Outliers (Huang et al., 2023) are eliminated, remedied, or looked into. Peaks in graphs are correctly aligned to account for little shifts brought on by equipment or human error. Ensure meaningful comparisons when data is collected in various batches, batch effects (Hu et al., 2023) are corrected. Duplicate entries or redundant data are removed. Cross-referencing with established standards verifies the accuracy of the cleaned data. To track more modifications, version control techniques are built toward the end.

5.2.4 NOISE REDUCTION

Noise refers to any unwanted data or records. Noise suppression in chromatographic data processing aims to enhance an analytical signal to produce a higher signal-to-noise ratio. Some common techniques used in noise reduction are given below.

5.2.4.1 MEDIAN FILTERING

Each data point in the spectra is replaced with a median value within a defined range to reduce the effect of outliers and random noise, thus preserving the relevant data. As depicted in Figure 5.4, the graphical representation of data points is uniformly organized after applying the median filter (Tay, 2021).

5.2.4.2 THRESHOLDING

A minimum threshold (Li and Mao, 2023) value is set, and the data points smaller in magnitude, than it is considered as "noise" and eliminated or replaced with an estimated value automatically.

5.2.4.3 BINNING

A large dataset is grouped into smaller bins (Chai et al., 2023). Bin averages are calculated to reduce the impact of small variations and random noise. Bins are more convenient to analyze, as depicted in Figure 5.5.

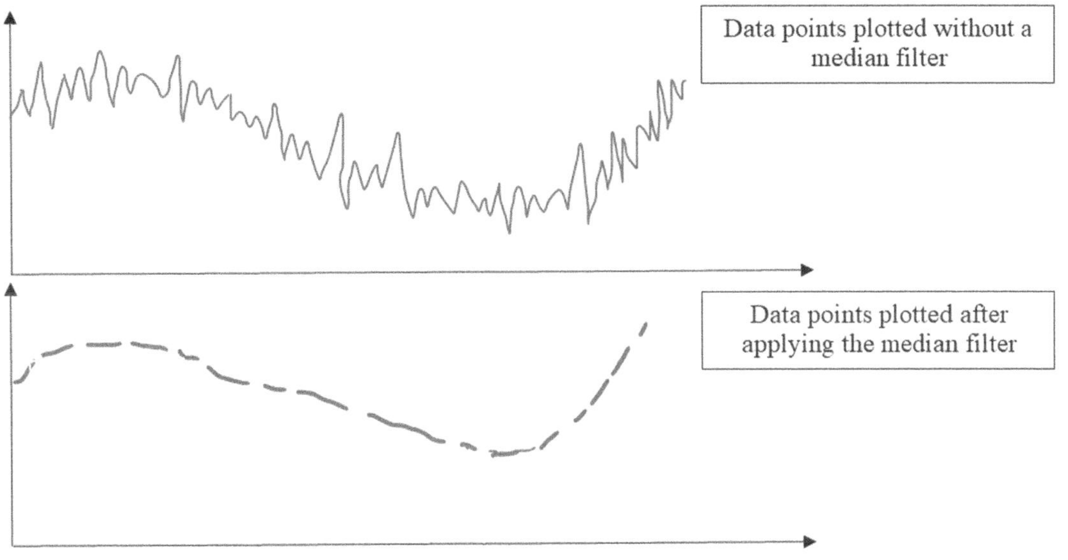

FIGURE 5.4 The first graph shows the data points plotted without a median filter, and the second graph shows the data points plotted after applying the median filter.

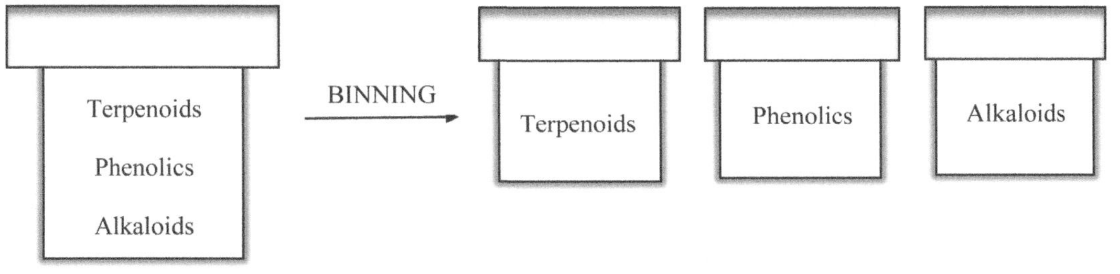

FIGURE 5.5 Binning of three plant metabolite datasets: (1) terpenoids (Williams et al., 2023), (2) phenolics (Medic et al., 2023), and (3) alkaloids (Li et al., 2023).

5.2.5 BASELINE CORRECTION METHODS

Baseline correction (Jiao et al., 2023) is necessary to reduce or remove unwanted background variations and separate peaks from the baseline of the spectral curve. Some common methods used for baseline correction are given below.

5.2.5.1 MINIMUM VALUES

Minimum intensity values are determined within specified boundaries on both sides of the peak and subtracted from the entire spectrum.

5.2.5.2 MEDIAN SMOOTHING

A median filter when applied to the spectrum reduces baseline fluctuations.

5.2.6 HANDLING MISSING VALUES AND OUTLIERS

If the data is insignificant, missing values (Yang et al., 2023) can be excluded. They can also be imputed with approximated values and replaced with predefined placeholders such as "N/A." Outliers can be visually identified by using graphical tools such as box plots and scatter plots (Figure 5.6). Replacement with nearest nonoutlier value reduces the overall impact of extreme values. Extreme values may also be removed beyond a certain threshold.

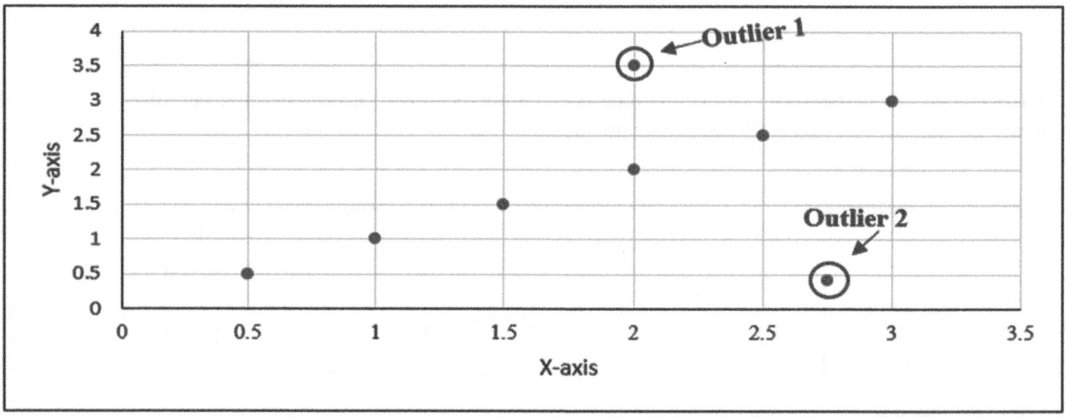

FIGURE 5.6 Position of two outliers in a graphical plot.

5.3 FEATURE EXTRACTION AND PEAK ANALYSIS

By turning raw data into useful metabolite information through feature extraction (Sachar and Kumar, 2021), insights into plant metabolism (Liu et al., 2023), responses (Baharudin and Osman, 2023), and pathways (Paul et al., 2023) are made possible. The key steps are listed as follows:

1. Once the raw data is preprocessed, local maxima/peaks are identified.
2. The highest point of the identified peak is located.
3. Its mass-to-charge ratio or time retention value is determined.
4. Calculate the area under the curve of each peak. The value, thus, obtained represents its intensity.
5. Align retention times across samples in chromatography-based techniques to correct for variations.

6. Check for characteristic isotopic patterns (Sarangi et al., 2022) if any.
7. Compare the peaks against known metabolite databases (Petrov et al., 2020) or spectral libraries (Choudhury et al., 2023).

5.4 METABOLITE ANNOTATION

The process of precisely identifying and characterizing metabolites is known as metabolite annotation. It entails comparing the experimental data to spectral libraries and reference databases that are already in existence. This offers molecular insights into the biochemistry of the plant metabolome as a whole as well as into the understanding of metabolic pathways.

The main steps involved in metabolite annotation are given as follows:

1. **Obtain mass spectra:** Mass spectra of the desired sample are obtained using different data acquisition techniques. The raw data is preprocessed, and the peak is identified.
2. **Compare with existing database:** Mass spectra, retention time, and other relevant information of detected peaks are compared with established metabolite databases and spectral libraries.
3. **Data matching:** The similarity in patterns of the detected experimental spectra and the reference spectra aids in further identification and confirmation.
4. **Retention time matching:** Compare the retention time of detected experimental peak and reference from a database or predicted values.
5. **Isotope analysis:** Compare the observed isotopic pattern with the theoretical isotopic distribution for a candidate metabolite.
6. **Assign chemical formula:** Deduce the possible chemical formula of the metabolite based on its mass and elemental composition.
7. **Tentative identification:** Label the identified metabolite as "tentative" to signify uncertainty.
8. **Adduct and ion annotation:** If any adduct or ionization state is contributing to the mass of the metabolite as observed in MS data, consider it.
9. **Biochemical context:** Take into consideration the biochemical pathways (Belcour et al., 2020) that the metabolite may be a part of.
10. **Literature search:** Go through scientific literature and databases to gather additional information about the identified metabolite.
11. ***In silico* tools:** Utilize computational tools (Martinez, 2013) and techniques for metabolite prediction.
12. **Confirmation:** Compare the mass spectra and retention time of the detected peaks with that of known standard metabolites.
13. **Validation:** Compare the spectra obtained by MS of the candidate metabolite with the reference spectra to validate the metabolite.
14. **Cross-validation:** Compare the identified metabolite across different analytical platforms or techniques.

15. **Quality score:** Assign a quality score or confidence level to each annotation based on matching criteria and validation results.
16. **Documentation:** The identified metabolite is documented in the database, including its name, chemical formula, retention times, mass spectra, and other associated metadata.
17. **Annotation updates:** As new data and better references are available, keep updating the annotation to ensure accuracy and comprehensiveness.

5.4.1 TOOLS

Many analytical tools are available to assist scientists and researchers in accurate identification of plant metabolites.

Some commonly used tools are given as follows:

1. **MZmine:** It is an open-source software for processing MS data and metabolite annotation (Zhou et al., 2021).
2. **MetaboAnalyst:** Web-based collection of tools for the analysis, interpretation, and annotation of metabolomic data (Kitamoto and Kitamoto, 2022).
3. **SIRIUS:** Java-based software for structural elucidation of metabolites. It integrates the isotope patterns and fragmentation patterns in MS spectra to produce detailed information about the metabolite of interest.

5.4.2 CHALLENGES

Metabolite annotation (Cai et al., 2023) is a challenging task due to the complexity and dynamic nature of plant metabolites. Some of them are given as follows:

1. **Spectral variability:** Variations arise in the spectrum due to human, instrumental, or environmental error. Some random errors may also complicate spectral data.
2. **Incomplete databases:** The reference database may not hold complete details of all the plant metabolites. This makes it difficult to establish the identity of the candidate metabolite.
3. **Adducts and derivatives:** Ionization of candidate metabolites can result in adduct formation and derivative compounds that may affect the mass spectral data.

5.4.3 POTENTIAL SOLUTIONS

To overcome challenges in metabolite annotation, a holistic approach integrating improved reference databases, state-of-the-art spectrum libraries, and consistent experimental procedures is required. High-resolution MS, tandem MS (Sabaragamuwa et al., 2022) with accurate mass measurements, specialized algorithms, and computational tools for spectrum matching can all help to increase the accuracy of annotations. The development of multidisciplinary collaboration and continuous development of annotation methods will aid in the advancement of metabolomics research.

5.5 PLANT METABOLOME DATABASES

Plant metabolome databases are digital repositories of plant metabolites. Such databases store information related to structure, biochemistry, and physiology (Liu et al., 2023) of metabolites. Plant metabolome data is stored in a well-organized manner to facilitate quick and efficient retrieval of information as and when required. With technological and scientific advancements, sophisticated tools and techniques are being used to diversify the applications of plant metabolome databases.

5.5.1 DATABASE STRUCTURE AND CONTENT

A typical plant metabolome database consists of the following:

1. **Metabolite entries:** Each entry is for a particular metabolite and contains information on that metabolite's name, chemical formula, molecular weight, structural details, and any relevant identifiers such as the International Chemical Identifier (InChI).
2. **Spectral data:** Spectral data obtained by MS, LC-MS, GC-MS, or NMR spectroscopy provide molecular insights into the structural and chemical characteristic features of metabolites.
3. **Taxonomic information:** Metabolites are categorized by plant species in database sections or fields, facilitating comparative investigations (Javed et al., 2021) of various plants.
4. **Concentration of metabolites:** Measurement of concentrations of metabolites in various plant tissues, growth stages, or under specific conditions provides insights into metabolic responses and involvement in different biochemical pathways.
5. **Associated genes and enzymes:** Metabolites are linked to the genes and enzymes necessary for their manufacture through databases of the plant metabolome. This offers insights into the molecular basis of metabolism (Liu et al., 2023).
6. **Literature references:** Plant metabolome databases cite all scientific publications where metabolite information was reported. This ensures the validity of information contained within the database and facilitates further explorations.
7. **Links to external resources:** Cross-references to other databases, genetic data, pathways, and ontologies enhance the data's integration and accessibility.

5.5.2 FEATURES OF DATABASES

A plant metabolome database offers multiple functionalities (Shawky, 2016) such as data retrieval (Lechtenberg et al., 2022), visualization (Huang et al., 2018), and data analysis (Zhang et al., 2023), as depicted in Figure 5.7.

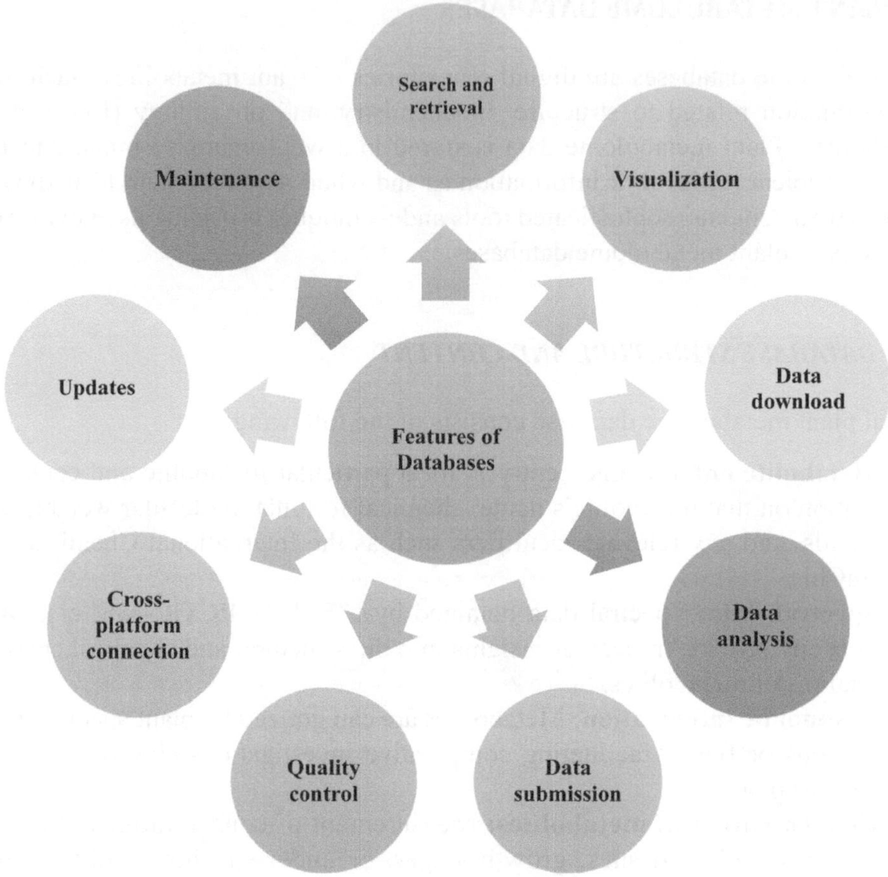

FIGURE 5.7 Features of databases.

5.5.3 OVERVIEW OF COMMON DATABASES (Table 5.2)

TABLE 5.2 An Overview of Some Commonly Used Plant Metabolome Databases for Scientific Studies and Research

Name of Database	Uniform Resource Locator	Description	References
Plant Metabolic Network (PMN)	https://pmn.plantcyc.org/	Information on chemicals, enzymes, and metabolic pathways is present in each organism-specific database housed by PMN. It provides interactive route maps that enable users to understand the relationships between various metabolic processes across a wide range of plant species.	(Kruger and Ratcliffe, 2009)
PlantCyc	https://plantcyc.org/	PlantCyc offers several databases, including curated metabolic pathways, enzymes, and compounds. Users can analyze metabolic networks and plant metabolites and look into relationships across metabolic pathways.	(Rafińska et al., 2022)

TABLE 5.2 *(Continued)*

Name of Database	Uniform Resource Locator	Description	References
MassBank	http://www.massbank.jp/	MassBank is a mass spectral database that has a wide range of mass spectral information for many metabolites from various plant species. Researchers can compare experimental mass spectra with reference spectra to identify and annotate metabolites.	(Vinaixa et al., 2016)
Global Natural Products Social Molecular Networking (GNPS)	https://gnps.ucsd.edu/	GNPS is a collaborative platform, which makes it simpler to identify metabolites by allowing spectrum matching and encouraging networking among researchers.	(Zhang et al., 2023)
Kyoto Encyclopedia of Genes and Genomes (KEGG)	https://www.genome.jp/kegg/	KEGG contains data on chemical properties, metabolic routes, and molecular interaction networks. It aids in the understanding of metabolic processes and the analysis of pathways for many different species, including plants.	(Huang et al., 2023)
Chemical Entities of Biological Interest (ChEBI)	https://www.ebi.ac.uk/chebi/	It offers a curated collection of molecules, including plant metabolites. By giving information on chemical structures, molecular properties, and biological activities, it aids researchers in the identification of key plant metabolites.	(Ibrahim et al., 2023)
PubChem	https://pubchem.ncbi.nlm.nih.gov/	PubChem offers information on a variety of chemicals, including plant metabolites. Researchers can download, search, and visualize data.	(Kim et al., 2022)
MetaCyc	https://metacyc.org/	For several species, including plants, MetaCyc offers a choice of well-selected enzymes and metabolic pathways. It facilitates the investigation of metabolic networks, the analysis of pathways, and the understanding of the operation of biochemical processes.	(Maranas and Burgard, 2001)
MASST	https://masst.unibas.ch/	MASST is a mass spectral database for specific plant metabolites. By encompassing a range of plant species and metabolite types, it aids researchers in the identification and annotation of plant metabolites.	(Venegas-Molina et al., 2021)
Plants Metabolomics Repository (PMR)	http://www.plantmetabolomics.org/	The PMR contains a variety of raw and processed metabolomics data related to plant studies. It promotes data sharing, collaboration, and reproducibility in the metabolomics industry.	(Ebbels et al., 2023)

5.5.4 APPLICATIONS OF PLANT METABOLOME DATABASES

As illustrated in Figure 5.8, the plant metabolome database can be used in a variety of applications (Fan et al., 2023; Millan et al., 2022; Letunic et al., 2008; Krayem et al., 2021; Easton et al., 2010; Shelke et al., 2023; Bagal et al., 2023; Zhang et al., 2020; McClements and McClements, 2023).

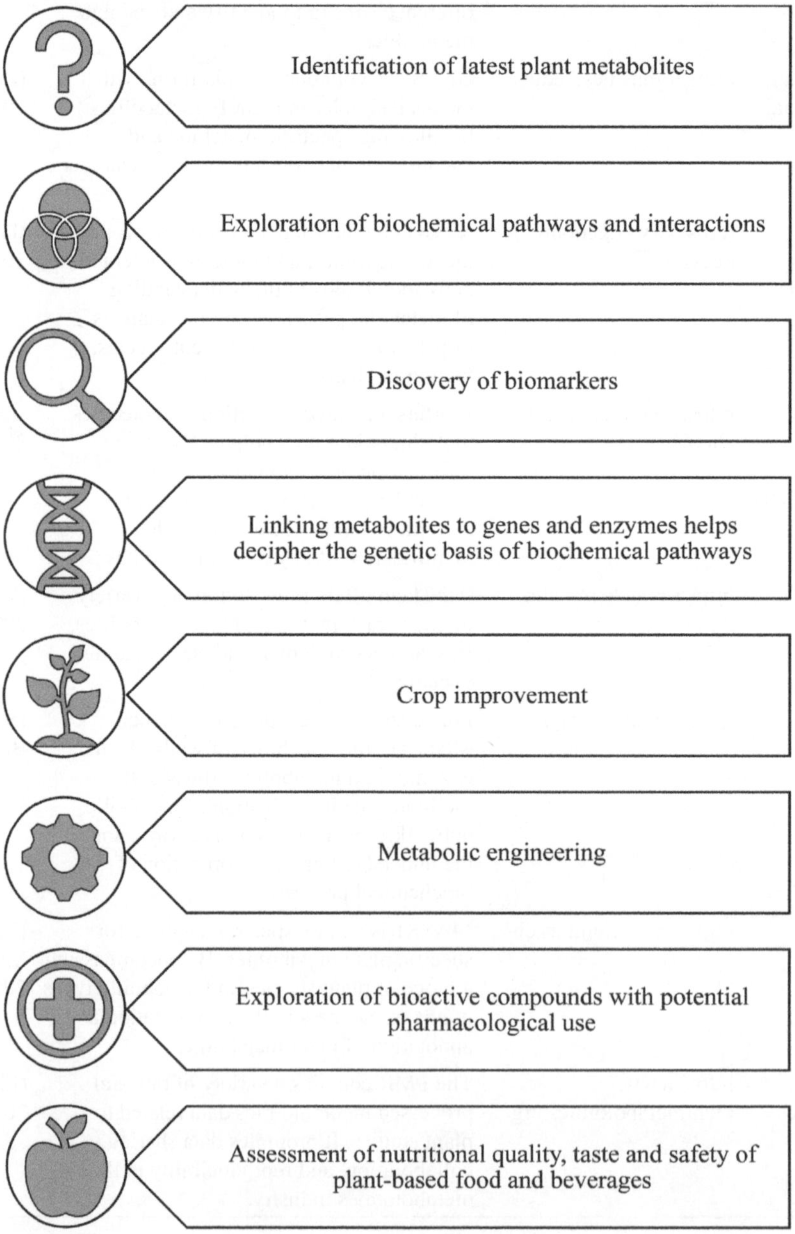

FIGURE 5.8 Applications of plant metabolome databases.

5.6 DATA SUBMISSION

Plant metabolome (Jo et al., 2022) comprises a wide range of metabolites (Wang et al., 2019). Once the metabolites are extracted (Liang et al., 2020), preprocessed (Zhou et al., 2019), and annotated (Chaleckis et al., 2019), the compiled data (Gupta et al., 2024) needs to be submitted to the respective plant metabolome databases.

The submission of plant metabolome data usually involves the following steps:

1. Registration:

A new account, providing basic details, needs to be registered in the desired plant metabolome database (Kosmacz et al., 2020) where the plant metabolome data is to be submitted.

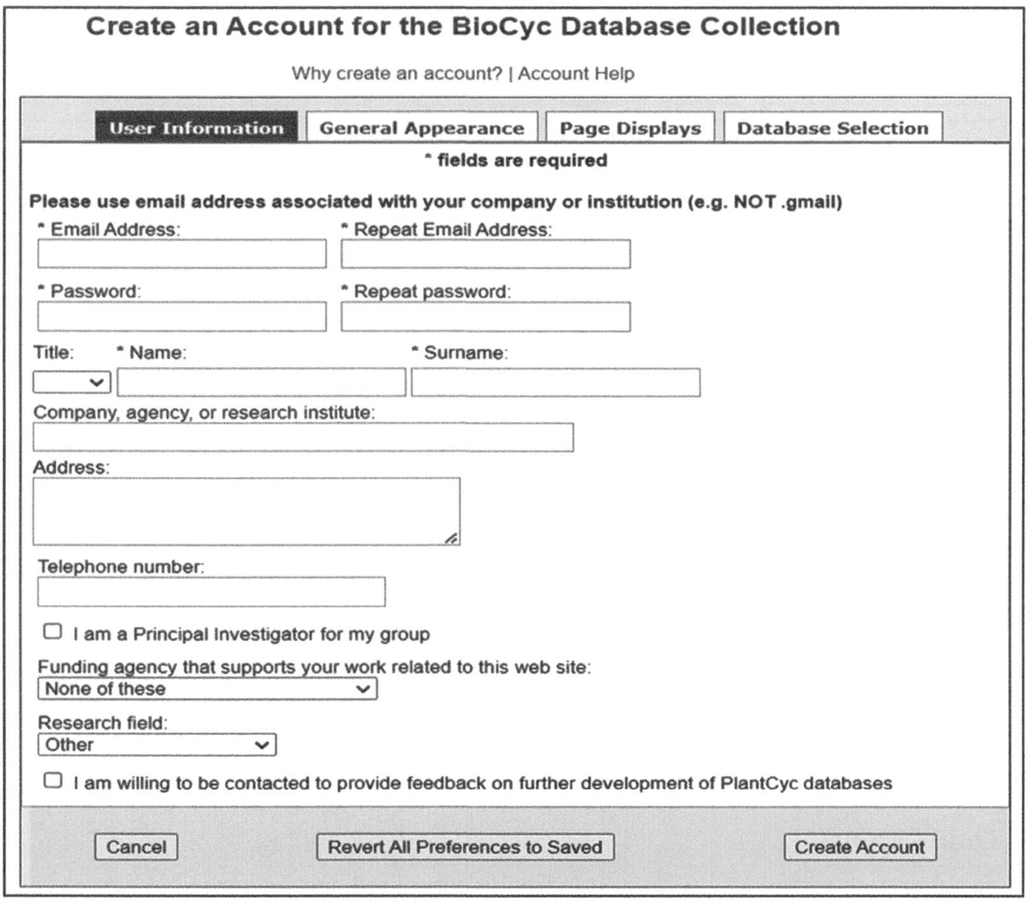

Source: https://pmn.plantcyc.org/preferences.html?status=new

2. Data preparation:

The plant metabolome data to be submitted is organized in a structured and data-relevant format. Experimental details may also be included for authenticity (Oms-Oliu et al., 2013).

3. Metadata annotation:

Details regarding the experimental setup, tools, and techniques, etc., may be provided to assist future researchers in understanding the context of the data.

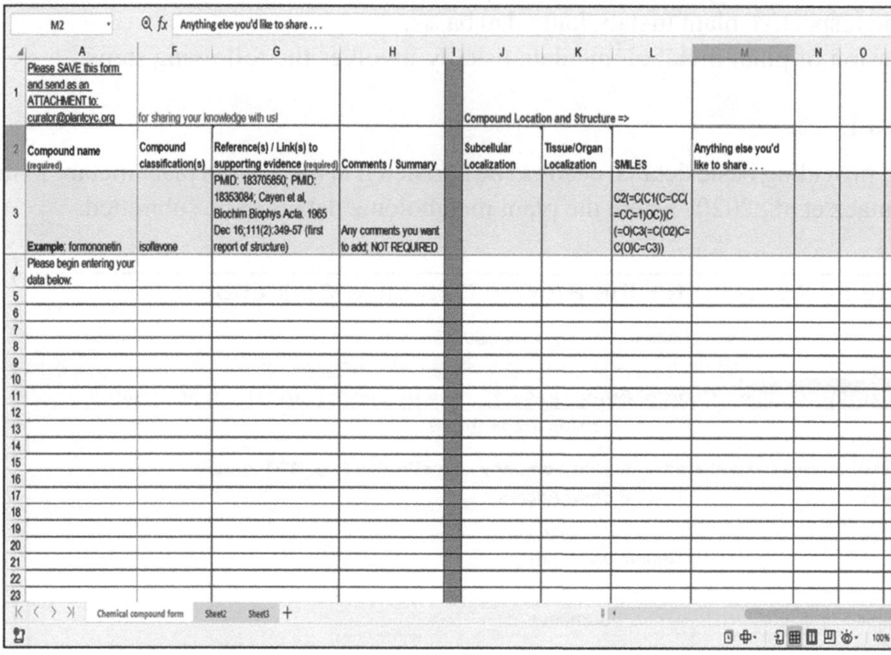

M2 Anything else you'd like to share . . .

	A	F	G	H	J	K	L	M
1	Please SAVE this form and send as an ATTACHMENT to: curator@plantcyc.org	for sharing your knowledge with us!			Compound Location and Structure =>			
2	Compound name (required)	Compound classification(s)	Reference(s) / Link(s) to supporting evidence (required)	Comments / Summary	Subcellular Localization	Tissue/Organ Localization	SMILES	Anything else you'd like to share . . .
3	Example: formononetin	isoflavone	PMID: 183705850; PMID: 18353084; Cayen et al, Biochim Biophys Acta. 1965 Dec 16;111(2):349-57 (first report of structure)	Any comments you want to add; NOT REQUIRED			C2(=C(C1(C=CC(=CC=1)OC))C(=O)C3(=C(O2)C=C(O)C=C3))	
4	Please begin entering your data below.							

Chemical compound form Sheet2 Sheet3

Source: https://carnegiedpbpmn.s3.amazonaws.com/s3fspublic/download/feedback/compound_form.xls

4. Data uploading:

After logging into the account, instructions are followed, and data is submitted in the assigned section (Zhu et al., 2017).

5. Review:

The submitted data and metadata (Leipzig et al., 2021) are assessed for quality and integrity before the final submission. Necessary corrections may be made as per the feedback (Suárez-Cabal et al., 2023) received after the review.

6. Completion:

After a thorough review of the data, it is finally submitted and integrated (Lv et al., 2022) in the database. It may take some time before it is visible publicly in the database.

Note: The terms and conditions vary among different databases. It is advisable to read them properly before finally submitting the plant metabolome (Hirai and Shiraishi, 2018) data.

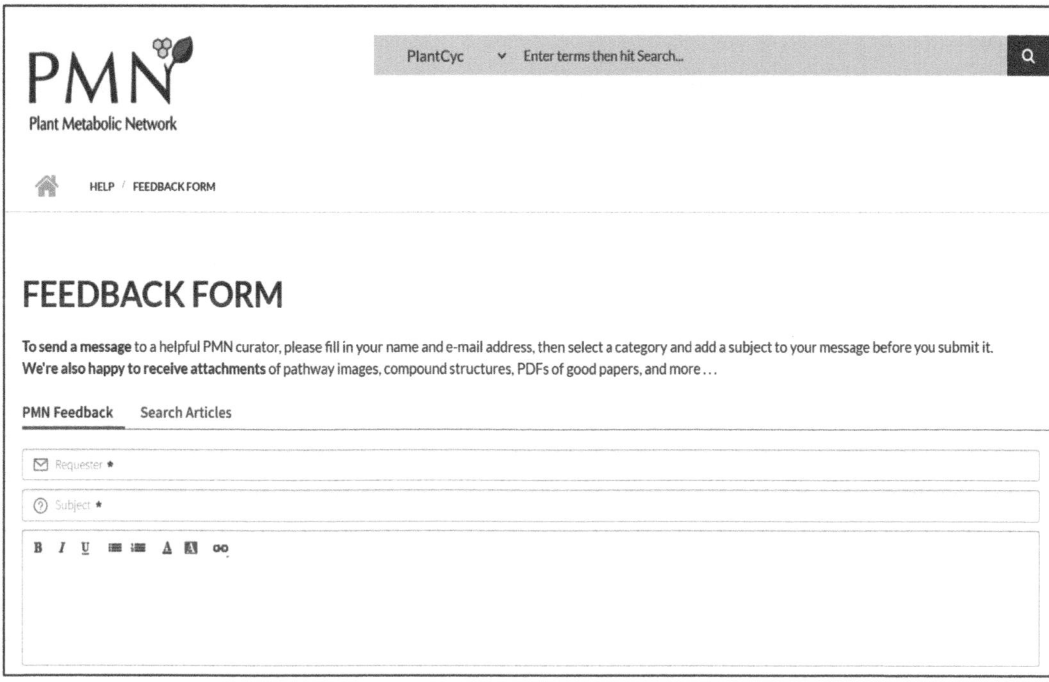

Source: https://plantcyc.org/feedback/data-submission

5.7 DATA ACCESS

Retrieval (Lechtenberg et al., 2021) of data from a plant metabolome database is a crucial step in research (Pang and Hu, 2023) and education (Hajjar et al., 2023). The following steps need to be followed for accessing data from a database:

1. Database search:

Visit the website of a plant metabolome database and search for specific metabolites, pathways, etc.

2. Browsing:

Explore the database by browsing through categories such as plant species (B R and Rani, 2023), metabolite classes (Adhikary and Dasgupta, 2023), metabolism, analysis (Li et al., 2023), or SmartTables.

3. Download:

One can download the chemical structure (Kliebenstein, 2013) or details of a metabolite (Bhinderwala et al., 2022) directly from the database.

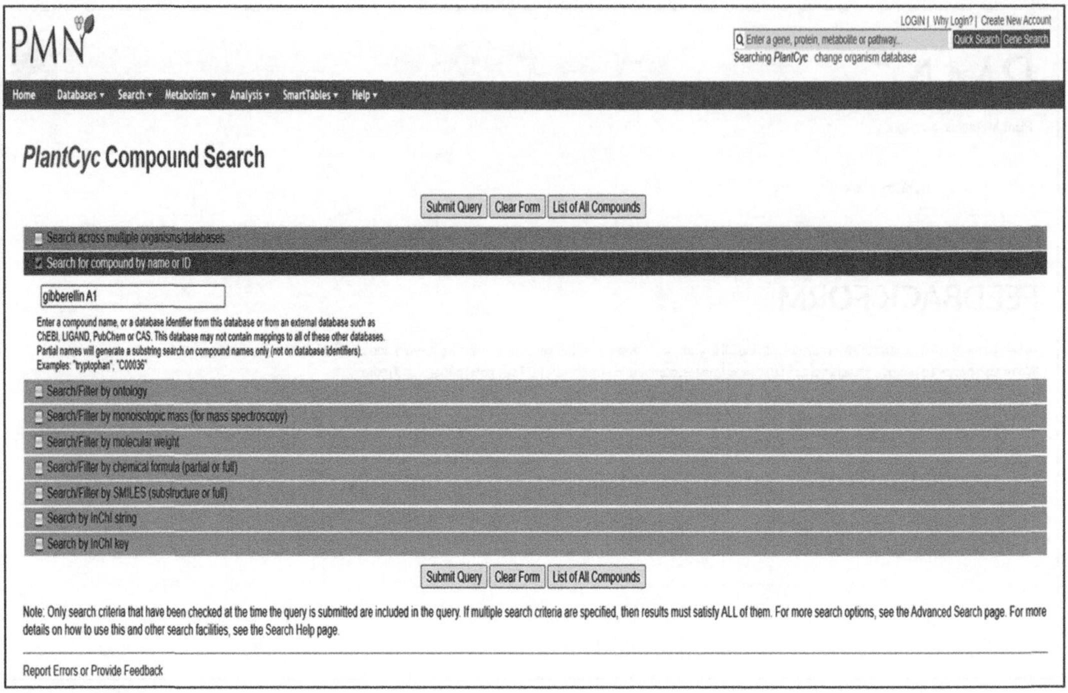

Source: https://pmn.plantcyc.org/cpd-search.shtml

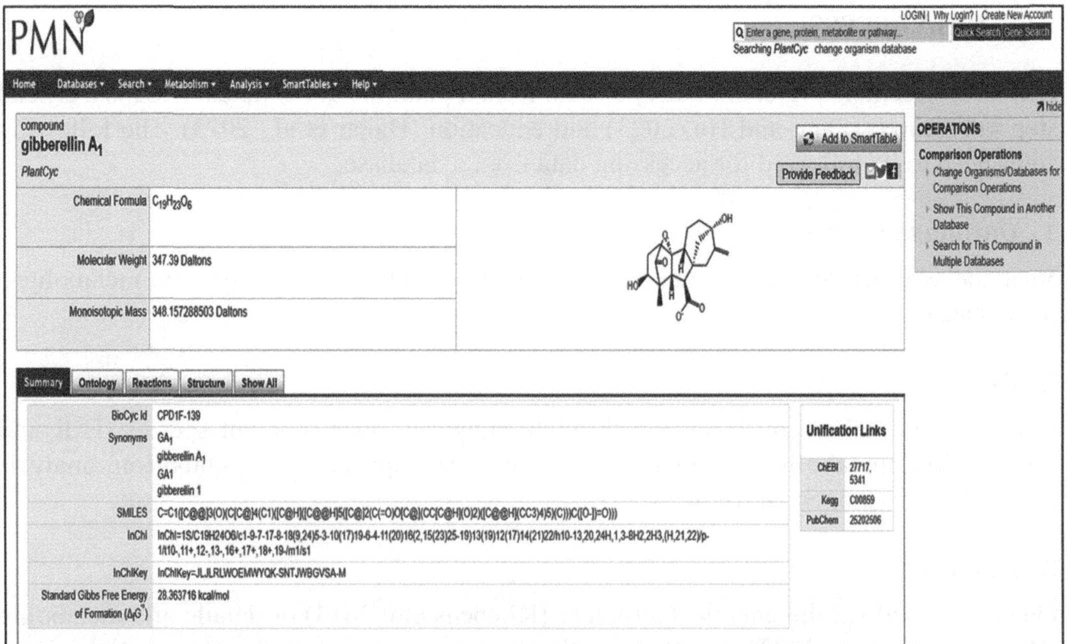

Source: https://pmn.plantcyc.org/compound?orgid=PLANT&id=CPD1F-139#

4. Citation

The source of data being must be properly cited as per the database guidelines. This ensures due acknowledgment of the efforts of the researchers and scientists who originally contributed to the data.

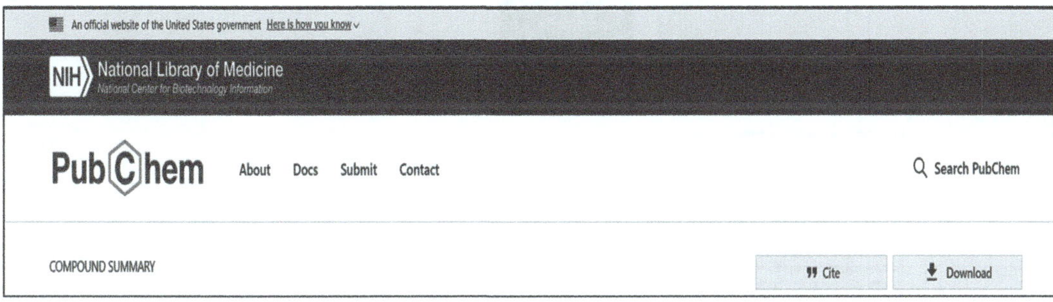

Source: https://pubchem.ncbi.nlm.nih.gov/compound/25202506

5. Tools and visualization:

Tools for visualization and analyses include metabolic pathway mapping (González-Rodríguez et al., 2023), interactive plots (Sarailidis et al., 2023), statistical analyses (Murari et al., 2023), etc.

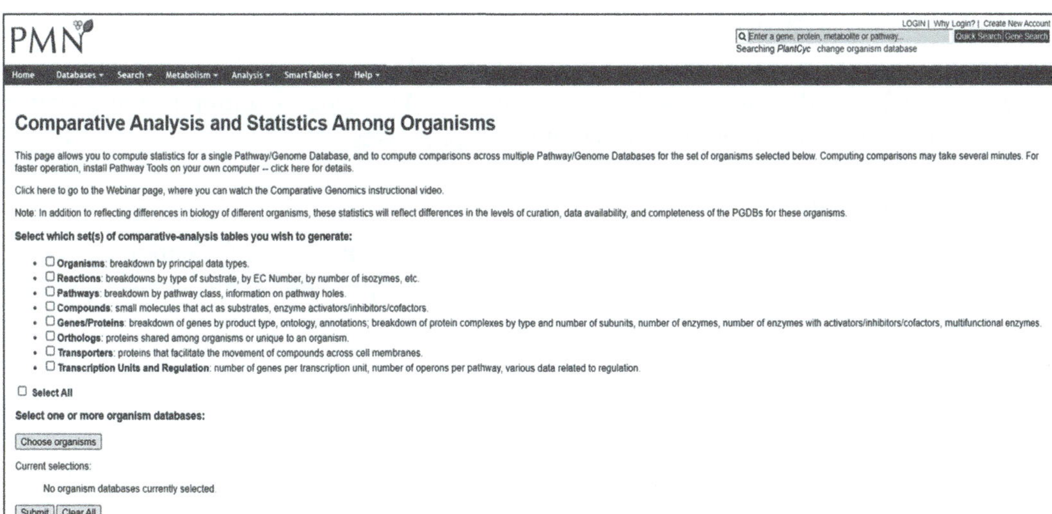

Source: https://pmn.plantcyc.org/comp-genomics

6. User support:

It includes frequently asked questions (FAQs), website user guides, tutorials, documentation, or contact information for help.

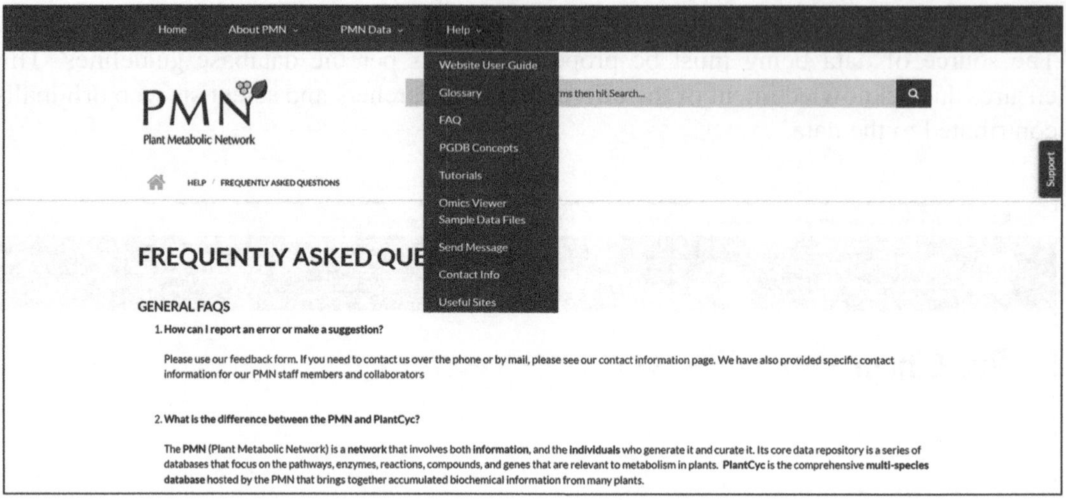

Source: https://plantcyc.org/help/faq?_ga=2.80574878.829198056.1692643980-709771005.1692445110

5.8 DATA INTEGRATION

Data integration (Silva et al., 2023) aids scientists in understanding and interconnecting the complex biochemical pathways (Xiong et al., 2019) and interactions within plant systems.

1. Multiomics integration:

Integration of plant metabolomics with genomics (Mandl et al., 2020), transcriptomics (Yue et al., 2023), and proteomics (Bykova et al., 2023) provides a comprehensive view of how genetic information translates into metabolic processes..

2. Sample annotation:

Integration of plant metabolome database with databases containing information on various compounds helps identify metabolites in the dataset and provides context regarding their biological functions (Zeng et al., 2021).

3. Pathway analysis:

Integrating plant metabolome data with metabolic pathway databases allows researchers to map metabolites onto known pathways. This helps in understanding the interrelation of metabolic reactions.

5.9 DATA VISUALIZATION

Data visualization in plant metabolome databases helps researchers and scientists identify patterns, trends, and insights that might not be interpreted from the raw data alone. Some common types of data visualization are given as follows:

1. **Heatmaps:** Display patterns and correlations in a grid of colored cells (Pati et al., 2023)
2. **Bar charts:** Compare quantities of different metabolites across samples or conditions (Elzer et al., 2011).
3. **Line charts:** Show changes in metabolite concentrations over time or experimental conditions.
4. **Scatter plots:** Visualize relationships between two metabolites or variables (Mohseni and Mokhtarzade, 2020)
5. **Network analysis:** Represents interactions between metabolites or metabolic pathways.

5.10 QUALITY CONTROL

Quality control is crucial for maintaining the integrity and reliability of plant metabolome databases. Ensuring high-quality data is crucial for accurate analysis and meaningful insights. Develop clear standard operating procedures (SOPs) for sample collection, processing, and data acquisition to ensure consistency. Automated checks are implemented for data integrity such as identifying missing values, outliers (Olteanu et al., 2023), and inconsistencies. Regular audits must be conducted to assess the quality and consistency of data across the databases. Domain experts may be involved in reviewing the quality of data. Users need to be encouraged to provide feedback. Quality control is a continuous process. Quality control procedures are revised based on user feedback, technological advancements, and the latest findings.

5.11 EMERGING TRENDS

With advancements in science and technology, plant metabolomic databases are offering a multitude of features. Some emerging trends in plant metabolome databases include:

1. **Metabolic pathway analysis:** Tools powered by artificial intelligence can predict metabolic connections and related pathways. The relationship between metabolites in a biological setting can be understood with the aid of network analysis and pathway prediction algorithms.
2. **Real-time data analysis:** Researchers can swiftly analyze and interpret metabolomics data through real-time data processing. This is helpful for dynamic studies and time-sensitive experiments (Neto et al., 2004).
3. **Personalized medicine:** According to a person's metabolic features, personalized metabolic profiles for individuals can direct treatment programs and dietary recommendations.

5.12 CHALLENGES

Due to genetic variations, growth circumstances, environmental changes, and technical limitations, plant metabolomics data can be highly variable. It is difficult to manage and normalize such a wide range of data. It is challenging to compare and integrate data effectively since different laboratories and studies lack consistent experimental techniques and data formats. To make data organization and retrieval easier, the development of standard and complete ontologies for plant metabolites, pathways, and biological processes is a continuous task.

5.13 CONCLUSION

The processing methods and establishment of plant metabolome databases play pivotal roles in advancing our understanding of plant metabolism and its intricate regulatory mechanisms. The intricate nature of metabolomics data demands sophisticated data processing techniques, including normalization, feature extraction, and statistical analysis, to uncover meaningful patterns and insights. Integration of machine learning and artificial intelligence further enhances data interpretation, enabling the extraction of intricate relationships and predictive models. Simultaneously, the development and maintenance of plant metabolome databases address the challenges of data organization, accessibility, and sharing. These databases provide researchers with centralized repositories enriched with curated data, metadata, and annotation, fostering cross-study comparisons and accelerating discoveries. While challenges such as data variability, annotation accuracy, and standardization persist, collaborative efforts between researchers, data scientists, and database administrators continue to refine processing methods and enhance the utility of plant metabolome databases. Ultimately, these resources serve as invaluable tools that drive advancements in plant science (Giacomello, 2021), agriculture (da Silveira et al., 2021), and biotechnology (Yuan et al., 2022), with far-reaching implications for sustainable crop improvement (Mohapatra et al., 2023) and ecosystem understanding.

KEYWORDS

- **metabolome**
- **Plant Metabolome Database**
- **metabolomics**
- **spectral databases**
- **feature matrix**

REFERENCES

Adamski, J. Introduction to metabolomics. *Metab. Biomed. Res.*, 2020, 1–15. https://doi.org/10.1016/B978-0-12-812784-1.00001-3.

Adhikary, S., Dasgupta, N. Role of secondary metabolites in plant homeostasis during biotic stress. *Biocatal. Agric. Biotechnol.*, 2023, 50, 102712. https://doi.org/10.1016/J.BCAB.2023.102712.

B R, P., Rani, N. S. DIMPSAR: Dataset for Indian medicinal plant species analysis and recognition. *Data Brief*, 2023, 49, 109388. https://doi.org/10.1016/J.DIB.2023.109388.

Bagal, D., Chowdhary, A. A., Mehrotra, S., Mishra, S., Rathore, S., Srivastava, V. Metabolic engineering in hairy roots: An outlook on production of plant secondary metabolites. *Plant Physiol. Biochem.*, 2023, 201, 107847. https://doi.org/10.1016/J.PLAPHY.2023.107847.

Baharudin, N. F., Osman, N. I. Plant development, stress responses, and secondary metabolism under ethylene regulation. *Plant Stress*, 2023, 7, 100146. https://doi.org/10.1016/J.STRESS.2023.100146.

Belcour, A., Girard, J., Aite, M., Delage, L., Trottier, C., Marteau, C., Leroux, C., Dittami, S. M., Sauleau, P., Corre, E., Nicolas, J., Boyen, C., Leblanc, C., Collén, J., Siegel, A., Markov, G. V. Inferring biochemical reactions and metabolite structures to understand metabolic pathway drift. *iScience*, 2020, 23(2), 100849. https://doi.org/10.1016/J.ISCI.2020.100849.

Bhinderwala, F., E. Roth, H., Noel, H., Feng, D., Powers, R. Chemical shift variations in common metabolites. *J. Magn. Reson.*, 2022, 345, 107335. https://doi.org/10.1016/J.JMR.2022.107335.

Bykova, N. V., Rampitsch, C., Igamberdiev, A. U. Plant proteomics: From the molecular basis of biological processes to a systems biology perspective. *J. Plant Physiol.*, 2023, 280, 153880. https://doi.org/10.1016/J.JPLPH.2022.153880.

Cabañas-García, E., Areche, C., Gómez-Aguirre, Y. A., Borquez, J., Muñoz, R., Cruz-Sosa, F., Balch, E. P. M. Biomass production and secondary metabolite identification in callus cultures of *Coryphantha macromeris* (Engelm.) Britton & Rose (Cactaceae), a traditional medicinal plant. *S. Afr. J. Bot.*, 2021, 137, 1–9. https://doi.org/10.1016/J.SAJB.2020.10.002.

Cai, Y., Zhou, Z., Zhu, Z. J. Advanced analytical and informatic strategies for metabolite annotation in untargeted metabolomics. *TrAC Trends Anal. Chem.*, 2023, 158, 116903. https://doi.org/10.1016/J.TRAC.2022.116903.

Chai, X., Liu, C., Fan, X., Huang, T., Zhang, X., Jiang, B., Liu, M. Combination of peak-picking and binning for NMR-based untargeted metabonomics study. *J. Magn. Reson.*, 2023, 351, 107429. https://doi.org/10.1016/J.JMR.2023.107429.

Chaleckis, R., Meister, I., Zhang, P., Wheelock, C. E. Challenges, progress and promises of metabolite annotation for LC–MS-based metabolomics. *Curr. Opin. Biotechnol.*, 2019, 55, 44–50. https://doi.org/10.1016/J.COPBIO.2018.07.010.

Chen, Y., Wang, F., Li, S., Xiao, B. J., Yang, F. The implementation of a data acquisition and service system based on HDF5. *Fusion Eng. Des.*, 2016, 112, 975–978. https://doi.org/10.1016/J.FUSENGDES.2016.02.047.

Choudhury, B. U., Narzari, R., Zafar, M., Singh, N., Mishra, V. K., Prabhakar, M. Spectral library of crops and discrimination of major vegetables grown in the eastern Himalayan ecosystem: A proximal hyperspectral remote sensing approach. *Ecol. Inform.*, 2023, 77, 102263. https://doi.org/10.1016/J.ECOINF.2023.102263.

da Silveira, F., Lermen, F. H., Amaral, F. G. An overview of agriculture 4.0 development: Systematic review of descriptions, technologies, barriers, advantages, and disadvantages. *Comput. Electron. Agric.*, 2021, 189, 106405. https://doi.org/10.1016/J.COMPAG.2021.106405.

De Souza Braga Neto, A., Rigaud, B., Mériguet, G., Rollet, A.-L., Sirieix-Plénet, J. Efficient method for in situ agitation of liquids directly inside NMR spectrometer. *MethodsX*, 2023, 11, 102254. https://doi.org/10.1016/J.MEX.2023.102254.

Deborde, C., Moing, A., Roch, L., Jacob, D., Rolin, D., Giraudeau, P. Plant metabolism as studied by NMR spectroscopy. *Prog. Nucl. Magn. Reson. Spectrosc.*, 2017, 102–103, 61–97. https://doi.org/10.1016/J.PNMRS.2017.05.001.

Dias, D. M., Coombes, S. R., Benstead, D., Whittaker, D. T. E., Ray, A., Xu, J. Advances in the specificity of mass spectrometry and nuclear magnetic resonance spectroscopy based structural characterisation methods for synthetic oligonucleotides. *J. Pharm. Sci.*, 2023, 112(9), 2524–2531. https://doi.org/10.1016/J.XPHS.2023.04.013.

Easton, J. M., Harris, L. M., Viant, M. R., Peet, A. C., Arvanitis, T. N. Linked metabolites: A tool for the construction of directed metabolic graphs. *Comput. Biol. Med.*, 2010, 40(3), 340–349. https://doi.org/10.1016/J.COMPBIOMED.2010.01.005.

Ebbels, T. M. D., van der Hooft, J. J. J., Chatelaine, H., Broeckling, C., Zamboni, N., Hassoun, S., Mathé, E. A. Recent advances in mass spectrometry-based computational metabolomics. *Curr. Opin. Chem. Biol.*, 2023, 74, 102288. https://doi.org/10.1016/J.CBPA.2023.102288.

Elzer, S., Carberry, S., Zukerman, I. The automated understanding of simple bar charts. *Artif. Intell.*, 2011, 175(2), 526–555. https://doi.org/10.1016/J.ARTINT.2010.10.003.

Fan, Y., Sun, L., Song, S., Sun, Y., Fan, X., Li, Y., Li, R., Sun, H. Integrated metabolome and transcriptome analysis of anthocyanin accumulation during the color formation of bicolor flowers in eustoma grandiflorum. *Sci. Hortic.*, 2023, 314, 111952. https://doi.org/10.1016/J.SCIENTA.2023.111952.

Giacomello, S. A new era for plant science: Spatial single-cell transcriptomics. *Curr. Opin. Plant Biol.*, 2021, 60, 102041. https://doi.org/10.1016/J.PBI.2021.102041.

González-Rodríguez, T., Pérez-Limón, S., Peniche-Pavía, H., Rellán-Álvarez, R., Sawers, R. J. H., Winkler, R. Genetic mapping of maize metabolites using high-throughput mass profiling. *Plant Sci.*, 2023, 326, 111530. https://doi.org/10.1016/J.PLANTSCI.2022.111530.

Gupta, S., Singh, A. P., Singh, G., Ding, X., Sharma, A. Plant based metabolomics: A future prospective and versatile tool for metabolite databases of curcuma longa. *J. Future Foods*, 2024, 4(1), 37–47. https://doi.org/10.1016/J.JFUTFO.2023.05.003.

Hajjar, G., Barros Santos, M. C., Bertrand-Michel, J., Canlet, C., Castelli, F., Creusot, N., Dechaumet, S., Diémé, B., Giacomoni, F., Giraudeau, P., Guitton, Y., Thévenot, E., Tremblay-Franco, M., Junot, C., Jourdan, F., Fenaille, F., Comte, B., Pétriacq, P., Pujos-Guillot, E. Scaling-up metabolomics: Current state and perspectives. *TrAC Trends Anal. Chem.*, 2023, 167, 117225. https://doi.org/10.1016/J.TRAC.2023.117225.

Hariharan, J., Ampatzidis, Y., Abdulridha, J., Batuman, O. An AI-based spectral data analysis process for recognizing unique plant biomarkers and disease features. *Comput. Electron. Agric.*, 2023, 204, 107574. https://doi.org/10.1016/J.COMPAG.2022.107574.

Hirai, M. Y., Shiraishi, F. Using metabolome data for mathematical modeling of plant metabolic systems. *Curr. Opin. Biotechnol.*, 2018, 54, 138–144. https://doi.org/10.1016/J.COPBIO.2018.08.005.

Hou, Z., Zhan, L., Cao, K., Luan, M., Wang, X., Zhang, B., Ma, L., Yin, H., Liu, Z., Liu, Y., Huang, G. Metabolite profiling and identification in living cells by coupling stable isotope tracing and induced electrospray mass spectrometry. *Anal. Chim. Acta*, 2023, 1241, 340795. https://doi.org/10.1016/J.ACA.2023.340795.

Hu, F., Chen, A. A., Horng, H., Bashyam, V., Davatzikos, C., Alexander-Bloch, A., Li, M., Shou, H., Satterthwaite, T. D., Yu, M., Shinohara, R. T. Image harmonization: A review of statistical and deep learning methods for removing batch effects and evaluation metrics for effective harmonization. *Neuroimage*, 2023, 274, 120125. https://doi.org/10.1016/J.NEUROIMAGE.2023.120125.

Huang, F., Fu, M., Li, J. R., Chen, L., Feng, K. Y., Huang, T., Cai, Y. D. Analysis and prediction of protein stability based on interaction network, gene ontology, and KEGG pathway enrichment scores. *Biochimica et Biophysica Acta (BBA)—Proteins and Proteomics*, 2023, 1871(3), 140889. https://doi.org/10.1016/J.BBAPAP.2023.140889.

Huang, J., Cheng, D., Zhang, S. A Novel outlier detecting algorithm based on the outlier turning points. *Expert Syst. Appl.*, 2023, 231, 120799. https://doi.org/10.1016/J.ESWA.2023.120799.

Huang, W., Luo, J., Bednarz, T., Duh, H. Making graph visualization a user-centered process. *J. Vis. Lang. Comput.*, 2018, 48, 1–8. https://doi.org/10.1016/J.JVLC.2018.07.001.

Ibrahim, T. N. B. T., Feisal, N. A. S., Kamaludin, N. H., Cheah, W. Y., How, V., Bhatnagar, A., Ma, Z., Show, P. L. Biological active metabolites from microalgae for healthcare and pharmaceutical industries: A comprehensive review. *Bioresour. Technol.*, 2023, 372, 128661. https://doi.org/10.1016/J.BIORTECH.2023.128661.

Javed, R., Ahmad, M. A., Gul, A., Ahsan, T., Cheema, M. Comparison of chemically and biologically synthesized nanoparticles for the production of secondary metabolites, and growth and development of plants. *Compr. Anal. Chem.*, 2021, 94, 303–329. https://doi.org/10.1016/BS.COAC.2021.02.002.

Jiao, Q., Guo, X., Liu, M., Kong, L., Hui, M., Dong, L., Zhao, Y. Deep learning baseline correction method via multi-scale analysis and regression. *Chemom. Intell. Lab. Syst.*, 2023, 235, 104779. https://doi.org/10.1016/J.CHEMOLAB.2023.104779.

Jo, J., Lee, J., Ahn, Y., Hwang, Y. S., Park, J., Lee, J., Choi, J. Metabolome and transcriptome analyses of plants grown in naturally attenuated soil after hydrogen fluoride exposure. *J. Hazard. Mater.*, 2022, 437, 129323. https://doi.org/10.1016/J.JHAZMAT.2022.129323.

Kim, S., Cheng, T., He, S., Thiessen, P. A., Li, Q., Gindulyte, A., Bolton, E. E. PubChem protein, gene, pathway, and taxonomy data collections: Bridging biology and chemistry through target-centric views of PubChem data. *J. Mol. Biol.*, 2022, 434(11), 167514. https://doi.org/10.1016/J.JMB.2022.167514.

Kitamoto, A., Kitamoto, T. Data on changes in lipid profiles during the differentiation and maturation of human subcutaneous white adipocytes analyzed using chromatographic and bioinformatic tools. *Data Brief*, 2022, 42, 108245. https://doi.org/10.1016/J.DIB.2022.108245.

Kliebenstein, D. J. Making new molecules — Evolution of structures for novel metabolites in plants. *Curr. Opin. Plant Biol.*, 2013, 16(1), 112–117. https://doi.org/10.1016/J.PBI.2012.12.004.

Kosmacz, M., Sokołowska, E. M., Bouzaa, S., Skirycz, A. Towards a functional understanding of the plant metabolome. *Curr. Opin. Plant Biol.*, 2020, 55, 47–51. https://doi.org/10.1016/J.PBI.2020.02.005.

Krayem, M., El Khatib, S., Hassan, Y., Deluchat, V., Labrousse, P. In search for potential biomarkers of copper stress in aquatic plants. *Aquat. Toxicol.*, 2021, 239, 105952. https://doi.org/10.1016/J.AQUATOX.2021.105952.

Kruger, N. J., Ratcliffe, R. G. Insights into plant metabolic networks from steady-state metabolic flux analysis. *Biochimie*, 2009, 91(6), 697–702. https://doi.org/10.1016/J.BIOCHI.2009.01.004.

Lechtenberg, F., Farreres, J., Galvan-Cara, A. L., Somoza-Tornos, A., Espuña, A., Graells, M. Information retrieval from scientific abstract and citation databases: A query-by-documents approach based on Monte-Carlo sampling. *Expert Syst. Appl.*, 2022, 199, 116967. https://doi.org/10.1016/J.ESWA.2022.116967.

Lechtenberg, F., Farreres, J., Somoza-Tornos, A., PachecoLópez, A., Espuña, A., Graells, M. Towards automated information retrieval of process data and knowledge from academic databases. *Comput. Aided Chem. Engineer.*, 2021, 50, 983–989. https://doi.org/10.1016/B978-0-323-88506-5.50152-2.

Leipzig, J., Nüst, D., Hoyt, C. T., Ram, K., Greenberg, J. The role of metadata in reproducible computational research. *Patterns*, 2021, 2(9), 100322. https://doi.org/10.1016/J.PATTER.2021.100322.

Letunic, I., Yamada, T., Kanehisa, M., Bork, P. IPath: Interactive exploration of biochemical pathways and networks. *Trends Biochem. Sci.*, 2008, 33(3), 101–103. https://doi.org/10.1016/J.TIBS.2008.01.001.

Li, C., Mao, Z. A label noise filtering method for regression based on adaptive threshold and noise score. *Expert Syst. Appl.*, 2023, 228, 120422. https://doi.org/10.1016/J.ESWA.2023.120422.

Li, J., Li, J.-X., Jiang, H., Li, M., Chen, L., Wang, Y.-Y., Wang, L., Zhang, N., Guo, H.-Z., Ma, K.-L. Phytochemistry and biological activities of corynanthe alkaloids. *Phytochemistry*, 2023, 213, 113786. https://doi.org/10.1016/J.PHYTOCHEM.2023.113786.

Li, Y. F., Chen, Y. Q., Chen, J. H., Shen, C. W. Flavonoid metabolites in tea plant (*Camellia sinensis*) stress response: Insights from bibliometric analysis. *Plant Physiol. Biochem.*, 2023, 202, 107934. https://doi.org/10.1016/J.PLAPHY.2023.107934.

Liang, X., Nielsen, N. J., Christensen, J. H. Selective pressurized liquid extraction of plant secondary metabolites: Convallaria majalis L. as a case. *Anal. Chim. Acta X*, 2020, 4, 100040. https://doi.org/10.1016/J.ACAX.2020.100040.

Liu, H., Van Oosterom, P., Hu, C., Wang, W. Managing large multidimensional array hydrologic datasets: A case study comparing NetCDF and SciDB. *Procedia Eng.*, 2016, 154, 207–214. https://doi.org/10.1016/J.PROENG.2016.07.449.

Liu, S., Rao, J., Zhu, J., Li, G., Li, F., Zhang, H., Tao, L., Zhou, Q., Tao, Y., Zhang, Y., Huang, K., Wei, C. Integrated physiological, metabolite and proteomic analysis reveal the glyphosate stress response mechanism in tea plant (*Camellia sinensis*). *J. Hazard. Mater.*, 2023, 454, 131419. https://doi.org/10.1016/J.JHAZMAT.2023.131419.

Liu, S., Zhang, Q., Kollie, L., Dong, J., Liang, Z. Molecular networks of secondary metabolism accumulation in plants: Current understanding and future challenges. *Ind. Crops Prod.*, 2023, 201, 116901. https://doi.org/10.1016/J.INDCROP.2023.116901.

Lv, J., Deng, M., Li, Z., Zhu, H., Wang, Z., Yue, Y., Yang, Z., Xu, J., Jiang, S., Zhao, W., Li, J., Zhao, K. Integrative analysis of the transcriptome and metabolome reveals the response mechanism to tomato spotted wilt virus. *Hortic. Plant J.*, 2022, 9(5), 958–970. https://doi.org/10.1016/J.HPJ.2022.12.008.

Ma, A., Qi, X. Mining plant metabolomes: Methods, applications, and perspectives. *Plant Commun.*, 2021, 2(5), 100238. https://doi.org/10.1016/J.XPLC.2021.100238.

Mandl, K. D., Glauser, T., Krantz, I. D., Avillach, P., Bartels, A., Beggs, A. H., Biswas, S., Bourgeois, F. T., Corsmo, J., Dauber, A., Devkota, B., Fleisher, G. R., Heath, A. P., Helbig, I., Hirschhorn, J. N., Kilbourn, J., Kong, S. W., Kornetsky, S., Majzoub, J. A., Marsolo, K., Martin, L. J., Nix, J., Schwarzhoff, A., Stedman, J., Strauss, A., Sund, K. L., Taylor, D. M., White, P. S., Marsh, E., Grimberg, A., Hawkes, C., Barkman, D., Borglund, E. M., Chakrabarty, R., Chandel, A., Degala, A. K., DeSain, T., Dexheimer, P., Divekar, P., Ellis, A., Furgason, M., Geehan, C., Guidetti, A. J., Gutierrez, A., Hallinan, B., Harper, B., Jalali, N., Khanna, J.,

Kirby, C., Korodi, G., Kouril, M., Kratchman, A., Kumar, R., Labilloy, G., Lee, I. H., Morgan, B., Morgan, J., Muglia, L. J., Nikitin, A., Pistone, M., Poduri, A., Rupert, A., Safier, K., Sliz, P., Stevenson, G., Geme, J. S., Thaker, V., Temporal, S., Velayutham, P., Wijesooriya, J., Wolf, B., Wooten, A., Yen, A., Zhang, Y. The genomics research and innovation network: Creating an interoperable, federated, genomics learning system. *Genet. Med.*, 2020, 22(2), 371–380. https://doi.org/10.1038/S41436-019-0646-3.

Maranas, C. D., Burgard, A. P. Review of EcoCyc and MetaCyc databases. *Metab. Eng.*, 2001, 3(2), 98–99. https://doi.org/10.1006/MBEN.2001.0180.

Martinez, M. From plant genomes to protein families: Computational tools. *Comput. Struct. Biotechnol. J.*, 2013, 8(10), e201307001. https://doi.org/10.5936/CSBJ.201307001.

McClements, I. F., McClements, D. J. Designing healthier plant-based foods: Fortification, digestion, and bioavailability. *Food Res. Int.*, 2023, 169, 112853. https://doi.org/10.1016/J.FOODRES.2023.112853.

Medic, A., Smrke, T., Hudina, M., Veberic, R., Zamljen, T. HPLC-mass spectrometry analysis of phenolics comparing traditional bilberry and blueberry liqueurs. *Food Res. Int.*, 2023, 173, 113373. https://doi.org/10.1016/J.FOODRES.2023.113373.

Millan, A. F. S., Gamir, J., Farran, I., Larraya, L., Veramendi, J. Identification of new antifungal metabolites produced by the yeast *Metschnikowia pulcherrima* involved in the biocontrol of postharvest plant pathogenic fungi. *Postharvest Biol. Technol.*, 2022, 192, 111995. https://doi.org/10.1016/J.POSTHARVBIO.2022.111995.

Mohapatra, B., Chamoli, S., Salvi, P., Saxena, S. C. Fostering Nanoscience's strategies: A new frontier in sustainable crop improvement for abiotic stress tolerance. *Plant Nano. Biol.*, 2023, 3, 100026. https://doi.org/10.1016/J.PLANA.2023.100026.

Mohseni, F., Mokhtarzade, M. A New soil moisture index driven from an adapted long-term temperature-vegetation scatter plot using MODIS data. *J. Hydrol. (Amst)*, 2020, 581, 124420. https://doi.org/10.1016/J.JHYDROL.2019.124420.

Murari, A., Spolladore, L., Rossi, R., Gelfusa, M. Combining dimensional and statistical analysis for efficient data driven modelling of complex systems. *Inf. Sci. (N Y)*, 2023, 644, 119243. https://doi.org/10.1016/J.INS.2023.119243.

Naithani, S., Van de Peer, Y. Genomic resources and databases for plant research community. *Curr. Plant Biol.*, 2016, 7–8, 1. https://doi.org/10.1016/J.CPB.2016.12.007.

Neto, P. F. R., Perkusich, M. L. B., Perkusich, A. Modelling and analysis of real-time databases for sensor networks using coloured petri nets. *IFAC Proc. Vol.*, 2004, 37(4), 653–658. https://doi.org/10.1016/S1474-6670(17)36189-X.

Olteanu, M., Rossi, F., Yger, F. Meta-survey on outlier and anomaly detection. *Neurocomputing*, 2023, 555, 126634. https://doi.org/10.1016/J.NEUCOM.2023.126634.

Oms-Oliu, G., Odriozola-Serrano, I., Martín-Belloso, O. Metabolomics for assessing safety and quality of plant-derived food. *Food Res. Int.*, 2013, 54(1), 1172–1183. https://doi.org/10.1016/J.FOODRES.2013.04.005.

Ozaki, M., Shimotsuma, M., Hirose, T. Separation of nicotinamide metabolites using a PBr column packed with pentabromobenzyl group modified silica gel. *Anal. Biochem.*, 2022, 655, 114837. https://doi.org/10.1016/J.AB.2022.114837.

Pang, H., Hu, Z. Metabolomics in drug research and development: The recent advances in technologies and applications. *Acta Pharm. Sin B*, 2023, 13(8), 3238–3251. https://doi.org/10.1016/J.APSB.2023.05.021.

Pati, S. K., Banerjee, A., Manna, S. Gene selection of microarray data using heatmap analysis and graph neural network. *Appl. Soft Comput.*, 2023, 135, 110034. https://doi.org/10.1016/J.ASOC.2023.110034.

Paul, M., Pandey, N. K., Banerjee, A., Shroti, G. K., Tomer, P., Gazara, R. K., Thatoi, H., Bhaskar, T., Hazra, S., Ghosh, D. An insight into omics analysis and metabolic pathway engineering of lignin-degrading enzymes for enhanced lignin valorization. *Bioresour. Technol.*, 2023, 379, 129045. https://doi.org/10.1016/J.BIORTECH.2023.129045.

Petrov, A. P., Sherman, L. M., Camden, J. P., Dovichi, N. J. Database of free solution mobilities for 276 metabolites. *Talanta*, 2020, 209, 120545. https://doi.org/10.1016/J.TALANTA.2019.120545.

Pokajewicz, K. Enhancing terpene and other plant volatiles analysis – A free spreadsheet tool "Retentify" for GC–MS data processing. *Microchem. J.*, 2023, 193, 108977. https://doi.org/10.1016/J.MICROC.2023.108977.

Qi, Z., Wang, H., He, T., Li, J., Gao, H. FRIEND: Feature selection on inconsistent data. *Neurocomputing*, 2020, 391, 52–64. https://doi.org/10.1016/J.NEUCOM.2020.01.094.

Rafińska, K., Wrona, O., Krakowska-Sieprawska, A., Walczak-Skierska, J., Kiełbasa, A., Rafiński, Z., Pomastowski, P., Kolankowski, M., Buszewski, B. Enzyme-assisted extraction of plant material—New functional aspects of the process on an example of Medicago sativa L. *Ind. Crops Prod.*, 2022, 187, 115424. https://doi.org/10.1016/J.INDCROP.2022.115424.

Rai, A., Kumari, K., Han, S. S. Polyphenolic profiling of *Victoria amazonica* using MRM LC-MS/MS: A comparative analysis of various plant parts. *Sci. Hortic.*, 2023, 320, 112206. https://doi.org/10.1016/J.SCIENTA.2023.112206.

Rozin, B. Towards solving the mystery of spiral phyllotaxis. *Prog. Biophys. Mol. Biol.*, 2023, 182, 8–14. https://doi.org/10.1016/J.PBIOMOLBIO.2023.04.004.

Sabaragamuwa, R., Perera, C. O., Fedrizzi, B. Ultrasound assisted extraction and quantification of targeted bioactive compounds of *Centella asiatica* (Gotu Kola) by UHPLC-MS/MS MRM tandem mass spectroscopy. *Food Chem.*, 2022, 371, 131187. https://doi.org/10.1016/J.FOODCHEM.2021.131187.

Sachar, S., Kumar, A. Survey of feature extraction and classification techniques to identify plant through leaves. *Expert Syst. Appl.*, 2021, 167, 114181. https://doi.org/10.1016/J.ESWA.2020.114181.

Sarailidis, G., Wagener, T., Pianosi, F. Integrating scientific knowledge into machine learning using interactive decision trees. *Comput. Geosci.*, 2023, 170, 105248. https://doi.org/10.1016/J.CAGEO.2022.105248.

Sarangi, V., Roy, S., Sanyal, P. Effect of burning on the distribution pattern and isotopic composition of plant biomolecules: Implications for paleoecological studies. *Geochim. Cosmochim. Acta*, 2022, 318, 305–327. https://doi.org/10.1016/J.GCA.2021.12.003.

Shawky, E. Multivariate analyses of NP-TLC chromatographic retention data for grouping of structurally-related plant secondary metabolites. *J. Chromatogr. B*, 2016, 1029–1030, 10–15. https://doi.org/10.1016/J.JCHROMB.2016.06.037.

Shelke, D. B., Islam, N. F., Chambhare, M. R., Sonawane, H. B., Patowary, R., Prasad, R., Sarma, H. Enhancing secondary metabolites and alleviating environmental stress in crops with mycogenic nanoparticles: A comprehensive review. *Biocatal. Agric. Biotechnol.*, 2023, 52, 102805. https://doi.org/10.1016/J.BCAB. 2023.102805.

Silva, B., Moreira, J., Costa, R. L. de C. Logical big data integration and near real-time data analytics. *Data Knowl. Eng.*, 2023, 146, 102185. https://doi.org/10.1016/J.DATAK.2023.102185.

Souihi, A., Mohai, M. P., Martin, J., Kruve, A. Mobile phase and column chemistry selection for high sensitivity non-targeted LC/ESI/HRMS screening of water. *Anal. Chim. Acta*, 2023, 1274, 341573. https://doi.org/10.1016/J.ACA.2023.341573.

Suárez-Cabal, M. J., Suárez-Otero, P., de la Riva, C., Tuya, J. MDICA: Maintenance of data integrity in column-oriented database applications. *Comput. Stand. Interfaces*, 2023, 83, 103642. https://doi.org/10.1016/J.CSI.2022.103642.

Tay, D. B. Sensor network data denoising via recursive graph median filters. *Signal Process.*, 2021, 189, 108302. https://doi.org/10.1016/J.SIGPRO.2021.108302.

Venegas-Molina, J., Molina-Hidalgo, F. J., Clicque, E., Goossens, A. Why and how to dig into plant metabolite–protein interactions. *Trends Plant Sci.*, 2021, 26(5), 472–483. https://doi.org/10.1016/J.TPLANTS.2020.12.008.

Vinaixa, M., Schymanski, E. L., Neumann, S., Navarro, M., Salek, R. M., Yanes, O. Mass spectral databases for LC/MS- and GC/MS-based metabolomics: State of the field and future prospects. *TrAC Trends Anal. Chem.*, 2016, 78, 23–35. https://doi.org/10.1016/J.TRAC.2015.09.005.

Wang, S., Alseekh, S., Fernie, A. R., Luo, J. The structure and function of major plant metabolite modifications. *Mol. Plant*, 2019, 12(7), 899–919. https://doi.org/10.1016/J.MOLP.2019.06.001.

Waschina, S., Seeger, K. Using In-tube extraction and slice selective NMR experiments allow imaging via statistical analysis of metabolic profiles. *Anal. Chim. Acta*, 2022, 1231, 340419. https://doi.org/10.1016/J.ACA.2022.340419.

Wilhelm, M., Kirchner, M., Steen, J. A. J., Steen, H. Mz5: Space- and time-efficient storage of mass spectrometry data sets. *Mol. Cell. Proteomics*, 2012, 11(1), O111.011379. https://doi.org/10.1074/MCP.O111.011379.

Williams, C., Stander, M. A., Medvedovici, A., Buica, A. Volatile terpenoid profiling in gin and beer – A targeted approach. *J. Food Compos. Anal.*, 2023, 118, 105178. https://doi.org/10.1016/J.JFCA.2023.105178.

Xiong, Q., Cao, C., Shen, T., Zhong, L., He, H. H., Chen, X. Comprehensive metabolomic and proteomic analysis in biochemical metabolic pathways of rice spikes under drought and submergence stress. *Biochimica et Biophysica Acta (BBA)—Proteins and Proteomics*, 2019, 1867(3), 237–247. https://doi.org/10.1016/J.BBAPAP.2019.01.001.

Yang, J., Yue, Z., Yuan, Y. Deep probabilistic graphical modeling for robust multivariate time series anomaly detection with missing data. *Reliab. Eng. Syst. Saf.*, 2023, 238, 109410. https://doi.org/10.1016/J.RESS.2023.109410.

Yuan, J. S., Pavlovich, M. J., Ragauskas, A. J., Han, B. Biotechnology for a sustainable future: Biomass and beyond. *Trends Biotechnol.*, 2022, 40(12), 1395–1398. https://doi.org/10.1016/J.TIBTECH.2022.09.020.

Yue, L., Liu, F., Hu, J., Yang, P., Wang, Y., Dong, J., Shu, W., Huang, X., Wang, S. A Guidebook of spatial transcriptomic technologies, data resources and analysis approaches. *Comput. Struct. Biotechnol. J.*, 2023, 21, 940–955. https://doi.org/10.1016/J.CSBJ.2023.01.016.

Zeng, L., Zhou, X., Liao, Y., Yang, Z. Roles of specialized metabolites in biological function and environmental adaptability of tea plant (*Camellia sinensis*) as a metabolite studying model. *J. Adv. Res.*, 2021, 34, 159–171. https://doi.org/10.1016/J.JARE.2020.11.004.

Zhang, F. X., Li, Z. T., Li, M., Yuan, Y. L. L., Cui, S. S., Wang, G. H., Li, R. man. An integrated strategy for revealing the pharmacological changes based on metabolites profiling and network pharmacology: Arctiin as an example. *J. Chromatogr. B*, 2020, 1157, 122270. https://doi.org/10.1016/J.JCHROMB.2020.122270.

Zhang, F., Ge, W., Huang, L., Li, D., Liu, L., Dong, Z., Xu, L., Ding, X., Zhang, C., Sun, Y., A, J., Gao, J., Guo, T. A Comparative analysis of data analysis tools for data-independent acquisition mass spectrometry. *Mol. Cell. Proteomics*, 2023, 22(9), 100623. https://doi.org/10.1016/J.MCPRO.2023.100623.

Zhang, S., Sun, L., Yan, F., Li, Q., Chen, R., Zhang, Z., Lai, X., Chen, Z., Li, Q., Yuan, E., Sun, S. Metabolomics analysis reveals the patterns and mechanisms of volatile metabolite in oolong tea during storage. *LWT*, 2023, 185, 115173. https://doi.org/10.1016/J.LWT.2023.115173.

Zhang, X., Qu, J., Dai, Z., Lin, Y., Lu, G., Yang, S., You, Y., Liu, H., Wu, Y., Jiang, G., Li, Y. Data-dependent acquisition based high-resolution mass spectrum for trace alternaria mycotoxin analysis and sulfated metabolites identification. *Food Chem.*, 2021, 364, 130450. https://doi.org/10.1016/J.FOODCHEM.2021.130450.

Zhang, Y., Lin, X., Gao, Z., Wang, T., Dong, K., Zhang, J. An omics data analysis method based on feature linear relationship and graph convolutional network. *J. Biomed. Inform.*, 2023, 104479. https://doi.org/10.1016/J.JBI.2023.104479.

Zheng, C., Zhou, Q., Shao, W., Zhang, J., Wang, J. Early identification of fungal leaf blight disease (Alternaria Alternate) on *Platycladus orientalis* plants by using gas chromatography-ion mobility spectrometry. *Microchem. J.*, 2022, 179, 107505. https://doi.org/10.1016/J.MICROC.2022.107505.

Zhou, J., Liu, F. J., Li, X. X., Li, P., Yang, H., Liu, Y. C., Chen, Y. H., dan Wei, C., Li, H. jun. A strategy for rapid discovery of traceable chemical markers in herbal products using MZmine 2 data processing toolbox: A case of Jing liqueur. *Chin. Herb. Med.*, 2021, 13(3), 430–438. https://doi.org/10.1016/J.CHMED.2021.05.004.

Zhou, L., Xu, J. Di, Zhou, S. S., Zhu, H., Kong, M., Shen, H., Zou, Y. T., Cong, L. J., Xu, J., Li, S. L. Independent or integrative processing approach of metabolite datasets from different biospecimens potentially affects metabolic pathway recognition in metabolomics. *J. Chromatogr. A*, 2019, 1587, 146–154. https://doi.org/10.1016/J.CHROMA.2018.12.024.

Zhu, Q., Uddin, M. Y. S., Qin, Z., Venkatasubramanian, N. Data collection and upload under dynamicity in smart community Internet-of-Things deployments. *Pervasive Mob. Comput.*, 2017, 42, 166–186. https://doi.org/10.1016/J.PMCJ.2017.10.003.

CHAPTER 6

Metabolomics-Assisted Breeding for Development of Stress-Tolerant Crops

ASHISH KUMAR* and MD. ZISHAN ANSARI

Department of Biotechnology, Guru Ghasidas Vishwavidyalaya (A Central University), Bilaspur, Chhattisgarh, India

*Corresponding authors

ABSTRACT

The demand for global food security has increased due to numerous biotic and abiotic stresses that limit crop growth and productivity. Plant metabolites are critical biomarkers of their physiological adaptations and response to different stresses. These metabolites may be detected, analyzed, and stored as datasets using modern analytical techniques such as mass spectrometry, liquid chromatography-mass spectrometry, gas chromatography-mass spectrometry, and nuclear magnetic resonance spectroscopy. Conventional breeding techniques can be time and resource intensive. By combining metabolomics data with traditional breeding techniques, metabolomic-assisted breeding makes it possible to identify key metabolites linked to stress tolerance. This knowledge can more precisely direct the selection of favorable metabolites in crop breeding programs. The conservation of genetic variety within farmed crops is further ensured, which enables the selection of plants with innate metabolic profiles that impart better stress tolerance. In addition, the integration of "omics" sciences such as transcriptomics, genomics, and proteomics will enable researchers and breeders to make informed decisions in selecting plants with superior stress responses. Metabolomics-assisted breeding signifies the crucial function of metabolomics in understanding stress response mechanisms and the pivotal role of multidisciplinary cooperation in tackling the global food security dilemma.

6.1 INTRODUCTION

In the aftermath of escalating global climate change and its effects on agriculture, the importance of crops capable of coping with multiple biotic and abiotic stresses has been rising faster than ever before. The quest for stress-tolerant yields has indulged scientists

Exploring Plant Stress Tolerance: A Metabolomics Approach.
Shuvasish Choudhury, Debojyoti Moulick, and Mirza Hasanuzzaman (Eds.)

DOI: 10.1201/9781779641731-6

and researchers across the globe in the exploration of innovative avenues. One such avenue recently gained substantial limelight is metabolomics-assisted stress-tolerant crop breeding (Fernie and Schauer, 2009).

Metabolomics, one of the latest additions to "omics" sciences, delves into the exploration of primary and secondary metabolites that play a pivotal role in biochemical and metabolic processes in plants. The studies of plant metabolites reveal their physiological adaptations and environment-related stress tolerance mechanisms (Gusain et al., 2023). Latest analytical techniques such as liquid chromatography-mass spectrometry (LC-MS) (Xiong et al., 2022), gas chromatography-mass spectrometry (GC-MS) (Abdulkader et al., 2022), and nuclear magnetic resonance (NMR) spectroscopy (Ng et al., 2023) enable extensive profiling of plant metabolites. Integrating metabolomics into crop breeding techniques is a novel strategy to study and discover a crop's metabolic responses to stress conditions. This enables researchers and crop breeders to bring forward some critical metabolites related to stress tolerance, thereby unraveling the fundamental molecular mechanisms underlying a crop's ability to adapt to demanding environmental conditions. Unlike conventional breeding techniques (Smith, 2021), metabolomics paves the way for molecular insights into the complex biochemical pathways within plants. Metabolomics-assisted breeding aims to identify crops with promising metabolic profiles that exhibit desirable attributes. This innovation expedites refining potential crop candidates for hybridization and subsequent cultivation, accelerating the crop breeding cycle. Metabolomics-accelerated crop breeding also leads to the discovery of biomarkers—the key metabolites closely associated with stress tolerance. These biomarkers guide light to restrict a plant's ability to thrive through unfavorable conditions.

Although faced with specific challenges such as functional validation of biomarkers and trait complexity, the amalgamation of metabolomics and crop breeding promises a future where the foundation of food security stands firm against the backdrop of numerous stress-related circumstances (Figure 6.1). This novel approach will revolutionize agricultural landscapes. The scientific study of stress tolerance at the molecular level, facilitated by metabolomics, overcomes the limitations osf traditional crop breeding practices. The synergistic interplay of metabolomics- and stress-resistant crop breeding is set to produce a wide variety (Derbyshire et al., 2022) of crops.

6.2 METABOLOMICS

Metabolomics has established itself as an impactful tool in stress-tolerant crop breeding. It offers molecular insights into metabolic biomarkers (Fu et al., 2017) that are critical in response to environmental stimuli and establishing the hereditary potential for enhanced crop varieties. This combination of metabolomics and crop breeding has revolutionized the traditional crop breeding practices and improvement. It opens doors for more efficient strategies to develop stress-tolerant (Arif et al., 2020) and high-yield crop cultivars. At its core, metabolomics involves a comprehensive analysis of plant metabolites necessary for the normal functioning of biochemical machinery. In the context of crop breeding, metabolomics emphasizes the possible interaction between genes and the environment. Advanced analytical techniques such as LC-MS (Xiong et al., 2022) and NMR spectroscopy (Ng et al., 2023) assist researchers and crop breeders to simultaneously detect and quantify numerous metabolites present within plant tissues.

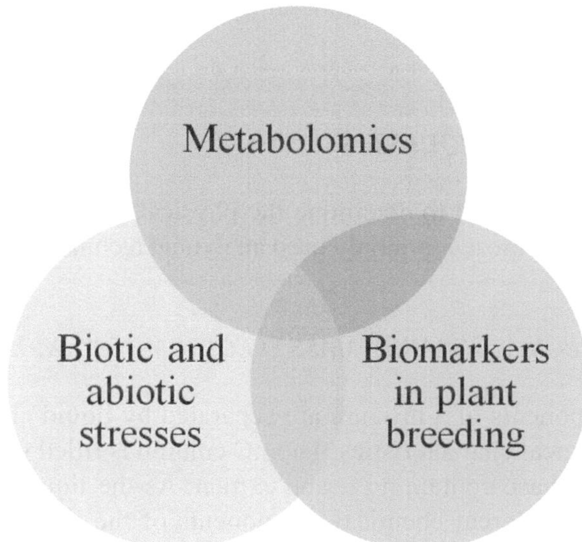

FIGURE 6.1 Venn diagram showing various components affecting the breeding of stress-tolerant crops.

Metabolomics focuses on phenotypic traits that determine a crop's growth and development (López-Bucio et al., 2022). It enables crop breeders to discover metabolic responses to stressors and the molecular mechanisms underlying such responses. Metabolic changes often precede visible symptoms of stress. Metabolomics facilitates the detection of small biochemical shifts, providing an early warning system to identify stress-tolerant genotypes before signs appear. Metabolomics contributes to trait mapping and identifying metabolite markers associated with specific characteristics by correlating metabolic profiles with desirable traits. Metabolomics enhances the ability to breed for quality traits such as flavor, aroma, and nutritional content. Metabolomics-assisted breeding allows for precise selection of parent plants based on their metabolic profiles. This targeted approach enables the development of hybrids with the desired traits. Metabolomics also delves into nonvisible characteristics such as nutrient utilization (Qiao et al., 2023), secondary metabolite production, and stress tolerance, which are often challenging to evaluate through traditional breeding practices. Metabolomics explains the complex interplay between genotypes and the environment. This understanding is crucial for developing crops that consistently withstand diverse conditions. Metabolomics allows the evaluation of large breeding populations, shortening the time required to identify promising candidates. Metabolomics aids in biofortification (Hefferon, 2023) efforts by identifying metabolites related to nutritional contents. This is pivotal for addressing malnutrition and enhancing the nutritive value of crop cultivars. Metabolomics bridges the gap between genotype and phenotype, facilitating more informed crop breeding decisions.

As technology advances and metabolomics databases expand, the integration of metabolomics into crop breeding becomes increasingly sophisticated. This interdisciplinary amalgamation holds promise not only for addressing immediate challenges posed by climate change but also for unlocking the potential of crops to meet the needs of an ever-evolving global population.

6.2.1 ANALYTICAL TECHNIQUES3–5

Analytical techniques are used to determine the physical and chemical characteristics of metabolites. Some of the most commonly used analytical techniques are mentioned below.

6.2.1.1 LIQUID CHROMATOGRAPHY-MASS SPECTROMETRY (XIONG ET AL., 2022)

The individual components of a mixture are separated by liquid chromatography based on their distinct chemical characteristics. The LC column is filled with a stationary phase and a mobile liquid phase containing tissue extract. As the liquid phase flows through the stationary phase, different chemical components of the tissue extract separate due to differences in their chemical reactivity. LC is combined with MS for structural and chemical characterization of crop metabolites (Anjali et al., 2023). The molecules to be analyzed are first ionized. The ionized particles are separated by applying electric and magnetic fields based on their mass-to-charge ratio (m/z). The mass spectrum, thus obtained, gives detailed information regarding the molecular composition of the metabolites (Figure 6.2).

6.2.1.2 GAS CHROMATOGRAPHY-MASS SPECTROMETRY (ABDULKADER ET AL., 2022)

The tissue extract is vaporized and introduced into the GC column. The components of the sample have different volatilities. More volatility, greater the extent of movement through the column. Consequently, the ingredients are separated as they move along the stationary phase. It has been widely applied in quantification (Kováčik, 2021) and selective detection of plant metabolites (Anjali et al., 2023) based on their volatility (Figure 6.3).

6.2.1.3 NUCLEAR MAGNETIC RESONANCE SPECTROSCOPY (NG ET AL., 2023)

The metabolite of interest is extracted from plant tissue and dissolved in an appropriate solvent. The mixture is loaded in an NMR tube and appropriately sealed. The tube is placed in an NMR spectrometer, and spectral data of the metabolite is generated, as depicted in Figure 6.4.

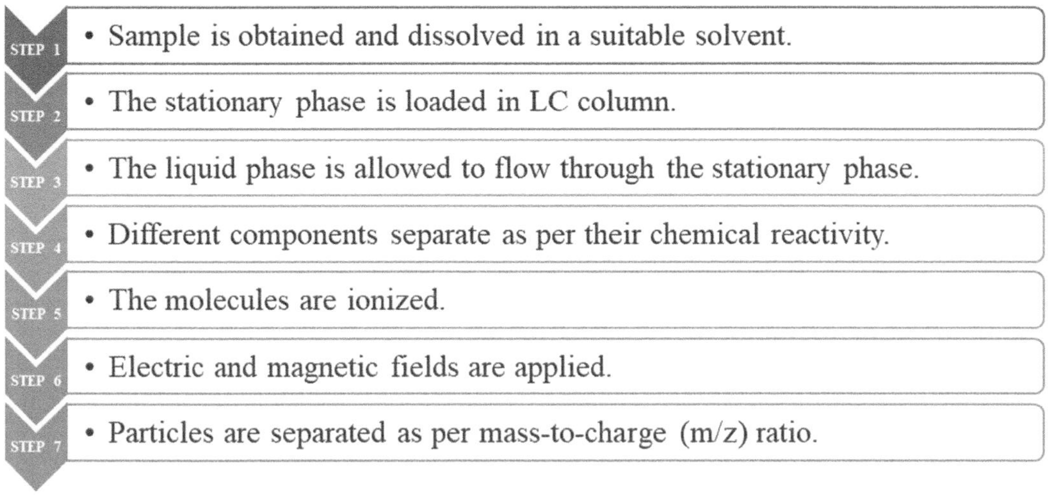

FIGURE 6.2 Summarized stepwise LC-MS.

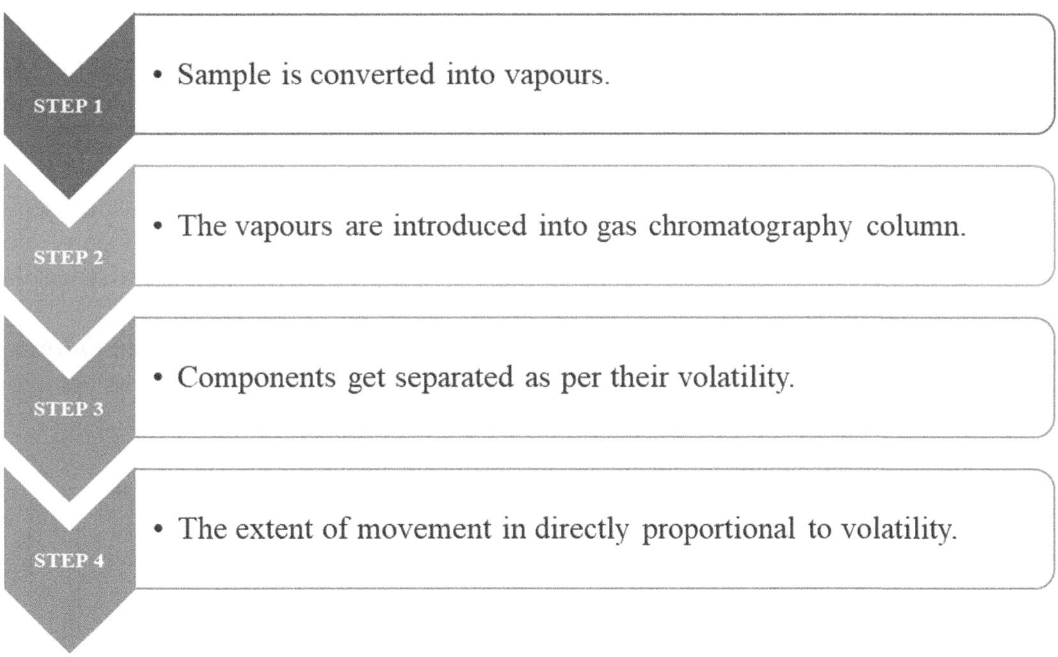

FIGURE 6.3 Summarized stepwise GC-MS.

6.3 BIOTIC STRESS

Biotic stress refers to the harmful effects of living organisms on crops, such as pathogens and pests. These directly affect field crops' growth, development, metabolism, and productivity. Some common biotic stresses are listed given as follows:

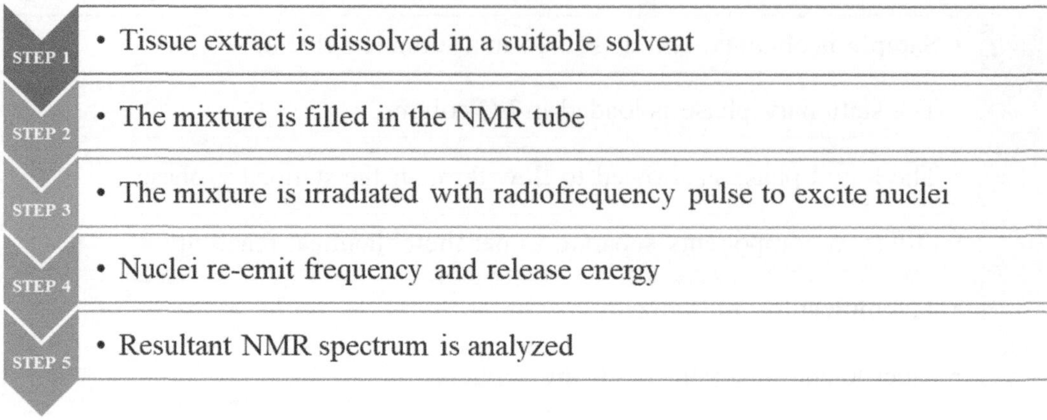

FIGURE 6.4 Summarized stepwise NMR spectroscopy (Ng et al., 2023).

1. **Microorganisms:** Harmful microorganisms include some bacteria, viruses, and fungi. For example, *Pseudomonas corrugata*, a Gram-negative plant pathogenic bacterium, causes pith necrosis in tomatoes (Su et al., 2023).
2. **Weeds:** These are undesirable plants that grow along with the principal crops. Competition for limited resources impedes the growth and productivity of main crops. For example, *Lepidium draba* is a significant threat to biodiversity (Sharma et al., 2021).
3. **Insects:** The exudates of plant pathogens usually stick to the body parts of insects. These insects may then carry pathogenic toxins to susceptible plant parts. For example, malaria involving *Plasmodium* is transmitted by Anopheles (Bordoloi and Agarwala, 2021).
4. **Fungus:** Pathogenic fungal species induce necrosis and abnormal growth. *Fusarium oxysporum*, *Alternaria*, and *Magnaporthe oryzae* are some of the most common fungi causing plant diseases (Collinge et al., 2022).
5. **Mites:** Mites suck out cellular contents from leaves. A high population of tiny mites is a considerable concern among crop breeders (Sousa et al., 2019).

6.4 ABIOTIC STRESS

Abiotic stress refers to the harmful effects of nonliving factors on plant parts, such as drought and other environmental perturbations. Some common crop abiotic stresses are given as follows:

a) **Salinity** refers to the total amount of salts dissolved in underground water. Any deviation from the optimum salinity level interferes with nitrogen uptake, plant reproduction, ionic imbalances, and damage to cellular machinery (Huang et al., 2023).
b) **Nutrient deficiency:** Nutrients are necessary for splitting water molecules during photosynthesis, opening/closing stomata, etc. Lack of micro- and macronutrients

such as nitrogen, phosphorus, potassium, iron, and manganese negatively affects plant growth and development (Tiziani et al., 2023).

c) **UV radiation:** Harmful ultraviolet (UV) radiations produce reactive oxygen species (ROS) that cause DNA damage and inactivation of key enzymes involved in photosynthesis (Ibnouf, 2021).

d) **Drought:** Insufficient moisture in soil may lead to drooping and senescence of leaves (Riyazuddin et al., 2023).

e) **Heavy metals:** Heavy metals such as Cd (cadmium), Zn (zinc), and Hg (mercury) can impair plant growth and lead to the accumulation of toxic compounds (Narayanan and Ma, 2023).

Figure 6.5 shows some of the effects of biotic and abiotic stresses on crops.

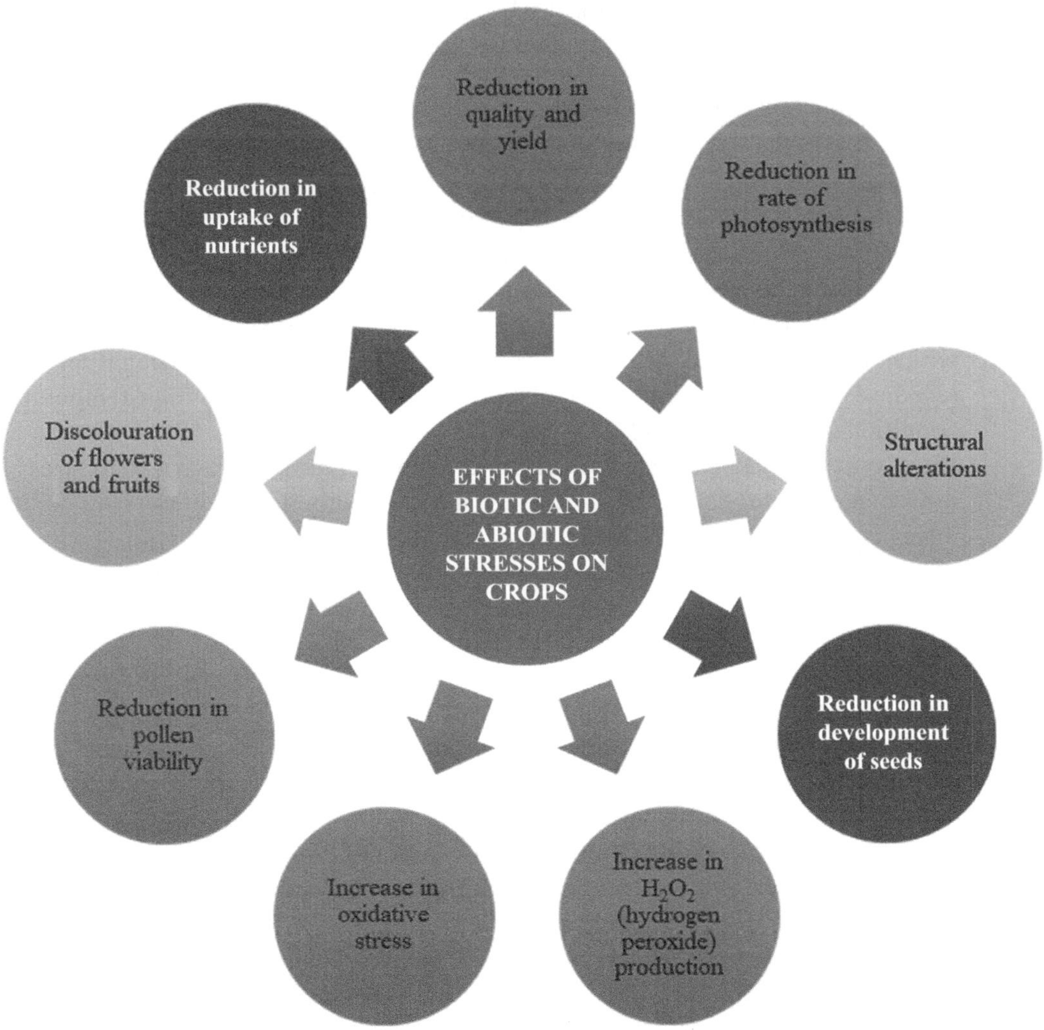

FIGURE 6.5 Effects of biotic and abiotic stresses on crops.

6.5 STRESS TOLERANCE MECHANISMS AND PATHWAYS

Field crops have evolved specific mechanisms and pathways to cope with biotic and abiotic stresses. Some of them are given as follows:

1. **Roots:** Crops may alter their root system architecture (RSA) in response to decreased moisture quantity. They tend to develop more profound and more extensive root systems to access water from deeper soil layers. Structural features such as number, length, and density of central and lateral roots may alter (Demissie et al., 2023).
2. **Cold adjustments:** When field crops are exposed to cold weather for an extended period, they increase the production of antifreeze proteins and change the composition of their membranes to maintain fluidity and prevent cold-induced damage (Ishimoto et al., 2021).
3. **Abscisic acid (ABA) production:** ABA is an essential hormone involved in plant stress responses. It regulates stomatal opening and closure to reduce water loss during stressful conditions and regulates the expression of various stress-responsive genes (Ding et al., 2022).
4. **Nutrient mobilization:** Plants can remobilize nutrients from older tissues to support the growth of younger and actively growing tissues at times of nutrient deficiency.
5. **Osmotic balance:** Plants regulate their internal osmotic potential by accumulating compatible solutes, such as proline and sugars, which help maintain water balance and prevent cellular dehydration during drought and salinity stress.
6. **Heat shock:** Heat stress causes the expression of heat shock proteins (HSPs), which help denatured proteins refold and prevent protein aggregation. This response aids in the protection of cellular functions during high-temperature conditions (Unel et al., 2023).
7. **Photosynthesis:** Plants can downregulate photosynthesis in response to stress to reduce energy and water consumption. This ultimately aids in preserving cellular integrity and preventing oxidative damage to the cellular machinery.
8. **Stomatal regulation:** Plants temporarily close their stomata during drought to reduce water loss through transpiration, thereby conserving water. ABA and other signaling pathways regulate stomatal opening and closure.
9. **RNA-mediated regulation:** microRNAs and siRNAs play a crucial role in post-transcriptional gene regulation during stress.
10. **Antioxidant machinery:** Plants have antioxidant defense systems that include enzymes such as superoxide dismutases, catalases, and peroxidases to neutralize ROS, thus preventing cell oxidative damage.
11. **Signal transduction pathways:** Plants use complex signaling networks involving proteins, hormones, and secondary messengers to transmit stress signals and activate appropriate responses.
12. **Epigenetic alterations:** Under stress conditions, epigenetic alterations such as DNA methylation and histone modifications can influence gene expression patterns, allowing field crops to adapt to any abrupt change in the environment (Harris et al., 2023). Figure 6.6 shows the summarized mechanism of stress tolerance.

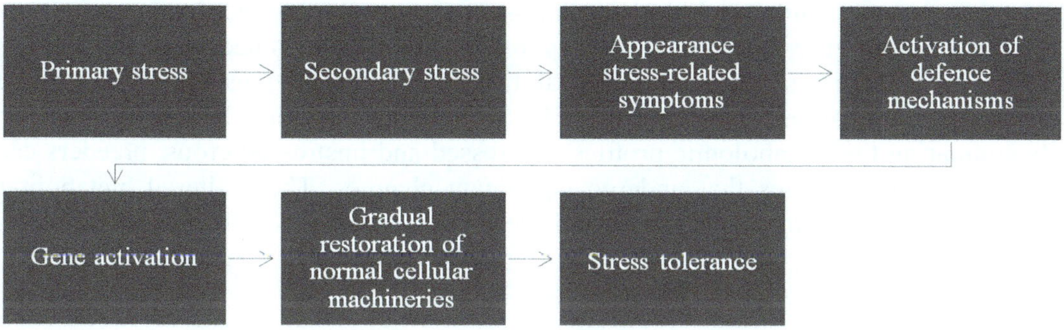

FIGURE 6.6 Flowchart summarizing stress tolerance mechanism.

6.6 METABOLITES AS STRESS RESPONSE INDICATORS

Crop stress encompasses diverse factors ranging from temperature (Samtani et al., 2022) to encounters with plant pathogens. In response to these stresses, plants have evolved to adapt, survive, and maintain homeostasis (Dhiman et al., 2023). A vital aspect of this adaptation is regulating metabolic processes coordinated by various metabolites. Metabolites are the intermediates and end-products of metabolic processes within the cells. They comprise a broad spectrum of molecules, including amino acids, carbohydrates, nucleotides, and secondary metabolites. Each metabolite contributes to vital cellular functions. Metabolites also emerge as potent indicators of a plant's stress responses. By keeping a record of fluctuations in metabolite levels, researchers gain insights into molecular mechanisms responsible for adaptation to stress conditions. Stress responses (Akpinar and Cansev, 2022) are complex and multifaceted reactions that plants undergo when encountering adverse conditions. These responses comprise physiological, biochemical, and behavioral changes to restore homeostasis and ensure average growth and developmental processes. To achieve this, plants must amend their internal machinery, allocating resources strategically and optimizing the cellular processes. When stress is encountered, cells must swiftly adapt to the new circumstances, requiring alterations in energy generation, biosynthetic pathways, and signaling. These adaptations also change the concentration of specific metabolites, as shown in Figure 6.7.

Metabolites reflect a plant's physiological state and adaptations (Francis et al., 2023). When biotic or abiotic stress hits the plant parts, the cellular machinery coordinates changes to optimize survival mechanisms. These changes manifest as shifts in metabolic pathways, altering the production and consumption of various metabolites. Consequently, the levels of specific metabolites can fluctuate significantly in response to stress, acting as quantitative indicators. Antioxidant metabolites, such as glutathione, are essential in maintaining cellular redox balance. At times of oxidative stress, the concentration of these metabolites may surge as cells attempt to counteract the increased levels of ROS (Tavanti et al., 2021). Simultaneously, the by-products of oxidation reactions may accumulate.

Nutrient limitation presents another stressful condition where metabolites reveal the intricate adaptation mechanism. When resources are scarce, cells must prioritize essential processes while sidelining the nonessential ones. For instance, amino acids may be diverted from growth-related pathways to biosynthetic routes necessary for survival during stress. Latest advancements in analytical techniques have ushered in the era of metabolomics, enabling the detailed study of metabolites within biological systems. By comparing the metabolomic profiles of stressed and unstressed crops, breeders can pinpoint the metabolites that undergo significant changes. These altered metabolites act as beacons, guiding scientists toward the pivotal metabolic pathways in the stress response.

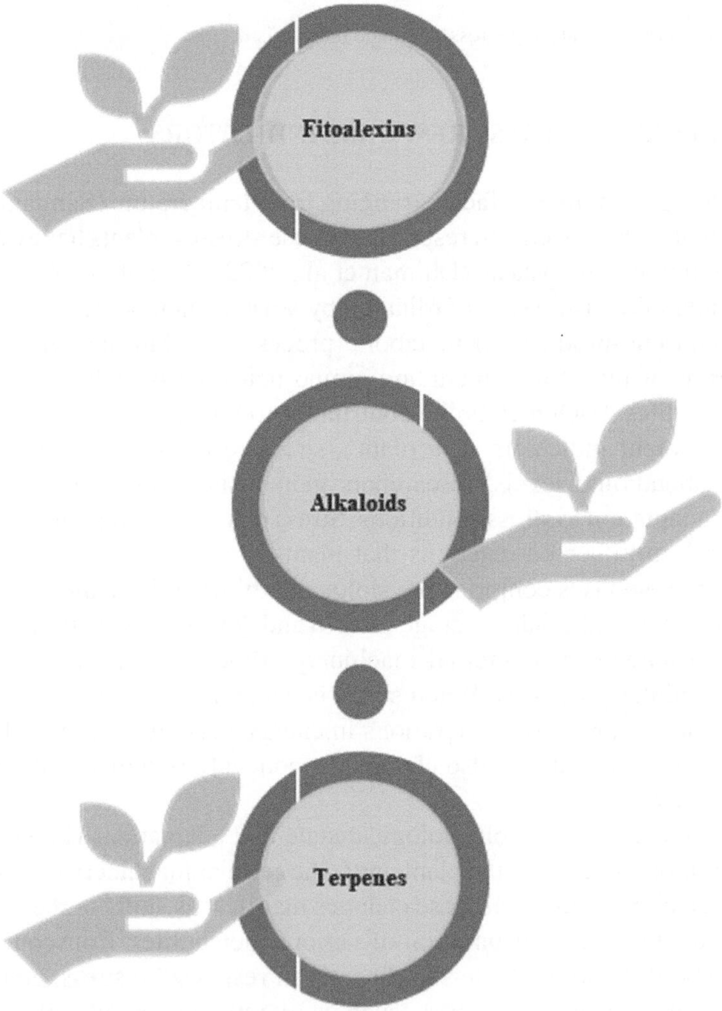

FIGURE 6.7 Examples of stress indicating metabolites in crops.

The profound implications of metabolites extend well beyond deciphering stress responses as they are also promising disease biomarkers (Hariharan et al., 2023).

Stress-induced metabolic alterations often precede clinical symptoms, offering a window of opportunity for early intervention. While the potential of metabolites as stress response indicators and disease biomarkers is intriguing, several challenges must also be addressed. The complexity of metabolite interactions (Yu et al., 2022), the dynamic nature of cellular responses, and the inherent variability in biological systems can complicate data interpretation. Moreover, identifying causal relationships between altered metabolite levels and downstream physiological outcomes requires meticulous experimentation and validation.

Future research will likely focus on the causal links between stress-induced metabolic changes and specific cellular outcomes. This could involve perturbation studies, where manipulating key metabolic pathways reveals their impact on stress resistance (Shah et al., 2023). Additionally, integrating metabolomics with other "omics" disciplines, such as genomics and proteomics, will provide a holistic understanding of stress responses and their implications. As researchers delve deeper into metabolites, their potential as diagnostic markers for stress-induced issues becomes increasingly apparent. By deciphering the metabolic fingerprints of plant diseases, we inch closer to preemptive interventions and crop-specific treatments. Metabolites are not mere bystanders; they are the messengers that reveal the stories of adaptation, resilience, and the delicate balance between cellular equilibrium (Dzinyela et al., 2023) and perturbations.

6.7 STRESS-RELATED METABOLOMIC PROFILING

Metabolomic profiling is a powerful scientific technique used to study the chemical characteristics and interaction among metabolites present in plant systems. Crop stress can arise from environmental changes (temperature, light, or humidity), pathogen attacks, nutrient deficiencies, or other disturbances. When plants are exposed to stress, their metabolism often adjusts to cope with the new conditions, leading to changes in the types and quantities of metabolites produced. The following are the steps involved in metabolomic profiling:

1. **Sample collection:** Plant tissues are collected from leaves, stems, roots, or any other plant part under controlled conditions and labeled as stressed and nonstressed (control) samples (Suh et al., 2018).
2. **Metabolite extraction:** Metabolites are extracted from the plant tissues using analytical techniques such as MS and LC-MS (Xiong et al., 2022). These extraction methods are used to capture a wide range of metabolites (Waschina and Seeger, 2022).
3. **Metabolite analysis:** Analytical techniques such as NMR (Ng et al., 2023) can identify and quantify various metabolites based on their molecular weight and chemical properties (Waschina and Seeger, 2022).
4. **Data processing and analysis:** The raw data obtained from metabolite analysis is processed and analyzed. This involves identifying peaks, eliminating outliers, metabolite annotation, and quantifying the abundance of different metabolites in the samples (Sun and Xia, 2023).

5. **Statistical analysis:** Statistical methods such as mean, standard deviation, and hypothesis testing are applied to the metabolomics data to identify significant differences between stressed and nonstressed samples. This helps in identifying those metabolites that are altered under stress conditions (Wang et al., 2022).
6. **Metabolite pathway analysis:** Plant metabolites are often involved in interconnected biochemical pathways. Pathway analysis tools help understand how the changes in metabolite concentrations may affect various metabolic pathways and the overall physiology.
7. **Biomarker identification:** Some metabolites can serve as biomarkers, indicating the presence or degree of stress. These biomarkers can be helpful for early stress detection and monitoring (Ben et al., 2023).
8. **Functional interpretation:** The identified changes in metabolite levels are then interpreted in the context of plant physiology and stress response mechanisms. This helps in gaining insights into how crops adapt to stress conditions (Osbourn, 2000).

6.8 METABOLOMICS-ASSISTED CROP BREEDING

Metabolomics-assisted breeding is one of the latest approaches in agriculture and crop breeding that utilizes the field of metabolomics to enhance the efficiency and precision of the breeding process. Metabolomics is a powerful tool for analyzing and quantifying the unique chemical fingerprints of plants in the context of crop breeding. Researchers can identify specific metabolites associated with desirable traits such as disease resistance, nutritional content, flavor, and stress tolerance by studying the metabolite profiles of different genotypes or varieties.

The key steps involved in metabolomics-assisted breeding are given as follows:

1. **Sample preparation:** Tissue samples are collected from different plant parts such as roots and leaves and prepared for metabolite extraction. Other isolation techniques are employed as per the type of sample and the metabolites of interest (Suh et al., 2018).
2. **Metabolite profiling:** High-throughput analytical techniques such as MS and NMR spectroscopy (Ng et al., 2023) are used to identify and quantify the metabolites present in the sample. This produces a detailed metabolite profile for each genotype (Waschina and Seeger, 2022).
3. **Data analysis:** The large datasets generated by metabolite profiling are processed using advanced computational tools. Statistical techniques and bioinformatics tools aid in identifying metabolites that are significantly related to the desired traits (Sun and Xia, 2023).
4. **Trait association:** Researchers and crop breeders can identify metabolites that are consistently associated with specific traits by comparing the metabolite profiles of different genotypes or populations. Specific metabolites, for example, could be linked to increased yield or drought tolerance.

5. **Marker development:** Metabolites with strong associations with desirable traits may be helpful as biomarkers. These biomarkers can further be used to create molecular markers (e.g., DNA markers) that can be screened in breeding populations quickly and accurately (Ben et al., 2023).
6. **Selection and breeding:** Metabolomics data can help breeders make informed parent plant selections for cross-breeding and hybridization. By concentrating on the metabolites contributing to those traits, new varieties with improved traits can be developed (Gerullis and Schulz, 2023).

6.8.1 BENEFITS

Figure 6.8 illustrates the benefits of metabolomics-assisted crop breeding (Munir et al., 2023; Mohapatra et al., 2023; Zhu et al., 2020).

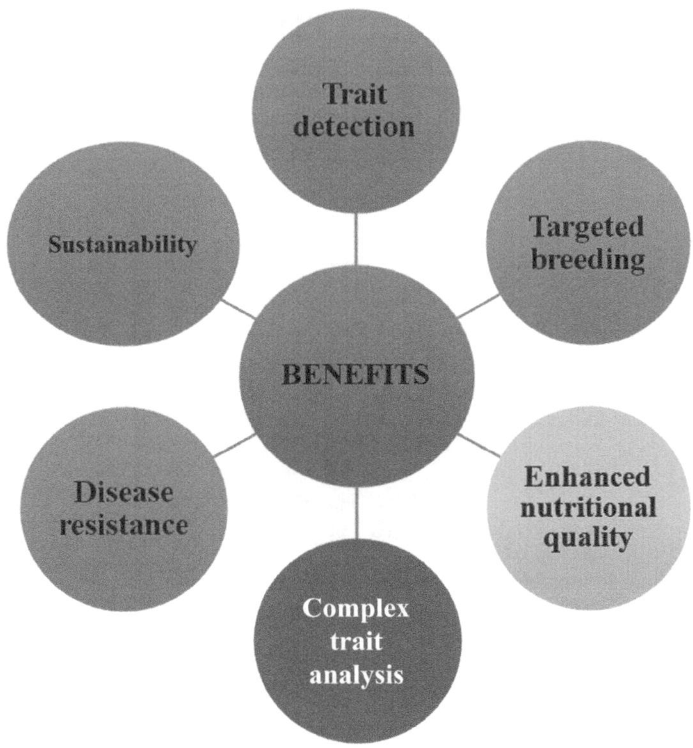

FIGURE 6.8 Benefits of metabolomics-assisted crop breeding.

6.8.2 LIMITATIONS

Figure 6.9 illustrates the limitations in metabolomics-assisted crop breeding (Sens et al., 2023; Ferrão et al., 2023; Weale, 2010).

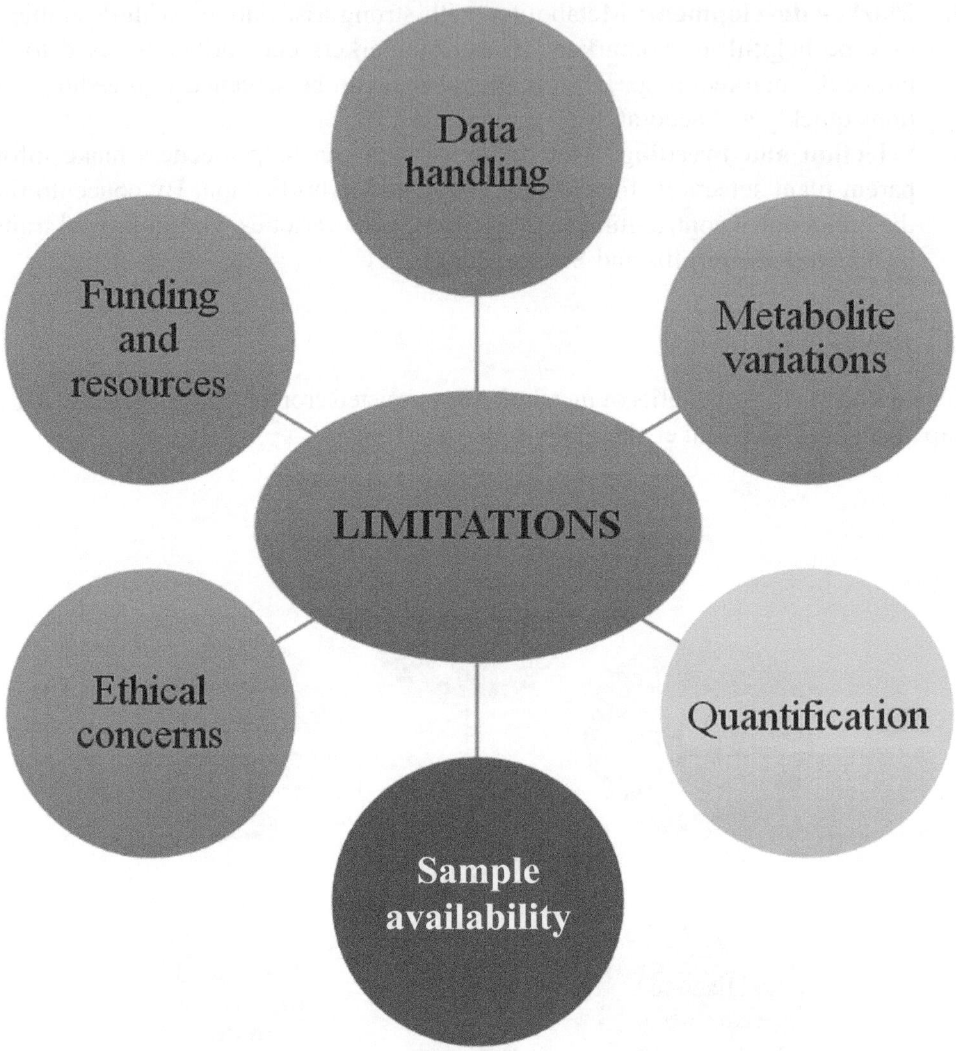

FIGURE 6.9 Limitations in metabolomics-assisted crop breeding.

6.9 SUCCESSFULLY DEVELOPED STRESS-TOLERANT CROPS

Researchers and breeders across the globe have employed the latest technology and analytical techniques to develop stress-tolerant crops. Table 6.1 summarizes the stress-tolerant crops that have been successfully developed.

6.10 ENHANCING GENETIC VARIABILITY

Enhancing plant genetic variability is crucial to modern agricultural and breeding practices. Genetic variability refers to the diversity of genetic traits within a plant species. This diversity

TABLE 6.1 Overview of Some Stress-Tolerant Crops

Crops	Stress tolerance	Description	References
Maize	Drought	Maize can withstand drought, making it a staple crop in water-stressed areas. Its deep root system aids in fetching water from deeper soil layers.	(Worku et al., 2020; Setimela et al., 2018)
Sorghum	Heat, drought	Sorghum is a drought-tolerant cereal crop that grows well in hot, dry climates. It can withstand high temperatures and a lack of water, making it suitable for harsh environments.	(Dev Sharma et al., 2006)
Amaranth	Heat, drought	Amaranth is a drought-tolerant crop. It can thrive in conditions of limited water availability and is adaptable to high-temperature environments.	(Qin et al., 2013)
Rice	Flood, salinity	Rice is renowned for its flood tolerance. It may thrive in coastal and brackish water settings because of its modest tolerance to salty soils.	(Aryal et al., 2022)
Potato	Cold, drought	Potatoes are hard and can withstand cold temperatures, allowing them to thrive in cooler climates. Because of their tuber storage ability, they are also drought tolerant.	(Alvarez-Morezuelas et al., 2022)
Quinoa	Drought, salinity	Quinoa can thrive in arid environments and requires little water to grow. It can also grow in saline environments due to its tolerance to soils with varying salt levels.	(Maestro-Gaitán et al., 2023)
Barley	Cold, salinity	Barley is known for its cold tolerance, making it ideal for cold climates. It can also tolerate soils with a moderate salt content, which is helpful in saline areas.	(Mostek et al., 2015)
Chickpea	Salinity	Chickpeas can grow in soils with high levels of salinity. They have some tolerance for saline conditions, allowing them to be grown in areas with salt-affected soils.	(Bhogireddy et al., 2023)
Pearl millet	Heat, drought	Pearl millet thrives in hot climates and water-stressed environments. It can thrive in areas with high temperatures and little rainfall, making it an important crop in arid regions.	(Sonkar et al., 2023)
Cowpea	Drought	Cowpea, also known as black-eyed peas, is drought resistant. Because of its ability to adapt to dry conditions, it is an essential crop in areas with erratic rainfall patterns.	(Olorunwa et al., 2023)

is the foundation for developing improved crops with enhanced traits, such as yield, disease resistance, nutritional content, and adaptability to varying environments. Figure 6.10 shows strategies and techniques used to enhance genetic variability in plants (Hoshikawa et al., 2023; da Silva et al., 2022; Weil and Martienssen, 2008; Kenzo et al., 2023; Angon et al., 2023).

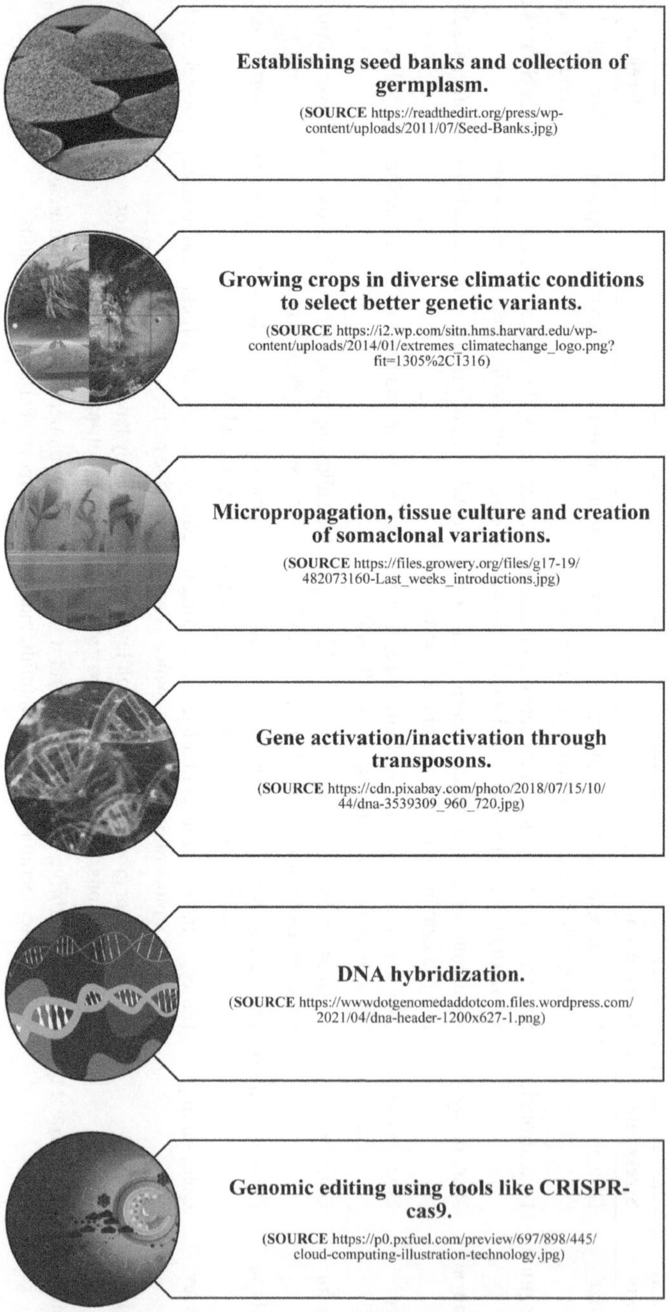

FIGURE 6.10 Techniques for enhancing genetic variability in field crops

6.11 CHALLENGES

Metabolomics studies generate complex datasets that require professional expertise to interpret and analyze. The lack of standard reference databases makes identification and annotation a challenging task. Environmental fluctuations (Zhang et al., 2023) considerably affect the concentration of metabolites and the trait of interest. Due to the inaccuracy of currently used analytical techniques, only limited quantitative analysis can be performed. Knowledge of bioinformatics (Mitra et al., 2022) and statistics is necessary to identify underlying issues while performing quantitative analysis. Successful integration of metabolomics with other "omics" sciences is a complex task. Ethical issues related to intellectual property rights (Kiran and Pandey, 2020) must be considered to prevent unprecedented consequences. To successfully develop stress-tolerant crop varieties (Liu et al., 2023), multiple generations of crop breeding may be required. This may also be time, energy, cost, and labor intensive. Currently, most crop breeders need to be made aware of the latest advances in crop breeding. They need to be skilled and educated about such tools and techniques. The access to data has also remained unclear. Many remotely located research institutions cannot access databases necessary for the development of stress-tolerant crops.

6.12 REGULATION

The formulation and strict implementation of guidelines and policies ensure safe and responsible use of metabolomics-assisted crop breeding. Regular risk assessments need to be done to identify and eliminate potential risk factors related to modified crops' environment, health, and socioeconomic impact. The quality of crops must be tested rigorously for toxicity and nutritional composition. A robust and safe transportation network will ensure easy trade and collaboration. To enhance transparency, all stakeholders' opinions must be taken into consideration. The final products should be adequately labeled before making available for public use. Ethical concerns related to genetic manipulation and its potential impacts on biodiversity, human health, and cultural practices should be addressed by regulations. Governments, scientific communities, industry stakeholders, and the general public must work together to create robust and efficient regulatory frameworks.

6.13 EMERGING TRENDS

Real-time data on plant development and stress responses are being gathered through remote sensing, drones, and high-resolution imaging techniques. Chemicals that have pharmacological potential (Roumani et al., 2020) are being explored. The synthesis of beneficial metabolites in crops, such as vitamins, antioxidants, and bioactive compounds, is being improved. Metabolomics examines how plants respond to climate change, including elevated CO_2 levels, varying temperature, and precipitation patterns, to assist breeders in developing resilient cultivars.

6.14 CONCLUSION

Metabolomics-assisted crop breeding offers a thorough knowledge of how genetic variants translate into perceptible phenotypes by delving deeper into the complex biochemical mechanisms that control crop growth, development, and responses to stressful conditions. Breeders may speed up the creation of superior crop varieties suited to many needs of our world in transition by using these insights to make informed choices. Cutting-edge technology and multidisciplinary collaborations (Zenda et al., 2023) are needed to overcome obstacles such as complex data processing and metabolite identification. Ethical issues, legal frameworks, and the requirement for long-term monitoring emphasize the significance of responsible implementation. We are paving the way for a more prosperous and sustainable agricultural future by leveraging the power of metabolite profiles and genetics, where crops are plentiful and carefully engineered to meet the dynamic requirements of our society.

KEYWORDS

- **global food security**
- **metabolomics-assisted breeding**
- **novel strategy**
- **stress-tolerant crops**
- **metabolic profiles**
- **multidisciplinary cooperation**

REFERENCES

Abdulkader, O. M., Sharaf, A. E. A., Fouda, H. M., Elhaw, M. H. Phytoconstitutes investigation of *Chenopodium ambrosioides* linn. and gas chromatography with mass spectroscopy material analysis. *Mater. Today Proc.*, 2022, 61, 992–997. https://doi.org/10.1016/J.MATPR.2021.10.280.

Akpinar, A., Cansev, A. Physiological and molecular responses of roots differ from those of leaves in spinach plants subjected to short-term drought stress. *S. Afr. J. Bot.*, 2022, 151, 9–17. https://doi.org/10.1016/J.SAJB.2022.09.032.

Alvarez-Morezuelas, A., Barandalla, L., Ritter, E., Lacuesta, M., Ruiz de Galarreta, J. I. Physiological response and yield components under greenhouse drought stress conditions in potato. *J. Plant Physiol.*, 2022, 278, 153790. https://doi.org/10.1016/J.JPLPH.2022.153790.

Angon, P. B., Mondal, S., Akter, S., Sakil, M. A., Jalil, M. A. Roles of CRISPR to mitigate drought and salinity stresses on plants. *Plant Stress*, 2023, 8, 100169. https://doi.org/10.1016/J.STRESS.2023.100169.

Anjali, Kumar, S., Korra, T., Thakur, R., Arutselvan, R., Kashyap, A. S., Nehela, Y., Chaplygin, V., Minkina, T., Keswani, C. Role of plant secondary metabolites in defence and transcriptional regulation in response to biotic stress. *Plant Stress*, 2023, 8, 100154. https://doi.org/10.1016/J.STRESS.2023.100154.

Arif, Y., Singh, P., Siddiqui, H., Bajguz, A., Hayat, S. Salinity induced physiological and biochemical changes in plants: An omic approach towards salt stress tolerance. *Plant Physiol. Biochem.*, 2020, 156, 64–77. https://doi.org/10.1016/J.PLAPHY.2020.08.042.

Aryal, J. P., Rahut, D. B., Acharya, S. Managing drought risks with drought-stress tolerant rice varieties and its impacts on yield and production risk: A case of Nepal. *Environ. Challenges*, 2022, 7, 100503. https://doi.org/10.1016/J.ENVC.2022.100503.

Ben, Y., Cheng, M., Liu, Y., Wang, X., Wang, L., Yang, Q., Huang, X., Zhou, Q. Biomarker changes and oxidative damage in living plant cells as new biomonitoring indicators for combined heavy metal stress assessment. *Ecol. Indic.*, 2023, 154, 110784. https://doi.org/10.1016/J.ECOLIND.2023.110784.

Bhogireddy, S., Kudapa, H., Bajaj, P., Garg, V., Chitikineni, A., Nayak, S., Varshney, R. K. Transcriptome analysis of chickpea during heat stress unveils the signatures of long intergenic non-coding RNAs (LincRNAs) and MRNAs in the Heat-QTL region. *Crop Design*, 2023, 2(1), 100026. https://doi.org/10.1016/J.CROPD.2023.100026.

Bordoloi, K. S., Agarwala, N. MicroRNAs in plant-insect interaction and insect pest control. *Plant Gene*, 2021, 26, 100271. https://doi.org/10.1016/J.PLGENE.2021.100271.

Collinge, D. B., Jensen, B., Jørgensen, H. J. Fungal endophytes in plants and their relationship to plant disease. *Curr. Opin. Microbiol.*, 2022, 69, 102177. https://doi.org/10.1016/J.MIB.2022.102177.

da Silva, L. M., Carvalho, V. S., Generoso, A. L., Miranda, D. P., Júnior, O. D. da C., Simioni, P. F., Santana, D. B., da Cunha, M., de Oliveira, J. G., Viana, A. P. Micropropagation of interspecific hybrids of *Vitis* spp. in microenvironments with different gas exchanges. *Sci. Hortic.*, 2022, 305, 111413. https://doi.org/10.1016/J.SCIENTA.2022.111413.

Demissie, H. S., Mindaye, T. T., Teklu, D. N., Kebede, F. G. Root system architecture analysis of sorghum genotypes and its effect on drought adaptation. *Rhizosphere*, 2023, 27, 100772. https://doi.org/10.1016/J.RHISPH.2023.100772.

Derbyshire, M. C., Batley, J., Edwards, D. Use of multiple 'omics techniques to accelerate the breeding of abiotic stress tolerant crops. *Curr. Plant Biol.*, 2022, 32, 100262. https://doi.org/10.1016/J.CPB.2022.100262.

Dev Sharma, A., Kumar, S., Singh, P. Expression analysis of a stress-modulated transcript in drought tolerant and susceptible cultivars of sorghum (Sorghum Bicolor). *J. Plant Physiol.*, 2006, 163(5), 570–576. https://doi.org/10.1016/J.JPLPH.2005.06.011.

Dhiman, S., Singh, A. D., Kour, J., Bhardwaj, T., Devi, K., Tikoria, R., Ohri, P., Bhardwaj, R., Sharma, P. Redox homeostasis in response to abiotic stresses in plants. *The Role of Growth Regulators and Phytohormones in Overcoming Environmental Tress*, 2023, 15–37. https://doi.org/10.1016/B978-0-323-98332-7.00014-7.

Ding, F., Zhang, B. L., Li, F., Li, Y. R., Li, J. H., Lu, Y. T. General control non-repressible 20 functions in the salt stress response of arabidopsis seedling by modulating ABA accumulation. *Environ. Exp. Bot.*, 2022, 198, 104856. https://doi.org/10.1016/J.ENVEXPBOT.2022.104856.

Dzinyela, R., Alhassan, A. R., Suglo, P., Movahedi, A. Advanced study of functional proteins involved in salt stress regulatory pathways in plants. *S. Afr. J. Bot.*, 2023, 159, 425–438. https://doi.org/10.1016/J.SAJB.2023.06.029.

Fernie, A. R., Schauer, N. Metabolomics-assisted breeding: A viable option for crop improvement? *Trends Genet.*, 2009, 25(1), 39–48. https://doi.org/10.1016/J.TIG.2008.10.010.

Ferrão, L. F. V., Dhakal, R., Dias, R., Tieman, D., Whitaker, V., Gore, M. A., Messina, C., Resende, M. F. R. Machine learning applications to improve flavor and nutritional content of horticultural crops through breeding and genetics. *Curr. Opin. Biotechnol.*, 2023, 83, 102968. https://doi.org/10.1016/J.COPBIO.2023.102968.

Francis, B., Aravindakumar, C. T., Brewer, P. B., Simon, S. Plant nutrient stress adaptation: A prospect for fertilizer limited agriculture. *Environ. Exp. Bot.*, 2023, 213, 105431. https://doi.org/10.1016/J.ENVEXPBOT.2023.105431.

Fu, H. Y., Hu, O., Zhang, Y. M., Zhang, L., Song, J. J., Lu, P., Zheng, Q. X., Liu, P. P., Chen, Q. S., Wang, B., Wang, X. Y., Han, L., Yu, Y. J. Mass-spectra-based peak alignment for automatic nontargeted metabolic profiling analysis for biomarker screening in plant samples. *J. Chromatogr. A*, 2017, 1513, 201–209. https://doi.org/10.1016/J.CHROMA.2017.07.044.

Gerullis, M. K., Schulz, W. Robustness of plant breeding systems under automated phenotyping. *Smart Agric. Technol.*, 2023, 5, 100225. https://doi.org/10.1016/J.ATECH.2023.100225.

Gusain, S., Joshi, S., Joshi, R. Sensing, signalling, and regulatory mechanism of cold-stress tolerance in plants. *Plant Physiol. Biochem.*, 2023, 197, 107646. https://doi.org/10.1016/J.PLAPHY.2023.107646.

Hariharan, J., Ampatzidis, Y., Abdulridha, J., Batuman, O. An AI-based spectral data analysis process for recognizing unique plant biomarkers and disease features. *Comput. Electron. Agric.*, 2023, 204, 107574. https://doi.org/10.1016/J.COMPAG.2022.107574.

Harris, C. J., Amtmann, A., Ton, J. Epigenetic processes in plant stress priming: Open questions and new approaches. *Curr. Opin. Plant Biol.*, 2023, 75, 102432. https://doi.org/10.1016/J.PBI.2023.102432.

Hefferon, K. L. Biofortification techniques to improve food security. *Sustainable Food Science—A Comprehensive Approach* 2023, 100–109. https://doi.org/10.1016/B978-0- 12-823960-5.00063-9.

Hoshikawa, K., Lin, Y. P., Schafleitner, R., Shirasawa, K., Isobe, S., Nguyen, D. C., Ohsawa, R., Yoshioka, Y. Genetic diversity analysis and core collection construction for amaranthus tricolor germplasm based on genome-wide single-nucleotide polymorphisms. *Sci. Hortic.*, 2023, 307, 111428. https://doi.org/10.1016/J.SCIENTA.2022.111428.

Huang, L., Zhong, S., Pan, C., Mo, H., Pan, Z., Huang, G., Mi, S., Gao, C., Liu, Y. Effects of acute salinity stress on physiology and immunoenzymatic activity in juvenile sea cucumber, *Stichopus monotuberculatus*. *Aquaculture*, 2023, 740094. https://doi.org/10.1016/J.AQUACULTURE.2023.740094.

Ibnouf, E. O. The effect of UV radiation on O-7 actinomycete in producing bioactive compounds in different growth conditions. *Saudi J. Biol. Sci.*, 2021, 28(8), 4619–4625. https://doi.org/10.1016/J.SJBS.2021.04.070.

Ishimoto, C., Waki, M., Soda, S. Adaptation of anammox granules in swine wastewater treatment to low temperatures at a full-scale simultaneous partial nitrification, anammox, and denitrification plant. *Chemosphere*, 2021, 282, 131027. https://doi.org/10.1016/J.CHEMOSPHERE.2021.131027.

Kenzo, T., Ichie, T., Norichika, Y., Kamiya, K., Inoue, Y., Min Ngo, K., Lum, S. K. Y. Drought tolerance in dipterocarp species improved through interspecific hybridization in a tropical rainforest. *Forest Ecol. Manag.*, 2023, 548, 121388. https://doi.org/10.1016/J.FORECO.2023.121388.

Kiran, U., Pandey, N. K. Transgenic food crops: Public acceptance and IPR. In: *Transgenic Technology Based Value Addition in Plant Biotechnology* (pp. 273–307), 2020. https://doi.org/10.1016/B978-0-12-818632-9.00012-5.

Kováčik, J. Basic physiology and biochemistry in environmental/experimental plant studies: How to quantify and interpret metabolites correctly. *Environ. Pollut.*, 2021, 286, 117192. https://doi.org/10.1016/J.ENVPOL.2021.117192.

Liu, S., Wang, H., Qin, F. Genetic dissection of drought resistance for trait improvement in crops. *Crop J.*, 2023, 11(4), 975–985. https://doi.org/10.1016/J.CJ.2023.05.002.

López-Bucio, J. S., Ravelo-Ortega, G., López-Bucio, J. Chromium in plant growth and development: Toxicity, tolerance and hormesis. *Environ. Pollut.*, 2022, 312, 120084. https://doi.org/10.1016/J.ENVPOL.2022.120084.

Maestro-Gaitán, I., Granado-Rodríguez, S., Poza-Viejo, L., Matías, J., Márquez-López, J. C., Pedroche, J. J., Cruz, V., Bolaños, L., Reguera, M. Quinoa plant architecture: A key factor determining plant productivity and seed quality under long-term drought. *Environ. Exp. Bot.*, 2023, 211, 105350. https://doi.org/10.1016/J.ENVEXPBOT.2023.105350.

Mitra, D., Mitra, D., Sabri Bensaad, M., Sinha, S., Pant, K., Pant, M., Priyadarshini, A., Singh, P., Dassamiour, S., Hambaba, L., Panneerselvam, P., Das Mohapatra, P. K. Evolution of bioinformatics and its impact on modern bio-science in the twenty-first century: Special attention to pharmacology, plant science and drug discovery. *Comput. Toxicol.*, 2022, 24, 100248. https://doi.org/10.1016/J.COMTOX.2022.100248.

Mohapatra, B., Chamoli, S., Salvi, P., Saxena, S. C. Fostering Nanoscience's strategies: A new frontier in sustainable crop improvement for abiotic stress tolerance. *Plant Nano Biol.*, 2023, 3, 100026. https://doi.org/10.1016/J.PLANA.2023.100026.

Mostek, A., Börner, A., Badowiec, A., Weidner, S. Alterations in root proteome of salt-sensitive and tolerant barley lines under salt stress conditions. *J. Plant Physiol.*, 2015, 174, 166–176. https://doi.org/10.1016/J.JPLPH.2014.08.020.

Munir, N., Gulzar, W., Abideen, Z., Hancock, J. T., El-Keblawy, A., Radicetti, E. Nanotechnology improves disease resistance in plants for food security: Applications and challenges. *Biocatal. Agric. Biotechnol.*, 2023, 51, 102781. https://doi.org/10.1016/J.BCAB.2023.102781.

Narayanan, M., Ma, Y. Mitigation of heavy metal stress in the soil through optimized interaction between plants and microbes. *J. Environ. Manage.*, 2023, 345, 118732. https://doi.org/10.1016/J.JENVMAN.2023.118732.

Ng, M. H., Nuzul Amri, I., Che Rahmat, C. M., Nu'man, A. H., Hasliyanti, A., Rohaya, M. H. Potential of nuclear magnetic resonance for the determination of organochlorine in edible oils. *J. Food Compos. Anal.*, 2023, 122, 105492. https://doi.org/10.1016/J.JFCA.2023.105492.

Olorunwa, O. J., Adhikari, B., Brazel, S., Bheemanahalli, R., Barickman, T. C., Reddy, K. R. Waterlogging stress reduces cowpea (*Vigna unguiculata* L.) genotypes growth, seed yield, and quality at different growth stages: Implications for developing tolerant cultivars under field conditions. *Agric. Water Manag.*, 2023, 284, 108336. https://doi.org/10.1016/J.AGWAT.2023.108336.

Osbourn, A. E. Plant secondary metabolites—A primary resource: Biochemistry of plant secondary metabolism and functions of plant secondary metabolites and their exploitation in biotechnology. *Trends Biotechnol.*, 2000, 18(7), 321–322. https://doi.org/10.1016/S0167-7799(00)01454-2.

Qiao, Y., Tie, J., Wang, X., Wei, B., Zhang, W., Liu, Z., Zhang, G., Lyu, J., Liao, W., Hu, L., Yu, J. Comprehensive evaluation on effect of planting and breeding waste composts on the yield, nutrient utilization, and soil environment of baby cabbage. *J. Environ. Manage.*, 2023, 341, 117941. https://doi.org/10.1016/J.JENVMAN.2023.117941.

Qin, L., Guo, S., Ai, W., Tang, Y., Cheng, Q., Chen, G. Effect of salt stress on growth and physiology in amaranth and lettuce: Implications for bioregenerative life support system. *Adv. Space Res.*, 2013, 51(3), 476–482. https://doi.org/10.1016/J.ASR.2012.09.025.

Riyazuddin, R., Singh, K., Iqbal, N., Labhane, N., Ramteke, P., Singh, V. P., Gupta, R. Unveiling the biosynthesis, mechanisms, and impacts of MiRNAs in drought stress resilience in plants. *Plant Physiol. Biochem.*, 2023, 202, 107978. https://doi.org/10.1016/J.PLAPHY.2023.107978.

Roumani, M., Duval, R. E., Ropars, A., Risler, A., Robin, C., Larbat, R. Phenolamides: Plant specialized metabolites with a wide range of promising pharmacological and health-promoting interests. *Biomed. Pharmacother.*, 2020, 131, 110762. https://doi.org/10.1016/J.BIOPHA.2020.110762.

Samtani, H., Sharma, A., Khurana, J. P., Khurana, P. Thermosensing in plants: Deciphering the mechanisms involved in heat sensing and their role in thermoresponse and thermotolerance. *Environ. Exp. Bot.*, 2022, 203, 105041. https://doi.org/10.1016/J.ENVEXPBOT.2022.105041.

Sens, A., Rischke, S., Hahnefeld, L., Dorochow, E., Schäfer, S. M. G., Thomas, D., Köhm, M., Geisslinger, G., Behrens, F., Gurke, R. Pre-analytical sample handling standardization for reliable measurement of metabolites and lipids in LC-MS-based clinical research. *J. Mass Spectrom. Adv. Clin. Lab.*, 2023, 28, 35–46. https://doi.org/10.1016/J.JMSACL.2023.02.002.

Setimela, P., Gasura, E., Thierfelder, C., Zaman-Allah, M., Cairns, J. E., Boddupalli, P. M. When the going gets tough: Performance of stress tolerant maize during the 2015/16 (El Niño) and 2016/17 (La Niña) season in Southern Africa. *Agric. Ecosyst. Environ.*, 2018, 268, 79–89. https://doi.org/10.1016/J.AGEE.2018.09.006.

Shah, S., Chen, C., Sun, Y., Wang, D., Nawaz, T., El-Kahtany, K., Fahad, S. Mechanisms of nitric oxide involvement in plant-microbe interaction and its enhancement of stress resistance. *Plant Stress*, 2023, 10, 100191. https://doi.org/10.1016/J.STRESS.2023.100191.

Sharma, G., Barney, J. N., Westwood, J. H., Haak, D. C. Into the weeds: New insights in plant stress. *Trends Plant Sci.*, 2021, 26(10), 1050–1060. https://doi.org/10.1016/J.TPLANTS.2021.06.003.

Smith, C. M. Conventional breeding of insect-resistant crop plants: Still the best way to feed the world population. *Curr. Opin. Insect. Sci.*, 2021, 45, 7–13. https://doi.org/10.1016/J.COIS.2020.11.008.

Sonkar, K. S., Achary, V. M. M., Sahoo, S., Reddy, M. K., Arockiasamy, A. Biochemical and structural characterization of a robust and thermostable ascorbate recycling monodehydroascorbate reductase (MDHAR) from stress adapted pearl millet. *Biochem. Biophys. Res. Commun.*, 2023, 662, 135–141. https://doi.org/10.1016/J.BBRC.2023.04.034.

Sousa, V. C., Zélé, F., Rodrigues, L. R., Godinho, D. P., Charlery de la Masselière, M., Magalhães, S. Rapid host-plant adaptation in the herbivorous spider mite tetranychus urticae occurs at low cost. *Curr. Opin. Insect. Sci.*, 2019, 36, 82–89. https://doi.org/10.1016/J.COIS.2019.08.006.

Su, Y., Wang, J., Gao, W., Wang, R., Yang, W., Zhang, H., Huang, L., Guo, L. Dynamic metabolites: A bridge between plants and microbes. *Sci. Total Environ.*, 2023, 899, 165612. https://doi.org/10.1016/J.SCITOTENV.2023.165612.

Suh, J. H., Han, S. B., Wang, Y. Development of an improved sample preparation platform for acidic endogenous hormones in plant tissues using electromembrane extraction. *J. Chromatogr. A*, 2018, 1535, 1–8. https://doi.org/10.1016/J.CHROMA.2017.12.068.

Sun, J., Xia, Y. Pretreating and normalizing metabolomics data for statistical analysis. *Genes Dis.*, 2023, 11(3), 100979. https://doi.org/10.1016/J.GENDIS.2023.04.018.

Tavanti, T. R., de Melo, A. A. R., Moreira, L. D. K., Sanchez, D. E. J., Silva, R. dos S., da Silva, R. M., dos Reis, A. R. Micronutrient fertilization enhances ROS scavenging system for alleviation of abiotic stresses in plants. *Plant Physiol. Biochem.*, 2021, 160, 386–396. https://doi.org/10.1016/J.PLAPHY.2021.01.040.

Tiziani, R., Pranter, M., Valentinuzzi, F., Pii, Y., Luigimaria, B., Cesco, S., Mimmo, T. Unraveling plant adaptation to single and combined nutrient deficiencies in a dicotyledonous and a monocotyledonous plant species. *Plant Sci.*, 2023, 335, 111793. https://doi.org/10.1016/J.PLANTSCI.2023.111793.

Unel, N. M., Baloglu, M. C., Altunoglu, Y. Ç. Comprehensive investigation of cucumber heat shock proteins under abiotic stress conditions: A multi-omics survey. *J. Biotechnol.*, 2023, 374, 49–69. https://doi.org/10.1016/J.JBIOTEC.2023.07.010.

Wang, X., Zhao, Y., Qi, X., Zhao, T., Wang, X., Ma, F., Zhang, L., Zhang, Q., Li, P. Quantitative analysis of metabolites in the aflatoxin biosynthesis pathway for early warning of aflatoxin contamination by UHPLC-HRMS combined with QAMS. *J. Hazard. Mater.*, 2022, 431, 128531. https://doi.org/10.1016/J.JHAZMAT.2022.128531.

Waschina, S., Seeger, K. Using In-tube extraction and slice selective NMR experiments allow imaging via statistical analysis of metabolic profiles. *Anal. Chim. Acta*, 2022, 1231, 340419. https://doi.org/10.1016/J.ACA.2022.340419.

Weale, A. Ethical arguments relevant to the use of GM crops. *N Biotechnol.*, 2010, 27(5), 582–587. https://doi.org/10.1016/J.NBT.2010.08.013.

Weil, C., Martienssen, R. Epigenetic interactions between transposons and genes: Lessons from plants. *Curr. Opin. Genet. Dev.*, 2008, 18(2), 188–192. https://doi.org/10.1016/J.GDE.2008.01.015.

Worku, M., De Groote, H., Munyua, B., Makumbi, D., Owino, F., Crossa, J., Beyene, Y., Mugo, S., Jumbo, M., Asea, G., Mutinda, C., Kwemoi, D. B., Woyengo, V., Olsen, M., Prasanna, B. M. On-farm performance and farmers' participatory assessment of new stress-tolerant maize hybrids in Eastern Africa. *Field Crops Res.*, 2020, 246, 107693. https://doi.org/10.1016/J.FCR.2019.107693.

Xiong, C. F., Bai, Y. L., Yin, X. M., Ye, T. T., Feng, Y. Q. Use of chemical labeling-assisted liquid chromatography-mass spectrometry for discovering derivatives of brassinosteroids. *J. Chromatogr. A*, 2022, 1685, 463639. https://doi.org/10.1016/J.CHROMA.2022.463639.

Yu, J., Tu, X., Huang, A. C. Functions and biosynthesis of plant signaling metabolites mediating plant–Microbe interactions. *Nat. Prod. Rep.*, 2022, 39(7), 1393–1422. https://doi.org/10.1039/D2NP00010E.

Zenda, T., Wang, N., Yan, X., Dong, A., Yang, Q., Zhong, Y., Duan, H. Opportunities and avenues for achieving crop climate resilience. *Environ. Exp. Bot.*, 2023, 213, 105414. https://doi.org/10.1016/J.ENVEXPBOT.2023.105414.

Zhang, L., Zhang, H., Zhu, Q., Niu, Y. Further investigating the performance of crop water stress index for maize from baseline fluctuation, effects of environmental factors, and variation of critical value. *Agric. Water Manag.*, 2023, 285, 108349. https://doi.org/10.1016/J.AGWAT.2023.108349.

Zhu, Q., Wang, B., Tan, J., Liu, T., Li, L., Liu, Y. G. Plant synthetic metabolic engineering for enhancing crop nutritional quality. *Plant Commun.*, 2020, 1(1), 100017. https://doi.org/10.1016/J.XPLC.2019.100017.

WEBLINKS

1. https://readthedirt.org/press/wp-content/uploads/2011/07/Seed-Banks.jpg
2. https://i2.wp.com/sitn.hms.harvard.edu/wpcontent/uploads/2014/01/extremes_climatechange_1ogo.png?fit=1305%2C1316
3. https://files.growery.org/files/g17-19/482073160-Last_weeks_introductions.jpg
4. https://cdn.pixabay.com/photo/2018/07/15/10/44/dna-3539309_960_720.jpg
5. https://wwwdotgenomedaddotcom.files.wordpress.com/2021/04/dna-header-1200x627-1.png
6. https://p0.pxfuel.com/preview/697/898/445/cloud-computing-illustration-technology.jpg

CHAPTER 7

Insight Into Metabolomics for Heavy Metal Stress Response and Tolerance in Plants†

NILOFER SHEIKH[1], UTSAB SINGHA[2], NAMITA DEVI[1], DHARMESWAR BARHOI[2], and SUNAYANA GOSWAMI[1*]

[1]*Biswanath College, Biswanath Chariali, Assam, India*

[2]*Department of Zoology, The Assam Royal Global University, Guwahati, Assam, India*

*Corresponding authors

ABSTRACT

Recent years have witnessed increased anthropogenic activities, rapid industrialization which has resulted in increased heavy-metal contamination in the environment which affect food quality, production, and security worldwide. Plants growth and yield are affected by the exposure of heavy metals. Intricate defense mechanisms have been developed by plants to protect them from the toxic effects of heavy metals. These defense mechanisms include compartmentalization and sequestration in cell organelles, complex formation is inactivated by organic ligands, exclusion with transporters, transcription factors etc. Omics technologies are cynical with feasible approaches for characterizing these roles which may then be employed to increase plant systems' resilience or their ability to withstand stress. The study of metabolomics has gain significant attention among all omics technologies due to technological advances in analytical biochemistry. The comprehensive analysis of the metabolic aspects will provide an overview of the metabolic responses of plants and strategies applied to mitigate the deleterious effects of stress conditions at a biochemical level. The current review summarizes the present status, challenges, and future prospects of metabolomics and related technological advancement in elucidating plant responses and various tolerance strategies to heavy-metal toxicity.

7.1 INTRODUCTION

Pollutants like heavy metals (HMs) are introduced into soil and water by a variety of unregulated human and natural activities, which has a negative impact on the health of living things. Several environmental stresses are experienced by plants during their lifecycle

†All authors of this chapter have contributed equally.

Exploring Plant Stress Tolerance: A Metabolomics Approach.
Shuvasish Choudhury, Debojyoti Moulick, and Mirza Hasanuzzaman (Eds.)

DOI: 10.1201/9781779641731-7

because of their sessile nature. HM stress has gain significant attention among all other stresses (Hussain et al., 2021; Riyazuddin et al., 2021; Rizwan et al., 2016). The group of HMs generally includes transition elements, metalloids, lanthanides and actinides, which are harmful even at least concentrations (Jamla et al., 2021). Since the plants are immobile, they cannot shift their position, as a result of which they are readily exposed to the HMs in the soil. About 53 elements are classified as HMs among all naturally existing elements, and a majority of them are not involved in any metabolic reaction in plants (Rahale et al., 2021). HMs are divided into essential and nonessential categories based on their biological relevance and possible impact on plant growth and yield (Rasheed et al., 2021). For instance, it has been revealed that a number of HMs such as nickel (Ni), manganese (Mn), cobalt (Co), and copper (Cu), play a critical role in a number of metabolic events in plants and, when present at an ideal concentration, have positive impacts on crop growth (Rahman et al., 2021). While it is possible to get the correct concentration of these vital nutrients for sustained plant development in the right climatic conditions, their intense bioavailability mostly causes oxidative bursts that cause various physiological and biochemical changes in plants. Nonetheless, plants have a tendency to scavenge the free radicals generated by antioxidants, which maintain the balance of reactive oxygen species (ROS). When grown in a HM toxic environment, plants produce a variety of responses, including decreased biomass, photosynthetic pigments, photosynthetic efficiency, water use efficiency, mineral nutrition imbalances, chromosomal aberrations, inhibition of cell divisions, chlorosis, and necrosis. However, few plants act as hyperaccumulators of the HMs and help in phytoremediation.

Under a variety of abiotic stress situations, plants experience metabolic disturbances. Because of their sessile nature, their metabolic networks constantly need to reconfigure themselves in response to environmental stimuli in order to preserve homeostasis and combat stress. Thus, a comprehensive analysis of these metabolic features will provide an overview to mitigate the deleterious effects of stress at a biochemical level. Recent years have seen a substantial increase in interest in metabolomics investigations because of the expanding technological advancements in analytical biochemistry (plant metabolomics). Sophisticated, advanced analytical techniques are needed for plant biochemical landscaping. Thus, some technological advancements in the field of metabolomics are necessary. Metabolomics offers a strong basis for a better understanding of plants' ability to modify their primary metabolic pathway and react to changing environmental conditions. The application of metabolomics in ecotoxicology and ecophysiology aids in the identification of metabolic pathways linked to stress responses, the assessment of possible hazards posed by the contaminants, and their mechanisms of action. One of the biggest environmental risks to plants' physiology and biochemical activities is HM stress. Hence, metabolomics technologies have been widely used in recent years to massively characterize plant molecular structure at different phases in order to comprehend the numerous components of cellular functioning systems that respond to HM stress and outline the application and advances of metabolomics in better understanding about how plants react to abiotic stress, with a particular emphasis on HM toxicity. In light of this, the current review highlights the latest developments in metabolomics in plants exposed to HM stresses and explores the potential for enabling efficient metabolite management techniques to mitigate the negative effects of HM pollution on plant development and productivity.

7.2 HEAVY METAL TOXICITY IN PLANTS

Among the naturally occurring HMs not all are biologically important. For instance, HMs like iron (Fe), molybdenum (Mo), and Mn can be utilized by plants as micronutrients while some others such as zinc (Zn), vanadium (V), Co, chromium (Cr), Cu, Ni, and so on, are significant as trace elements. HMs that confer toxicity to plants and microorganisms and have no noteworthy role as nutrients include uranium (U), antimony (Sb), lead (Pb), cadmium (Cd), silver (Ag), mercury (IIg), and arsenic (As). Among them, As, Pb, Cd, Ni, and Hg are ubiquitous in nature and known to affect vegetable growth is initiated at low concentrations but restricts growth at higher concentrations (Clemens and Ma, 2016). Table 7.1 illustrates some of the sources of toxic HMs and metalloids input in the soil due to agricultural practices.

TABLE 7.1 Sources of Heavy Metals in Agricultural Soil Owing to Various Anthropogenic Activities

Sources of Heavy Metal	Heavy Metal/Metalloids Input in Soil	References
Smelting, mining, and burning of fossil fuels	Copper (Cu), zinc (Zn), arsenic (As), lead (Pb), mercury (Hg), nickel (Ni), and cadmium (Cd), etc.	He et al. (2005), Wang et al. (2011), Wuana and Okieimen (2011), Li et al. (2014), Ali et al. (2019), Kohzadi et al. (2019), Tang et al. (2019), Afzaal et al. (2022), Adnan et al. (2024)
Use of chemical fertilizers	Arsenic (As), lead (Pb), cadmium (Cd), mercury (Hg), chromium (Cr), cobalt (Co), iron (Fe), manganese (Mn), etc.	Gimeno-García et al. (1996), Wuana and Okieimen (2011), Mar et al. (2012), Li et al. (2014), Wei et al. (2020), Zhuang et al. (2020), Khatun et al. (2022), Liu et al. (2023), Perčin et al. (2023), Orellana-Mendoza et al. (2024)
Use of agricultural pesticides	Arsenic (As), cobalt (Co), cadmium (Cd), mercury (Hg), chromium (Cr), nickel (Ni), lead (Pb), copper (Cu), etc.	Singh and Walker (2006), Arao et al. (2010), Rao et al. (2011), Lewis et al. (2016), Defarge et al. (2018), Abdallah Alnuwaiser (2019), Tonelli and Tonelli (2020), Alengebawy et al. (2021), Rey-Martínez et al. (2022), Adil et al. (2023)
Use of biosolids and manures	Zinc (Zn), arsenic (As), copper (Cu), lead (Pb), nickel (Ni), cadmium (Cd), chromium (Cr), etc.	Benitez et al. (2001), Sharma et al. (2017), Urra et al. (2019), Taghipour and Jalali (2019), Gianico et al. (2021), Poornima et al. (2022), Marchuk et al. (2023)
Wastewater disposal: Industrial Sewage Domestic	Copper (Cu), cadmium (Cd), mercury (Hg), nickel (Ni), chromium (Cr), lead (Pb), etc.	Sörme and Lagerkvist (2002), Zwolak et al. (2019), Geng et al. (2020), Elgarahy et al. (2021), Nunes et al. (2021), Rafique et al. (2022)
Through atmospheric deposition	Mercury (Hg), cadmium (Cd), lead (Pb), etc.	Pacyna et al. (2009), Suvarapu and Baek (2017), Cooke et al. (2020), Vithanage et al. (2022), Kumar et al. (2023)

HMs deposition and their sources are countless, starting from fertilizer formulation, pesticide formulation, combustion of fossil fuels, various technological processes, road construction, and many more that can cause significant threats, especially to the soil environment (Zwolak et al., 2019). Below are described some abundant sources of HMs and their subsequent hazards to agricultural field crops.

7.2.1 CADMIUM (Cd) AS HEAVY METALS

The HM, cadmium (Cd), is one of the most bio-persistent and toxic HMs can cause serious damage and inhibit the growth and development of agricultural field crops. Though Cd enters the environment through various natural processes, anthropogenic activities like the metal industry, traffic, landfills, mining, combustion emissions, and sewage sludges, and so on, can augment Cd pollution furthermore (Bigalke et al., 2017). Extensive use of synthetic phosphate-based fertilizers on field crops plays a prominent role in the accumulation of HMs like Cd that strongly alter soil chemistry, thereby influencing soil microbiota as well as agricultural crops therein (Haider et al., 2021). Traces of Cd are also used as an impurity in the production of petroleum and its products (Robbins and Walker, 1975). Major contributors of Cd pollution are considered to be the Cd–Ni-based batteries used in electronic gadgets like cell phones, photovoltaic cells of solar panels, metal smelting, synthetic pigment production, metal coatings, stabilizers in plastic products, and so on, that discharge large quantities of Cd each year globally (Järup et al., 1998; Bhattacharyya et al., 2000; Wuana and Okieimen, 2011). Other modes of Cd contamination come from leachate of municipal sewages, farmyard manure, discarded mine deposits, discarded parts of automobiles, and many more (Khan et al., 2017).

Moreover, Cd is mostly available in the soil in inorganic form and in one oxidation state (+2). Hence, bioavailability of Cd largely depends on the soil condition and composition, as it is more pronounced in aerobic soils than in anaerobic soils because of the formation of insoluble cadmium sulfide (CdS) in the latter (Clemens and Ma, 2016).

7.2.2 ARSENIC (As) AS HEAVY METALS/METALLOIDS

Arsenic (As), being the 20th abundant naturally occurring environmental contaminant on Earth's crust, has the ability to shape the biodiversity, health, and structure in the affected areas. Therefore, understanding the underlying effects of As, a potentially toxic metalloid released into the soil as a result of various anthropogenic processes, is of paramount importance in view of stress-related issues and subsequent tolerance among plants.

The ingression of As into the environment, especially into the soil and water bodies, largely depends on several natural as well as manmade alterations. Though As can find its way into the soil and aquatic environment through various geogenic processes like volcanic eruptions, rock weathering, and geothermal/hydrothermal processes; many other anthropogenic sources such as wood preservatives, mining, coal burning, smelting, excessive use of As-based fertilizers and pesticides, and extensive irrigation of ground water in agricultural fields can strengthen their concentrations many folds (Webster and Nordstrom, 2003; Basu et al., 2014; Bhat et al., 2021; Sharma et al., 2020; Shabbir et al., 2021).

Among the various forms of As, two inorganic forms, namely arsenite and arsenate, are taken up by the plant roots. Inorganic arsenite (+3) is the most toxic form; in addition, arsenate gets converted into arsenite once inside the cell, thus acting as a double-edged sword (Finnegan and Chen, 2012). Arsenic also exists in nature as methylated organic metabolites such as trimethylarsine oxide, monomethylarsonic acid, and dimethylarsinic acid. Arsenic in soils has been reported to be found at a wide concentration range from a few µg/kg to extremely high concentrations like 250,000 mg/kg (Mikutta et al., 2014). Arsenic rarely occurs as a free element, rather compounded with other minerals like sulfides, oxides, arsenates, arsenites, and arsenides. Moreover, they also associate with other HMs like Pb, Ni, Zn, Hg, Ag, Co, and so on; thus forming various arsenical compounds such as scorodite ($FeAsO_4{\cdot}2H_2O$), sulfide mineral arsenopyrite (FeAsS), realgar (AsS), oxide mineral diarsenic pentoxide (As_2O_5), claudetite (As_2O_3), arsenolite (As_4O_6), and orpiment (As_2O_3). Among them, arsenopyrite is the most common mineral (Sarkar and Paul, 2016).

7.2.3 LEAD (Pb) AS HEAVY METALS

Lead (Pb) is the 37th most abundant HMs on Earth's crust and considered as one of the most hazardous pollutants (after arsenic) by the Agency of Toxic Substance and Disease Registry (Pourrut et al., 2011). Both natural and anthropogenic sources of Pb contamination are very common and widespread across the globe. The natural sources may include weathering of rocks, volcanic activities, occasional forest fire, and so on, that contribute a considerably low level of Pb into the environment (Gadd, 2010). On the other hand, mainly the anthropogenic sources like mining, smelting of metallic ores, industrial applications, vehicle exhaust, and so on, may introduce Pb pollution several folds higher than the environmental relevant concentrations (3–94 µg/L in aquatic environment; 15–40 mg/kg on Earth's crust); evidently detrimental to biotic community, especially the plants (Wuana and Okieimen, 2011; Kataba et al., 2020; Fahr et al., 2013; Amari et al., 2017). Figure 7.1 schematically represents the various anthropogenic sources of Pb contamination that damage both the terrestrial as well as aquatic ecosystem health structures.

The distribution and accumulation of Pb in agricultural field crops varies widely and largely depends on the nearby point and nonpoint sources (Chen et al., 2018). Furthermore, the rate and amount of Pb absorption and bioaccumulation also depend on the soil texture, seasonal variations, and species of the plant concerned (Sharma and Dubey, 2005; Bidar et al., 2009).

7.2.4 MERCURY (Hg) AS HMS

The role of mercury (Hg) as HMs is of great concern as they are omnipresent in the Earth's biogeochemical cycle, contributed by both natural as well as anthropogenic sources of release. The primary sources of Hg flux in the environment are contributed mostly by the volcanic eruption and related geological processes (Gustin et al., 2008); though other sources like natural wildfires, weathering of Hg-containing rocks, meteorological conditions, and so on may contribute significantly to the global pool of HM pollution in the environment over a period of time (Ferrara et al., 1998; Pirrone et al., 2001; Obrist et al., 2018). Various man-made activities, including fossil fuel combustion, mining and mine waste deposition, leather tannery, disposal of Hg-based batteries, Hg-coated energy-efficient lamps, as well as other industrial activities compounded with indiscriminate use of Hg in several other sectors, have amplified the level of Hg pollution many fold higher (Xu et al., 2015; Bueno et al., 2009; Zhu et al., 2018; Tasca and Puccini, 2019). Figure 7.2 summarizes some of the major sectors that can power the anthropogenic Hg contamination in the Indian subcontinent.

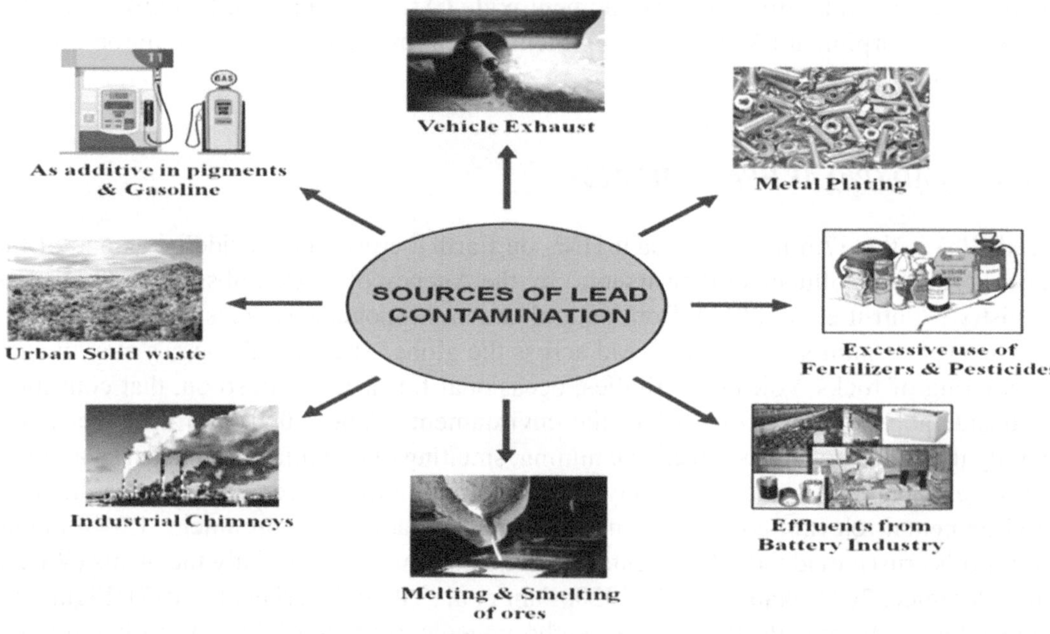

FIGURE 7.1 Major anthropogenic sources of lead contamination in the environment.

In nature, Hg exists in various forms, namely, elemental mercury (Hg^0), inorganic mercurous (Hg^{1+}), and mercuric (Hg^{2+}), as well as organic forms like methyl mercury (CH_3Hg) (Zahir et al., 2005; Clarkson and Magos, 2006). All these can interchange from one form to another through a complex web of reactions between soil, air, and water, mostly carried out by ambient temperature, pressure, and soil microbiota present therein (Beckers and Rinklebe, 2017). Figure 7.3 depicts the various biochemical events that occur in the biogeochemical cycling of Hg.

In soil, Hg is generally immobile due to high binding affinity with various organic substrates and sulfur ligands present in the soil. Hence, elevated levels of Hg can be found in the soil having high organic content. However, Hg gets released from terrestrial surfaces to the atmosphere under high temperature and pressure (Neculita et al., 2005).

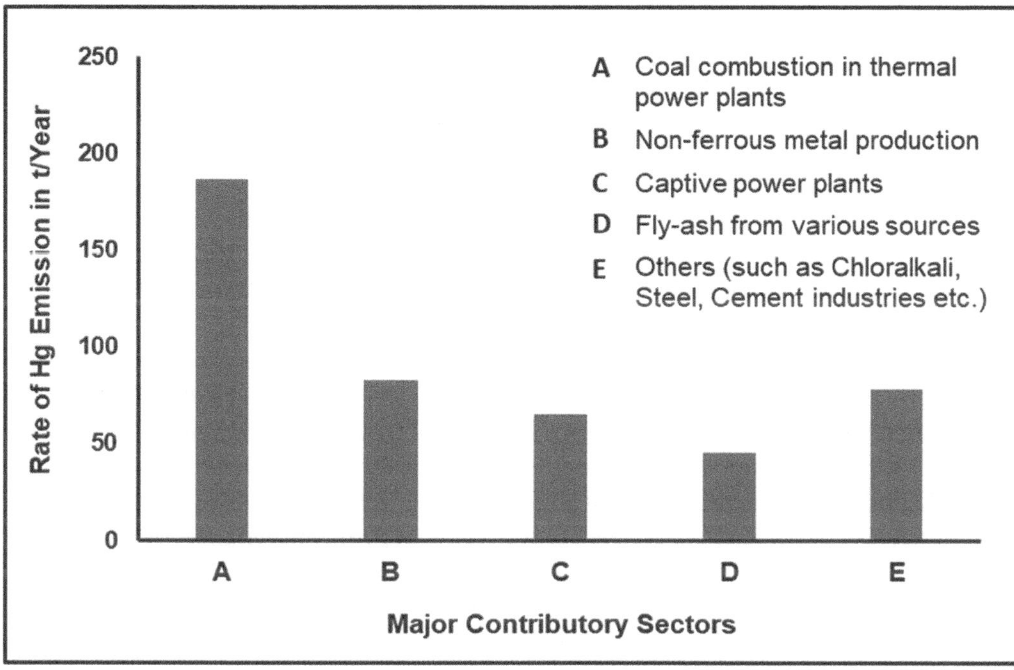

FIGURE 7.2 Major contributory sectors of mercury emissions in India, till 2019.
Source: Adapted from Sahu et al., 2024

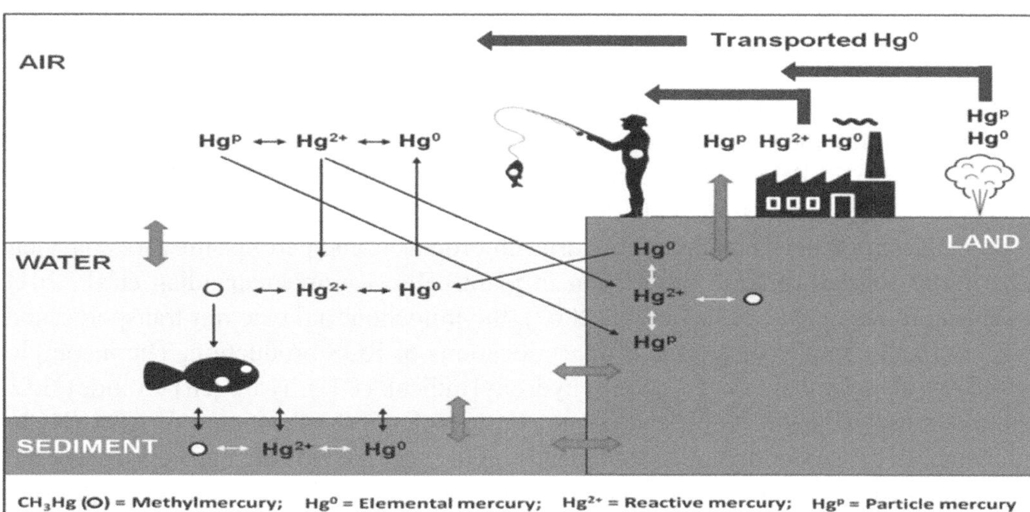

FIGURE 7.3 Biogeochemical cycling of mercury (Hg) between air, water, and soil.

7.3 HEAVY METAL AND ROS PRODUCTION

HMs (Zn, Pb, Hg, Cr, Cu, Cd, As, etc.) are one of the contaminants found in wastewater. Generally, elements having an atomic number greater than 20 are classified as HMs

(Lambert et al., 2000; Lasat, 1999). It is released to the environment both by geogenic processes like leaching and weathering and anthropogenic resources like mining, smelting, pesticides, and so on (Kumar and Trivedi, 2016). When found in low amounts, these metals serve as vital for maintaining an array of biological and physiological functions in organisms; however, once these concentrations exceed a certain threshold, they become toxic to plants, animals, and humans (Jaishankar et al., 2014). Plant growth is also restricted by pollutants containing HMs. Higher concentrations of metals in the soil have an impact on the physiological, biochemical, genetic, and seed germination processes, as well as plant growth and productivity (Bharti and Sharma, 2022).

HMs can be separated into redox-active and redox-inactive. The redox-active HMs through the Haber–Weiss and Fenton reactions directly contribute to the redox reaction in cells, which forms $O_2^{\bullet-}$ and then leads to the production of H_2O_2 and •OH. Plants exposed to redox-inactive HMs can also experience indirect mechanisms of oxidative stress, such as disruption of the electron transport chain, interaction with the antioxidant defense system, or induction of lipid peroxidation (Prasad et al., 1999; Schutzendubel and Polle, 2002).

Although essential HMs are needed for plant growth and maintenance, excessive concentrations of them can harm the plant by inhibiting the cytoplasmic enzymes and by damaging the cell structure caused by oxidative stress (Assche and Clijsters, 1990). ROS production is one of the effects observed when plants are exposed to the toxic concentration of HMs (Tamás et al., 2017). ROS are produced in plants both naturally as well as by the impact of HM. ROS are produced more frequently in plants when modifications are made to their proteins, nucleic acids, cellular structures, or enzymes. Plants have ROS scavenging pathways that include both enzymatic and nonenzymatic mechanisms that regulate ROS production (Mansoor et al., 2022). Thus, higher plants are induced to produce ROS under unfavorable conditions in order to support a wide range of physiological variations. Drought, salinity, HM toxicity, and other factors cause the breakdown of antioxidants through the expression of genes related to the antioxidative response causes the generation of ROS. Moreover, ROS have the ability to harm macromolecules like DNA and lipids (lipid peroxidation) (Mansoor et al., 2023).

Some plant organelles like mitochondria, chloroplasts, and peroxisomes are considered to be major sources of ROS production in plant cells (Karuppanapandian et al., 2011). According to Barranco-Medina et al. (2007), the mitochondrial electron transport chain's complex I and complex II are the primary locations of ROS production. The incomplete breakdown of molecular oxygen, such as hydroxyl radicals (OH), hydrogen peroxide (H_2O_2), superoxide radical anion ($O_2^{\bullet-}$), and ozone (O_3), results in the formation of ROS (Nathan and Ding, 2010). Production of ROS depends on the specific HM element. Excessive ROS are formed inside the cell as a result of HM toxicity; Cu can directly produce ROS, while Cd is a redox-inactive HM and can only produce ROS indirectly through the inactivation of enzymes, the induction of lipoxygenase gene expression in plant tissues, and the oxidation of polyunsaturated fatty acids (Cho and Seo, 2005). ROS are also produced when HMs interact with various components, mainly in mitochondria and chloroplasts, altering the electron transport chain's activity (Suzuki et al., 2012). An imbalance in the membrane potential brought on by a rise in ROS levels encourages ion channel leakage, lipid peroxidation, and macromolecule destruction. Additionally, the length of the stress, the developmental stage, the dose and concentration of the particular HM, and the plant organ in question can all affect how toxic HMs are to subcellular organelles that release ROS (Mansoor et al., 2023).

By directly binding and consuming glutathione (GSH) and its derivatives, which are necessary to scavenge any ROS generated, HMs also increase the production of ROS (Lee et al., 2003). Furthermore, oxidative stress caused by HMs is linked to plasma-membrane-bound NADPH oxidase (Potocký et al., 2012). Cytosolic NADPH can be used by plasma membrane-bound NADPH oxidases to produce $O_2^{\bullet-}$, which superoxide dismutase (SOD) swiftly converts to H_2O_2 (Pourrut et al., 2008). Outside the plasma membrane, where the pH is often lower than inside the cell, is where the ROS produced by NADPH oxidase are found (Sagi and Fluhr, 2006). Furthermore, it has been reported that Ca^{2+} and protein kinases lead to the formation of ROS induced by HMs by activating NADPH oxidase. NADPH oxidase-induced production of ROS by HMs was observed in Cd-treated *Citrus sativum*, Ni-managed *Durum triticum*, and Pb-treated *Vicia faba* (Shahid et al., 2014).

7.4 PLANT RESPONSE TO HEAVY METAL TOXICITY

Plants are continuously faced with various environmental constraints due to the discharge of hazardous chemicals, including HMs, into the environment. Hence, due to the high concentration of HMs in soil, plants show potential toxic effects on growth and development in plants. Plants adapt itself at the anatomical, physiological, biochemical, and molecular levels under HM stress (Figure 7.4) to combat the toxic effects of HM under stress.

7.4.1 ANATOMICAL RESPONSE

When applied separately or in combination, Cu or Cd significantly decreased the diameter of the root, the breadth and thickness of the leaf midrib, the diameter of the xylem vessels of all the seedling organs, the parenchyma cell area in the stem, the leaf midrib, the pith, and the cortex of the root, the size of the stem vascular bundles, the number of xylem arms in the root, and the frequency of stomata on the abaxial leaf surface (Kasim, 2006). Under HM stress, it was also reported that the number of xylem tissues increased due to a higher rate of transpiration and thicker cell wall, the number of cells in the cortical region in the roots decreased as a result of HM accumulation in the cell wall of the roots and the vacuoles of the cortical cells (Rauser and Ackerley, 1987). Actin is present in plant cells in three different forms that can change mutually. These include a pool of monomeric (unpolymerized) actin, a thin near-membrane network formed by polymeric actin, and tight bundles of longitudinal filaments connected to one another into a single network with the cell membrane. Actin filament organization and the ratio of polymeric to monomeric forms of actin are frequently altered in conjunction with alterations in cell metabolism. It was demonstrated that aluminum (Al) encouraged the production of tight bundles of filaments of actin in the meristematic cells of the *Triticum turgidum* root, while

Cd caused the disarray and microfilaments and microtubules destruction in the green alga *Spirogyra*. Additionally, it was shown that the tobacco mutants' resistance to Al was linked to a significant rise in actin content overall, particularly in the percentage of tightly wound actin filament bundles in leaf cells. The role of heat shock proteins (HSPs), or their counterparts, as molecular chaperons in maintaining the native structure of proteins and preventing denaturation and aggregation is widely recognized. When plants are exposed to HMs, the interaction between actin and HSPs crucial for maintaining the organization of their cytoskeleton and ensuring their survival. Rape plants treated for a short period of time with 50 μM $CuSO_4$ showed significant increases in polymeric actin content in root cells, as well as an increase in the quantity of HSP70 and HSP60 interacting with actin filaments. According to the theory, HSP70 and HSP60 serve as molecular chaperons that shield the actin cytoskeleton from harm when they interact with the polymeric actin when exposed to HMs (Kulikova et al., 2009).

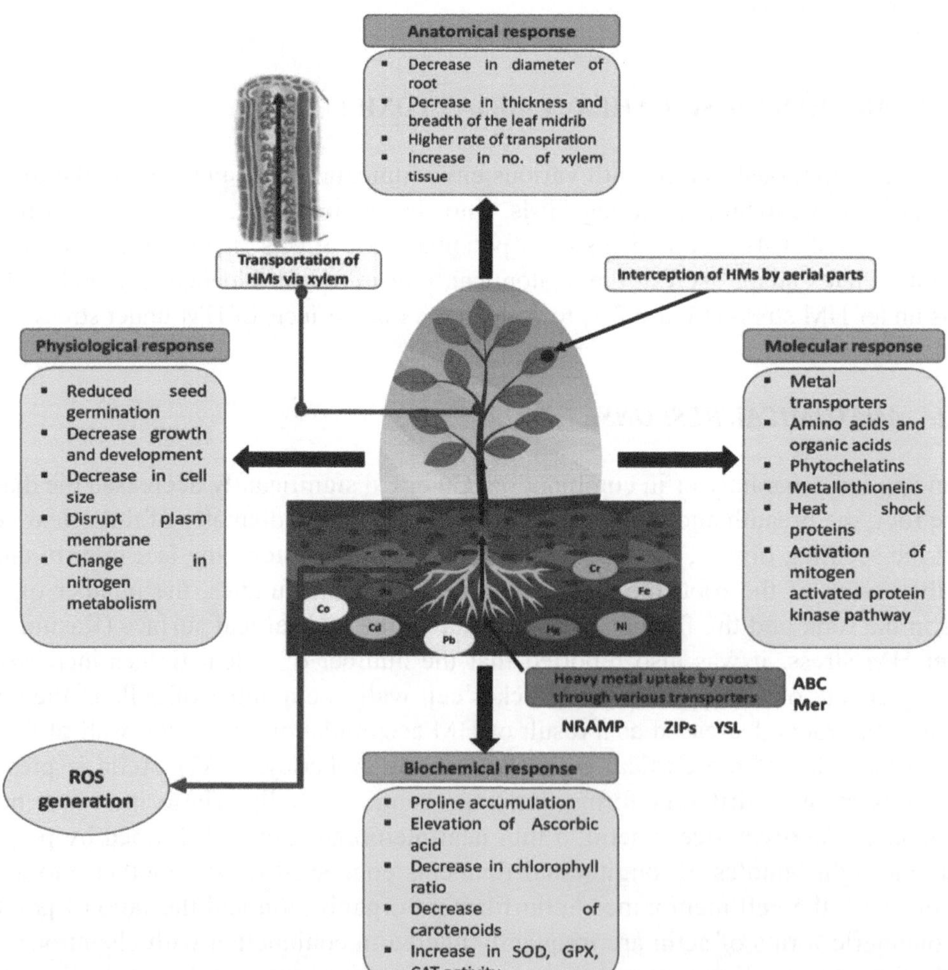

FIGURE 7.4 Schematic diagram of plant responses to various heavy metal toxicities.

7.4.2 PHYSIOLOGICAL RESPONSE

Many physiological reactions in plants are activated and altered by HM toxicity. For instance, a number of HMs, including Cr, Cu, Cd, As, Hg, and Co, have been shown to decrease seed germination and plant development as well as cause a variety of morphological changes in plants. Furthermore, Pb inhibited the growth and development of *Oryza sativa* and *Brassica juncea* L. (Bassegio et al., 2020). Similarly, roots and shoots of fresh and dry biomass of *Triticum aestivum* were deceased simultaneously with the increase of Cd concentrations from 50 to 200 μM. Marrugo-Negrete et al. (2016) observed that an increased amount of Hg in nutrient solution from 5 to 80 μg/mL reduced the root and shoot biomass of *Jatropha curcas*. According to Ashraf et al. (2011), Ni toxicity has reduced the amount of protease and α-amylase in sunflower seeds, which has lowered seed germination. Furthermore, when Mn concentration increased from 2 to 1000 μM, detrimental effects of Mn toxicity were seen in a growing trend on the biomass of roots and shoots, chlorophyll biosynthesis, photochemical ability of PSII, and stomatal conductance of *Vaccinium corymbosum* L. (Millaleo et al., 2020). *Brassica napus* root fresh and dry biomass, plant height, root length, stomatal conductance, water use efficiency, and net photosynthesis were all significantly decreased by Cr toxicity in soils (Zaheer et al., 2020). Roots are the main tissues that interact with HM toxicity in soils; prior research has shown a significant decrease in root elongation, diameter, and root hair formation, as well as an improvement in root dieback (Rahman et al., 2021; Rucińska-Sobkowiak, 2016). Roots of *Lolium perenne* and *Festuca arundinacea* were reported to harden their cell walls and hinder their elongation by avoiding the penetration of Cu (Zhao et al., 2010). Moreover, HMs that have penetrated root cells are sequestered into vacuoles by the cells in order to prevent interference with the metabolic processes of organelles. For instance, the involvement of root HMs relieving and avoiding was highlighted by Cr deposition and compartmentation into the vacuole and cell wall of *Arabidopsis thaliana* (Eleftheriou et al., 2015; Yasir et al., 2021). It has been shown that the cell plasma membrane controls, shields, and neutralizes undesirable substances from entering the cell organelles and mitigates the negative effects of various toxicities. Nevertheless, it has been shown that HMs like Cd can damage the structure and function of the plasma membrane, which can result in cell death (Janicka-Russak et al., 2012). When some undesirable changes occur, such as when the supply of water and nutrients is cut off, these negative consequences might intensify. The plant does not exhibit any of the symptoms associated with Cd stress, such as reduced nutrient intake (Ghnaya et al., 2007), altered water balance (Wang et al., 2008), suppression of stomatal opening, or alterations in nitrogen metabolism. Furthermore, by altering the composition of lipids and some membrane-associated enzymes, including H^+ ATPase (Djebali et al., 2005), Cd^{2+} ions may change the permeability of membranes (Fodor et al., 1995). Additionally, net photosynthesis is susceptible to Cd and has a direct impact on the manufacture of chlorophyll (Ekmekçi et al., 2008; Li et al., 2008) and the healthy growth of chloroplasts (Jin et al., 2008). Significant decreases in both the quantity and diameter of xylem vessels were also brought about by Cd. Cd reduced the size of the cells in *Phaseolus vulgaris* (Barceló et al., 1986); nevertheless, it increased the diameters of the mesophyll cells in *B. napus* and *Pisum sativum* (Baryla et al., 2001). It was discovered that the aquatic plant *Elodea canadensis* submerged leaves exposed to Cd inhibited cell division but only significantly enlarged one of the two cell layers that make up the leaf blade (Dalla Vecchia et al., 2005). This implies that rather than being caused by a disruption in water balance, the alterations in leaf cell enlargement brought on by Cd might be the result of particular morphogenetic effects (Kovačević et al., 1999). Under metal stress circumstances, *P. sativum* further demonstrated reduced photosynthesis as a result of lower PSII activity (Chugh and Sawhney, 1999). The leaves with high cell density have less photosynthetic reaction, which affects the intensity of gas exchange.

7.4.3 BIOCHEMICAL RESPONSE

Many transporters, such as cation diffusion facilitator, HM-transporting ATPases, natural resistance-associated macrophage proteins, multidrug and toxin efflux, and zinc–iron permease proteins, allow HMs to enter plant cells and activate several biochemical responses by replacing/displacing essential ions or stopping multiple functional groups. They were found to compete with vital nutrients at metabolic catalyzing enzymes' active binding sites, altering their structure and ultimately impairing the enzymes' ability to operate (Ghori et al., 2019). Plants possess intricate molecular and cellular mechanisms for scavenging ROS. These processes reduce oxidative damage to cells and boost tolerance to HMs by inhibiting or slowing the oxidative chain reaction and oxidation of biomolecules (Michalak, 2006). Monodehydroascorbate reductase, dehydroascorbate reductase, glutathione reductase (GR), catalase (CAT), peroxidase (POX), and SOD are examples of antioxidant enzymes found in plants. Low-molecular weight quenchers include cysteine, ascorbic acid, thiols, proline (Singh and Sinha, 2005), nitrogen compounds, GSH, carotenoids, phenolic, and α-tocopherol (Michalak, 2006). The biochemical response of *Vigna radiata* cultivated in soil containing 50% Cr was studied by Rath et al. (2019), and the findings showed a notable decrease in the levels of starch, amino acids (AAs), and chlorophyll. Due to biochemical stress stimuli, several ornamental plants exposed to Cd toxicity demonstrated a notable reduction in their overall photosynthesis and root elongation; concurrently, however, the levels of soluble sugar, membrane stability index, proline concentration of leaf, osmotic adjustment, metal–proline complexes and membrane damage increased (Song et al., 2020). The overproduction of MDA (malondialdehyde), EL (electrolyte leakage), and H_2O_2 caused by the Cu-mediated biochemical response in *Hibiscus cannabinus* caused damaging effects in roots that led to shoot tissues. However, the harmful effects were countered by the upregulation of CAT, SOD, POD, GSH, and GPX, like antioxidants are antagonists with metal contents in nutrient solution (Saleem et al., 2020). According to research done on *Sesuvium portulacastrum* exposed to increased concentrations of Cu, plants boosted antioxidants that fought against the overproduction of MDA and ROS (Aljarba et al., 2021; Lokhande et al., 2020). Typically, these antioxidants control, balance, and restore the proper amount of ROS (Hasanuzzaman et al., 2020; Ur Rahman et al., 2021b). Helaoui et al. (2020) investigated the Ni-mediated biochemical reaction, and the findings showed that higher concentrations of POD (peroxidase) and GST (glutathione S-transferase) were necessary to reverse ROS-mediated damage. The results on antioxidants corroborated the upregulation of the *Prx1C* gene family, suggesting that plants have a better biochemical reaction as an effective defense strategy. Moreover, HM toxicities have a major impact on carbon fixation, distribution, and assimilation. For instance, in *T. aestivum*, toxicities from Cd and Zn impeded CO_2 assimilation (Paunov et al., 2018). According to Son et al. (2014), HM toxicity decreased the expression of Rubisco, a crucial CBB cycle enzyme, and controlled CO_2 fixation by catalyzing the carboxylation of RuBP. This may have been caused by HMs interacting with the thiol group, which changed the structure of the enzymes. Under the greater concentration of HMs, nitrogen fixation, metabolism, and carbon fixation are all negatively impacted. For example, during Cr toxicity in *Cyamopsis tetragonoloba*, the expression of the NO_3^- and NH_4^+ fixing enzymes (nitrogenase and urease) was reduced by increasing structure destruction through augmenting protease enzymes (Sangwan et al., 2014).

7.4.4 MOLECULAR RESPONSE

Plants reacted to HM toxicity by impairing mitotic activity, which in turn delayed or temporarily delayed root formation (Campos et al., 2018). In Ni-stressed *Plantago lanceolate*, decreased meristem mitosis was shown to result in limited root development and elongation (Pavlova, 2017). Similarly, in Cr-stressed rice seedlings, Cr delayed and prolonged the cell cycle, inhibiting mitotic cell division (Sundaramoorthy et al., 2010). Plants generate a variety of chelating agents, multiple metal-binding structures, or decontamination methods such as AAs, organic acids (OAs), metallothionein, phytochelatins (PCs), and so on, to detoxify the harmful effects of HMs (Raza et al., 2022). Even though plants have a number of mechanisms in place to keep HMs out of the nucleus, the penetration of HMs somehow causes DNA mutations (additions, deletions, and modifications), protein–DNA cross-associations, changes to the DNA's primary structure, or breaks in DNA strands (Emamverdian et al., 2015). They also damage chlorophyll, upset lipid homeostasis, interfere with ETC and ATP production, and prevent ATP biomolecules from integrating, which results in cell death (Parmar et al., 2013). By modifying auxin signaling, chelation and antioxidant response, and cytokinin signaling, microRNAs have been shown a significant role in plant HMs adaptation and resistance in the context of molecular response (Ding et al., 2020). Casarrubia et al. (2020) found that microRNAs significantly influence how sensitive the *Vaccinium myrtillus* plant is to Cd exposure, working in concert with the mycorrhizal link. Enhancing Al and Cd tolerance in *Nicotiana tabacum* has been found to significantly benefit from increased microRNA transcription, as indicated by the biosynthesis/biogenesis of different alternative metabolites (Saenz-de la O et al., 2020).

7.5 METABOLOMICS APPROACHES FOR HEAVY METAL TOXICITY IN PLANTS

Metabolomics is a branch of "omics" science that involves the profiling of the metabolome (Deshmukh et al., 2014). Metabolomics is the study of metabolites through nuclear magnetic resonance (NMR) spectroscopy, while metabolomics is associated with mass spectrometry (GC–MS or LC–MS). Metabolomics is utilized for both quantitative and qualitative analysis of metabolites and sensitively reflects subtle metabolic changes in organisms under adverse stress, also studies abnormalities by observing the changes in metabolites (Akash et al., 2023). Plants achieve HM tolerance through intracellular chelation of HMs with metal-binding ligands, OAs, AAs, or GSH. The metabolites can be detected from the sample after derivatization, including sugars, AAs, and OAs, and so on.

7.5.1 AMINO ACID

Amino acids (AAs) are signaling molecules for plant growth and development; the accumulation of AAs signifies various roles of plants, such as modulation of stomatal opening, its role as osmolytes, regulation of ion transport, or HM detoxification by binding metal ions in several plant tissues. For instance, proline and histidine have been observed to bind metal ions inside the xylem sap (Rai, 2002). The metabolites like proline and oligopeptides, such as betaine, polyamines, and nicotianamine, are synthesized under the stress of Cd, Cu, Ni, and Zn (Singh et al., 2015). These molecules have three major functions, namely, chelation of metals, antioxidant defense, and signaling. Histidine, other AAs, GSH, and PCs play an important role in metal binding, whereas polyamines function as signaling molecules and antioxidants. Their accumulation is considered as an active response of stress defense instead of metabolic deregulation. In transgenic plants and algae, strong correlative and mechanistic evidence indicates the involvement of proline in metal stress defense. The experimentation of Cd-stress on *Lactuca sativa (L.)* by Costa and Morel explained that proline oxidation inhibition is the main reason for higher root proline levels (Sharma and Dietz, 2006). Upregulation of proline under HM stress is encountered in plants (algae to gymnosperms) (Sharma and Dietz, 2006; Costa and Morel, 1994). Metabolites, such as glutamine and glycine, are associated with the biosynthesis of GSH, and their concentration increased in the Enrei cultivar but decreased in the Harosoy cultivar but in the case of proline during Cd stress, the up- and downregulation act in opposite manner in both the cultivars (Siripornadulsil et al., 2002). To mention, in Cd detoxification, methionine, apart from proline, is a noted precursor of various secondary metabolites decreases at least twofold in both cultivars. A specific case of upregulation of free proline up to >20-fold is seen in the leaves of *Silene vulgaris*, and this proline-inducing ability in a growth medium is most effective for Cu, followed by Cd and Zn (Singh et al., 2015). Proline content in other plant species increases by metal induction is often in *S. vulgaris*, the proline level is lower even if exposed to HM concentrations. The increase in proline content is higher in the shoot than in the roots, and the bulk amount of the metal-induced proline is restricted to the leaves in *S. vulgaris*. This higher accumulation of proline in leaves on exposure to HMs is mainly due to deterioration of the plant–water balance (Ahsan et al., 2012). The water deficit is also evident in *L. sativa*, where an increase in Cd concentration at 100 μM causes a 13-fold rise in root proline content (Sharma and Dietz, 2006). The correlation mechanism of water deficit condition and metal-dependent proline induction was established by Schat and coworkers in the year 1997, where proline accumulation was almost completely inhibited by suppression of transpiration (placing plants under transparent polyethylene covers) at higher relative humid condition (98%), whereas it causes increase of leaf proline by 20-fold level at 75% relative humidity in uncovered control plants. Report states that in Cd-stressed *T. aestivum*, silicon-induced proline content hinders osmotic stress within a cell, resists denaturation of protein, undergoes antioxidant stabilization of enzymes, and also reduces the overproduction of ROS (Kastori et al., 1992). A synergic correlation to Cd concentration is found to rely on the use and unuse of jasmonic acid and gibberellic acids in roots and shoots of *Cicer arietinum* (Howladar et al., 2018).

The proline content in the case of Cu stressed-detached rice leaves increases rapidly as well, probably due to proteolysis and increased activities of D1-pyrroline-5-carboxylate reductase or ornithine-D-amino-transferase (Ahmad et al., 2021). *Armeria maritima*, populated in Cu-containing bog in Wales, possesses high Cu-tolerance with increased proline content in the roots; however, in a noncopper site, it lacks significant elevation of proline (Chen et al., 2001). A similar case of Zn-tolerance in *Deschampsia cespitosa* shows fourfold increased levels of proline content in roots compared to nontolerant matches (Farago and Mullen, 1979). Unlike Zn and Cd, Hg, although belongs to the same group in the periodic table, the metabolomics studies on crops under Hg stress are lacking. It is reported that inorganic Hg probably impacts soluble compartments, while methyl-Hg mainly impacts the nonsoluble ones (Smirnoff and Stewart, 1987). Only a few reports under Hg stress show increased content of tryptophan, threonine, and phenylalanine in *Elodea nuttallii* shoots.

Another essential AA, histidine, plays a critical role in HM detoxification in plants (Cosio and Renault, 2020). The metal scavenging role of histidine in the metal-contaminated environment is mainly dependent on the type and concentration of HM and the affected plant species (Dutta et al., 2018). During Ni-stress, a His–Ni complex is easily formed in Ni-contaminated nutrient solution, and this complexation restricts the mobility and availability of Ni in the nutrient, which otherwise can cause Ni-induced oxidative stress. This is how histidine protects AA-treated *T. aestivum* (Ozen and Yaman, 2016). A similar L-histidine-mediated HM scavenging mechanism was recently reported by Kozhevnikova and group, which showed Ni-complexation and compartmentation in the root cell sap of Ni-stressed *Noccaea caerulescens* (Dalir and Khoshgoftarmanesh, 2014). A study on equilibrium in xylem fluids by metal complex of Zn treated soybean and tomato show that major part of Cu and Ni was bound to asparagine and histidine (Kozhevnikova et al., 2020).

7.5.2 ORGANIC ACID

Detoxification of HMs through chelation by low molecular weight organic acids (LMWOAs) is a defensive mechanism in plants (Kramer et al., 1996). OAs (citrate, oxalate, and malate) as metal transporters in xylem sap have also been involved in metal sequestration in the vacuole (Hall, 2002). Under HM stress especially in Cd tolerance, rice root exudates could bear more OAs [oxalic acid ($H_2C_2O_4$), tartaric acid (HOOCCH(OH)CH(OH)COOH), malic acid (HOOC–CH_2–CH(OH)–COOH), citric acid ($C_6H_8O_7$), succinic acid (COOH–CH_2–CH_2–COOH), malonic acid (HOOC–CH_2–COOH), and acetic acid (CH_3COOH)] and AAs (including lysine, glycine, alanine, methionine, glutamic acid and histidine), among which tartaric acid and histidine were the most abundant (Wang et al., 2015). Mwamba et al. (2020) studied the metabolomics responses between low and high Cd cumulative genotypes and found that the antioxidant capacity of tolerant Cd-accumulating genotypes (including ascorbic acid (AsA), serotonin, phytosteroids, monoterpenes, carotenoids, and lignans) was enhanced. Citrate also plays a central role in Al stress tolerance. A higher concentration of Pb suppresses the AA synthesis in radish roots as reported by Wang et al. (2015). The metabolism study of roots of the radish plant under Cd and Pb stress shows involvement of citrate, malate, and gluconate. The content of gluconate was enhanced in both treatments, with citrate upregulation and specific downregulation of malate under Pb stress. OAs like hexadecanoic acid, linoleic acid, α-linolenic acid, and octadecanoic acid are also accumulated under metal stress, though not involved in primary metabolism. These results indicate that LMWOAs are associated in the case of radish in Cd stress, and carbon flow is mainly concentrated in gluconate, which justifies that the radish roots are possibly more likely to use gluconate to relieve the Cd toxicity.

The report also shows an increase in the fructose, glucose, and isocaproate contents in roots of halophyte *Suaeda salsa*, whereas a subsequent decrease in sucrose, citrate, and malate concentration under Pb stress (Wu et al., 2012). Plants synthesize more AAs and OAs under HM stress helps to improve their tolerance capability. Stress of Cu in cucumber induces accumulation of threonic acid, tyrosine, phenylalanine, and polyphenol (Zhao et al., 2018). A similar report on the detoxification of excess Cu in *Celosia argentea* by OAs such as proline, glyceric acid, succinic acid, and AAs (glutamic acid, serine, aspartic acid, and threonine) is also known (Wang et al., 2020). Pidatala et al. (2018) found that the levels of OAs, AAs, and coenzymes, including asparagine, proline, histidine, tryptophan, valine, isoleucine, threonine, nicotinamide, and methionine, were increased in vetiver under Pb stress.

Malate has maximum affinity for Zn ions and is primarily engaged in solubilizing cytosolic ions (Podar and Maathuis, 2022). Turhadi et al. (2019) reported metabolite response of rice to Fe stress in terms of decreased concentration of linolenic acid and increased content of elaidic acid.

7.5.3 PHENOL

The phenol content in plant tissues is subjected to a tolerance range of plants to the various stress factors, especially increased biosynthesis of phenolic compounds helps to protect plants from oxidative stress. For example, flavonoids lessen the effect of harmful hydroxyl radicals through metal chelation before they damage plant cells. An experiment by Chrzanowski and group (2022) in hybrid *Triticum aestivum* L. showed the accumulation effect of Pb and Cu salts on seed germination; the authors analyzed enzyme activity in the phenylalanine ammonia-lyase pathway (PAL) and tyrosine ammonia-lyase (TAL) with the total flavonoid and phenol content (Jańczak-Pieniążek et al., 2023). Several researchers reported the activation of phenolic compounds, particularly PAL and chalcone synthase, in numerous plant species like *Carthamus tinctorius*, *Malus pumila*, *Solanum lycopersicum*, *Vitis vinifera*, and *Viola tricolor* L. under several HMs toxicities (Rahman et al., 2022). The phenolic metabolism and how it manages the stress tolerance of Cd and Zn in *Kandelia obovata* was reported by Chen et al. (2009). The report showed improvement of total phenolic compounds in roots and leaves (3.6–44.6% in the roots, and by 0.4–126.6% in the leaves) with increased Cd or Zn contents, whereas biomass and chlorophyll a content in the leaves were suppressed (Chen et al., 2019). The metal toxicity and phenol production are correlated as carboxyl and hydroxyl groups undergo metal chelation. The use of Pb, Cu, and Cd increases the total phenolics of the leaves of the corn, for example, chlorogenic acid and rutin are the main phenolics and they may affect the changes in their total concentration whereas contents of caffeic acid, ferulic acid, and vanillic acid are decreased subsequently (Kisa et al., 2016). An antagonistic correlation between phenolic application and As toxicity is known in As-stressed *Zea mays*, while the synergetic correlation was found between phenolic production and As concentration (Kısa et al., 2016). The Cu metal-driven phenolic concentration in leaf tissues of flax plants further improved with Si-application to detoxify Cu-mediated ROS (El-Beltagi et al., 2020). A phenolic compound named tannin, reported in *Camellia sinensis* plants in maximum quantity, is highly resistant to Mn and protects tea plants from Mn stress through its complexation and compartmentation in several parts of the plant (Farzadfar et al., 2017).

7.5.4 GLUTATHIONE AND A-TOCOPHEROL

Glutathione (GSH) is a hydrophilic low molecular weight tripeptide thiol (c–Glu–Cys–Gly), and due to its cellular defensive strategy, it is significantly involved in the detoxification of HMs. Mechanistically, the S–S bond in glutathione disulfide is reduced to GSH and further reduced into PCs through a transpeptidation reaction promoted by the catalyst GR (Kumar et al., 2012). The thiol moiety (GSH) plays a vital role in metal stress tolerance as it resists metal binding with cysteine-rich proteins, which are responsible for upregulation of plant growth, reproduction, and defense system (Liu et al., 2017). The PCs, however, form metal chelates and restrict the bioavailability and mobility of HMs in transport medium (Xie et al., 2019). *N. caerulescens* is the first hyperaccumulator reported in the late 1980s; hyperaccumulators are plants that absorb a greater concentration of HMs than the soil, often used in phytoremediation technologies. Many Cd hyperaccumulators, namely *Thlaspi caerulescens* and *Solanum nigrum* L., have been studied and over the years researchers tried to establish the mechanisms of phytoremediation efficiency by Cd enrichment in plants (Zeng et al., 2009). This phytoremediation of Cd toxicity is through metal chelation with PCs and other ligands and further compartmentalization in the vacuole. When compared to other HM ions, Cd^{2+} is the strongest inducer of PCs synthesis, a descendent of GSH and mainly obtained as a stable metal-PCs coordination complex. Mechanism shows, gene transplantation of phytochelatin synthase or ɣ-ECS, a key enzyme for production of GSH, resulted in the overexpression of PCs and GSH. Synergic correlation of GR expression and As toxicity in shoots of As-stressed *Isatis cappadocica* desv., explains GR's protective role against the toxic effects of HMs (Hossain et al., 2012). Mengdi Xie and group investigated the physiological responses of *Amaranthus hypochondriacus* to Cd stress with LCMS/MS-based metabolomics and obtained thiol compounds through HPLC analysis (Xie et al., 2019). The leaves of *A. hypochondriacus* under high Cd stress accumulate PCs 40 times than leaves that are not under Cd stress conditions. Being a central molecule of both the antioxidant defense system and the glyoxalase system is involved in both direct and indirect control of ROS and thus protects the plant from HM-induced oxidative damage.

An increased level of GSH biosynthesis plays a central role in Ni tolerance in Ni hyperaccumulators (*Thlaspi* spp.) (Hossain et al., 2012). Apart from GSH, the nonprotein thiol PCs are used in uptake and complexation in the Zn/Cd hyperaccumulator *Arabis paniculata*, showing no indication toward metal-induced ROS metabolism (Zeng et al., 2009). The roots of Cd-treated (50 μM, 21 d) plants had higher APX, CAT, GR, GST, and GPX activities and AsA and GSH content. Activity of increased antioxidant enzymes in Cd-treated plants suggests that metal tolerance in *Phragmites* plants might be associated with the efficient regulation of AsA and GSH and their metabolizing enzymes (Hossain et al., 2012). *Sedum alfredii*, a new Zn hyperaccumulator and Pb accumulator, showed an increase in GSH content, and compared to PCs, they are more likely to HM transport, hyperaccumulation/accumulation, and tolerance. GSH protects proteins against denaturation caused by stress, as well as plays an indirect role in protecting membranes by maintaining α-tocopherol and zeaxanthin in the reduced state.

α-Tocopherol, a highly reactive form of vitamin E mainly produced in the plastid cells of higher plants, is necessary for plant physiological and metabolic processes associated with plant growth, blooming, photoprotection, senescence, and longevity of plants. This byproduct of vitamin E also encounters HM toxicity by alleviating the overproduction of ROS and MDA by reducing O^{2-} in the plastids and plays a crucial role in HM toxicity by protecting the cellular damage of plants (Sadiq et al., 2019). Numerous researchers have observed the altered expression of α-tocopherol in HM toxicity with different concentrations. It has been observed that an increasing trend in α-tocopherol concentration under Cd toxicities and concluded that an increasing gene expression trend directly correlated to its biological production under HM toxicity (Rahman et al., 2022). In Arabidopsis plant species under Cu and Cd-stressed conditions, the upregulation of vitamin E-mediated enzymes (VTE2- and VTE5-mRNA), which regulate the oxidative burst by enhancing HPPD (hydroxyphenylpyruvate dioxygenase) and CPs proteins contents to minimize mobility and bioavailability of both Cu and Cd in the root cell sap (Rahman et al., 2022).

7.6 APPLIED METABOLOMICS PERSPECTIVE OF PLANT RESPONSE AND TOLERANCE TO HM TOXICITY

Applying novel techniques like metabolomics may be the key to mitigating the harmful impacts of the heavy pollutants that are hurting agricultural production and yields due to the increase in anthropogenic activities that cause environmental contamination. Metabolomics is the study of plant metabolite profiles, involving the detection, identification, characterization, and quantification of these chemicals, as well as the production of a cellular chemical fingerprint of regulatory processes. According to Razzaq et al. (2019), the term "metabolome" refers to the study of the entire metabolite pool that is active or present in an organism at a given time under particular circumstances. By making it easier to profile and characterize functional primary and specialized secondary metabolites globally in plants expressed under a variety of environmental conditions, especially in abiotic stress response, resistance, and tolerance, metabolomics analysis has completely changed the phytochemical landscape (Isah, 2019; Dawid and Hille, 2018). A comprehensive study of metabolic pathways, metabolite production, and bioactivity under various environmental conditions has been made possible over time by metabolomics-based phytochemistry (Figure 7.5).

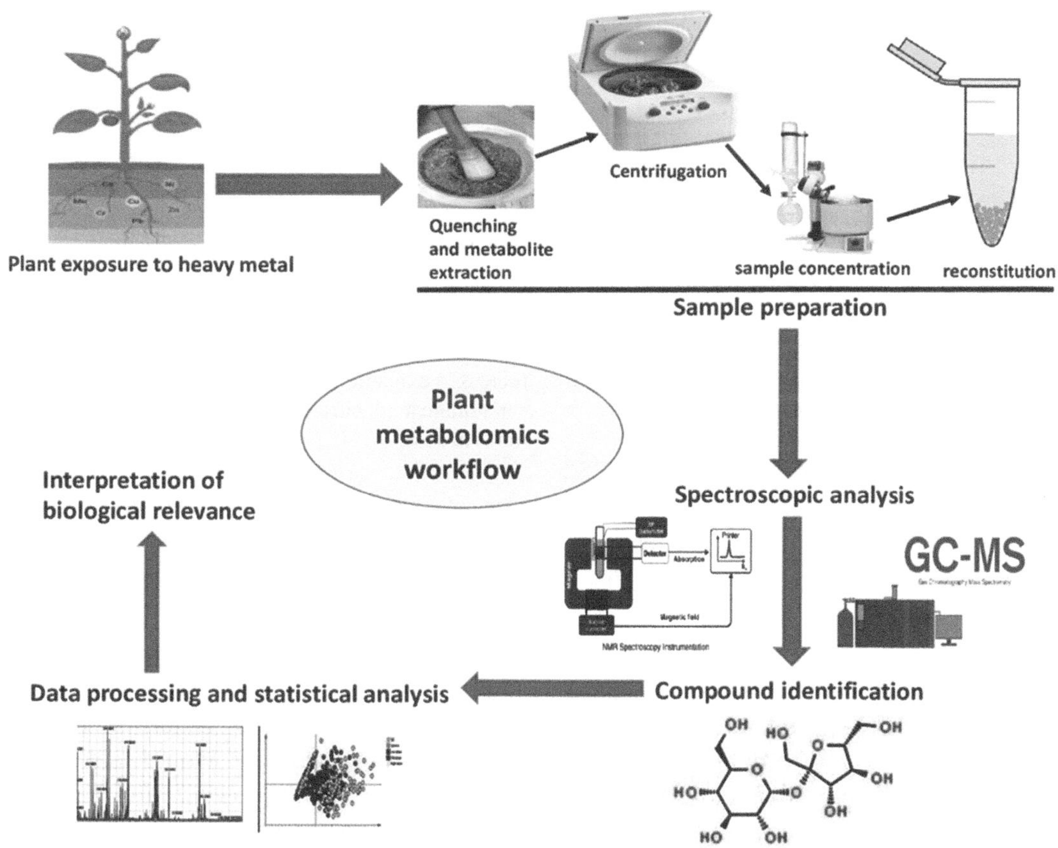

FIGURE 7.5 Schematic diagram of general plant metabolomics workflow.

Understanding the adaptive mechanisms could be much enhanced by the ability to monitor the entire range of metabolites. The goal of metabolomics research is to offer a comprehensive understanding of an organism's functional state, greatly advancing the field of study on plant stress biology (Table 7.2). Specific approaches can be used in metabolomics investigations, such as metabolic fingerprinting, metabolic profiling, and targeted analysis. A variety of analytical techniques, such as GC–MS, LC–DAD–MS, FT–IR, and NMR, are successfully employed for metabolic fingerprinting and profiling. The large volumes of data generated by these approaches require advanced multivariate statistical analysis. Hence, metabolomics data handling has been improved by available bioinformatics tools and databases. According to Nanjo et al. (2012) and Swigonska and Weidner (2013), contemporary bioinformatics and genomic sequencing employing complemented metabolomics and proteomics data provide a powerful tool for the characterization and identification of novel proteins that will comparatively allow plant species to adapt to harsh environmental settings.

TABLE 7.2 Metabolomics Study of Toxic Metals/Metalloids Toxicity of Different Plant Species

Sl No.	Heavy Metal	Plants Species	Data Analysis	Key Observations	References
1.	Pb	*Whitmania pigra*	MS(UPL C–Q/TOF–MS)	24 metabolites changes (included lipid, nucleotides, dipeptides, sphingolipid, and nucleotides)	Luo et al. (2020)
2.	Cd	Tobacco plant	UHPL C–Q–TOF–MS	Alteration in 150 and 76 metabolites from roots and leaves	Zou et al. (2022)
3.	Cd	*Solanum lycopersicum*	^{1}H NMR and HPLC–PDA	Significant increase in soluble carbohydrates and sugar content	Hediji et al. (2010)
4.	Cd	*Oryza sativa* L.	UPLC–ESI–Q–TOF–MS	Upregulation of lipid and fatty acids in low Cd-accumulating cultivar (TY816)	Liu et al. (2021)
5	Cd, Cu, Zn	*Poa annua* L., *Medicago polymorpha* L., *Malva sylvestris*	HPLC–PDA	Increased exudation of oxalic acid, malic acid, citric acid, and fumaric acid	Marshabela et al. (2022), Jaiswal et al. (2018)
6.	Pb and Cd	Reddish root	GC–MS	Metabolic profiling	Wang et al. (2015)
7.	Al^{3+}	*Triticum aestivum*	UHPL C–QTOF–MS	Higher amount of flavonoid	Marshabela et al. (2022)

According to Kráľová et al. (2012), plants growing in soil contaminated with HMs can be identified by symptoms like chlorosis, stunted development, halted nutrient content, water translocation, and altered photosynthesis and water relations. Determining these impacts and the plant-defense mechanisms against HM toxicity would aid in the development of predictive strategies to increase plant tolerance. *Whitmania pigra* exposed to Pb was collected by Luo et al. (2020) and subjected to an ultra-performance quadrupole time-of-flight mass spectrometry (UPLC–Q/TOF–MS) investigation. In response to Pd exposure were revealed by metabolomics data. Lipids, nucleotides, and dipeptides were among the altered metabolites that aided in the metabolism of glycerophospholipids, sphingolipids, and nucleotides. Most recently, major changes in 150 and 76 metabolites under Cd stress from roots and leaves were found using UHPLC–QTOF–MS profiling of tobacco plants exposed to Cd toxicity (Zou et al., 2022). The results of the study demonstrated differential control of metabolites linked to AA biosynthesis, arginine and proline metabolism, nicotinate and nicotinamide metabolism, and flavonoid biosynthesis. It has previously been demonstrated that proline reduces oxidative stress, nutritional absorption, and photosynthetic characteristics in *Cajanus cajan* L. exposed to Cd stress (Hayat et al., 2021). Concurrently, it was discovered that buckwheat cultivars with higher flavonoid content were more resistant to Pb^{2+} and Cd^{2+} than cultivars with lower flavonoid content (Horbowicz et al., 2013). Similar findings were reported by Mashabela et al. (2022b), who found that varieties of wheat (*Triticum aestivum* L.) resistant to aluminum

(Al^{3+}) had higher amounts of flavonoids following a UHPLC–QTOF–MS-based metabolic characterization. Using ^{1}H NMR and HPLC–PDA, the reaction of tomato plants (*S. lycopersicum*) to prolonged Cd exposure was assessed (Hediji et al., 2010). Leaves of tomato exposed to 20 μM of Cd, absolute metabolite quantification revealed significant increases in carbohydrate content. These findings were observed to correlate with the increase in sucrose content with a decrease in glucose and fructose level in the same leaves, which was suggested to be due to Cd inhibition of invertase activity (Devi et al., 2013). Furthermore, the investigation revealed elevated concentrations of α-tocopherol and ascorbate in leaves exposed to Cd; these metabolites play a role in plant antioxidant defense, preventing scavenging lipid peroxyl radicals by lipid peroxidation in thylakoid membranes, thereby enhancing plant resistance to stress. In HM-contaminated soils, roots are the main point of interaction with metals, which makes HM more bioavailable and easier for HM accumulator plants to absorb for bioremediation. Rice cultivar response under Cd stress conditions was assessed using metabolomics analysis (Liu et al., 2021). The study conducted metabolic profiling analysis of Cd-treated cultivars using untargeted UPLC–ESI–Q–TOF–MS. The results showed that lipids and fatty acids were upregulated in the low-accumulating cultivar (TY816) as opposed to the high-accumulator (JY841). Liu et al. (2021) claimed that HM toxicity causes host plants to emit different amounts of lipids and fatty acids as a cellular membrane-mediated defensive strategy against Cd stress. Here, oxidative stress brought on by the perception of Cd causes membrane lipids to desaturate, increasing the amount of fatty acids that are exuded from roots into the soil (rhizosphere); this process also strengthens the roots' ability to withstand damage to their membranes (Liu et al., 2021). LMWOAs were shown to be a crucial plant response mechanism to phytotoxicity brought on by HMs (Montiel-Rozas, and Modejon, 2016). Treatment of plants with Cd and Zn, an HPLC–PDA analysis of root exudates from *Poa annua* L. (Poaceae), *Medicago polymorpha* L. (Leguminosae), and *Malva sylvestris* L. (Malvaceae) plants with increased levels of exudations like oxalic acid, malic acid, citric acid, and fumaric acid. In the rhizosphere, OAs with metal-chelating properties, such as fumarate, oxalate, citrate, and malate, can form nontoxic complexes with HM substrates to decrease HM uptake and accumulation in plants, hence reducing phytotoxicity (Mashabela et al., 2022a; Jaiswal et al., 2018). Proline is a widely recognized and studied plant metabolite due to its diverse functions, such as its role in metal chelation and the activation of antioxidant enzymes. As a result, it is utilized in the manufacture of transgenic plants that are resistant to metals (Khalid et al., 2019). According to Sharma and Dietz (2006), putrescine was also utilized to give crops resistance to HM stress. According to Khalid et al. (2019), there was an upregulation of α-tocopherol levels in arabidopsis during Cd stress, which was thought to reduce stress. Another metabolite that appears to have a role in metal toxicity is ascorbate and GSH, which are significantly implicated in the ROS stress response. By triggering antioxidant enzymes and lowering the production of reactive species when HMs build up in plant cells from their surroundings, certain secondary metabolites, such as carotenoids, flavonoids, phenols, and tannins, modulate stress responses. According to Khalid et al. (2019), these metabolites are also known to prevent ROS formation, therefore safeguarding the photosynthetic apparatus. Zhang et al. (2019) observed increased buildup of shoot Cd chelates and increased root-pectin demethylation in the tolerant cultivar of oilseed rape under Cd stress using high-resolution metabolic profiling. According to Ali et al. (2020), tocopherol in oilseed rape roots inhibits both the absorption of Cd in the roots and its translocation along the shoot. When GSH and tocopherol accumulation were found, *Lemna minor* was observed to demonstrate lower hyper buildup of Hg and less oxidative damage caused by Hg (Zhang et al., 2017). The function of NO in controlling Cd-induced stress in Fescue, which was investigated by Zhu et al. (2020), by a comprehensive transcriptome and metabolomics investigation. Exogenous sodium nitroprusside treatment was found to reduce the accumulation of Cd and 15 metabolites in this NO-regulated Cd-stress tolerance pathway, including arginine, proline, GSH, and others. According to Wang et al. (2015), a metabolic profile of radish roots treated with Cd and Pb revealed changes in sugar, AA, and secondary metabolites, as well as variable rates of metabolism in the GSH, energy, carbohydrate, and oxidative phosphorylation pathways. Technological developments in recent times have contributed to an even greater surge in metabolite profiling research. Characterization and quantification of these metabolites have advanced significantly as a consequence of equipment tinkering and application of fundamental concepts of tools such as mass spectrometry and chromatography. Combinational techniques such as capillary electrophoresis MS, direct flow injection MS, gas chromatography–MS (GC–MS), and liquid chromatography mass spectrometry (LC–MS) are currently being developed for the profiling of plant metabolomes. Depending on the requirements and kind of characterization required, other instruments like high-performance thin-layer chromatography, ultraperformance liquid chromatography (UPLC), high-resolution MS, matrix-assisted laser desorption/ionization–MS imaging, and ion mobility–MS are also frequently used. Using GC–MS, Wang et al. (2015) performed metabolomic profiling, revealing the metabolites and metabolic pathways involved in radish root Pb and Cd stress responses. Similar to this, Zhu et al. (2020) combined transcriptome and metabolomic analysis to determine the critical elements behind the tolerance and stress responses to Cd in tall fescue plants, which are mostly controlled by nitric oxide. A network of metabolites under stress (biotic and abiotic) can be created with the use of metabolomics technique, and this network can then be investigated for crop development initiatives. This information can be applied to transgenic plants, phenotyping, drug discovery and extraction, mutant characterization, metabolomics-assisted breeding, and related fields (Razzaq et al., 2019). A plant that is under stress deviates from its natural mechanisms by halting or modifying its growth and development processes and concentrating more on reaction and survival. The quantity, content, and kind of metabolites generated vary as a result of transcriptional reprogramming or altered gene expression patterns.

7.7 CONCLUSION

The environment has become contaminated with toxic concentrations of HMs, particularly Cd, As, Al, Hg, and Pb, due to rapid industrialization, soil mineralization, rigorous weathering of rocks, and anthropogenic intrusions. Regrettably, the ambient soil bioavailability of these HMs is gradually increasing globally. Through the production of ROS, lipid peroxidation, DNA damage, sudden changes in the cell cycle, nutritional imbalance, and toxicities, these HMs negatively affect cells and the components that make up cells. These toxic HMs have also been shown to alter the expression of chaperones and proteases, which impacts protein metabolism and may even prevent the synthesis of new proteins. The toxic concentration of HMs poses a threat to plant growth, development, and crop yields, as well as to the food chains, because plants frequently absorb and store even minute but hazardous concentrations of these HMs in their seeds and edible sections. The scientific community has been interested in studying the crucial elements of HM toxicity and cultivating tolerance in crop plants because of the significance of HMs. The mechanistic insights of HM absorption, transport, translocation, and sequestration/chelation in cellular compartments have attracted ongoing research. However, more thorough research is required to find novel transporters, uncover retrograde signaling networks and cross-talks, identify potential hyperaccumulators in plants, and understand the implications of couptake of metals in future perspectives. The Metabolomics technique, which is part of the "New Age Omics" paradigm, necessitates managing vast amounts of data generated during quantification and analysis. In order to perform data mining, it is necessary to create and maintain enormous libraries and databases for each species, as well as characterize these metabolites according to factors including mechanism, size, activity, and reactions. With its broader use, higher level of sophistication, more popularity, and larger database creation, metabolomics offers a wealth of information about how plants function internally and react to varying environmental conditions. It is possible to observe and evaluate the dynamic nature of plants' responses to both stressful and normal circumstances. Hence, this knowledge can then be applied to quantify and describe the genetic and environmental responses of the plant.

KEYWORDS

- **heavy metal toxicity**
- **plants**
- **metabolomics**
- **stress tolerance**

REFERENCES

Abdallah Alnuwaiser, M. An analytical survey of trace heavy elements in insecticides. *Int. J. Anal. Chem.*, 2019(1), 8150793.

Adil, N., Ashraf, K., Munir, M., Mohiuddin, M., Abbasi, A., Riaz, U., Aslam, A., Marey, S. A., Hatamleh, A. A., Zaman, Q. U. Pesticides, heavy metals and plasticizers: Contamination and risk assessment of drinking-water quality. *Sustainability*, 2023, 15(17), 13263.

Adnan, M., Xiao, B., Ali, M. U., Xiao, P., Zhao, P., Wang, H., Bibi, S. Heavy metals pollution from smelting activities: A threat to soil and groundwater. *Ecotoxicol. Environ. Saf.*, 2024, 274, 116189.

Afzaal, M., Hameed, S., Liaqat, I., Ali Khan, A. A., abdul Manan, H., Shahid, R., Altaf, M. Heavy metals contamination in water, sediments and fish of freshwater ecosystems in Pakistan. *Water Pract. Technol.*, 2022, 17(5), 1253–1272.

Ahmad, P., Raja, V., Ashraf, M., Wijaya, L., Bajguz, A., Alyemeni, M. N. Jasmonic acid (JA) and gibberellic acid (GA3) mitigated Cd-toxicity in chickpea plants through restricted Cd uptake and oxidative stress management, *Sci. Rep.*, 2021, 11, 19768.

Ahsan, N., Nakamura, T., Komatsu, S. Differential responses of microsomal proteins and metabolites in two contrasting cadmium (Cd)-accumulating soybean cultivars under Cd stress. *Amino Acids*, 2012, 42, 317.

Akash, M. S. H., Yaqoob, A., Rehman, K., Imran, M., Assiri, M. A., Al-Rashed, F., Al-Mulla, F., Ahmad, R., Sindhu, S. Metabolomics: A promising tool for deciphering metabolic impairment in heavy metal toxicities. *Front. Mol. Biosci.*, 2023, 10.

Alengebawy, A., Abdelkhalek, S. T., Qureshi, S. R., Wang, M. Q. Heavy metals and pesticides toxicity in agricultural soil and plants: Ecological risks and human health implications. *Toxics*, 2021, 9(3), 4 2.

Ali, E., Hassan, Z., Irfan, M., Hussain, S., Rehman, H. U., Shah, J. M., Shahzad, A. N., Ali, M., Alkahtani, S., Abdel-Daim, M. M., Bukhari, S. A. H. Indigenous tocopherol improves tolerance of oilseed rape to cadmium stress. *Front. Plant Sci.*, 2020, 11, 547133.

Ali, H., Khan, E., Ilahi, I. Environmental chemistry and ecotoxicology of hazardous heavy metals: Environmental persistence, toxicity, and bioaccumulation. *J. Chem.*, 2019, 2019(1), 6730305.

Aljarba, N. H., Ali, H., Alkahtani, S. Synergistic dose permutation of isolated alkaloid and sterol for anticancer effect on young Swiss Albino Mice. *Drug Des., Dev. Ther.*, 2021, 15, 4043.

Amari, T., Ghnaya, T., Abdelly, C. Nickel, cadmium and lead phytotoxicity and potential of halophytic plants in heavy metal extraction. *S. Afr. J. Bot.*, 2017, 111, 99–110.

Arao, T., Ishikawa, S., Murakami, M., Abe, K., Maejima, Y., Makino, T. Heavy metal contamination of agricultural soil and countermeasures in Japan. *Paddy Water Environ.*, 2010, 8, 247–257.

Ashraf, M. Y., Sadiq, R., Hussain, M., Ashraf, M., Ahmad, M. S. A. Toxic effect of nickel (Ni) on growth and metabolism in germinating seeds of sunflower (Helianthus annuus L.). *Biol. Trace Elem. Res.*, 2011, 143, 1695–1703.

Assche, F., Clijsters, H. Effects of metals on enzyme activity in plants. *Plant Cell Environ.*, 1990, 24, 1–15

Barceló, J., Poschenrieder, C., Andreu, I., Gunse, B. Cadmium-induced decrease of water stress resistance in bush bean plants (Phaseolus vulgaris L. cv. Contender) I. Effects of Cd on water potential, relative water content, and cell wall elasticity. *J. Plant Physiol.*, 1986, 125(1–2), 17–25.

Barranco-Medina, S., Krell, T., Finkemeier, I., Sevilla, F., Lázaro, J. J., Dietz, K. J. Biochemical and molecular characterization of the mitochondrial peroxiredoxin PsPrxII F from Pisum sativum. *Plant Physiol. Biochem.*, 2007, 45(10–11), 729–739.

Baryla, A., Carrier, P., Franck, F., Coulomb, C., Sahut, C., Havaux, M. Leaf chlorosis in oilseed rape plants (Brassica napus) grown on cadmium-polluted soil: Causes and consequences for photosynthesis and growth. *Planta*, 2001, 212, 696–709.

Bassegio, C., Campagnolo, M. A., Schwantes, D., Gonçalves Junior, A. C., Manfrin, J., Schiller, A. D. P., Bassegio, D. Growth and accumulation of Pb by roots and shoots of Brassica juncea L. *Int. J. Phytorem.*, 2020, 22(2), 134–139.

Basu, A., Saha, D., Saha, R., Ghosh, T., Saha, B. A review on sources, toxicity and remediation technologies for removing arsenic from drinking water. *Res. Chem. Intermed.*, 2014, 40, 447–485.

Beckers, F., Rinklebe, J. Cycling of mercury in the environment: Sources, fate, and human health implications: A review. *Crit. Rev. Environ. Sci. Technol.*, 2017, 47(9), 693–794.

Benitez, E., Romero, E., Gomez, M., Gallardo-Lara, F., Nogales, R. Biosolids and biosolids-ash as sources of heavy metals in a plant-soil system. *Water Air Soil Pollut.*, 2001, 132, 75–87.

Bharti, R., Sharma, R. Effect of heavy metals: An overview. *Mater. Today: Proc.*, 2022, 51, 880–885.

Bhat, J. A., Ahmad, P., Corpas, F. J. Main nitric oxide (NO) hallmarks to relieve arsenic stress in higher plants. *J. Hazard. Mater.*, 2021, 406, 124289.

Bhattacharyya, M. H., Wilson, A. K., Rajan, S. S., Jonah, M. Biochemical pathways in cadmium toxicity. *Mol. Biol. Toxicol. Metals*, 2000, 34–74.

Bidar, G., Pruvot, C., Garçon, G., Verdin, A., Shirali, P., Douay, F. Seasonal and annual variations of metal uptake, bioaccumulation, and toxicity in Trifolium repens and Lolium perenne growing in a heavy metal-contaminated field. *Environ. Sci. Pollut. Res.*, 2009, 16, 42–53.

Bigalke, M., Ulrich, A., Rehmus, A., Keller, A. Accumulation of cadmium and uranium in arable soils in Switzerland. *Environ. Pollut.*, 2017, 221, 85–93.

Bueno, P. C., Bellido, E., Rubí, J. M., Ballesta, R. J. Concentration and spatial variability of mercury and other heavy metals in surface soil samples of periurban waste mine tailing along a transect in the Almadén mining district (Spain). *Environ. Geol.*, 2009, 56, 815–824.

Campos, N. V., Arcanjo-Silva, S., Freitas-Silva, L., de Araújo, T. O., Souza-Fernandes, D. P., Azevedo, A. A. Arsenic hyperaccumulation in Pityrogramma calomelanos L. (Link): Adaptive traits to deal with high metalloid concentrations. *Environ. Sci. Pollut. Res.*, 2018, 25(11), 10720–10729.

Casarrubia, S., Martino, E., Daghino, S., Kohler, A., Morin, E., Khouja, H.-R., Murat, C., Barry, K. W., Lindquist, E. A., Martin, F. M. Modulation of plant and fungal gene expression upon cd exposure and symbiosis in ericoid mycorrhizal Vaccinium myrtillus. *Front. Microbiol.* 2020, 11, 341

Chen, C. T., Chen, L. M., Lin, C. C., Kao, C. H. Regulation of proline accumulation in detached rice leaves exposed to excess copper, *Plant Sci..*, 2001, 160, 283.

Chen, T., Liu, X., Li, X., Zhao, K., Zhang, J., Xu, J., Shi, J., Dahlgren, R. A. Heavy metal sources identification and sampling uncertainty analysis in a field-scale vegetable soil of Hangzhou, China. *Environ. Pollut.*, 2009, 157(3), 1003–1010.

Chen, S. B., Meng, W. A. N. G., Li, S. S., Zhao, Z. Q., Wen-di, E. Overview on current criteria for heavy metals and its hint for the revision of soil environmental quality standards in China. *J. Integr. Agric.*, 2018, 17(4), 765–774.

Chen, S., Wang, Q., Lu, H., Li, J., Yang, D., Liu, J., Yan, C. Phenolic metabolism and related heavy metal tolerance mechanism in Kandelia Obovata under Cd and Zn stress. *Ecotoxicol. Environ. Saf.*, 2019, 169, 134.

Cho, U. H., Seo, N. H. Oxidative stress in Arabidopsis thaliana exposed to cadmium is due to hydrogen peroxide accumulation. *Plant Sci.*, 2005, 168(1), 113–120.

Chugh, L. K., Sawhney, S. K. Photosynthetic activities of Pisum sativum seedlings grown in presence of cadmium. *Plant Physiol. Biochem.*, 1999, 37(4), 297–303.

Clarkson, T. W., Magos, L. The toxicology of mercury and its chemical compounds. *Crit. Rev. Toxicol.*, 2006, 36(8), 609–662.

Clemens, S., Ma, J. F. Toxic heavy metal and metalloid accumulation in crop plants and foods. *Annu. Rev. Plant Biol.*, 2016, 67, 489–512.

Cooke, C. A., Martínez-Cortizas, A., Bindler, R., Gustin, M. S. Environmental archives of atmospheric Hg deposition–a review. *Sci. Total Environ.*, 2020, 709, 134800.

Cosio, C., Renault, D. Effects of cadmium, inorganic mercury and methylmercury on the physiology and metabolomic profiles of shoots of the macrophyte Elodea nuttallii, *Environ. Pollut.* 2020, 257, 113557.

Costa, G., Morel, J.-L. Water relations, gas exchange and amino acid content in Cd-treated lettuce. *Plant Physiol. Biochem.*, 1994, 32, 561.

Dalir, N., Khoshgoftarmanesh, A. H. Symplastic and apoplastic uptake and root to shoot translocation of nickel in wheat as affected by exogenous amino acids, *J. Plant Physiol.*, 2014, 171, 531.

Dalla Vecchia, F., La Rocca, N., Moro, I., De Faveri, S., Andreoli, C., Rascio, N. Morphogenetic, ultrastructural and physiological damages suffered by submerged leaves of Elodea canadensis exposed to cadmium. *Plant Sci.*, 2005, 168(2), 329–338.

Dawid, C., Hille, K. Functional metabolomics—A useful tool to characterize stress-induced metabolome alterations opening new avenues towards tailoring food crop quality. *Agronomy*, 2018, 8(8), 138.

Defarge, N., De Vendômois, J. S., Séralini, G. E. Toxicity of formulants and heavy metals in glyphosate-based herbicides and other pesticides. *Toxicol. Rep.*, 2018, 5, 156–163.

Deshmukh, R., Sonah, H., Patil, G., Chen, W., Prince, S., Mutava, R., Vuong, T., Valliyodan, B., Nguyen, H. T. Integrating omic approaches for abiotic stress tolerance in soybean, Front, *Plant Sci..*, 2014, 5, 244.

Devi, R., Munjral, N., Gupta, A. K., Kaur, N. Effect of exogenous lead on growth and carbon metabolism of pea (Pisum sativum L.) seedlings. *Physiol. Mol. Biol. Plants*, 2013, 19, 81–89.

Ding, Y., Ding, L., Xia, Y., Wang, F., Zhu, C. Emerging roles of microRNAs in plant heavy metal tolerance and homeostasis. *J. Agric. Food Chem.* 2020, 68(7), 1958–1965.

Djebali, W., Zarrouk, M., Brouquisse, R., El Kahoui, S., Limam, F., Ghorbel, M. H., Chaïbi, W. Ultrastructure and lipid alterations induced by cadmium in tomato (Lycopersicon esculentum) chloroplast membranes. *Plant Biol.*, 2005, 7(04), 358–368.

Dutta, S., Mitra, M., Agarwal, P., Mahapatra, K., De, S., Sett, U., Roy, S. Oxidative and genotoxic damages in plants in response to heavy metal stress and maintenance of genome stability, *Plant Signal. Behav.* 2018, 13, e1460048.

Ekmekçi, Y., Tanyolac, D., Ayhan, B. Effects of cadmium on antioxidant enzyme and photosynthetic activities in leaves of two maize cultivars. *J. Plant Physiol.*, 2008, 165(6), 600–611.

El-Beltagi, H. S., Sofy, M. R., Aldaej, M. I., Mohamed, H. I. Silicon alleviates copper toxicity in flax plants by up-regulating antioxidant defense and secondary metabolites and decreasing oxidative damage. *Sustainability*, 2020, 12, 4732.

Eleftheriou, E. P., Adamakis, I.-D. S., Panteris, E., Fatsiou, M. Chromium-induced ultrastructural changes and oxidative stress in roots of Arabidopsis thaliana. *Int. J. Mol. Sci.*, 2015, 16(7), 15852–15871.

Elgarahy, A. M., Elwakeel, K. Z., Mohammad, S. H., Elshoubaky, G. A. A critical review of biosorption of dyes, heavy metals and metalloids from wastewater as an efficient and green process. *Clean. Engineer. Technol.*, 2021, 4, 100209.

Emamverdian, A., Ding, Y., Mokhberdoran, F., Xie, Y. Heavy metal stress and some mechanisms of plant defense response. *Sci. World J.*, 2015, 2015, 756120.

Fahr, M., Laplaze, L., Bendaou, N., Hocher, V., Mzibri, M. E., Bogusz, D., Smouni, A. Effect of lead on root growth. *Front. Plant Sci.*, 2013, 4, 175.

Farago, M. E., Mullen, W. A. Plants which accumulate metals. IV. A possible copper–proline complex from the roots of Armeria maritime, *Inorganica Chim. Acta*, 1979, 32, L93.

Farzadfar, S., Zarinkamar, F., Hojati, M. Magnesium and manganese affect photosynthesis, essential oil composition and phenolic compounds of Tanacetum parthenium, *Plant Physiol. Biochem.*, 2017, 112, 207.

Ferrara, R., Mazzolai, B., Edner, H., Svanberg, S., Wallinder, E. Atmospheric mercury sources in the Mt. Amiata area, Italy. *Sci. Total Environ.*, 1998, 213(1–3), 13–23.

Finnegan, P. M., Chen, W. Arsenic toxicity: The effects on plant metabolism. *Front. Physiol.*, 2012, 3, 182.

Fodor, E., Szabó-Nagy, A., Erdei, L. The effects of cadmium on the fluidity and H+-ATPase activity of plasma membrane from sunflower and wheat roots. *J. Plant Physiol.*, 1995, 147(1), 87–92.

Gadd, G. M. Metals, minerals and microbes: Geomicrobiology and bioremediation. *Microbiology*, 2010, 156(3), 609–643.

Geng, H., Xu, Y., Zheng, L., Gong, H., Dai, L., Dai, X. An overview of removing heavy metals from sewage sludge: Achievements and perspectives. *Environ. Pollut.*, 2020, 266, 115375.

Ghnaya, T., Slama, I., Messedi, D., Grignon, C., Ghorbel, M. H., Abdelly, C. Effects of Cd2+ on K+, Ca^{2+} and N uptake in two halophytes Sesuvium portulacastrum and Mesembryanthemum crystallinum: Consequences on growth. *Chemosphere*, 2007, 67(1), 72–79.

Ghori, N.-H., Ghori, T., Hayat, M., Imadi, S., Gul, A., Altay, V., Ozturk, M. Heavy metal stress and responses in plants. *Int. J. Environ. Sci. Technol.*, 2019, 16(3), 1807–1828.

Gianico, A., Braguglia, C. M., Gallipoli, A., Montecchio, D., & Mininni, G. Land application of biosolids in Europe: Possibilities, constraints and future perspectives. *Water*, 2021, 13(1), 103.

Gimeno-García, E., Andreu, V., Boluda, R. Heavy metals incidence in the application of inorganic fertilizers and pesticides to rice farming soils. *Environ. Pollut.*, 1996, 92(1), 19–25.

Gustin, M. S., Lindberg, S. E., Weisberg, P. J. An update on the natural sources and sinks of atmospheric mercury. *Appl. Geochem.*, 2008, 23(3), 482–493.

Haider, F. U., Liqun, C., Coulter, J. A., Cheema, S. A., Wu, J., Zhang, R., Wenjun, M., Farooq, M. Cadmium toxicity in plants: Impacts and remediation strategies. *Ecotoxicol. Environ. Saf.*, 2021, 211, 111887.

Hall, J. Cellular mechanisms for heavy metal detoxification and tolerance, *J. Exp. Bot.*, 2002, 53, 1.

Hasanuzzaman, M., Bhuyan, M. B., Zulfiqar, F., Raza, A., Mohsin, S. M., Mahmud, J. A., Fotopoulos, V. Reactive oxygen species and antioxidant defense in plants under abiotic stress: Revisiting the crucial role of a universal defense regulator. *Antioxidants*, 2020, 9(8), 681.

Hayat, K., Khan, J., Khan, A., Ullah, S., Ali, S., Salahuddin, Fu, Y. Ameliorative effects of exogenous proline on photosynthetic attributes, nutrients uptake, and oxidative stresses under cadmium in pigeon pea (Cajanus cajan L.). *Plants*, 2021, 10, 796

He, Z. L., Yang, X. E., Stoffella, P. J. Trace elements in agroecosystems and impacts on the environment. *J. Trace Elem. Med. Biol.*, 2005, 19(2–3), 125–140.

Hediji, H., Djebali, W., Cabasson, C., Maucourt, M., Baldet, P., Bertrand, A. Effects of long-term cadmium exposure on growth and metabolomic profile of tomato plants. *Ecotoxicol. Environ. Saf.*, 2010, 73, 1965–1974

Helaoui, S., Boughattas, I., Hattab, S., Mkhinini, M., Alphonse, V., Livet, A., Bousserrhine, N., Banni, M. Physiological, biochemical and transcriptomic responses of Medicago sativa to nickel exposure. *Chemosphere*, 2020, 249, 126121.

Horbowicz, M., Dębski, H., Wiczkowski, W., Szawara-Nowak, D., Koczkodaj, D., Mitrus, J., Sytykiewicz, H. The impact of short-term exposure to Pb and Cd on flavonoid composition and seedling growth of common buckwheat cultivars. *Polish J. Environ. Stud.*, 2013, 22(6), 1723–1730.

Hossain, M. A., Piyatida, P., da Silva, J. A. T., Fujita, M. Molecular mechanism of heavy metal toxicity and tolerance in plants: Central role of glutathione in detoxification of reactive oxygen species and methylglyoxal and in heavy metal chelation, *J. Bot.*, 2012, 2012, 872875.

Howladar, S. M., Al-Robai, S. A., Al-Zahrani, F. S., Howladar, M. M., Aldhebiani, A. Y., Silicon and its application method effects on modulation of cadmium stress responses in Triticum aestivum (L.) through improving the antioxidative defense system and polyamine gene expression, *Ecotoxicol. Environ. Saf.* 2018, 159, 143.

Hussain, B., Ashraf, M. N., Abbas, A., Li, J., Farooq, M. Cadmium stress in paddy fields: Effects of soil conditions and remediation strategies. *Sci. Total Environ.*, 2021, 754, 142–188.

Isah, T. Stress and defense responses in plant secondary metabolites production. *Biol. Res.*, 2019, 52, 39.

Jaishankar, M., Mathew, B. B., Shah, M. S., Gowda, K. R. S. Biosorption of few heavy metal ions using agricultural wastes. *J. Environ. Pollut. Hum. Health*, 2014, 2(1), 1–6.

Jaiswal, S. K., Naamala, J., Dakora, F. D. Nature and mechanisms of aluminium toxicity, tolerance and amelioration in symbiotic legumes and rhizobia. *Biol. Fert. Soils*, 2018, 54, 309–318.

Jamla, M., Khare, T., Joshi, S., Patil, S., Penna, S., Kumar, V. Omics approaches for understanding heavy metal responses and tolerance in plants. *Curr. Plant Biol.*, 2021, 27, 100213.

Jańczak-Pieniążek, M., Cichoński, J., Michalik, P., Chrzanowski, G. Effect of heavy metal stress on phenolic compounds accumulation in winter wheat plants, *Molecules*, 2023, 28, 241.

Janicka-Russak, M., Kabała, K., Burzynski, M. Different effect of cadmium and′ copper on H+-ATPase activity in plasma membrane vesicles from Cucumis sativu s roots. *J. Exp. Bot.*, 2012, 63(11), 4133–4142.

Järup, L., Berglund, M., Elinder, C. G., Nordberg, G., Vanter, M. Health effects of cadmium exposure–a review of the literature and a risk estimate. *Scand. J. Work Environ. Health*, 1998, 24, 1–51.

Jin, X., Yang, X., Islam, E., Liu, D., Mahmood, Q. Effects of cadmium on ultrastructure and antioxidative defense system in hyperaccumulator and non-hyperaccumulator ecotypes of Sedum alfredii Hance. *J. Hazard. Mater.*, 2008, 156(1–3), 387–397.

Karuppanapandian, T., Moon, J. C., Kim, C., Manoharan, K., Kim, W. Reactive oxygen species in plants: Their generation, signal transduction, and scavenging mechanisms. *Aust. J. Crop Sci.*, 2011, 5(6), 709–725.

Kasim, W. A. Changes Induced by Copper and Cadmium Stress in the Anatomy and Grain Yield of Sorghum bicolor (L.) Moench. *Int. J. Agr. Biol.*, 2006, 8, 123.

Kastori R, Petrovic M, Petrovic N. Effect of excess lead, cadmium, copper and zinc on water relations in sunflower, *J. Plant Nutr.*, 1992, 15, 2427.

Kataba, A., Botha, T. L., Nakayama, S. M., Yohannes, Y. B., Ikenaka, Y., Wepener, V., Ishizuka, M. Acute exposure to environmentally relevant lead levels induces oxidative stress and neurobehavioral alterations in larval zebrafish (Danio rerio). *Aquat. Toxicol.*, 2020, 227, 105607.

Khalid, N., Aqeel, M., Noman, A. System biology of metal tolerance in plants: An integrated view of genomics, transcriptomics, metabolomics, and phenomics. In: *Plant Metallomics and Functional Omics: A System-Wide Perspective* (pp. 107–144), 2019.

Khan, M. A., Khan, S., Khan, A., Alam, M. Soil contamination with cadmium, consequences and remediation using organic amendments. *Sci. Total Environ.*, 2017, 601, 1591–1605.

Khatun, J., Intekhab, A., Dhak, D. Effect of uncontrolled fertilization and heavy metal toxicity associated with arsenic (As), lead (Pb) and cadmium (Cd), and possible remediation. *Toxicology*, 2022, 477, 153274.

Kısa, D., Elmastaş, M., Öztürk, L., Kayır, Ö. Responses of the phenolic compounds of Zea mays under heavy metal stress. *Appl. Biol. Chem.*, 2016, 59, 813.

Kohzadi, S., Shahmoradi, B., Ghaderi, E., Loqmani, H., Maleki, A. Concentration, source, and potential human health risk of heavy metals in the commonly consumed medicinal plants. *Biol. Trace Elem. Res.*, 2019, 187, 41–50.

Kovačević, G., Kastori, R., Merkulov, L. J. Dry matter and leaf structure in young wheat plants as affected by cadmium, lead, and nickel. *Biologia Plantarum*, 1999, 42, 119–123.

Kozhevnikova, A., Seregin, I., Schat, H. Accumulation of nickel by excluder Thlaspi arvense and hyperaccumulator Noccaea caerulescens upon short-term and long-term exposure, *Russ. J. Plant Physiol.*, 2020, 67, 303.

Kráľová, K., Jampílek, J., Ostrovský, I. Metabolomics-useful tool for study of plant responses to abiotic stresses. *Ecol. Chem. Eng. S*, 2012, 19(2), 133–161.

Kramer, U., Cotter-Howells, J. D., Charnock, J. M., Baker, A. J. M., Smith, J. A. C. Free histidine as a metal chelator in plants that accumulate nickel, *Nature*, 1996, 379, 635.

Kulikova, A. L., Kholodova, V. P., Kuznetsov, V. V. Actin is involved in early plant responses to heavy metal stress and associates with molecular chaperons in stress environments. *Doklady Biol. Sci.*, 2009, 424, 49.

Kumar, A., Prasad, M. N. V., Sytar, O. Lead toxicity, defense strategies and associated indicative biomarkers in Talinumtriangularegrown hydroponically, *Chemosphere*, 2012, 89, 1056.

Kumar, H., Singh, G., Mishra, V. K., Singh, R. P., Singh, P. Airborne heavy metals deposition and contamination to water resources. In: *Metals in Water* (pp. 155–173). Elsevier, 2023.

Kumar, S., Trivedi, P. K. Heavy metal stress signaling in plants. In: *Plant Metal Interaction* (pp. 585–603). Elsevier, 2016.

Lambert, M., Leven, B. A., Green, R. M. New methods of cleaning up heavy metal in soils and water. *Environ. Sci. Technol. Briefs Citizens*, 2000, 7(4), 133–163.

Lasat, M. M. Phytoextraction of metals from contaminated soil: A review of plant/soil/metal interaction and assessment of pertinent agronomic issues. *J. Hazard. Subst. Res.*, 1999, 2(1), 5.

Lee, S., Moon, J. S., Ko, T. S., Petros, D., Goldsbrough, P. B., Korban, S. S. Overexpression of Arabidopsis phytochelatin synthase paradoxically leads to hypersensitivity to cadmium stress. *Plant Physiol.*, 2003, 131(2), 656–663.

Lewis, K. A., Tzilivakis, J., Warner, D. J., Green, A. An international database for pesticide risk assessments and management. *Hum. Ecol. Risk Assess.*, 2016, 22(4), 1050–1064.

Li, M., Zhang, L. J., Tao, L., Li, W. Ecophysiological responses of Jussiaea rapens to cadmium exposure. *Aquatic Bot.*, 2008, 88(4), 347–352.

Li, Z., Ma, Z., van der Kuijp, T. J., Yuan, Z., Huang, L. A review of soil heavy metal pollution from mines in China: Pollution and health risk assessment. *Sci. Total Environ.*, 2014, 468–469, 843–53.

Liu, C., Lan, M. M., He, E. K., Yao, A. J., Wang, G.-B., Tang, Y. T., Qui, R. L. Phenomic and metabolomic responses of roots to cadmium reveal contrasting resistance strategies in two rice cultivars (Oryza sativa L.). *Soil Ecol. Lett.*, 2021, 3, 220–229.

Liu, X., Zhang, H., Jiao, H., Li, L., Qiao, X., Fabrice, M. R., Wu, J., Zhang, S. Expansion and evolutionary patterns of cysteine-rich peptides in plants. *BMC Genom.*, 2017, 18(1), 1–14.

Liu, Z., Bai, Y., Gao, J., Li, J. Driving factors on accumulation of cadmium, lead, copper, zinc in agricultural soil and products of the North China Plain. *Sci. Rep.*, 2023, 13(1), 7429.

Lokhande, V. H., Patade, V. Y., Srivastava, S., Suprasanna, P., Shrivastava, M., Awasthi, G. Copper accumulation and biochemical responses of Sesuvium portulacastrum (L.). *Mater. Today.: Proc.* 2020, 31, 679–684.

Luo, X., Meng, J., Chen, X., Cheng, L., Yan, S., Gao, L., Xue, M., Yang, Y. Metabolomics-based study reveals the effect of lead (Pb) in the culture environment on Whitmania pigra. *Sci. Rep.*, 2020, 10(1), 4794.

Mansoor, S., Ali Wani, O., Lone, J. K., Manhas, S., Kour, N., Alam, P., Ahmad, P. Reactive oxygen species in plants: From source to sink. *Antioxidants*, 2022, 11(2), 225.

Mansoor, S., Ali, A., Kour, N., Bornhorst, J., AlHarbi, K., Rinklebe, J., Chung, Y. S. Heavy metal induced oxidative stress mitigation and ROS scavenging in plants. *Plants*, 2023, 12(16), 3003.

Mar, S. S., Okazaki, M., Motobayashi, T. The influence of phosphate fertilizer application levels and cultivars on cadmium uptake by Komatsuna (Brassica rapa L. var. perviridis). *Soil Sci. Plant Nutrit.*, 2012, 58(4), 492–502.

Marchuk, S., Tait, S., Sinha, P., Harris, P., Antille, D. L., McCabe, B. K. Biosolids-derived fertilisers: A review of challenges and opportunities. *Sci. Total Environ.*, 2023, 875, 162555.

Marrugo-Negrete, J., Durango-Hernández, J., Pinedo-Hernández, J., Enamorado-Montes, G., Díez, S. Mercury uptake and effects on growth in Jatropha curcas. *J. Environ. Sci.*, 2016, 48, 120–125.

Mashabela, M. D., Piater, L. A., Dubery, I. A., Tugizimana, F., Mhlongo, M. I. Rhizosphere tripartite interactions and PGPR-mediated metabolic reprogramming towards ISR and plant priming: A metabolomics review. *Biology*, 2022a, 11, 346.

Mashabela, M. D., Piater, L. A., Steenkamp, P. A., Dubery, I. A., Tugizimana, F., Mhlongo, M. I. Comparative metabolite profiling of wheat cultivars (Triticum aestivum) reveals signatory markers for resistance and susceptibility to stripe rust and aluminium (Al3+) toxicity. *Metabolites*, 2022b, 12, 98.

Michalak, A. Phenolic compounds and their antioxidant activity in plants growing under heavy metal stress. *Polish J. Environ. Stud.*, 2006, 15(4), 523–530.

Mikutta, C., Mandaliev, P. N., Mahler, N., Kotsev, T., Kretzschmar, R. Bioaccessibility of arsenic in mining-impacted circumneutral river floodplain soils. *Environ. Sci. Technol.*, 2014, 48(22), 13468–13477.

Millaleo, R., Alvear, M., Aguilera, P., González-Villagra, J., de la Luz Mora, M., Alberdi, M., Reyes-Díaz, M. Mn toxicity differentially affects physiological and biochemical features in highbush blueberry (Vaccinium corymbosum L.) cultivars. *J. Soil Sci. Plant Nutrit.*, 2020, 20(3), 795–805.

Montiel-Rozas, M. M., Modejon, E. M. P. Effect of heavy metals and organic matter on root exudates (low molecular weight organic acids) of herbaceous species: An assessment in sand and soil conditions under different levels of contamination. *Environ. Pollut.*, 2016, 216, 273–281.

Mwamba, T. M., Islam, F., Ali, B., Lwalaba, J. L. W., Gill, R. A., Zhang, F., Farooq, M. A., Ali, S., Ulhassan, Z., Huang, Q., Zhou, W. Comparative metabolomic responses of low- and high-cadmium accumulating genotypes reveal the cadmium adaptive mechanism in Brassica napus, *Chemosphere*, 2020, 250, 126308.

Nanjo, Y., Skultety, L., Uváčková, L. U., Klubicová, K., Hajduch, M., Komatsu, S. Mass spectrometry-based analysis of proteomic changes in the root tips of flooded soybean seedlings. *J. Proteome Res.*, 2012, 11(1), 372–385.

Nathan, C., Ding, A. SnapShot: Reactive oxygen intermediates (ROI). *Cell*, 2010, 140(6), 951–951.

Neculita, C. M., Zagury, G. J., Deschênes, L. Mercury speciation in highly contaminated soils from chlor-alkali plants using chemical extractions. *J. Environ. Qual.*, 2005, 34(1), 255–262.

Nunes, N., Ragonezi, C., Gouveia, C. S., Pinheiro de Carvalho, M. Â. Review of sewage sludge as a soil amendment in relation to current international guidelines: A heavy metal perspective. *Sustainability*, 2021, 13(4), 2317.

Obrist, D., Kirk, J. L., Zhang, L., Sunderland, E. M., Jiskra, M., Selin, N. E. A review of global environmental mercury processes in response to human and natural perturbations: Changes of emissions, climate, and land use. *Ambio*, 2018, 47, 116–140.

Orellana-Mendoza, E., Camel, V., Yallico, L., Quispe-Coquil, V., Cosme, R. Effect of fertilization on the accumulation and health risk for heavy metals in native Andean potatoes in the highlands of Perú. *Toxicol. Rep.*, 2024, 12, 594–606.

Ozen, S. A., Yaman, M. Phytoextraction of lead and its relationship with histidine in six plant species using ICP-MS and HPLC-MS, *Chem. Ecol.*, 2016, 32, 346.

Pacyna, J. M., Pacyna, E. G., Aas, W. Changes of emissions and atmospheric deposition of mercury, lead, and cadmium. *Atmos. Environ.*, 2009, 43(1), 117–127.

Parmar, P., Dave, B., Sudhir, A., Panchal, K., Subramanian, R. Physiological, biochemical and molecular response of plants against heavy metals stress. *Int. J. Curr. Res.*, 2013, 5(1), 80–89

Paunov, M., Koleva, L., Vassilev, A., Vangronsveld, J., Goltsev, V. Effects of different metals on photosynthesis: Cadmium and zinc affect chlorophyll fluorescence in durum wheat. *Int. J. Mol. Sci.*, 2018, 19(3), 787.

Pavlova, D. Nickel effect on root-meristem cell division in Plantago lanceolata (Plantaginaceae) seedlings. *Aust. J. Bot.*, 2017, 65(5), 446–452.

Perčin, A., Zgorelec, Ž., Karažija, T., Kisić, I., Župan, N., Šestak, I. Metals contained in various formulations of mineral nitrogen fertilizers determined using portable X-ray fluorescence. *Agronomy*, 2023 13(9), 2282.

Pidatala, V. R., Li, K., Sarkar, D., Wusirika, R., Datta, R. Comparative metabolic profiling of vetiver (Chrysopogon zizanioides) and maize (Zea mays) under lead stress, *Chemosphere*, 2018, 193, 903.

Pirrone, N., Costa, P., Pacyna, J. M., Ferrara, R. Mercury emissions to the atmosphere from natural and anthropogenic sources in the Mediterranean region. *Atmos. Environ.*, 2001, 35(17), 2997–3006.

Podar, D., Maathuis, F. J. The role of roots and rhizosphere in providing tolerance to toxic metals and metalloids. *Plant, Cell Environ.*, 2022, 45, 719.

Poornima, R., Suganya, K., Sebastian, S. P. Biosolids towards Back–To–Earth alternative concept (BEA) for environmental sustainability: A review. *Environ. Sci. Pollut. Res.*, 2022, 29(3), 3246–3287.

Potocký, M., Pejchar, P., Gutkowska, M., Jiménez-Quesada, M. J., Potocká, A., de Dios Alché, J., Kost, B., Žárský, V. NADPH oxidase activity in pollen tubes is affected by calcium ions, signaling phospholipids and Rac/Rop GTPases. *J. Plant Physiol.*, 2012, 169(16), 1654–1663.

Pourrut, B., Perchet, G., Silvestre, J., Cecchi, M., Guiresse, M., Pinelli, E. Potential role of NADPH-oxidase in early steps of lead-induced oxidative burst in Vicia faba roots. *J. Plant Physiol.*, 2008, 165(6), 571–579.

Pourrut, B., Shahid, M., Dumat, C., Winterton, P., Pinelli, E. Lead uptake, toxicity, and detoxification in plants. *Rev. Environ. Contam. Toxicol. Vol.,* 2011, 213, 113–136.

Prasad, M. N. V., Hagemeyer, J., Dietz, K. J., Baier, M., Krämer, U. Free radicals and reactive oxygen species as mediators of heavy metal toxicity in plants. In: *Heavy Metal Stress in Plants: From Molecules to Ecosystems* (pp. 73–97), 1999.

Rafique, M., Hajra, S., Tahir, M. B., Gillani, S. S. A., Irshad, M. A review on sources of heavy metals, their toxicity and removal technique using physico-chemical processes from wastewater. *Environ. Sci. Pollut. Res.*, 2022, 29(11), 16772–16781.

Rahale, C. S., Lakshmanan, A., Sumithra, M., Kumar, E. R. Humic acid involved chelation of ZnO nanoparticles for enhancing mineral nutrition in plants. *Solid State Commun..*, 2021, 333, 114355.

Rahman, S. U., Nawaz, M. F., Gul, S., Yasin, G., Hussain, B., Li, Y., Cheng, H., State-of-the-art OMICS strategies against toxic effects of heavy metals in plants: A review, *Ecotoxicol. Environ. Saf.*, 2022, 242, 113952.

Rahman, S. U., Xuebin, Q., Kamran, M., Yasin, G., Cheng, H., Rehim, A., Riaz, L., Rizwan, M., Ali, S., Alsahli, A. A. Silicon elevated cadmium tolerance in wheat (Triticum aestivum L.) by endorsing nutrients uptake and antioxidative defense mechanisms in the leaves. *Plant Physiol. Biochem.*, 2021, 166, 148–159.

Rai, V. K. Role of amino acids in plant responses to stresses. *Biol. Plant.*, 2002, 45, 481.

Rao, M. M., Kumar Meena, A., Galib. Detection of toxic heavy metals and pesticide residue in herbal plants which are commonly used in the herbal formulations. *Environ. Monit. Assess.*, 2011, 181, 267–271.

Rasheed, A., Hassan, M. U., Fahad, S., Aamer, M., Batool, M., Ilyas, M., Shang, F., Wu, Z., Li, H. Heavy metals stress and plants defense responses. In: *Sustainable Soil and Land Management and Climate Change* (pp. 57–82). 2021, CRC Press.

Rath, B. P., Hota, S., Subhadarshini, S., Dash, D., Das, P. K. Consequence of chromium-tainted soil on physical and biochemical responses of Vigna radiata *L. J. Appl. Biol. Biotechnol.* 2019, 7(1), 3–1

Rauser, W. E., Ackerley, C. A. Localization of cadmium in granules within differentiating and mature root cells. *Can. J. Bot.*, 1987, 65, 643.

Raza, A., Tabassum, J., Zahid, Z., Charagh, S., Bashir, S., Barmukh, R., Khan,R. S. A.,Barbosa Jr, F., Zhang, C., Chen, H. Advances in "Omics" approaches for improving toxic metals/metalloids tolerance in plants. *Front. Plant Sci.*, 2022, 2949.

Razzaq, A., Sadia, B., Raza, A., Khalid Hameed, M., Saleem, F. Metabolomics: A way forward for crop improvement. *Metabolites*, 2019, 9(12), 303.

Rey-Martínez, N., Guisasola, A., Baeza, J. A. Assessment of the significance of heavy metals, pesticides and other contaminants in recovered products from water resource recovery facilities. *Resour. Conserv. Recycl.*, 2022, 182, 106313.

Riyazuddin, R., Nisha, N., Ejaz, B., Khan, M. I. R., Kumar, M., Ramteke, P. W., Gupta, R. A Comprehensive review on the heavy metal toxicity and sequestration in plants. *Biomolecules*, 2021, 12(1), 43.

Rizwan, M., Meunier, J. D., Davidian, J. C., Pokrovsky, O., Bovet, N., Keller, C. Silicon alleviates Cd stress of wheat seedlings (Triticum turgidum L. cv. Claudio) grown in hydroponics. *Environ. Sci. Pollut. Res.*, 2016, 23(2), 1414–1427.

Robbins, W. K., Walker, H. H. Analysis of petroleum for trace metals. Determination of trace quantities of cadmium in petroleum by atomic absorption spectrometry. *Anal. Chem.*, 1975, 47(8), 1269–1275.

Rucińska-Sobkowiak, R. Water relations in plants subjected to heavy metal stresses. *Acta Physiologiae Plantarum*, 2016, 38, 1–13.

Sadiq, M., Akram, N. A., Ashraf, M., Al-Qurainy, F., Ahmad, P. Alpha-tocopherol-induced regulation of growth and metabolism in plants under non-stress and stress conditions. *J. Plant Growth Regul.* 2019, 38, 1325.

Saenz-de la O, D., Cedillo-Jimenez, C. A., García-Ortega, L. F., Martínez-Resendiz, M., Arn´e-Robles, D., Cruz-Hernandez, A., Guevara-Gonzalez, R. G. Response of transgenic tobacco overexpressing the CchGLP gene to cadmium and aluminium: Phenotypic and microRNAs expression changes. *Physiol. Mol. Biol. Plants*, 2020, 26(1), 3–13.

Sagi, M., Fluhr, R. Production of reactive oxygen species by plant NADPH oxidases. *Plant Physiol.*, 2006, 141(2), 336–340.

Sahu, S. K., Mishra, M., Mishra, A., Mangaraj, P., Beig, G. Quantification and assessment of hazardous mercury emission from industrial process and other unattended sectors in India: A step towards mitigation. *J. Hazard. Mater.*, 2024, 470, 134103.

Saleem, M. H., Fahad, S., Rehman, M., Saud, S., Jamal, Y., Khan, S., Liu, L. Morpho-physiological traits, biochemical response and phytoextraction potential of short-term copper stress on kenaf (Hibiscus cannabinus L.) seedlings. *PeerJ*, 2020, 8, e8321

Sangwan, P., Kumar, V., Joshi, U. Effect of chromium (VI) toxicity on enzymes of nitrogen metabolism in clusterbean (Cyamopsis tetragonoloba L.). *Enzym. Res.*, 2014, 2014, 784036.

Sarkar, A., Paul, B. The global menace of arsenic and its conventional remediation-A critical review. *Chemosphere*, 2016, 158, 37–49.

Schutzendubel, A., Polle, A. Plant responses to abiotic stresses: Heavy metal-induced oxidative stress and protection by mycorrhization. *J. Exp. Bot.*, 2002, 53(372), 1351–1365.

Shabbir, A., Saqib, M., Murtaza, G., Abbas, G., Imran, M., Rizwan, M., Naeem, M. A., Ali, S., Javeed, H. M. R. Biochar mitigates arsenic-induced human health risks and phytotoxicity in quinoa under saline conditions by modulating ionic and oxidative stress responses. *Environ. Pollut.*, 2021, 287, 117348.

Shahid, M., Pourrut, B., Dumat, C., Nadeem, M., Aslam, M., Pinelli, E. Heavy-metal-induced reactive oxygen species: Phytotoxicity and physicochemical changes in plants. *Rev. Environ. Contamination Toxicol. Vol.*, 2014, 232, 1–44.

Sharma, B., Sarkar, A., Singh, P., Singh, R. P. Agricultural utilization of biosolids: A review on potential effects on soil and plant grown. *Waste Manage. (Oxford)*, 2017, 64, 117–132.

Sharma, P., Dubey, R. S. Lead toxicity in plants. *Brazil. J. Plant Physiol.*, 2005, 17, 35–52.

Sharma, P., Monika, Goyal, K., Kumar, T., Chauhan, N. S. Inimical effects of arsenic on the plant physiology and possible biotechnological solutions to mitigate arsenic-induced toxicity. In: *Contaminants in Agriculture: Sources, Impacts and Management* (pp. 399–422), 2020.

Sharma, S. S., Dietz, K.-J., The significance of amino acids and amino acid-derived molecules in plant responses and adaptation to heavy metal stress, *J. Exp. Bot.*, 2006, 57, 711–726

Singh, B. K., Walker, A. Microbial degradation of organophosphorus compounds. *FEMS Microbiol. Rev.*, 2006, 30(3), 428–471.

Singh, S., Parihar, P., Singh, R., Singh, V. P., Prasad, S. M. Heavy metal tolerance in plants: Role of transcriptomics, proteomics, metabolomics, and ionomics. *Front. Plant Sci.*, 2015, 6, 1143.

Singh, S., Sinha, S. Accumulation of metals and its effects in Brassica juncea (L.) Czern. (cv. Rohini) grown on various amendments of tannery waste. *Ecotoxicol. Environ. Saf.*, 2005, 62(1), 118–127.

Siripornadulsil, S., Traina, S., Verma, D. P. S., Sayre, R. T. Molecular mechanisms of proline-mediated tolerance to toxic heavy metals in transgenic microalgae, *Plant Cell.*, 2002, 14, 2837.

Smirnoff, N., Stewart, G. R. Nitrogen assimilation and zinc toxicity to zinc-tolerant and non-tolerant clones of Deschampsia cespitosa (L.) Beauv. *New Phytol..*, 1987, 107, 671.

Son, J. A., Narayanankutty, D. P., Roh, K. S. Influence of exogenous application of glutathione on rubisco and rubisco activase in heavy metal-stressed tobacco plant grown in vitro. *Saudi J. Biol. Sci.*, 2014, 21(1), 89–97.

Song, X., Zhang, C., Chen, W., Zhu, Y., Wang, Y. Growth responses and physiological and biochemical changes in five ornamental plants grown in urban lead-contaminated soils. *Plant. Interact.*, 2020, 1(1), 29–47.

Sörme, L., Lagerkvist, R. J. S. O. T. T. E. Sources of heavy metals in urban wastewater in Stockholm. *Sci. Total Environ.*, 2002, 298(1–3), 131-145.

Sundaramoorthy, P., Chidambaram, A., Ganesh, K. S., Unnikannan, P., Baskaran, L. Chromium stress in paddy:(i) nutrient status of paddy under chromium stress; (ii) phytoremediation of chromium by aquatic and terrestrial weeds. *Comptes Rendus Biol.*, 2010, 333(8), 597–607.

Suvarapu, L. N., Baek, S. O. Determination of heavy metals in the ambient atmosphere: A review. *Toxicol. Ind. Health*, 2017, 33(1), 79–96.

Suzuki, N., Koussevitzky, S. H. A. I., Mittler, R. O. N., Miller, G. A. D. ROS and redox signalling in the response of plants to abiotic stress. *Plant, Cell Environ.*, 2012, 35(2), 259–270.

Swigonska, S., Weidner, S. Proteomic analysis of response to long-term continuous stress in roots of germinating soybean seeds. *J. Plant Physiol.*, 2013, 170(5), 470–479.

Taghipour, M., Jalali, M. Impact of some industrial solid wastes on the growth and heavy metal uptake of cucumber (Cucumis sativus L.) under salinity stress. *Ecotoxicol. Environ. Saf.*, 2019, 182, 109347.

Tamás, L., Mistrík, I., Zelinová, V. Heavy metal-induced reactive oxygen species and cell death in barley root tip. *Environ. Exp. Bot.*, 2017, 140, 34–40.

Tang, J., Zhang, J., Ren, L., Zhou, Y., Gao, J., Luo, L., Yang, Y., Peng, Q., Huang, H., Chen, A. Diagnosis of soil contamination using microbiological indices: A review on heavy metal pollution. *J. Environ. Manage.*, 2019, 242, 121–130.

Tasca, A. L., Puccini, M. Leather tanning: Life cycle assessment of retanning, fatliquoring and dyeing. *J. Cleaner Prod.*, 2019, 226, 720–729.

Tonelli, F. C. P., Tonelli, F. M. P. Causes and effects of pesticide and metal pollution on different ecosystems. In: Bhat, R., Hakeem, K., Dervash, M. (Eds.), *Bioremediation and Biotechnology* (Vol 2). Springer: Cham, 2020.

Turhadi, T., Hamim, H., Ghulamahdi, M., Miftahudin, M. Iron toxicity-induced physiological and metabolite profile variations among tolerant and sensitive rice varieties, *Plant Signal. Behav.* 2019, 14, 1682829.

Ur Rahman, S., Xuebin, Q., Zhao, Z., Du, Z., Imtiaz, M., Mehmood, F., Hongfei, L., Hussain, B., Ashraf, M. N. Alleviatory effects of Silicon on the morphology, physiology, and antioxidative mechanisms of wheat (Triticum aestivum L.) roots under cadmium stress in acidic nutrient solutions. *Sci. Rep.*, 2021b, 11(1), 1–12.

Urra, J., Alkorta, I., Garbisu, C. Potential benefits and risks for soil health derived from the use of organic amendments in agriculture. *Agronomy*, 2019, 9(9), 542.

Vithanage, M., Bandara, P. C., Novo, L. A., Kumar, A., Ambade, B., Naveendrakumar, G., Ranagalage, M., Magana-Arachchi, D. N. Deposition of trace metals associated with atmospheric particulate matter: Environmental fate and health risk assessment. *Chemosphere*, 2022, 303, 135051.

Wang, L., Zhou, Q., Ding, L., Sun, Y. Effect of cadmium toxicity on nitrogen metabolism in leaves of Solanum nigrum L. as a newly found cadmium hyperaccumulator. *J. Hazard. Mater.*, 2008, 154(1–3), 818–825.

Wang, S., He, T., Xu, F., Li, X., Yuan, L., Wang, Q., Liu, H. Analysis of physiological and metabolite response of Celosia argentea to copper stress, *Plant Biol.*, 2020, 23, 391.

Wang, W., Wang, S., Ma, X., Gong, J. Recent advances in catalytic hydrogenation of carbon dioxide. *Chem. Soc. Rev.*, 2011, 40(7), 3703–3727.

Wang, Y., Xu, L., Shen, H., Wang, J., Liu, W., Zhu, X., Wang, R., Sun, X., Liu, L. Metabolomic analysis with GC-MS to reveal potential metabolites and biological pathways involved in Pb & Cd stress response of radish roots, *Sci. Rep.*, 2015, 5, 18296.

Webster, J. G., Nordstrom, D. K. Geothermal arsenic. In: Welch, A. H., Stollenwerk, K. G. (Eds.), *Arsenic in Ground Water: Geochemistry and Occurrence* (pp. 101–126). Kluwer, Boston, 2003.

Wei, B., Yu, J., Cao, Z., Meng, M., Yang, L., Chen, Q. The availability and accumulation of heavy metals in greenhouse soils associated with intensive fertilizer application. *Int. J. Environ. Res. Public Health,* 2020. 17(15), 5359.

Wu, H., Liu, X., Zhao, J., Yu, J., Pang, Q., Feng, J. Toxicological effects of environmentally relevant lead and zinc in halophyte Suaeda salsa by NMR-based metabolomics. *Ecotoxicol*, 2012, 21, 2363.

Wuana, R. A., Okieimen, F. E. Heavy metals in contaminated soils: A review of sources, chemistry, risks and best available strategies for remediation. *Int. Scholarly Res. Notices*, 2011, 2011(1), 402647.

Xie, M., Chen, W., Lai, X., Dai, H., Sun, H., Zhou, X., Chen, T. Metabolic responses and their correlations with phytochelatins in Amaranthus hypochondriacus under cadmium stress, *Environ. Pollut.*, 2019, 252, 1791.

Xu, J., Bravo, A. G., Lagerkvist, A., Bertilsson, S., Sjöblom, R., Kumpiene, J. Sources and remediation techniques for mercury contaminated soil. *Environ. Int.*, 2015, 74, 42–53.

Yasir, M. W., Siddique, M. B. A., Shabbir, Z., Ullah, H., Riaz, L., Shah, A. A. Biotreatment potential of co-contaminants hexavalent chromium and polychlorinated biphenyls in industrial wastewater: Individual and simultaneous prospects. *Sci. Total Environ.*, 2021, 779, 146345.

Zaheer, I. E., Ali, S., Saleem, M. H., Imran, M., Alnusairi, G. S., Alharbi, B. M., Soliman, M. H. Role of iron–lysine on morpho-physiological traits and combating chromium toxicity in rapeseed (Brassica napus L.) plants irrigated with different levels of tannery wastewater. *Plant Physiol. Biochem.*, 2020, 155, 70–84.

Zahir, F., Rizwi, S. J., Haq, S. K., Khan, R. H. Low dose mercury toxicity and human health. *Environ. Toxicol. Pharmacol.*, 2005, 20(2), 351–360.

Zeng, X., Ma, L. Q., Qiu, R., Tang, Y. Responses of nonprotein thiols to Cd exposure in Cd hyperaccumulator Arabis paniculata Franch, *Environ. Exp. Bot.*, 2009, 66, 242.

Zhang, T., Lu, Q., Su, C., Yang, Y., Hu, D., Xu, Q. Mercury induced oxidative stress, DNA damage, and activation of antioxidative system and Hsp70 induction in duckweed (Lemna minor). *Ecotoxicol. Environ. Saf.*, 2017, 143, 46–56.

Zhang, Z. H., Zhou, T., Tang, T. J., Song, H. X., Guan, C. Y., Huang, J. Y., Hua, Y. P. A multiomics approach reveals the pivotal role of subcellular reallocation in determining rapeseed resistance to cadmium toxicity. *J. Exp. Bot.*, 2019, 70(19), 5437–5455.

Zhao, L., Huang, Y., Paglia, K., Vaniya, A., Wancewicz, B., Keller, A. A. Metabolomics reveals the molecular mechanisms of copper induced cucumber leaf (cucumis sativus) senescence, *Environ. Sci. Technol.*, 2018, 52, 7092.

Zhao, S., Liu, Q., Qi, Y., Duo, L. Responses of root growth and protective enzymes to copper stress in turfgrass. *Acta Biol. Crac. S. Bot.*, 2010, 52(2)

Zhu, H., Ai, H., Hu, Z., Du, D., Sun, J., Chen, K., Chen, L. Comparative transcriptome combined with metabolome analyses revealed key factors involved in nitric oxide (NO)-regulated cadmium stress adaptation in tall fescue. *Bmc Genomics [Electronic Resource]*, 2020, 21, 1–13.

Zhu, W., Li, Z., Li, P., Yu, B., Lin, C. J., Sommar, J., Feng, X. Re-emission of legacy mercury from soil adjacent to closed point sources of Hg emission. *Environ. Pollut.*, 2018, 242, 718–727.

Zhuang, Z., Mu, H. Y., Fu, P. N., Wan, Y. N., Yu, Y., Wang, Q., Li, H. F. Accumulation of potentially toxic elements in agricultural soil and scenario analysis of cadmium inputs by fertilization: A case study in Quzhou county. *J. Environ. Manage.*, 2020, 269, 110797.

Zou, C., Lu, T., Wang, R., Xu, P., Jing, Y., Wang, R., Xu, J., Wan, J. Comparative physiological and metabolomic analyses reveal that Fe3O4 and ZnO nanoparticles alleviate Cd toxicity in tobacco. *J. Nanobiotechnol.*, 2022, 20(1), 302.

Zwolak, A., Sarzyńska, M., Szpyrka, E., Stawarczyk, K. Sources of soil pollution by heavy metals and their accumulation in vegetables: A review. *Water Air Soil Pollut.*, 2019, 230, 1–9.

CHAPTER 8

Plant–Pathogen Interactions: What Metabolomics Can Contribute?

SAYANTI MANDAL

Department of Chemistry and Biochemistry, Sharda School of Engineering & Sciences, Sharda University, Greater Noida, Uttar Pradesh, India

ABSTRACT

Plants within their ecological environment engage in interactions with a diverse array of living species and must adapt to a wide range of abiotic and biotic stress. For decades, it has been known that plants can synthesize hundreds of thousands of chemical compounds, but only a handful have been completely identified as being involved in defensive responses. Even though these metabolites are very important for plants, we still know nothing about how they are synthesized and what they do. Metabolomics has been used to identify the metabolites that are crucial to plant–microbe (including pathogen) interactions. The primary chemical that plants utilize to actively modify their microbiome are metabolites. In addition, this emerging discipline has received considerable attention over the past decade for deciphering metabolites under abiotic stress in plants. Moreover, in contrast to genomic and transcriptomic, this metabolomics application and its tools are still underutilized in plant–microbe interaction research. Consequently, it is essential to apply metabolomics to the study of plant resistance/susceptibility. This chapter looks at metabolomics research that correlates plant resistance to fungal, microbial, nematode, and viral diseases with changes in primary or specialized metabolism. Lastly, it is explored how combining metabolomics with other -omics might benefit plant-pathology research.

8.1 INTRODUCTION

Analyzing biological systems' metabolites and their dynamic reactions to the challenging environment changes both qualitatively and quantitatively is the primary goal of metabolomics (Liu et al., 2021; Marchev et al., 2021). The fundamental components of any metabolomics technique are metabolite analysis, identification, and detection. In the 1970s, metabolite profiling gave rise to the discipline of metabolomics. The term "metabolomics" was coined by

Exploring Plant Stress Tolerance: A Metabolomics Approach.
Shuvasish Choudhury, Debojyoti Moulick, and Mirza Hasanuzzaman (Eds.)

DOI: 10.1201/9781779641731-8

Fiehn et al. (2000) and is described as "a thorough and quantitative analysis of all metabolites in a biological system." (Chen et al., 2019; Fiehn et al., 2000). Metabolomics, which is often used to investigate plant and microbial systems, focuses on all small-molecule components and the changes in particular cells, cell types, tissue types, or organs. Metabolomics has emerged as a powerful tool for studying plant and microbial systems since it zeroes in on all components of small molecules as well as changes in specific cells, cell types, tissues, or organs. Metabolomics is an expanding discipline within the area of omics sciences. It primarily focuses on obtaining rapid and comprehensive snapshots of metabolomes (Shafi et al., 2021). Plants are home to a diverse range of microbiota, which includes actinomycetes, bacteria, fungi, algae, and protozoa (Pirzadah et al., 2021). Consequently, because of this variability, it is challenging to generalize plant physiology. Bacteria are present in the majority among these organisms (Glick, 2012). Plant growth-promoting bacteria are mostly found in the rhizosphere and/or roots of plants, and they may be very helpful to the growth and development of such plants (Adedayo and Babalola., 2023; Mandal et al., 2023). Fungi, nematodes, and bacteria also have a significant detrimental impact on agricultural and natural environments. They interact with both aboveground and belowground plant tissues, and they play a significant role in the recycling and decomposition of organic compounds (Zeilinger et al., 2016). From parasitism to mutualism, there are many different results from these intricate interactions between plants and fungi. The utilization of metabolomics not only offers an exhaustive depiction of the metabolic pathways implicated in the examination of interactions between different species but also elucidates the underlying mechanisms that govern interactions between hosts (plants) and microbes (Khanijou et al., 2022; Zhang et al., 2023). The fascinating topic of "interactions between pathogen and their hosts" studies how different phytopathogens interact with different types of plants (Chen et al., 2019; Choudhury et al., 2021). Figure 8.1 depicts the effect of phytopathogens (biotic stress) on plant growth and development.

Many integrated technologies and approaches are used for the large-scale investigation of very complex plant extracts, including mass spectrometry (MS)-based techniques (Patel et al., 2021). A variety of integrated technologies and methodologies are utilized in the extensive investigation of exceedingly intricate plant extracts, including approaches founded on MS. These consist of capillary-based MS, gas chromatography–mass spectrometry (GC-MS), liquid chromatography-mass spectrometry (LC-MS), NMR, matrix-assisted laser desorption/ionization (MALDI), and other MS-based methods (Zhang et al., 2023; Patel et al., 2021). Furthermore, the development of genome editing technologies has made it possible for plant scientists to precisely and effectively modify specific genes and metabolic pathways in a broad range of plant species (Razzaq et al., 2022). The application of these contemporary instruments has significantly facilitated the growth of crop improvement programs through the enhancement of quality attributes such as flavonoid, folate, and protein composition. This thorough analysis focuses on the most recent studies on plant metabolomics and how it might be used to enhance crops. With an emphasis on plant–bacterial and plant–fungal interactions, this chapter aims to provide an overview of the benefits of metabolomics analytical platforms and their applications in examining the interactions of various species.

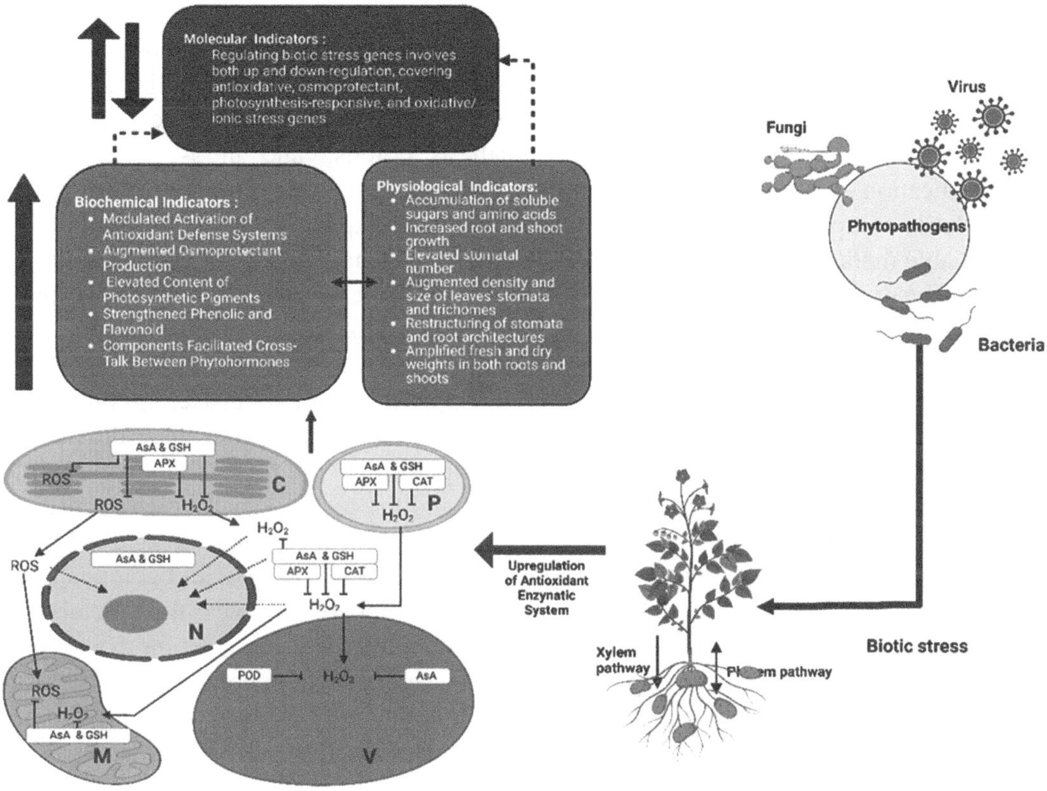

FIGURE 8.1 Schematic illustration of biotic responses in plants. M: Mitochondria, V: Vacuole, N: Nucleus, C: Chloroplast, P: Peroxisome. The figure was created with BioRender.com.

8.2 METABOLOMICS: A COMPREHENSIVE OVERVIEW

In order to better understand biological systems, metabolomics use analytical chemistry techniques to quantify a broad variety of tiny molecules produced by metabolic activities, which are often classified as fewer than 1500 Daltons (Da) (Villas-Bôas et al., 2005). This covers the response of biochemical pathways to therapies, species-specific differences in biochemical processes, and the identification of new metabolites that may function as biomarkers of certain biological conditions, pathways, and reactions. Metabolomics is particularly well-suited for research that generates hypotheses and lacks knowledge of specific molecular targets associated with a given scientific question. Two distinct plant diseases, for example, may result in different symptoms in the same plant, but the molecular basis for this variation remains unclear. In this light, it represents a significant possibility to develop nontraditional biological markers relevant for biosecurity. The last component of the genes, mRNA, proteins, posttranslational modification, and metabolite cascade is metabolites, which represent the interplay of environment, genetics, and the resultant phenotype (Fiehn, 2002). The functions of primary metabolites, which include fatty acids, amino acids, and intermediates in energy metabolism, are numerous in biology. These roles include those of structure and building blocks for other cofactors, macromolecules, and substrates for enzymes, regulators of gene expression, energy metabolism, and waste products. Thousands of secondary metabolites are also found in plants; these metabolites serve a variety of biological purposes, such as defense and intra- or interspecies communication, but they are not necessary for growth and development. Polyphenols and volatile compounds are two examples of them. The quantity of the metabolome (all quantifiable metabolites in an organism, including exogenous metabolites) of species that are significant to biosecurity is poorly estimated. Many agriculturally significant plants, like *Solanum tuberosum* (4470), *Actinidia deliciosa* (4062), and *Solanum lycopersicum* (4194), have thousands of metabolites included in the food metabolome database (www.foodb.ca, accessed on May 23, 2022). The human metabolome database, in contrast, has 217,920 items, indicating that plant metabolomes are really far larger (albeit the database also includes substances of plant and synthetic origin if they are found in human samples) (Wishart et al., 2021). Depending on the sample and the techniques used to quantify the metabolites, the number of metabolites that may be identified and detected using any approach in practice ranges from 50 to 1000. Metabolite production and composition in a sample are influenced by both genetic and environmental variables, thus metabolome measurements of a given sample may comprise many metabolomes. As an example, the metabolome of an insect will include not just the insect's own metabolites but also those of its host plant, its gut microbiome, any parasites it may have, and its external microbiome. The term "exposome" refers to the group of metabolites from external sources, which includes dietary chemicals, insecticides, pollutants, and microbial compounds (Orešič et al., 2020). There are many challenges that come up because of this high level of complexity and variety. Nevertheless, this also signifies that the relative proportions of metabolites may vary depending on circumstances such as the host–plant relationship, species and subspecies, or geographic location. A substantial influence of the environment on the expression of metabolites in plants is suggested by estimates of metabolomic heritability, which is the percentage of the metabolome that is stable from generation to generation. These estimates range from 5% to 47% based on genome-wide association studies of metabolite genes in *Oryza sativa* and *S. tuberosum* (Soltis and Kliebenstein, 2015). Additionally, there is a great deal of variance across plant species. For example, in *Arabidopsis thaliana* and *O. sativa*, a single genetic locus may account for as much as 90% of the metabolite variation observed, but in maize, it can only account for up to 62% (Soltis and Kliebenstein, 2015). Metabolomic trends are hard to understand because metabolites are not always specific to one species or route. Amino acids, for example, are found in all biological samples, and their amounts are controlled by several routes, limiting the value of specific amino acids as biomarkers in many circumstances. Utilizing the chemometric (i.e., data-driven analysis of chemical information) and bioinformatics developments of the last decade to process and interpret metabolomics data in order to acquire additional knowledge about the organism or system of interest is possible despite the difficulty posed by this high degree of complexity (Chong et al., 2019).

8.3 METHODS FOR PERFORMING METABOLOMICS

The two primary methods used to measure the metabolome are MS and NMR spectroscopy. The basis for NMR metabolite detection is the resonance of certain atoms' nuclei—such as hydrogen or carbon-13—under an electromagnetic field. The various resonances, or chemical shifts, are employed in metabolomics to distinguish between metabolites in intricate combinations. Solid samples (referred to as "solid-state NMR") may be subjected to NMR, which often needs little sample preparation. One advantage of this method is its quantitative nature, as the intensity of a signal is directly proportional to the quantity of molecules generating that signal. NMR's sensitivity in comparison to MS, size, and cost are the primary barriers to its broader use in metabolomics, while smaller, less expensive devices are available at the penalty of reduced sensitivity and resolution (Draper and McCarney, 2021).

As an alternative, the mass of charged molecules is measured by MS. It identifies molecules by combining their fragmentation pattern and molecular weight, which is determined by their mass-to-charge ratio. In metabolomics, it is customary to couple mass spectrometers with chromatography in order to separate metabolites prior to analysis (Gupta et al., 2020). This is done to prevent the ionization process, in which a charge is introduced to the metabolites prior to detection (a phenomenon called "ion suppression"), from being overwhelmed. Retention time, or the amount of time a metabolite takes to go through a chromatography column, is another tool used to help identify compounds. The two most common kinds of chromatography are LC and GC. Nonetheless, other techniques for separation are also used, such as supercritical CO_2 chromatography and capillary electrophoresis.

Compared to NMR, mass spectrometry is more sensitive and capable of detecting a higher quantity of metabolites (Meredith et al., 2023; Choudhury et al., 2022). It is less costly than NMR and often needs a smaller laboratory setup. The primary drawbacks include the fact that it is not quantifiable without standards since different molecules ionize at different intensities, and that sample extraction is required prior to analysis, which may introduce artifacts into the studies and take a long time, particularly for solid materials (Barding et al., 2013). Research in the area is ongoing to address these issues and enhance the quality and throughput of metabolomics analysis, as is the case with any limits. Ambient ionization mass spectrometry (AI-MS) is a rapidly developing tool with applications in biosecurity diagnostics. The term AI-MS describes a mass spectrometer that has an interface that makes it possible to analyze a sample with little to no sample preparation; after a few seconds of examination, the device generates a mass spectrum "fingerprint." A variety of AI-MS models operate based on various ideas for loading samples into a mass spectrometer. Direct analysis in real-time mass spectrometry (Beyramysoltan et al., 2018) and rapid evaporative ionization mass spectrometry (Balog et al., 2010, 2013) are two examples of this. Compared to other metabolomics instruments, AI-MS can prepare and analyze samples rapidly, which is a need that is easily satisfied. The primary limitation of AI-MS is its restricted capacity to recognize the target compounds. This presents a challenge when the main goal is to comprehend the biology of the sample, even if all that is needed for diagnosis is a strong fingerprint of the metabolites that are there (Ross et al., 2020). AI-MS may be used to tentatively identify metabolites based on the kind of mass spectrometer used, such as mass fragmentation or high-resolution mass. Metabolomics depends on technology as well as data processing and interpretation.

8.4 METABOLOMICS: IMPROVED TOOLS FOR INTERPRETING PATHOGEN–PLANT INTERACTIONS

Metabolomics is the thorough, unbiased, high-throughput examination of every metabolite in a biological system. It started with metabolite profiling, which was made possible by the groundbreaking research of the Baylor College of Pharmacy's Horning group (Horning and Horning, 1970). Oliver et al. (1998) proposed the idea of the "metabolome," which emerged with the emergence and fast development of mass spectrometry-based analytical methods (Oliver et al., 1998). The idea of "metabonomics," which denotes tracking metabolic alterations in reaction to genetic mutation or pathogenic stressors, was first put out in 1999 (Nicholson et al., 1999). Eventually, Fiehn put out the first, more thorough explanation of metabolomics and its uses in plant biology studies (Fiehn, 2002). Many studies have been carried out in the field of utilizing metabolomics in plant biology research for metabolic engineering, crop improvement, and food evaluation, as well as in clinical research for illness diagnosis, medication screening, and therapy. Metabolomics acts as a bridge between genotypes and phenotypes (Pang et al., 2019; Salem et al., 2020; Choudhury et al., 2021). As a functional genomics method, metabolomics has become more popular for deciphering intricate biological system interactions. Metabolites represent the final products of cellular metabolism; therefore, their concentrations can be regarded as the biological system's definitive reaction to external stimuli (Fiehn, 2002). Therefore, in contrast to genomes or proteomics, which may not always have a clear correlate to biochemical phenotypes, metabolomic investigations directly reflect the phenotypic changes of a given system (Feussner and Polle, 2015). Modifications to a certain cell's transcriptome, proteome, or genome will undoubtedly result in modifications to the metabolome (Gomez-Casati et al., 2013). Metabolomics is the highest degree of postgenomic analysis since it considers both transcriptional and posttranscriptional regulation. Because metabolites have a different chemistry compared to transcripts and proteins, determining their presence requires the use of advanced instruments like MS and NMR (Broeckling et al., 2004; Salem, 2020). Both targeted and untargeted methods are often used in metabolomics research (Salem, 2020). Untargeted metabolomics aims to identify as many compounds from examined samples as possible. This strategy generates hypotheses and is ideal for unexplained discoveries. The measurement of a predetermined set of certain metabolites or a particular class of metabolites is the focus of focused metabolomics, on the other hand, which is a biased and hypothesis-led method. Experiment design, sample preparation, data collection, data processing, and biological interpretation are often included in a metabolomics study (Figure 8.2). Experiment design has to be carefully examined in order to get significant findings. A variety of criteria, including the kind of tissue, number of groups, number of repetitions per condition, and time of harvesting, must be carefully chosen in order to statistically represent the validity of the experimental findings (Chen et al., 2019). It is essential to collect samples appropriately, quench them quickly, and extract them directly in order to prevent metabolite deterioration. In metabolomics research, the extraction of metabolites is a crucial stage in sample preparation that should be carefully handled, as it will determine which analytes are subsequently identified depending on the solvent type. A major factor influencing the final metabolite matrix is the extraction solvent and technique used. Extracting all of the metabolites from a sample using a single extraction technique is challenging since various metabolites have varying solubilities (Salem et al., 2020). Because of this, the extraction technique must be selected in accordance with the experiment's goals and the spectrum of metabolites to be studied. Organic acids, carbohydrates, and amino acids may all be effectively extracted by the all-purpose solvent water. Single solvents like methanol or blends of immiscible organic solvents may be utilized for secondary metabolites (Salem et al., 2016). We concentrate on the developments in data analysis and analytical methods in the following sections.

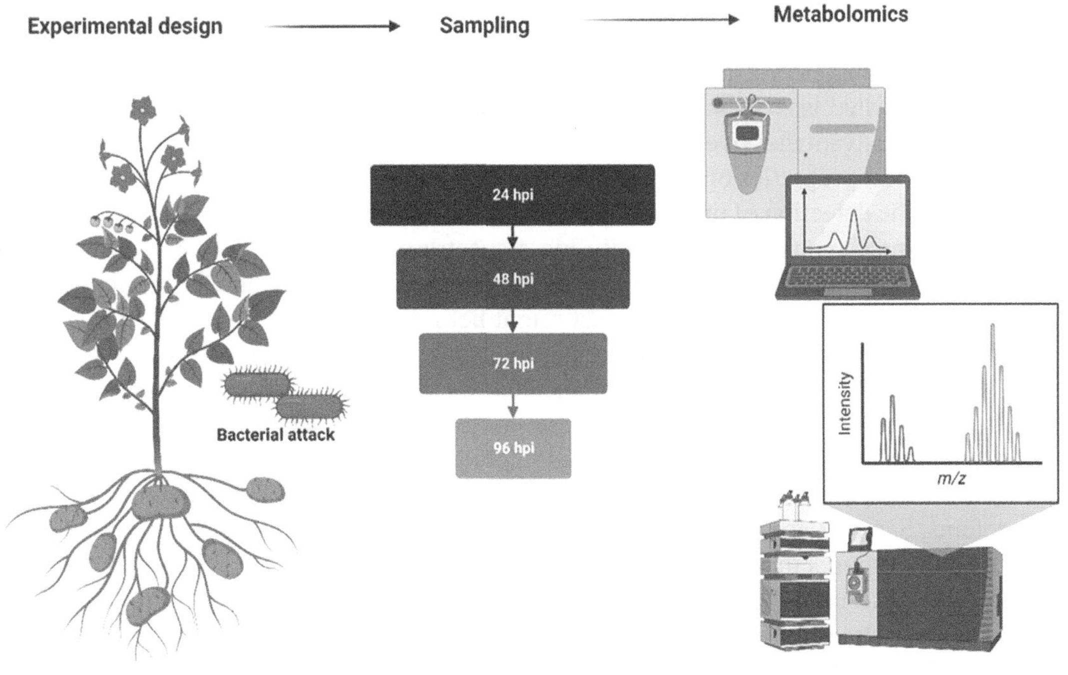

FIGURE 8.2 Design of experiments in plant–pathogen metabolomics. Through inoculation with the pathogenic microbe (a bacterium that induces necrotic lesions) or exposure to other biotic stresses, plant tissues are infected. Typically, subsequent sampling of infected material is conducted in accordance with a time course denoted in hours postinoculation (hpi). Snap-freezing the samples in liquid nitrogen halts their metabolism. The utilization of targeted metabolomics enables the quantification of particular compounds (in this case, the defense hormone). The image is modified from Allwood et al. (2021). The figure was created with BioRender.com.

8.5 PLATFORMS FOR METABOLOMICS AND COMPREHENSIVE METABOLITE DATABASES

A vibrant and expanding field, metabolomics provides a thorough understanding of the metabolic properties of living systems. Using high-throughput analytical methods, metabolomics is the systematic study of the metabolome of biofluids, cells, tissues, or organisms with the goal of identifying and quantifying metabolite alterations associated with illness. Because of the complexity of the metabolome and the wide variety of physicochemical characteristics of the metabolites, many analytical approaches are necessary. The analytical platforms that are most often used include mass spectrometry, NMR, LC-MS, and GC-MS. These methods allow for the creation of large amounts of data and improved chemometric analysis, which provide fundamental. The field of metabolomics is burgeoning and dynamic, offering a full understanding of the metabolic properties of living systems (Misra, 2021). The methodical study of a cell, biofluid, tissue, or organism's metabolome using high-throughput analytical tools to detect and quantify metabolite alterations associated with diseases is known as metabolomics. Because mass spectrometry is more sensitive than NMR, researchers may cover the metabolome data in a systematic manner (Misra, 2021). Consequently, new chemicals and metabolic biomarkers that might help with the reconstruction of metabolic networks were discovered by researchers. Mass spectrometry's accuracy has increased thanks to recent advancements in ionization methods, including electrospray ionization (ESI), air pressure chemical ionization (APCI), and matrix-assisted laser desorption/ionization time-of-flight (MALDI-TOF) (Issaq et al., 2009). NMR's advantages over MS information about metabolites still include the ability to identify physical properties of ligands, binding sites on proteins, direct binding of the target protein, and the detection of protein–ligand complex structures, despite its lower sensitivity and large sample requirement. Because mass spectrometry is more sensitive than NMR, researchers may cover the metabolome data in a systematic manner. Consequently, new chemicals and metabolic biomarkers that might help with the reconstruction of metabolic networks were discovered by researchers. Advancements in ionization methods, including ESI, APCI, and MALDI-TOF, have enhanced mass spectrometry's precision (Issaq et al., 2009). NMR's benefits over MS remain the same, despite its larger sample size required and poorer sensitivity. These advantages include its ability to detect protein–ligand complex structures, determine physical characteristics of ligands, and locate binding sites on proteins. Samples are derivatized for the GC-MS platform, which renders the chemicals volatile and obscures the presence of underivatized molecules (apart from hydrocarbons) in the study. The introduction of GC × GC-TOF-MS (gas chromatography time of flight mass spectrometry) has enabled higher sample throughput and coeluting peak separation (Ralston-Hooper et al., 2011). The main tools used by LC-MS to detect both primary and secondary metabolites of greater mass are ESI and APCI, which are widely used for both targeted and nontargeted techniques (Turner et al., 2016). Apart from these platforms, focused metabolomics focuses on sensitive and untargeted metabolomics investigations that analyze all detectable metabolites, including unidentified substances, as well as the examination of certain categories of metabolites with high selectivity. With capillary electrophoresis–mass spectrometry (CE-MS), different analyte groups (charged, neutral, polar, and hydrophobic) may be separated with great resolution (Ramautar and de Jong, 2014). Additionally, FAIMS (field asymmetric waveform ion mobility spectrometry), an ion mobility-based electrophoretic technique, is combined with MS. The FAIMS technique is used to identify biological materials, such as volatile compounds released during bacterial growth (Zrodnikov and Davis, 2012). High-throughput technology spectrometry (CE-MS) provides high-resolution separation of different analyte groups (charged, neutral, polar, and hydrophobic). A few examples of the data processing platforms used to process the large data sets generated by this technology are MET-Align, MET-COFEA, Chroma TOF, and MET-XAlign (Ramautar and de Jong, 2014). Additionally, MS is used with FAIMS, an ion mobility-based electrophoretic technique. The FAIMS technique is used to identify biological materials, such as volatile compounds released during bacterial growth (Zrodnikov and Davis, 2012). The massive data sets generated by the aforementioned high-throughput technologies are processed using a variety of data processing systems, including MET-COFEA, ChromaTOF, MET-Align, and MET XAlign (Misra and van der Hooft, 2016; Perez de Souza et al., 2017). This included baseline correction, alignment, separation of co-eluting peaks, and normalization before the compounds were identified. To find metabolites, use databases such as National Institute of Standards and Technology, metabolite and tandem mass spectrometry, Golm Metabolome Database, and other metabolome databases (Johnson et al., 2015). In addition, statistical analysis is performed on the data using web-based tools and software such as MetaboAnalyst 5.0, Cytoscape 3.10.1, and other statistical analysis tools to identify the metabolites (Tsugawa et al., 2015; Lei et al., 2023). Such investigations facilitate the identification of metabolic indicators associated with different attributes. Researchers will benefit from the sharing and analysis of data made possible by initiatives such as the Metabolomics Workbench and the Plant Metabolic Network, which will provide centralized databases of plant metabolites, pathways, and associated information.

8.6 METABOLITES SERVE AS SIGNALS IN THE PLANT SYMBIOSIS RELATIONSHIP

Plant roots are not isolated subterranean structures; a diverse range of microorganisms inhabit their rhizosphere. The rhizosphere serves as a focal point for plant–microbe interactions due to the fact that root exudates provide the primary carbon source for microbe proliferation and influence the composition of the microbial community (Jacoby et al., 2017; Mendes et al., 2013; Sasse et al., 2018). There are two well-studied symbiont groups in the rhizosphere: mycorrhizal fungi, which create a symbiotic relationship with plant roots, and rhizobia, which form a nitrogen-fixing symbiosis in legumes (Smith and Smith, 2012; Udvardi and Poole, 2013). Due to its ability to produce nitrogen fertilizer without using up limited fossil fuel resources or emitting greenhouse gases, biological nitrogen fixation is a significant source of nitrogen for plants and has a significant influence on agricultural practices. Flavonoids secreted from plant roots and nod factors generated by Rhizobia facilitate this functional symbiosis. The secretion of flavonoids from plant roots and the generation of nod factors by Rhizobia enable this functional symbiotic relationship (Bladergroen and Spaink, 1998). They stimulate rhizobacterial nod genes by binding to the membrane-associated nodD protein, which forms a complex with the flavonoids and acts as a transcriptional regulator of other nodulation genes involved in rhizobia production of nod factors (Schlaman et al., 1989). Plant roots may detect signals in the form of lipo-chitooligosaccharides, commonly sulfated, called nod factors (Long, 1996). They initiate a series of signaling events that stimulate changes in the structure of root hairs and the division of cell walls. This process makes it easier for rhizobia to get trapped and guides them to the roots, where they can form nodules where symbiotic interactions and nitrogen fixation take place (Cullimore et al., 2001; Long, 1996; Oldroyd et al., 2011). Because nod factors are so distinct, nodulation between a certain plant species and rhizobial strain must be strictly specified (Oldroyd et al., 2011). However, flavonoids are more involved in the symbiotic relationship between plants and rhizobia. They are necessary for the beginning of nodules because they control auxin transport (Wasson et al., 2006). Furthermore, in the actinorhizal symbiosis with Frankia, flavonoids and isoflavonoids are crucial for the other form of nodulation (Abdel-Lateif et al., 2013; Auguy et al., 2011). The phylum Glomeromycota's arbuscular mycorrhizal fungi (AMF) are the primary source of mycorrhiza for herbaceous plants. These fungi infiltrate several commercially significant crops, such as vegetables, horticultural plants, and grains (Smith and Smith, 2012). Mycorrhiza reduces abiotic stress, provides nutrition, and increases crop output. The phosphorus (P) that AMF mycelium efficiently scavenges improves plant P absorption, which in turn promotes favorable host plant development. AMF may even enhance plant development in P-deficient environments (Harrison et al., 2002; Smith et al., 2003). Significantly more resistant to a variety of abiotic stimuli, such as salt, drought, heat, cold, nutrient shortage, and metal toxic stress, are host plants when treated with AMF (Jung et al., 2012). Similar to nodulation, the signaling leading to AMF symbiosis is initiated by a plant signal that causes fungal hyphae development and the manufacture of a "myc" factor. This signal, together with other signals, sets off a symbiotic signaling cascade that has many characteristics of nodulation signaling (Schmitz and Harrison, 2014). Some of the most important signals sent by plants are strigolactones, which are a type of sesquiterpene; phytohormones that control the branching of shoots and roots; and Striga, which starts the growth of parasitic plants (Waters et al., 2017). Strigolactones cause significant branching in the AMF and are released from plant roots, especially when phosphate is limited (Schmitz and Harrison,2014). Nonetheless, other plant metabolic signals are also required for the symbiosis to occur, such as flavonoids, which stimulate spore germination, and C16 aliphatic fatty acids linked to cutin, which are essential for the formation of appressoria and plant cell penetration (Wang et al., 2012). Additionally, two sets of metabolites—lipo-chitooligosaccharides and chitin oligomers—have been shown to have the same impact on plant roots as mycorrhiza fungi, which explains the long-elusive myc signal (Genre et al., 2013). Chitin oligomers are often found in a variety of fungal infections, while myc-lipo-chitooligosaccharides are structurally related to nod factors and frequently sulphated (Schmitz and Harrison, 2014). As fungal exudates have extremely low abundances of different compounds, it is currently unclear how these signals are differentiated, as well as how many and how specific they are. The discovery of hydroxy- and carboxyblumenol glucosides in mycorrhized plant leaves, whose accumulation varies quantitatively with colonization degree, is an interesting new breakthrough (Wang et al., 2018). Many plant species have been shown to produce the derivatives of blumenol, which were induced by a variety of mycorrhiza fungus. For example, in genetic screens, they might prove to be an invaluable instrument for quickly and accurately estimating the amounts of mycorrhiza colonization (Wang et al., 2018).

8.7 METABOLITE'S ROLE IN THE INTERACTION BETWEEN PLANT–PATHOGENS

Some well-known soil-borne plant pathogens include nematodes, fungi, oomycetes, and bacteria (Back et al., 2002; Kamoun et al., 2015; Mansfield et al., 2012). Root rot, root blackening, wilting, stunting, or seedling damping-off are symptoms of soil-borne pathogen infection in plants. In the coniferous forests of western North America, for instance, *Phellinus weirii*, a common native fungal infection, may kill certain robust conifer trees and induce laminated root rot in a number of conifer species (Hansen and Goheen, 2000). Phloem and cambium nutrients are used by *P. weirii* when it enters roots. Therefore, even if the tree seems to have green foliage, the roots are irrevocably harmed. In the conifer forest west of the Cascade Mountains, this fungus is thought to be the disease that causes the most economic harm (Hansen and Goheen, 2000). The fungal pathogen *Fusarium* spp., which colonizes roots and produces root rot symptoms as well as plant wilting, is another common disease that affects soybeans, cereals, tomatoes, and other crops (Arias et al., 2013; Gordon, 2017). For root pathogens to identify and infect root tissues, chemical signals present in the rhizosphere are also crucial. To control its development, the oomycete disease *Phytophthora sojae*, for instance, uses the tiny amounts of compounds generated by soybeans, called isoflavones, such as genistein and daidzein, to guide its hyphae and zoospores (Morris et al., 1998). Cochliophilin A is one of the flavonoids found in spinach root exudates that attracts *Aphanomyces cochlioides*, the bacterium that causes root rot disease in spinach (Tahara et al., 2001). Generally speaking, root exudates are crucial in drawing pathogens to the roots and either inducing or discouraging microbiome activities that aid in the defense of plants against their invasion (Haney et al., 2015; Jacoby et al., 2020; Sasse et al., 2018). The plant immune response includes a number of metabolites that shield plants against different types of pests. The majority of what we know about these compounds comes from studying leaf pathogens in tests. The defensive chemicals in Arabidopsis that are produced from tryptophan, including indolic glucosinolates and camalexin, are among the greatest examples. The "mustard oil bomb" is made up of glucosinolates, which are sulfur-rich secondary compounds of the Caparales (Bones and Rossiter, 1996). When tissue is damaged, the enzyme myrosinase comes into contact with the glucosinolates that are stored in the vacuoles. This causes the enzyme to cleave a glucose residue, and the resultant unstable intermediate rearranges into organic isothiocyanates, nitriles, or thiocyanates (Halkier and Gershenzon, 2006). In addition to discouraging herbivores, these strong volatiles boost defense against bacterial and fungal infections (Bednarek et al., 2009; Clay et al., 2009). The primary amino acid sources for the synthesis of glucosinolates are methionine for aliphatic glucosinolates, phenylalanine for aromatic glucosinolates, and tryptophan for indolic glucosinolates (Halkier and Gershenzon, 2006). The phytoalexin camalexin, one of the Brassicaceae family's defensive chemicals, is a precursor of tryptophan. First identified and investigated in Arabidopsis, where it is generated under infection with fungal pathogens, some bacterial strains, and abiotic conditions, camalexin was originally derived from the plant camelina (*Camelina sativa*), a member of the Brassicaceae family (Glawischnig, 2007). The resistance of leaves of Arabidopsis accessions to the necrotrophic disease *Botrytis cinerea* has been shown to be primarily regulated by variations in the induction of camalexin production (Rowe and Kliebenstein, 2008). It has only been recently discovered that both of these chemicals play significant roles in interacting with microorganisms that promote plant development in infected roots and root exudates (Iven et al., 2012; Koprivova et al., 2019). Nevertheless, nothing is known regarding the degree of significance of indolic phytoalexins in root pathogen defense, save from the fact that camalexin clearly inhibits the development of the root fungal disease *Verticillium longisporum*. The range of secondary changes that indolic chemicals generated from plants go through presents a barrier in understanding their modes of action. Many hydroxylation, methylation, and glycosylation processes attest to the great structural variety seen in plant roots in indolic glucosinolates and other indole chemicals (Stahl et al., 2016). Consequently, the creation of novel research pipelines to precisely identify and quantify the structures of these molecules is one potential for high-throughput plant metabolomics. However, phytoalexins are not only indoles and glucosinolates. For instance, it has been shown that flavonoids are crucial in shielding a variety of plant species against *Fusarium* and other fungi (Mierziak et al., 2014). Mainly, the difference between hops cultivars resistant to soilborne pathogens and those vulnerable to them was the accumulation of phenolic chemicals, *Verticillium nonalfalfae* (Kunej et al., 2020). Priming, or the enhancing of immune response by prior stimuli, is an intriguing facet of plant immunity research. For instance, callose-rich cell wall depositions are stimulated by β-aminobutyric acid (BABA), a priming agent in Arabidopsis that boosts plants' tolerance to the necrotrophic fungus *Plectosphaerella cucumerina* (Ton and Mauch-Mani, 2004). Arabidopsis's BABA-induced resistance may be based on the creation of stable hormone conjugates, which may quickly hydrolyze into active signals that trigger the synthesis of defensive chemicals in response to an assault (Dietz et al., 2000). The identification of these hormone conjugates is crucial since they play a role in the plant immune system. Plant immunity may be studied using salicylic acid, jasmonate acid, abscisic acid, and indole derivatives; a focused metabolomics approach can assist in determining which of these compounds mediates the priming response. An excellent example for more research is the discovery of indole-3-carboxylic acid accumulation in BABA-primed Arabidopsis plants and its role in defense against *P. cucumerina* (Gamir et al., 2012). Identifying the whole range of bioactive compounds generated by plant pathogens is a crucial problem for metabolomic studies in plant–microbe interactions. Additionally, some of these metabolites are hazardous to humans. These compounds may be used as nonprotein effectors to assault the plant (Pusztahelyi et al., 2015; Adeniji et al., 2020). Aflatoxin, toxoflavin, alternariol, fumonisin, or coronatine are among the chemically varied range of toxins that are generated by pathogenic species of *Alternaria*, *Aspergillus*, *Burkholderia*, *Pseudomonas*, and *Fusarium*, respectively (Adeniji et al., 2020; Lee and Ryu, 2017). It is believed that mycotoxins, which have detrimental effects on human health, infect 25% of crops worldwide. With a sensitivity of 0.03 $\mu g kg^{-1}$, metabolomics, and specifically LC-MS technology, can identify mycotoxins in plant material and food (Eshelli et al., 2018). Metabolomics has great promise for shedding light on pathogen biology; nevertheless, more methodological development is required to decrease matrix effects and limit interference from other plant metabolites (Eshelli et al., 2018).

8.8 CHARACTERIZATION OF GENE FUNCTIONS THROUGH METABOLOMICS

Endogenous metabolites from plants are essential to both humans and the plants themselves. Therefore, the use of comparative metabolomics has been shown to be very effective in highlighting the metabolic variations across various plant species as well as in cross-referencing metabolite quantitative trait loci (mQTL) among populations of the same plant species (Wang et al., 2014a; Wen et al., 2014). Additionally, metabolomics makes it possible to identify plant metabolites that are essential for biotic stress resistance or abiotic stress tolerance. By combining metabolomics with other -omics technologies, it is possible to identify the genes responsible and explore the metabolic route further. This is becoming more crucial for metabolic engineering and marker-assisted breeding, which aim to target vital plant processes. According to earlier research, it is possible to find novel metabolites by multiomics analysis utilizing natural variations and near-isogenic lines of a particular plant (Tsugawa et al., 2021). The majority of agricultural attributes are polygenic, meaning that quantitative genomics techniques may easily clarify their underlying genetic basis. Over the last 15 years, metabolomics and genomics-related technologies have developed rapidly. As a result, metabolite genome-wide association studies (mGWAS) and mQTL have become widely used to analyze the genetic architectures underlying the variety of metabolites found in plants. This process creates links or associations between chromosomal locations and metabolite contents. Then, in several plant species with annotated genomes that have undergone a thorough evaluation, the potential genes found inside the genomic interval or close to the corresponding marker loci are identified (Fang and Luo, 2019). Comparative metabolomics and mGWAS analysis were recently used by Liang et al. (2021a) to find 10 candidate genes significantly correlated with the abundances of 37 metabolites, which are biomarkers for salt stress tolerance (Liang et al., 2021a). Using a wheat mGWAS analysis, Chen et al. (2020a) find 26 candidate genes and verify two genes involved in the flavonoid decorating process of wheat kernels (Chen et al., 2020a). Using a similar multiomics integration technique, various research studies have also been published to describe genes and analyze the genetics of metabolic variables in other crops, including cucumber, rice, tomato, and maize (Wen et al., 2014, 2016; Alseekh et al., 2015; Zhu et al., 2018a; Chen et al., 2016). Moreover, in order to confirm the significance of a number of potential rice genes, the researchers studied integrated metabolomics, genomics, and other technologies (Liu et al., 2020). Second, the integration of metabolomics and transcriptomics is a strategy that also helps to reveal the relationship between an organism's genotype and phenotype. Examples of these relationships include the identification of the function of core genes involved in specific metabolic pathways, such as the biosynthesis of proanthocyanidin in *A. thaliana*, noscapine, tanshione in *Salvia miltiorrhiza*, favonoids and amino acids in *Camellia sinensis* L., and anthocyanin in *Triticum aestivum* L. (Huang et al., 2018; Wang et al., 2021b).

To illustrate the control of plant hormones and carbon metabolism in *Vigna radiata* throughout postgermination seedling development, for instance, we have combined metabolomics and transcriptomics (Wang et al., 2021a). The same multiomics approach was used in several earlier research studies to identify metabolic changes or metabolic regulation that took place throughout plant growth or to comprehend the global metabolic responses to various stimuli (Asfaw Degu et al., 2014; Oates et al., 2015; Agarrwal et al., 2016). Furthermore, the integration of metabolomics and proteomics has been employed to illustrate changes in cellular composition, including proteins and metabolism, that are necessary for plants to exhibit phenotypic plasticity. For instance, this has been demonstrated in response to various stresses and perturbations in glucosinolate biosynthesis (Chen et al., 2012; Kushalappa and Gunnaiah, 2013). Amiour et al. (2012) used metabolomics, transcriptomics, and proteomics to uncover important processes in maize nitrogen metabolism. Dai et al. (2021) recently studied the phytohormone, amino acid, and energy metabolism responses to winter adaptation in the winter and spring young shoots of a unique ever-growing tea tree by comparative metabolomics, proteomics, and transcriptomics investigations. Through the integration of metabolomics and proteomics datasets for the purpose of studying banana carotenoid content, we discovered that elevated levels of carbohydrate accumulation and increased abundance of proteins associated with carotenogenesis contribute to the high carotenoid content of banana pulp during development. This suggests that a multitarget approach is required to improve the carotenoid content of banana (Heng et al., 2019).

8.9 LIMITATIONS OF METABOLOMICS

Despite the many uses of metabolomics in the identification of antimicrobial drug leads that we have just discussed, this technology is not without its drawbacks. While *in vitro* testing is the focus of most research on bacteria's reaction to plant secondary metabolites, understanding the *in vivo* mechanism of action is essential for the creation of novel medications (Domínguez et al., 2017). This condition may be explained by a number of factors. The first is that samples acquired *in vitro* have a greater degree of homogeneity than those collected *in vivo*. Both types of samples ought to represent the metabolome in its real state, although differences in technique might affect the results. Moreover, conditions *in vitro* are much more consistent and do not differ across animals (Halouska et al., 2013). The ability to distinguish between metabolites from the host and bacteria is the other factor. Correctly assigning the observed alterations may be difficult if determining the route of action necessitates analyzing metabolites generated by both the bacterium and the host. The ability to distinguish between metabolites from the host and bacteria is the other factor. Correctly assigning the observed alterations may be difficult if determining the route of action necessitates analyzing metabolites generated by both the bacterium and the host. Consequently, in order to get over this restriction, metabolomics must be complemented with further "OMICS" in order to investigate the *in vivo* action of antibiotics. Aside from the issue with *in vitro*/*in vivo* testing, metabolomics may not always be enough to provide a good explanation for antibiotics work mechanism. Basic chemometrics and molecular networking, for instance, are useful in establishing connections between the chemical makeup of crude plant mixes and their antibacterial activity, but they are unable to elucidate their mode of action (Choudhury et al., 2021). These techniques provide an approximate identification of the active compounds, but more focused isolation and definitive identification are required to fully comprehend their impact on bacteria. Then, the metabolome of the bacteria under the effect of the active compounds must be monitored (Allard et al., 2017). Comparably, a more thorough gene-metabolite method may provide more information on the regulation of bacteria's response to stressors than can pathway analysis alone (Domínguez et al., 2017). Another problem that should be addressed in the future is the scarcity of bacterial model organisms generally accessible for evaluating metabolic alterations. *Bacillus subtilis*, *Escherichia coli*, *Thermotoga maritima*, *Streptococcus pyogenes* M1 476 (serotype M1), *Synechococcus elongatus* PCC7942, *Klebsiella pneumoniae* MGH 78,578 (serotype K52), *Mesorhizobium japonicum* MAFF 303,099, and *K. variicola* At-22 are some of the most studied models. However, in certain situations, this is insufficient, and data must be examined in the phylogenetically closest organisms (Pang et al., 2020). Although it is acknowledged as a limitation of metabolomics research that not all metabolites can be identified in all sample types, it is nevertheless vital to map metabolites onto metabolic pathways for a large number of model species (Edison et al., 2016).

8.10 CONCLUSION

Plants depend on a complex and layered metabolic defense mechanism to protect themselves and effectively retaliate against the detrimental impacts of both biotic and abiotic stressors. Extensive metabolomic analysis of various plant species exposed to biotic stress, with a specific focus on pathogen infection, demonstrated that pathogen infection caused changes in the levels of several primary metabolites, including carboxylic compounds, glycolysis, pentose phosphate, and shikimate pathways. Additionally, secondary metabolites such as terpenoid/isoprenoid, phenolic compounds, and nitrogen/sulfur-containing compounds were also affected. The outcome of host–pathogen interactions is frequently influenced by metabolic alterations, which are typically linked to the availability and flux of nutrients within the infected plant. Basically, these metabolic changes might have a role in the host and pathogen interaction, either directly or indirectly, to make the host more robust and resistant to phytopathogens' negative effects. However, phytopathogens themselves have the ability to exploit a variety of metabolites as virulence factors or to compromise host immunity. To unravel the complicated interactions between the plant and phytopathogenic microorganisms, a deeper knowledge of metabolic reprogramming and a complete view of how metabolic networks are controlled throughout the host–pathogen interaction is required. In conclusion, we think that the current topic "Metabolic-based host–pathogen interactions" opens up intriguing possibilities for better understanding host–pathogen interactions at the metabolic level and the possible role(s) of host/microbe primary and secondary metabolites against pathogens and abiotic stress.

KEYWORDS

- **plant-microbe interaction**
- **biotic stress**
- **phytopathogen**
- **specialized metabolites**
- **metabolic reprogramming**
- **metabolomics**

REFERENCES

Abdel-Lateif, K., Vaissayre, V., Gherbi, H., Verries, C., Meudec, E., Perrine-Walker, F., et al. Silencing of the chalcone synthase gene in Casuarina glauca highlights the important role of flavonoids during nodulation. *New Phytol.*, 2013, 199(4), 1012–1021. https://doi.org/10.1111/nph.12326.

Adedayo, A. A., Babalola, O. O. The potential of biostimulants on soil microbial community: A review. *Front. Indust. Microbiol.*, 2023, 1, 1308641. https://doi.org/10.3389/finmi.2023.1308641

Adeniji, A. A., Babalola, O. O., Loots, D. T. Metabolomic applications for understanding complex tripartite plant-microbes interactions: Strategies and perspectives. *Biotechnology Reports (Amsterdam, Netherlands)*, 2020, 25, e00425. https://doi.org/10.1016/j. btre.2020.e00425.

Agarrwal, R., Padmakumari, A. P., Bentur, J. S., Nair, S. Metabolic and transcriptomic changes induced in host during hypersensitive response mediated resistance in rice against the Asian rice gall midge. *Rice*, 2016, 9, 5.

Allard, P. M., Genta-Jouve, G., Wolfender J-L. Deep metabolome annotation in natural products research: towards a virtuous cycle in metabolite Identification. *Curr. Opin. Chem. Biol.*, 2017, 36, 40–49.

Allwood, J. W., Williams, A., Uthe, H., van Dam, N. M., Mur, L. A. J., Grant, M. R., Pétriacq, P. Unravelling plant responses to stress—The importance of targeted and untargeted metabolomics. *Metabolites*, 2021, 11(8), 558. https://doi.org/10.3390/metabo11080558.

Alseekh, S., Tohge, T., Wendenberg, R., Scossa, F., Omranian, N., Li, J., et al. Identification and mode of inheritance of quantitative trait loci for secondary metabolite abundance in tomato. *Plant Cell.*, 2015, 27, 485–512.

Amiour, N., Imbaud, S., Clément, G., Agier, N., Zivy, M., Valot, B., et al. The use of metabolomics integrated with transcriptomic and proteomic studies for identifying key steps involved in the control of nitrogen metabolism in crops such as maize. *J. Exp. Bot.*, 2012, 63(2), 695–709.

Arias, M. M., Leandro, L. F., Munkvold, G. P. Aggressiveness of Fusarium species and impact of root infection on growth and yield of soybeans. *Phytopathology*, 2013, 103(8), 822–832. https://doi.org/10.1094/PHYTO-08-12-0207-R.

Asfaw Degu, U. H., Sikron, N., Venturini, L., Buson, G., Ghan, R., Plaschkes, I., et al. Metabolite and transcript profiling of berry skin during fruit development elucidates differential regulation between Cabernet Sauvignon and Shiraz cultivars at branching points in the polyphenol pathway. *BMC Plant Biol.*, 2014, 14, 188.

Auguy, F., Abdel-Lateif, K., Doumas, P., Badin, P., Guerin, V., Bogusz, D., et al. Activation of the isoflavonoid pathway in actinorhizal symbioses. *Funct. Plant Biol.*, 2011, 38(9), 690–696. https://doi.org/10.1071/FP11014.

Back, M. A., Haydock, P. P. J., Jenkinson, P. Disease complexes involving plant parasitic nematodes and soilborne pathogens. *Plant Pathol.*, 2002, 51(6), 683–697. https://doi.org/10.1046/j.1365-3059.2002.00785.x.

Balog, J., Sasi-Szabó, L., Kinross, J., Lewis, M., Muirhead, L., Veselkov, K., Mirnezami, R., Dezső, B., Damjanovich, L., Darzi, A., et al. Intraoperative tissue identification using rapid evaporative ionization mass spectrometry. *Sci. Transl. Med.*, 2013, 5, 194ra193.

Balog, J., Szaniszlo, T., Schaefer, K.-C., Denes, J., Lopata, A., Godorhazy, L., Szalay, D., Balogh, L., Sasi-Szabo, L., Toth, M., et al. Identification of biological tissues by rapid evaporative ionization mass spectrometry. *Anal. Chem.*, 2010, 82, 7343–7350.

Barding, G. A., Jr., Béni, S., Fukao, T., Bailey-Serres, J., Larive, C. K. Comparison of GC-MS and NMR for metabolite profiling of rice subjected to submergence stress. *J. Proteome Res.*, 2013, 12, 898–909.

Bednarek, P., Pislewska-Bednarek, M., Svatos, A., Schneider, B., Doubsky, J., Mansurova, M., et al. A glucosinolate metabolism pathway in living plant cells mediates broad-spectrum antifungal defense. *Science*, 2009, 323(5910), 101–106. https://doi.org/10.1126/science.1163732.

Beyramysoltan, S., Giffen, J., Rosati, J., Musah, R. A. Direct analysis in real time-mass spectrometry and Kohonen artificial neural networks for species identification of larva, pupa and adult life stages of carrion insects. *Anal. Chem.*, 2018, 90, 9206–9217.

Bladergroen, M. R., Spaink, H. P. Genes and signal molecules involved in the rhizobia-leguminoseae symbiosis. *Curr. Opin. Plant Biol.*, 1998, 1(4), 353–359. https://doi.org/10.1016/1369-5266(88)80059-1.

Bones, A. M., Rossiter, J. T. The myrosinase-glucosinolate system, its organisation and biochemistry. *Physiol. Plant.*, 1996, 97, 194–208.

Broeckling, C. D., Huhman, D. V., Farag, M. A., Smith, J. T., May, G. D., Mendes, P., Dixon, R. A., Sumner, L. W. Metabolic profiling of Medicago truncatula cell cultures reveals the effects of biotic and abiotic elicitors on metabolism. *J. Exp. Bot.*, 2004, 56, 323–336.

Chen, W., Wang, W., Peng, M., Gong, L., Gao, Y., Wan, J., et al. Comparative and parallel genome-wide association studies for metabolic and agronomic traits in cereals. *Nat. Commun.*, 2016, 7, 12767.

Chen, Y. Z., Pang, Q. Y., He, Y., Zhu, N., Branstrom, I., Yan, X. F., et al. Proteomics and metabolomics of Arabidopsis responses to perturbation of glucosinolate biosynthesis. *Mol. Plant.*, 2012, 5, 1138–50.

Chen, F., Ma, R., Chen, X. Advances of metabolomics in fungal pathogen–plant interactions. *Metabolites*, 2019, 9(8), 169. https://doi.org/10.3390/metabo9080169

Chong, J., Yamamoto, M., Xia, J. MetaboAnalystR 2.0: From raw spectra to biological insights. *Metabolites*, 2019, 9, 57.

Choudhury, S., Mazumder, M. K., Moulick, D., Sharma, P., Tata, S. K., Ghosh, D., Ali, H. M., Siddiqui, M. H., Brestic, M., Skalicky, M., Hossain, A. A computational study of the role of secondary metabolites for mitigation of acid soil stress in cereals using dehydroascorbate and mono-dehydroascorbate reductases. *Antioxidants*, 2022, 11(3), 458. https://doi.org/10.3390/antiox11030458

Choudhury, S., Sharma, P., Moulick, D. et al. Unrevealing metabolomics for abiotic stress adaptation and tolerance in plants. *J. Crop Sci. Biotechnol.*, 2021, 24, 479–493. https://doi.org/10.1007/s12892-021-00102-8

Clay, N. K., Adio, A. M., Denoux, C., Jander, G., Ausubel, F. M. Glucosinolate metabolites required for an Arabidopsis innate immune response. *Science*, 2009, 323(5910), 95–101. https://doi.org/10.1126/science.1164627.

Cullimore, J. V., Ranjeva, R., Bono, J. J. Perception of lipo-chitooligosaccharidic nod factors in legumes. *Trends Plant Sci.*, 2001, 6(1), 24–30. https://doi.org/10.1016/s1360-1385(00)01810-0.

Dai, Z., Huang, H., Zhang, Q., Bei, J., Chen, Z., Liu, Q., et al. Comparative multi-omics of tender shoots from a novel ever growing tea cultivar provide insight into the winter adaptation mechanism. *Plant Cell Physiol.*, 2021, 62, 366–77.

Dietz, K. J., Sauter, A., Wichert, K., Messdaghi, D., Hartung, W. Extracellular beta-glucosidase activity in barley involved in the hydrolysis of ABA glucose conjugate in leaves. *J. Exp. Bot.*, 2000, 51(346), 937–944.

Domínguez, Á., Muňoz, E., López, M. C., Cordero, M., Martínez, J. P., Viňas, M. Transcriptomics as a tool to discover new antibacterial targets. *Biotechnol. Lett.*, 2017, 39, 819–828

Draper, S. L., McCarney, E. R. Benchtop nuclear magnetic resonance spectroscopy in forensic chemistry. *Org. Magn. Reson.*, 2021, 61, 106–129.

Edison, A. S., Hall, R. D., Junot, C., Karp, P. D., Kurland, I. J., Mistrik, R., Reed, L. K., Saito, K., Salek, R. M., Steinbeck, C., Sumner, L. W., Viant, M. R. The time is right to focus on model organism metabolomes. *Metabolites*, 2016, *6*(1), 8, https://doi.org/10. 3390/metabo6010008

Eshelli, M., Qader, M. M., Jambi, E. J., Hursthouse, A. S., Rateb, M. E. Current status and future opportunities of Omics tools in Mycotoxin research. *Toxins (Basel)*, 2018, 10(11), 433. https://doi.org/10.3390/toxins10110433.

Fang, C., Luo, J. Metabolic GWAS-based dissection of genetic bases underlying the diversity of plant metabolism. *Plant J.*, 2019, 97, 91–100.

Feussner, I., Polle, A. What the transcriptome does not tell—Proteomics and metabolomics are closer to the plants' patho phenotype. *Curr. Opin. Plant Biol.*, 2015, 26, 26–31.
Fiehn, O. Metabolomics—The link between genotypes and phenotypes. *Plant Mol. Biol.*, 2002, 48, 155–171.
Fiehn, O., Kopka, J., Dörmann, P., Altmann, T., Trethewey, R. N. and Willmitzer, L. Metabolite profiling for plant functional genomics. *Nat. Biotechnol.*, 2000, 18, 1157–1161 https://doi.org/10.1038/81137
Gamir, J., Pastor, V., Cerezo, M., Flors, V. Identification of indole-3-carboxylic acid as mediator of priming against *Plectosphaerella cucumerina*. *Plant Physiol. Biochem.*, 2012, 61, 169–179. https://doi.org/10.1016/j.plaphy.2012.10.004.
Genre, A., Chabaud, M., Balzergue, C., Puech-Pages, V., Novero, M., Rey, T., et al. Short-chain chitin oligomers from arbuscular mycorrhizal fungi trigger nuclear Ca2+ spiking in Medicago truncatula roots and their production is enhanced by strigolactone. *New Phytol.*, 2013, 198(1), 190–202. https://doi.org/10.1111/nph.12146.
Glawischnig, E. Camalexin. *Phytochemistry*, 68(4), 401–406. https://doi.org/10.1016/j.phytochem. 2007, 2006. 12.005.
Glick, B. R. Plant growth-promoting bacteria: mechanisms and applications. *Scientifica*, 2012, 963401 https://doi.org/10.6064/2012/963401
Gomez-Casati, D. F., Zanor, M. I., Busi, M. V. Metabolomics in plants and humans: Applications in the prevention and diagnosis of diseases. *Biomed. Res. Int.*, 2013, 2013, 792527.
Gordon, T. R. *Fusarium oxysporum* and the Fusarium wilt syndrome. *Annu. Rev. Phytopathol.*, 2017, 55,23–39. https://doi.org/10.1146/annurev-phyto-080615-095919.
Gupta, P., Gupta, S., Pruthi, V. Techniques for detection and extraction of metabolites. In: Singh, V., Singh, A., Bhargava, P., Joshi, M., Joshi, C. (Eds.), *Engineering of Microbial Biosynthetic Pathways*. Springer: Singapore, 2020. https://doi.org/10.1007/978-981-15-2604-6_3
Halkier, B. A., Gershenzon, J. Biology and biochemistry of glucosinolates. *Annu. Rev. Plant Biol.*, 2006, 57, 303–333. https://doi.org/10.1146/annurev.arplant.57.032905.105228.
Halouska, S., Zhang, B., Gaupp, R., Lei, S., Snell, E., Fenton, R. J., Barletta, R. G., Somerville, G. A., Powers, R. Revisiting protocols for the NMR analysis of bacterial metabolomes. *J. Integr. OMICS.*, 2013, 3, 120–137
Haney, C. H., Samuel, B. S., Bush, J., Ausubel, F. M. Associations with rhizo sphere bacteria can confer an adaptive advantage to plants. *Nat. Plants*, 2015, 1(6), 15051. https://doi.org/10.1038/nplants.2015.51.
Hansen, E. M., Goheen, E. M. *Phellinus Weirii* and other native root pathogens as determinants of Forest structure and process in Western North America. *Annu. Rev. Phytopathol.*, 2000, 38, 515–539. https://doi.org/10.1146/annurev.phyto.38.1.515.
Harrison, M. J., Dewbre, G. R., Liu, J. A phosphate transporter from *Medicago truncatula* involved in the acquisition of phosphate released by arbuscular mycorrhizal fungi. *Plant Cell*, 2002, 14(10), 2413–2429.
Heng, Z., Sheng, O., Huang, W., Zhang, S., Fernie, A. R., Motorykin, I., et al. Integrated proteomic and metabolomic analysis suggests high rates of glycolysis are likely required to support high carotenoid accumulation in banana pulp. *Food Chem.*, 2019, 297, 125016.
Horning, E. C., Horning, M. G. Metabolic profiles: Chromatographic methods for isolation and characterization of a variety of metabolites in man. *Methods Med. Res.*, 1970, 12, 369–371.
Huang, H., Yao, Q., Xia, E., Gao, L. Metabolomics and transcriptomics analyses reveal nitrogen influences on the accumulation of flavonoids and amino acids in young shoots of tea plant (*Camellia sinensis* L.) associated with tea flavor. *J. Agric. Food Chem.*, 2018, 66, 9828–38.
Issaq, H. J., Van, Q. N., Waybright, T. J., Muschik, G. M., Veenstra, T. D. Analytical and statistical approaches to metabolomics research. *J. Sep. Sci.*, 2009, 32, 2183–2199.
Iven, T., Konig, S., Singh, S., Braus-Stromeyer, S. A., Bischoff, M., Tietze, L. F., et al. Transcriptional activation and production of tryptophan-derived secondary metabolites in Arabidopsis roots contributes to the defense against the fungal vascular pathogen *Verticillium longisporum*. *Mol. Plant*, 2012, 5(6), 1389–1402. https://doi.org/10.1093/mp/sss044.
Jacoby, R. P., Chen, L., Schwier, M., Koprivova, A., Kopriva, S. Recent advances in the role of plant metabolites in shaping the root microbiome. *F1000Res*, 2020, 9, F1000, https://doi. org/10.12688/f1000research.21796.1.
Jacoby, R. P., Peukert, M., Succurro, A., Koprivova, A., Kopriva, S. The role of soil microorganisms in plant mineral nutrition-current knowledge and future directions. *Front. Plant Sci.*, 2017, 8, 1617. https://doi.org/10.3389/fpls.2017.01617.

Johnson, C. H., Ivanisevic, J., Benton, H. P., Siuzdak, G. Bioinformatics: The next frontier of metabolomics. *Anal. Chem.*, 2015, 87, 147–156.

Jung, S. C., Martinez-Medina, A., Lopez-Raez, J. A., Pozo, M. J. Mycorrhiza induced resistance and priming of plant defenses. *J. Chem. Ecol.*, 2012, 38(6), 651–664. https://doi.org/10.1007/s10886-012-0134-6.

Kamoun, S., Furzer, O., Jones, J. D., Judelson, H. S., Ali, G. S., Dalio, R. J., et al. The top 10 oomycete pathogens in molecular plant pathology. *Mol. Plant Pathol.*, 2015, 16(4), 413–434. https://doi.org/10.1111/mpp.12190.

Khanijou, J. K., Kulyk, H., Bergès, C., Khoo, L. W., Ng, P., Yeo, H. C., Helmy, M., Bellvert, F., Chew, W., Selvarajoo, K. Metabolomics and modelling approaches for systems metabolic engineering. *Metab. Eng. Commun.*, 2022, 15, e00209. https://doi.org/10.1016/j.mec.2022.e00209

Koprivova, A., Schuck, S., Jacoby, R. P., Klinkhammer, I., Welter, B., Leson, L., et al. Root-specific camalexin biosynthesis controls the plant growth-promoting effects of multiple bacterial strains. *Proc. Nat. Acad. Sci. U.S.A.*, 2019, 116(31), 15735–15744. https://doi.org/10.1073/pnas.1818604116.

Kunej, U., Mikulic-Petkovsek, M., Radisek, S., Stajner, N. Changes in the phenolic compounds of hop (*Humulus lupulus* L.) induced by infection with Verticillium non-alfalfae, the causal agent of hop Verticillium wilt. *Plants (Basel)*, 2020, 9(7), 841. https://doi. org/10.3390/plants9070841.

Kushalappa, A. C., Gunnaiah, R. Metabolo-proteomics to discover plant biotic stress resistance genes. *Trends Plant Sci.*, 2013, 18, 522–31.

Lee, H. J., Ryu, D. Worldwide occurrence of Mycotoxins in cereals and cereal derived food products: Public health perspectives of their co-occurrence. *J. Agric. Food Chem.*, 2017, 65(33), 7034–7051. https://doi.org/ 10.1021/acs.jafc. 6b04847.

Lei, C., Liu, Y., Ma, Q., Chen, Z., Kong, Q., Jia, Y., Wang, X., Li, X., Xue, Z., Chen, J. Metabolomics-based alleviation of depression by Xiaoyaosan through regulation of XDH and GRIA2. *Pharmacological Research—Modern Chinese Medicine*, 2023, 8, 100284. https://doi.org/10.1016/j.prmcm.2023.100284

Liang, X., Liu, S., Wang, T., Li, F., Cheng, J., Lai, J., et al. Metabolomics-driven gene mining and genetic improvement of tolerance to salt-induced osmotic stress in maize. *New Phytol.*, 2021a, 230, 2355–70.

Liu, X., Qin, D., Piersanti, A., Zhang, Q., Miceli, C., Wang, P. Genome-wide association study identifies candidate genes related to oleic acid content in soybean seeds. *BMC Plant Biol.*, 2020, 20, 399.

Liu, R., Bao, X., Zhao, J., Li, H. Advances in the study of metabolomics and metabolites in some species interactions. *Molecules*, 2021, 26(11). https://doi.org/10.3390/molecules26113311

Long, S. R. Rhizobium symbiosis: Nod factors in perspective. *Plant Cell*, 1996, 8(10), 1885–1898. https://doi. org/10.1105/tpc.8.10.1885.

Mandal, S., Anand, U., López-Bucio, J., Kumar, M., Lal, M. K., Tiwari, R. K., Dey, A. Biostimulants and environmental stress mitigation in crops: A novel and emerging approach for agricultural sustainability under climate change. *Environ. Res.*, 2023, 233, 116357. https://doi.org/10.1016/j.envres.2023.116357

Mansfield, J., Genin, S., Magori, S., Citovsky, V., Sriariyanum, M., Ronald, P., et al. Top10 plant pathogenic bacteria in molecular plant pathology. *Mol. Plant Pathol.*, 2012, 13(6), 614–629. https://doi.org/10.1111/j. 1364-3703.2012.00804.x.

Marchev, A. S., Vasileva, L. V., Amirova, K. M. et al. Metabolomics and health: from nutritional crops and plant-based pharmaceuticals to profiling of human biofluids. *Cell. Mol. Life Sci.*, 2021, 78, 6487–6503. https://doi. org/10.1007/s00018-021-03918-3

Mendes, R., Garbeva, P., Raaijmakers, J. M. The rhizosphere microbiome: Significance of plant beneficial, plant pathogenic, and human pathogenic microorganisms. *FEMS Microbiol. Rev.*, 2013, 37(5), 634–663. https://doi. org/10.1111/1574 6976.12028.

Meredith, L. K., Ledford, S. M., Riemer, K., Geffre, P., Graves, K., Honeker, L. K., LeBauer, D., Tfaily, M. M., Krechmer, J. Automating methods for estimating metabolite volatility. *Front. Microbiol.*, 2023, 14, 1267234. https://doi.org/10.3389/fmicb.2023.1267234

Mierziak, J., Kostyn, K., Kulma, A. Flavonoids as important molecules of plant interactions with the environment. *Molecules*, 2014, 19(10), 16240–16265. https://doi.org/10.3390/molecules191016240.

Misra, B. B. New software tools, databases, and resources in metabolomics: updates from 2020. *Metabolomics*, 2021, 17, 49. https://doi.org/10.1007/s11306-021-01796-1

Misra, B. B., van der Hooft, J. J. Updates in metabolomics tools and resources: 2014–2015. *Electrophoresis*, 2016, 37, 86–110.

Morris, P. F., Bone, E., Tyler, B. M. Chemotropic and contact responses of *phytophthora sojae* hyphae to soybean isoflavonoids and artificial substrates. *Plant Physiol.*, 1998, 117(4), 1171–1178. https://doi.org/10.1104/pp.117.4.1171.

Nicholson, J. K., Lindon, J. C., Holmes, E. 'Metabonomics': Understanding the metabolic responses of living systems to pathophysiological stimuli via multivariate statistical analysis of biological NMR spectroscopic data. *Xenobiotica*, 1999, 29, 1181–1189.

Oates, C. N., Kulheim, C., Myburg, A. A., Slippers, B., Naidoo, S. The transcriptome and terpene profile of eucalyptus grandis reveals mechanisms of defense against the insect pest, leptocybe invasa. *Plant Cell Physiol.*, 2015, 56, 1418–28.

Oldroyd, G. E., Murray, J. D., Poole, P. S., Downie, J. A. The rules of engagement in the legume-rhizobial symbiosis. *Annu. Rev. Genet.*, 2011, 45, 119–144. https://doi. org/10.1146/annurev-genet-110410-132549.

Oliver, S. G., Winson, M. K., Kell, D. B., Baganz, F. Systematic functional analysis of the yeast genome. *Trends Biotechnol.*, 1998, 16, 373–378.

Orešič, M., McGlinchey, A., Wheelock, C. E., Hyötyläinen, T. Metabolic signatures of the exposome—Quantifying the impact of exposure to environmental chemicals on human health. *Metabolites*, 2020, 10, 454.

Pang, Z., Chong, J., Li, S., Xia, J. MetaboAnalystR 3.0: toward an optimized workflow for global metabolomics. *Metabolites*, 2020, 10(5), 186, https://doi.org/10.3390/metabo10050186

Pang, H., Jia, W., Hu, Z. Emerging applications of metabolomics in clinical pharmacology. *Clin. Pharmacol. Ther.*, 2019, 106, 544–556.

Patel, M. K., Pandey, S., Kumar, M., Haque, M. I., Pal, S., Yadav, N. S. Plants metabolome study: Emerging tools and techniques. *Plants*, 2021, 10(11), 2409. https://doi.org/10.3390/plants10112409

Perez de Souza, L., Naake, T., Tohge, T., Fernie, A. R. From chromatogram to analyte to metabolite. How to pick horses for courses from the massive web resources for mass spectral plant metabolomics. *Gigascience*, 2017, 6, gix037.

Pirzadah, T. B., Malik, B. and Hakeem, K. R. *Plant-Microbe Dynamics: Recent Advances for Sustainable Agriculture*. CRC Press: London, 2021.

Pusztahelyi, T., Holb, I. J., Pocsi, I. Secondary metabolites in fungus-plant inter actions. *Front. Plant Sci.*, 2015, 6, 573. https://doi.org/10.3389/fpls.2015.00573.

Ralston-Hooper, K. J., Adamec, J., Jannash, A., Mollenhauer, R., Ochoa-Acuña, H., Sepúlveda, M. S. Use of GC GC/TOF MS and LC/TOF-MS for metabolomic analysis of *Hyalella azteca* chronically exposed to atrazine and its primary metabolite, desethylatrazine. *J. Appl. Toxicol.*, 2011, 31, 399–410.

Ramautar, R., de Jong, G. J. Recent developments in liquid-phase separation techniques for metabolomics. *Bioanalysis*, 2014, 6, 1011–1026.

Razzaq, A., Wishart, D. S., Wani, S. H., Hameed, M. K., Mubin, M., Saleem, F. Advances in metabolomics-driven diagnostic breeding and crop improvement. *Metabolites*, 2022, 12(6). https://doi.org/10.3390/metabo12060511

Ross, A., Brunius, C., Chevallier, O., Dervilly, G., Elliott, C., Guitton, Y., Prenni, J. E., Savolainen, O., Hemeryck, L., Vidkjær, N. H., et al. Making complex measurements of meat composition fast: Application of rapid evaporative ionisation mass spectrometry to measuring meat quality and fraud. *Meat Sci.*, 2020, 181, 108333.

Rowe, H. C., Kliebenstein, D. J. Complex genetics control natural variation in Arabidopsis thaliana resistance to *Botrytis cinerea*. *Genetics*, 2008, 180(4), 2237–2250. https://doi.org/10.1534/genetics.108.091439.

Salem, M. A., Juppner, J., Bajdzienko, K., Giavalisco, P. Protocol: A fast, comprehensive and reproducible one-step extraction method for the rapid preparation of polar and semi-polar metabolites, lipids, proteins, starch and cell wall polymers from a single sample. *Plant Methods*, 2016, 12, 45.

Salem, M. A., Perez de Souza, L., Serag, A., Fernie, A. R., Farag, M. A., Ezzat, S. M., Alseekh, S. Metabolomics in the context of plant natural products research: From sample preparation to metabolite analysis. *Metabolites*, 2020, 10, 37.

Sasse, J., Martinoia, E., Northen, T. Feed your friends: Do Plant exudates shape the root microbiome? *Trends Plant Sci.*, 2018, 23(1), 25–41. https://doi.org/10.1016/j. tplants.2017.09.003.

Schlaman, H. R., Spaink, H. P., Okker, R. J., Lugtenberg, B. J. Subcellular localization of the nodD gene product in *Rhizobium leguminosarum*. *J. Bacteriol.*, 1989, 171(9), 4686–4693. https://doi.org/10.1128/jb.171.9.4686-4693.1989.

Schmitz, A. M., Harrison, M. J. Signaling events during initiation of arbuscular mycorrhizal symbiosis. *J. Integr. Plant Biol.*, 2014, 56(3), 250–261. https://doi. org/10.1111/jipb.12155.

Shafi, A., Zahoor, I. and Habib, H. Omics technologies to unravel plant-microbe interactions. In: Pirzadah, T. B., Malik, B., Hakeem, K. R. (Eds.), *Plant-Microbe Dynamics: Recent Advances for Sustainable Agriculture* (pp. 201–220). CRC Press: Boca Raton, FL, USA, 2021.

Smith, S. E., Smith, F. A. Fresh perspectives on the roles of arbuscular mycorrhizal fungi in plant nutrition and growth. *Mycologia*, 2012, 104(1), 1–13. https://doi.org/10.3852/11-229.

Smith, S. E., Smith, F. A., Jakobsen, I. Mycorrhizal fungi can dominate phosphate supply to plants irrespective of growth responses. *Plant Physiol.*, 2003, 133(1), 16–20.

Soltis, N. E., Kliebenstein, D. J. Natural variation of plant metabolism: Genetic mechanisms, interpretive caveats, and evolutionary and mechanistic insights. *Plant Physiol.*, 2015, 169, 1456–1468.

Stahl, E., Bellwon, P., Huber, S., Schlaeppi, K., Bernsdorff, F., Vallat-Michel, A., et al. Regulatory and functional aspects of indolic metabolism in plant systemic acquired resistance. *Mol. Plant*, 2016, 9(5), 662–681. https://doi.org/10.1016/j.molp.2016.01.005.

Tahara, S., Ohkawa, K., Takayama, T., Ogawa, Y. The third naturally occurring attractant toward zoospores of phytopathogenic *Aphanomyces cochlioides* from the Spinacia oleracea host plant. *Biosci. Biotechnol. Biochem.*, 2001, 65(8), 1755–1760. https://doi.org/10.1271/bbb.65.1755.

Ton, J., Mauch-Mani, B. Beta-amino-butyric acid-induced resistance against necrotrophic pathogens is based on ABA-dependent priming for callose. *Plant, J.*, 2004, 38(1), 119–130. https://doi.org/10.1111/j.1365-313X.2004.02028.x.

Tsugawa, H., Rai, A., Saito, K., Nakabayashi, R. Metabolomics and complementary techniques to investigate the plant phytochemical cosmos. *Nat Prod Rep.*, 2021, 38, 1729–59.

Tsugawa, H., Cajka, T., Kind, T., Ma, Y., Higgins, B., Ikeda, K., Kanazawa, M., Vander Gheynst, J., Fiehn, O., Arita, M. MS-DIAL: Data-independent MS/MS deconvolution for comprehensive metabolome analysis. *Nat. Methods*, 2015, 12, 523–526.

Turner, M. F., Heuberger, A. L., Kirkwood, J. S., Collins, C. C., Wolfrum, E. J., Broeckling, C. D., Prenni, J. E., Jahn, C. E. Non-targeted metabolomics in diverse sorghum breeding lines indicates primary and secondary metabolite profiles are associated with plant biomass accumulation and photosynthesis. *Front. Plant Sci.*, 2016, 7, 953.

Udvardi, M., Poole, P. S. Transport and metabolism in legume-rhizobia symbi oses. *Annu. Rev. Plant Biol.*, 2013, 62(64), 781–805. https://doi.org/10.1146/annurev arplant-050312-120235.

Villas-Bôas, S. G., Rasmussen, S., Lane, G. A. Metabolomics or metabolite profiles? *Trends Biotechnol.*, 2005, 23, 385–386.

Wang, L., Czedik-Eysenberg, A., Mertz, R. A., Si, Y., Tohge, T., Nunes-Nesi, A., et al. Comparative analyses of C(4) and C(3) photosynthesis in developing leaves of maize and rice. *Nat. Biotechnol.*, 2014a, 32, 1158–1165.

Wang, X., Jin, S., Chang, X., Li, G., Zhang, L., Jin, S. Two interaction proteins between *AtPHB6* and *AtSOT12* regulate plant salt resistance through ROS signaling. *Plant Physiol. Biochem.*, 2021a, 169, 70–80.

Wang, E., Schornack, S., Marsh, J. F., Gobbato, E., Schwessinger, B., Eastmond, P., et al. A common signaling process that promotes mycorrhizal and oomycete colonization of plants. *Curr. Biol.*, 2012, 22(23), 2242–2246. https://doi.org/10.1016/j.cub.2012.09.043.

Wang, M., Schafer, M., Li, D., Halitschke, R., Dong, C., McGale, E., et al. Blumenols as shoot markers of root symbiosis with arbuscular mycorrhizal fungi. *eLife*, 2018, 7, e37093. https://doi.org/10.7554/eLife.37093.

Wang, X., Yao, X., Zhao, A., Yang, M., Zhao, W., LeTourneau, M. K., et al. Phosphoinositide-specific phospholipase C gene involved in heat and drought tolerance in wheat (*Triticum aestivum* L.). *Genes Genomics.*, 2021b, 43, 1167–77.

Wasson, A. P., Pellerone, F. I., Mathesius, U. Silencing the flavonoid pathway in Medicago truncatula inhibits root nodule formation and prevents auxin transport regulation by rhizobia. *Plant Cell*, 2006, 18(7), 1617–1629. https://doi.org/10.1105/tpc.105.038232.

Waters, M. T., Gutjahr, C., Bennett, T., Nelson, D. C. Strigolactone signaling and evolution. *Annu. Rev. Plant Biol.*, 2017, 62(68), 291–322. https://doi.org/10.1146/annurev-arplant-042916-040925.

Wen, W., Li, D., Li, X., Gao, Y., Li, W., Li, H., et al. Metabolome-based genome-wide association study of maize kernel leads to novel biochemical insights. *Nat. Commun.*, 2014, 5, 3438.

Wen, W., Liu, H., Zhou, Y., Jin, M., Yang, N., Li, D., et al. Combining quantitative genetics approaches with regulatory network analysis to dissect the com plex metabolism of the maize kernel. *Plant Physiol.*, 2016, 170, 136–46.

Wishart, D. S., Guo, A., Oler, E., Wang, F., Anjum, A., Peters, H., Dizon, R., Sayeeda, Z., Tian, S., Lee, B. L., et al. HMDB 5.0: The human metabolome database for 2022. *Nucleic Acids Res.*, 2021, 50, D622–D631.

Zeilinger, S., Gupta, V. K., Dahms, T. E., Silva, R. N., Singh, H. B., Upadhyay, R. S. et al. Friends or foes? Emerging insights from fungal interactions with plants. *FEMS Microbiol. Rev.*, 2016, 40, 182–207 https://doi.org/10.1093/femsre/fuv045

Zhang, Y., Chen, R., Zhang, D., Qi, S., Liu, Y. Metabolite interactions between host and microbiota during health and disease: Which feeds the other? *Biomed. Pharmacother.*, 2023, 160, 114295. https://doi.org/10.1016/j.biopha.2023.114295

Zhu, G., Wang, S., Huang, Z., Zhang, S., Liao, Q., Zhang, C., et al. Rewiring of the fruit metabolome in tomato breeding. *Cell*, 2018a, 172, 249–261 e212.

Zrodnikov, Y., Davis, C. E. The highs and lows of FAIMS: Predictions and future trends for high field asymmetric waveform ion mobility spectrometry. *J. Nanomed. Nanotechnol.*, 2012, 3, 109e.

CHAPTER 9

Metabolite Diversity and Its Function in Stress Responses and Tolerance

AMIT KUMAR PRADHAN[1], MANABENDRA NATH[4], UMAKANTA CHOWRA[2], UDDIPTA BORTHAKUR[4], NIBEDITA SARMA[4], DIKSHIT GOSWAMI[4], KHIROD KALITA[4], BIKASH KUMAR KUNDU[4], SUBRAT KAKATI[4], ZINA MONI SHANDILYA[3], PREETOM REGON[5], and BHABEN TANTI[4*]

[1]*Department of Botany, Pragjyotish College, Guwahati, Assam, India*

[2]*Department of Botany, Guwahati College, Guwahati, Assam, India*

[3]*Department of Botany, Nowgong Girls College, Nagaon, Assam, India*

[4]*Department of Botany, Gauhati University, Guwahati, Assam, India*

[5]*Department of Entomology, Agricultural Research Organization, The Volcani Institute, Rishon LeZion, Israel*

*Corresponding authors

ABSTRACT

The increasing impact of climate change and environmental stressors on crop production pose a serious threat to global food security. Although numerous organisms with enhanced stress tolerance have been created through breeding and genetic engineering, their practical application is often delayed due to lengthy biological cycles and legal considerations. Among the various omics technologies, metabolomics emerges as the most versatile, flexible, and broadly applicable tool, playing a crucial role in deciphering the molecular complexities of plant stress responses. Metabolites serve as integrative indicators of biological systems, reflecting the dynamic interactions of genes, proteins, and regulatory networks. Through metabolomic profiling, distinctive metabolic patterns associated with stress tolerance can be identified, offering valuable insights into plant adaptation strategies. Its widespread utility transcends species boundaries, making it a powerful tool for understanding and improving stress resilience across diverse crops. Moreover, exploring how plants modulate primary and secondary metabolites in response to stress provides crucial molecular insights that inform sustainable agricultural practices. Overall, identifying the presence or absence of specific metabolites, along with quantifying their relative accumulation levels in relation

Exploring Plant Stress Tolerance: A Metabolomics Approach.
Shuvasish Choudhury, Debojyoti Moulick, and Mirza Hasanuzzaman (Eds.)

DOI: 10.1201/9781779641731-9

to gene expression data, provides greater accuracy in selecting tolerant crop varieties for breeding programs

9.1 INTRODUCTION

In recent years, the global agricultural systems have been confronted with unprecedented challenges originating from climate change, anthropogenic activities, and accelerated depletion of natural resources (Fadiji et al., 2022). These adversities present approaching threats to worldwide food resilience, demanding a thorough cognition of plant stress responses and the advancement of resilient agricultural practices (Zhao et al., 2017). In particular, the increasing impact of climate adversity leading to erratic weather patterns has led to the appearance of severe environmental stresses, such as salinity, water deficit, and extreme temperature, all of which have harsh consequences for crop efficiency and agricultural output (Fadiji et al., 2022).

The implications of elevation in planetary temperature on staple crops such as maize, sorghum, wheat, rice, and soybean are concerning. As per the report, with each 1°C increase in the global average temperature, these crops suffer a substantial reduction in yield, ranging from 3% to 8% over a 29-year warming trend (Priya and Singh, 2024; Zhao et al., 2017). Additionally, the impact of both drought and salinity, intensified by climate change, contributes to 50% of the world's cultivated and irrigated agricultural land (Orimoloye, 2022).

The effects of these abiotic stresses go beyond their direct impact on plant growth, as they also heighten the occurrence of biotic stressors, such as bacteria, fungi, herbivores, and insects. Alarming statistics reveal that up to 40% of global agricultural output is adversely affected by pests and diseases, with their prevalence rising due to the changing climate (Savary et al., 2019).

With the current progress in scientific understanding, a range of strategies has been developed to improve stress tolerance in crops. The strategies encompass: (1) the development of genetically modified organisms, including transgenic, cis-genic, and intragenic crops; (b) the implementation of traditional breeding programs; (c) the application of new breeding technologies like CRISPR/Cas editing (FAO, 2022); and (d) the utilization of plant bio-stimulants (Pradhan et al., 2023a; Regon et al., 2022).

Beyond genetic approaches, nongenetic aspects to enhance crop productivity under stress include the external induction of biostimulants, such as natural plant and fungal extracts, in addition to microbial inoculants. The incorporation of sugars, peptide constituents, and bioactive specialized metabolites and synthetic complexes has proven to be a practical and easily implementable method for enhancing plant stress tolerance in the short term (Khan et al., 2020). The application of endogenous metabolites has also been reported to enhance plant stress responses, with the advantage that these compounds help combat various types of stress (Iqbal et al., 2011). Phytohormones, vital regulators in plant advancement, have been utilized in crops to enhance their resistance to environmental stress. The outward implementation of phytohormones—the vital component for plant advancement has been effective in reducing the effects of environmental stress, along with genetic modification methods to enhance internal production of growth factors (Priya and Singh, 2024; Swain et al., 2023).

Various findings about the utilization and diversity of metabolites have been reported as consequences of abiotic impact (Jorge et al., 2016; Kopka et al., 2004). In response to abiotic factors like salinity, drought, UV light, temperature fluctuations, and cold stress, plants initiate the synthesis of diverse secondary metabolites (SMs). These metabolites, including polyamines, flavonoids, jasmonic acid, methyl-jasmonate, and glycine betaine (GB), are species-specific and their production is reliant on environmental cues. They play a crucial protective role against abiotic stressors. To mitigate the impact of such stressors, plants employ a range of strategies such as scavenging reactive oxygen species (ROS), synthesizing antioxidants, preserving membrane integrity, and regulating the accumulation of compatible solutes (Slama et al., 2015; Mata-Pérez et al., 2016).

Plants, fungi, and microbes produce around 200,000 metabolites, classified as primary and SMs, in response to diverse environmental stimuli. Primary metabolites, which comprise proteins, lipids, and carbohydrates, are produced by all living things and are necessary for central, critical functions, including growth and development. Plants produce SMs in various cellular compartments. SMs majorly comprise of three types, such as phenolics, terpenoids, and alkaloids. Among its many uses is defense against biotic and abiotic stress. The creation of these compounds occurs via a variety of metabolic pathways, including the tricarboxylic acid (TCA) cycle, the acetate–malonate pathway, and also the shikimate pathway (Figure 9.1).

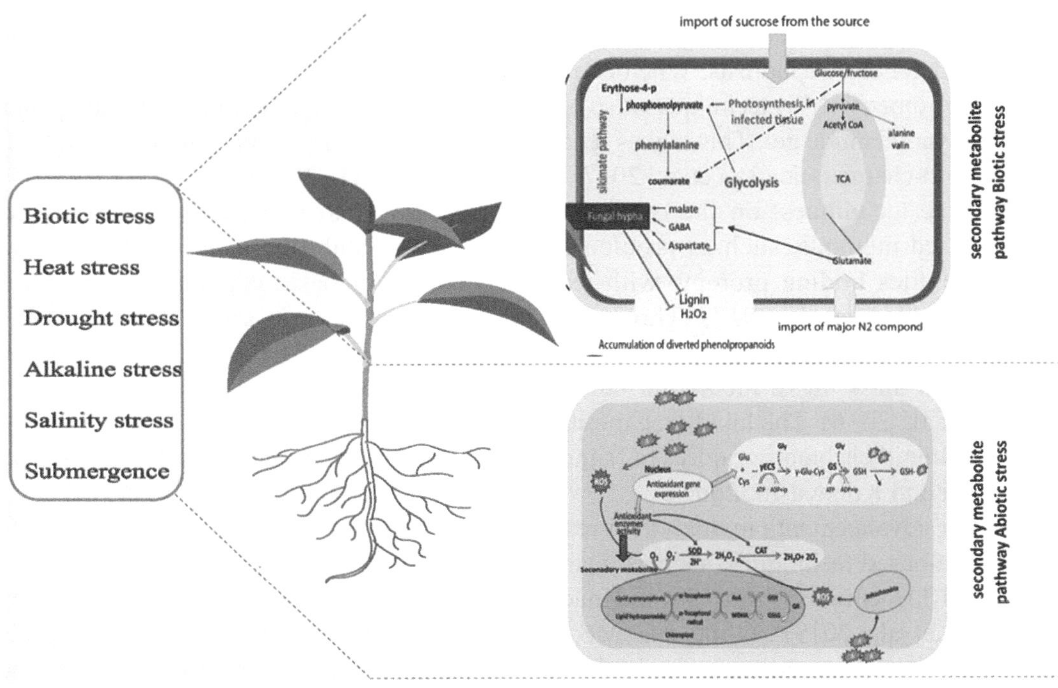

FIGURE 9.1 Schematic diagram representing secretion of secondary metabolites by plants following different metabolomic pathways against biotic and abiotic stress conditions.

Specialized metabolites are synthesized by plants when they undergo stress (Marone et al., 2022; Ding et al., 2021). These metabolites, though require a considerable amount of energy for their production, offer an effective defense mechanism of plants against both biotic and abiotic stressors (D'Amelia et al., 2021; Marone et al., 2022). Beyond their application in stress adaptability, the metabolites are also utilized in diverse fields of nutrition and drug industries, due to their nutritional and therapeutic values (Chen et al., 2021).

The exploration of plant specialized metabolism is benefited by advancements in analytical chemistry and the rapid expansion of -omics technologies (Ding et al., 2021). Metabolomics, genomics, transcriptomics, and proteomics have become fundamental tools, enabling researchers to decipher the complex relations between stress, genetic responses, and the production of specialized metabolites. However, as reported, different obstacles emerge in understanding the exact roles of these metabolites in ecological interactions and identifying the genetic elements involved in the production and stimulation (D'Amelia et al., 2021).

9.2 PLANT METABOLOMICS AND PROTEOMICS: ANALYTICAL METHODS AND TECHNIQUES

The implementation of proteomics and metabolomics covers a wide range of methodologies to investigate the metabolites, both essential and secondary generated in the plants. Proteomics research, for example, can use gel-based or gel-free techniques. Comprehensive protein studies are usually done using gel-based techniques, including 2D gel electrophoresis and differential gel electrophoresis. When combined with these gel-based techniques, advanced mass spectrometry (MS) techniques also allow detection of a large number of proteins on a single polyacrylamide gel. This makes it easier to examine their posttranslational alterations and mass-to-charge ratio (Tan et al., 2017; Vanderschuren et al., 2013).

Gel-free techniques, on the other hand, were created to overcome the drawbacks of gel-based methods, such as problems with bias, repeatability, technical know-how, and difficulties finding proteins with extremely acidic or basic characteristics or low abundance (Tan et al., 2017). Three main identifying protocols are employed in this approach: tag-based labeling (e.g., isobaric tags for relative and absolute quantification and tandem mass tags), metabolic labeling, and label-free methods (Tan et al., 2017; Ludwig et al., 2016). The label-free methods utilize techniques such as multidimensional capillary liquid chromatography (LC) and sequential window attainment of mass spectra (Fukusaki and Kobayashi, 2005).

Recent advancements in analytical research have resulted in the expansion of several metabolite-based techniques, such as gas chromatography (GC), capillary electrophoresis (CE), and LC coupled with MS and nuclear magnetic resonance (NMR) spectroscopy (Piasecka et al., 2019). While GC–MS is a popular technique for measuring organic chemicals that are volatile or semivolatile, LC–MS provides a more thorough method that enables the measurement of a wide variety of metabolites from crude extracts. Through ultraperformance, LC combined with high-resolution MS techniques, including time-of-flight, Fourier transform infrared spectroscopy, and orbitrap-based MS, LC–MS is standardized over time to improve the collection of metabolomics data (Salem et al., 2020).

Due to its complex and varied extraction requirements, CE–MS is a powerful technology that is rarely employed for plant metabolite research. However, it shows great sensitivity, especially for active compounds (Fukusaki and Kobayashi, 2005). Known for its independence from ionization and perceived as less biased, NMR spectroscopy can discover novel chemicals and is highly repeatable with a very less amount of prepared sample needed (Valentino et al., 2020). NMR spectroscopy is a useful tool for studying the metabolomics of rice in accordance to biotic stress, despite its infrequent application (Salem et al., 2020).

In order to effectively identify and categorize proteins and metabolites according to their distinct roles and associated pathways, proteomics and metabolomics provide distinct hints in the application of analytical methods and databases as prerequisites. As a result, several well-established analytical tools and databases are commonly utilized, including the Kyoto Encyclopedia of Genes and Genomes, RiceCyc, OryzaCyc, the Gene Ontology knowledgebase, and the NCBI protein database (Piasecka et al., 2019).

9.3 UTILIZING PLANT METABOLOMICS FOR ENHANCING TOLERANCE TO BIOTIC STRESS

Plants have evolved a wide range of innate and adaptive defensive systems in response to threats from other living entities, enabling them to defend themselves against harsh conditions generated by invasive disorders. The first line of protection comprises structural hardness and defense mechanisms like cell walls, barks, and waxy epidermal cuticles. The "zigzag" paradigm explains how inducible defense mechanisms, specifically effector-triggered immunity and pathogen-associated molecular pattern-triggered immunity, function as adaptive responses. Notably, in reaction to biotic stress, plants activate complex signaling networks that involve the emergence of ROS and hormone signaling. SMs like phytoalexins and phenolics are produced as a result of these cascades that control kinase signaling, which in turn activate transcription factors (TFs) and defense-related genes. Derived from several mechanisms, SMs serve as pathogen virulence factors, damage-associated molecular patterns, and antibiotic substances. Certain metabolites, including glucosinolates, benzoxazinoids, phenylpropanoid phytoalexins, and volatile indole, are implicated in rice's defense against biotic stress.

Thereby, understanding the intricate responses of plants to various biotic stresses necessitates an in-depth understanding of proteins and metabolites. Proteomics focuses on identifying and analyzing proteins to uncover specific alterations occurring during distinct developmental phases or as a consequence of stressful conditions. In contrast, metabolomics plays a vital role in elucidating the molecular mechanisms underlying plant diversity.

9.3.1 METABOLITE-INDUCED STRESS TOLERANCE AGAINST BACTERIAL INFECTION

Bacterial blight, a significant rice disease caused by the pathogen *Xanthomonas oryzae* pv. *oryzae* (Xoo), has been extensively studied to understand the global changes in rice proteins during infection. Research has shown a reduction in central carbon catabolism and an increase in signal transduction related to pathogenesis, disease resistance, and cellular metabolic regulation. As per reports, different probable receptor-like kinases, possible resistance (R) genes, and pathogenesis-related proteins have been found to be elevated in this context (Li et al., 2012; Sana et al., 2010; Mahmood et al., 2006, 2009). Furthermore, metabolomics research in susceptible and resistant phenotypes showed variations in metabolite buildup prior to and following Xoo infection (Sana et al., 2010). Notably, even without treatment, XA21-expressing plants exhibited distinct metabolite profiles, characterized by elevated levels of carbon-containing alcohols, the Kerb's cycle intermediates, and various other chemicals. More sensitive metabolites were found in XA21 plants following treatment, which may indicate a role for these plants in alkaloid metabolism and polyamine production. Notably, XA21 exhibited decreased acetophenone, a virulence indicator.

Plant growth-promoting rhizobacteria (PGPR) facilitate nutrient uptake and phytohormone production in plants by being stimulated by root exudates. Rice cultivars infected with rice-associated *Azospirillum* strains showed alterations in phenolic derivatives, including bioflavonoids and hydroxycinnamic derivatives. These alterations were influenced by the interaction between the cultivar and the PGPR strain, as revealed by metabolite profiling initiated in 2013 (Chamam et al., 2013). Common metabolomics patterns were seen in *nipponbare* inoculated with different strains of PGPR, including decreased levels of alkyl resorcinol and selective inoculation of antimicrobial derivatives, *N-p*-coumaroyl putrescine, and *N*-feruloyl putrescine (Valette et al., 2020). High-performance liquid chromatography showed that salicylic acid (SA) was produced in rice roots and root exudates by *Pseudomonas putida*, a biocontrol agent against rice illnesses (Kandaswamy et al., 2019). Based on its ability to induce systemic acquired resistance, *Pseudomonas aeruginosa* has been found to create siderophores, antibacterial chemicals, and defense-related enzymes in rice plants. According to this result, *P. aeruginosa* might be able to improve rice plant resistance to pathogens (Yasmin et al., 2017).

9.3.2 METABOLITE-INDUCED STRESS TOLERANCE AGAINST FUNGAL INFECTION

Phytopathogenic fungi, which are fundamentally divided into three categories: necrotrophs, hemibiotrophs, and biotrophs, are one of the primary infectious agents in plants (Table 9.1). Fungi and plants contain thousands of SMs. Necrotrophic fungi are dispersed more broadly among hosts than biotrophs, and they often employ toxins and enzymes that degrade cell walls, which can be small peptides or SMs (Howlett, 2006) (Table 9.1). The four primary chemical types of fungal signaling metabolites are polyketides, terpenoids, chemicals derived from shikimic acid, and nonribosomal peptides (Figure 9.1). Furthermore, meroterpenoids—a fusion of terpenes and polyketides—are a common example of hybrid metabolites made up of components from several families. Due to the significant agricultural losses caused by *Magnaporthe oryzae*, the inducing biotic pathogen, extensive research into rice responses has been conducted through proteomic and metabolomic analysis, which have clarified metabolic variations and the function of SMs in defense mechanisms. Metabolomic assays are utilized to identify certain chemicals, like the rice saponin bayogenin 3-*O*-cellobioside, which has an association with blast resistance. Different changes are seen in photosynthesis, sugar metabolism, and defense-related proteins when the necrotrophic fungus *Rhizoctonia solani* causes sheath blight. Increased antifungal protein levels, ROS-scavenging machinery, and altered energy metabolism are all present in resistant cultivars.

TABLE 9.1 List of Fungal Diseases in Addition to Their Individual Host Plants

Sl. No	Disease	Host
1	Powdery mildew	Various plants, including roses, cucurbits (e.g., squash, cucumber), grapes, beans, and many ornamental plants
2	Downy mildew	Grapes, cucurbits, lettuce, onions, and other crops
3	Rust	Wheat, barley, oats, soybeans, coffee, and various ornamental plants
4	Late blight	Tomatoes and potatoes
5	Early blight	Tomatoes, potatoes, and plants of the Solanaceae family
6	Anthracnose	Beans, cucurbits, tomatoes, peppers, mangoes, and many fruit and vegetable crops
7	*Fusarium* wilt	Various plants, including tomatoes, cucurbits, melons, and bananas
8	*Verticillium* wilt	Tomatoes, potatoes, strawberries, eggplants, and many other fruits and vegetable crops
9	Botrytis blight (gray mold)	Strawberries, grapes, tomatoes, lettuce, and various ornamental plants
10	Sooty mold	Ornamental plants, citrus trees, and other plants infested by honeydew-producing insects
11	Dutch elm disease	Elm trees
12	Citrus canker	Citrus trees
13	Apple scan	Apple trees
14	Peach leaf curl	Peach and nectarine trees
15	Coffee leaf rust	Coffee plants
16	Smuts	Corn, wheat, barley, oats, and other grasses
17	Leaf-spot diseases	Various plants such as tomatoes, roses, beans, and ornamental plants
18	*Septoria* leaf spots	Tomatoes, pepper, and other Solanaceous plants
19	White mold	Beans, peas, lettuce, and other leguminous and leafy crops
20	Root rot	Various plants, including vegetables, ornamentals, and trees
21	Fire blight	Apple and pear trees
22	Black sigatoka	Banana Plants
23	Sclerotinia	Soybeans, sunflowers, lettuce, and other crops
24	Cedar apple rust	Apple and cedar trees
25	Wheat stripe rust	Wheat and barley
26	Bean rust	Various beans, including common beans and soybeans
27	Pine rust	Pine trees
28	Peach brown rot	Peaches, plums, and other stone fruit trees
29	Cabbage black rot	Cruciferous vegetables, including cabbage and broccoli

Notes: Fungal diseases in plants are quite diverse and can affect various plant species.

9.3.3 METABOLITE-INDUCED STRESS TOLERANCE AGAINST VIRAL INFECTION

According to Alexander and Cilia (2016), viruses that infect rice plants or insects cause physiological changes, including the suppression of photosynthesis and the induction of chlorosis. Carbon metabolism is the main effect of viral infections carried by insects as vectors. After viral infection, the glycolytic enzyme glyceraldehyde-3-phosphate dehydrogenase expressed is more widely in different plant species and has multiple functions in nonmetabolic processes (Chen et al., 2021). The rice black-streaked dwarf virus decreases photosynthesis by increasing H_2O_2 generation, which alters light absorption. RBSDV, for example, promotes ABA and cytokinin, decreases gibberellins and indole-3-acetic acid, and regulates phytohormones. Susceptible and resistant cultivars differ in their levels of ROS, and some antiviral therapies cause defense-related enzymes to become overexpressed (Alexander and Cilia, 2016). Li et al.'s (2019) research, for example, showed a correlation between greater viral resistance and higher flavonoid synthesis, underscoring the importance of these SMs as important players in viral stress tolerance. The Jan et al. (2021) study offers insight into how terpenoid biosynthesis is altered during viral stress, indicating its potential for use as a defensive tactic. One of the most distinctive reactions to viral stress is the reprogramming of energy metabolism, which represents the plant's attempts to reallocate resources for defense. Research by Dhal and Pal (2022) provides insights into the regulation of glycolytic pathways during viral stress, revealing the complex regulatory mechanisms involved. The study by Nováková et al. (2020) explores the modulation of amino acid metabolism during cucumber mosaic virus infection, revealing significant alterations in key amino acids such as glutamine and arginine.

9.3.4 METABOLITE-INDUCED STRESS TOLERANCE AGAINST INSECTS' INFECTION

In order for an invasion to be successful, insects must employ sophisticated defense systems in their structures and sophisticated invasion methods in their plants. SMs are compounds that plants use to ward off herbivores, draw in natural enemies, and cause herbivore death (Dong et al., 2017). According to Lu et al. (2018), insects can overcome these protections by employing SMs or secreting effectors in salivary proteins (Peng et al., 2016). While plant defense tactics against various insects can differ, detoxification, cell wall changes, photosynthesis, phytohormones, jasmonic acid (JA) signaling, and defensive SMs are common defense mechanisms (Zogli et al., 2020; Ling et al., 2019).

The study on the impact of rice by brown plant hopper (BPH) infection, using proteomics and metabolomics, has uncovered significant changes in various aspects of rice physiology (Agarrwalet al., 2014; Wei et al., 2009). BPH, a prevalent vascular feeder, impacts lipid transport and metabolism, the biosynthesis of SMs, amino acid transport and metabolism, and phytohormone signaling, in both prone and tolerant cultivars (Zha and You, 2020; Zhang et al., 2018; 2019). Time-course investigations have shown metabolic changes in susceptible cultivars, including increased β-oxidation, glyoxylate shunt, gluconeogenesis, and GABA shunt. Glycolysis is enhanced in resistant cultivars, assisting the shikimate pathway in SM production (Liu et al., 2010). According to Dong et al. (2017), resistant varieties respond in the early stage by upregulating the machinery that removes ROS and SA, which reduces tissue damage. The *Bph6* gene transmits resistance through wax production, including methyl esters of fatty acids (Zhang et al., 2018). All cultivars exhibit early-stage inhibition of primary metabolism, which recovers in the late stage; resistant cultivars have steady levels of fatty acids, organic acids, and amino acids. Resistant rice induces higher amounts of flavonoid glycosides than susceptible rice (Kang et al., 2019; Uawisetwathana et al., 2019). By enhancing both SA and jasmonic acid response pathways, the endophytic bacterium *Bacillus velezensis* YC7010 provides tolerance to BPH pathogenicity in rice (Liu et al., 2016).

Certain interactions have also been identified by studies on how cereals react to different insects. The induction of light-dependent reactions of photosynthesis, along with the synthesis of protein constituents and other secondary bioactive molecules, is the mechanism through which the rice cultivar *Qingliu* resists the rice leafroller (*Cnaphalocrocis medinalis*). In addition, fatty acid changes both prior to infection and are linked to resistance of rice gall midge, whereas 2-amino-4-carbamoylbutanoic acid and 23-keto-docosanoic acid are linked to vulnerability (Cheah et al., 2020).

9.3.5 METABOLITE-INDUCED STRESS TOLERANCE AGAINST NEMATODE INFECTION

Nematodes, prevalent in nature, include parasitic species that affects plants like rice (Sato et al., 2019). Alterations in SMs, such as chlorogenic esters and the phytoalexin, are predominantly present in resistant rice. Notably, proteins associated with stress, metabolic pathways, and SM biosynthesis get accumulated in the early infection stage, persisting into later stages. Transcript analyses showed that in the susceptible cultivar, four notable peptide molecules related to α-linolenic acid bioactivity, glutathione (GSH) metabolism, phenyl propanoid development, and plant–pathogen interaction pathways were subsequently suppressed or reduced. Conversely, the synthesis of these proteins was enhanced in the tolerant varieties (Xiang et al., 2020; Li et al., 2012).

9.4 INDUCTION OF PLANT METABOLOMICS IN ENVIRONMENTAL STRESS TOLERANCE

Understanding the complex biochemical reactions of plants to abiotic stresses, such as drought, salt, high temperatures, and nutrient shortages, are made possible through the study of plant metabolomics. Metabolomics provides important insights into the metabolic signaling routes and regulatory networks involved in stress tolerance mechanisms by thoroughly examining the changes in metabolite profiles during stress. With this knowledge, stress-responsive metabolites, biomarkers, and important metabolic processes that plants employ to adapt to harsh environmental circumstances can be identified. Additionally, by specifically altering metabolic pathways, metabolomics makes it easier to assess stress-reduction tactics and create crop kinds that can withstand stress. A comprehensive understanding of plant stress responses is made possible by integration with other omics techniques, which eventually helps to sustainably upgrade environmental stress tolerance in crops, which is essential to guaranteeing food security worldwide.

9.4.1 METABOLITE-INDUCED STRESS TOLERANCE AGAINST DROUGHT/ SUBMERGENCE STRESS

Water is necessary for the metabolism and physiology of plants. It is necessary for the movement of nutrients and metabolites throughout the plant. Plants experience drought stress due to high transpiration rates and low water supply, which causes a drop in cell pressure and hydraulic potential. As a result, several essential physiological processes are altered, including the biosynthesis of SMs (Ashraf et al., 2018). Plants can develop drought resilience thanks to this changed SM synthesis (Verma and Shukla, 2015). In order to speed up energy production and support the glutamic acid-mediated proline biosynthesis pathway, which is necessary to improve osmotic management in plants, SMs were also utilized to improve the TCA cycle and glycolysis (Qu et al., 2019).

Several plants showed elevated levels of endogenous SMs in response to drought. Compared to Q2, which is the most vulnerable to drought with 33.9 mg of gallic acid equivalent and 27.4 mg of rutin equivalent, the rice cultivar Q8 was identified as the most tolerant among the 20 cultivars as reported by Quan et al.(2016). The levels of thymol, γ-terpinene, proline, and carbohydrates in both seedlings and calli increased following mannitol-induced drought stress, helping the plants withstand the stress (Razavizadeh and Komatsu, 2018; Ormeno et al., 2007). Plants are able to withstand drought stress due to the biosynthesis of SMs, phytohormones, osmolytes, proline, and some adaptation processes.

Terpenes and their production are stimulated in drought-stricken environments (Turtola et al., 2003). By activating the ROS-scavenging mechanism, diterpene gives salvia plants drought resistance (Munn′e-Bosch et al., 2001). Relative humidity's effects on apple plants were examined, and the results indicated that terpenes like (E,E)-α-farnesene, limonene, β-pinene, camphene, and α-pinene are released when the humidity is low (Vallat et al., 2005). Drought negatively impacts the output of monoterpenes in spearmint and rosemary (Delfine et al., 2005). It has been demonstrated that TFs related to the production of terpenes and flavonoids in potatoes and cotton are changed by drought (Payton et al., 2011; Zhang et al., 2014). Drought damage control has been linked to the release of monoterpenes, glucosides, terpenoids, and carotenoids in rosemary and grapes that are caused by drought (Liu et al., 2014). Different types of terpene hydrocarbons, a sesquiterpene called (E,E)-α-farnesene, and a terpenoid called camphor were discovered to be elevated in rapeseed when the amount of water was lower and herbivory. It is interesting to note that in risky circumstances like herbivore assaults or drought stress, plants release volatile terpenes or isoprenoids to warn nearby tissue or plants (Savoi et al., 2016; Loreto et al., 2014).

When there is a water deficit, phenolic acid and alkamide accumulate strongly in cornflower (Rani et al., 2009; Gray et al., 2003). Tea plants' ability to withstand drought is associated with a greater accumulation of polyphenols, such as catechins and bioflavonoids (Cheruiyot et al., 2008). Similar to this, in drought-resistant and sensitive cultivars of wheat, peanuts, and sesame, differential increases in flavonoids and polyphenols or expression of their biosynthesis-related genes were also noted in the event of a water deficit (Ma et al., 2014; Kermani et al., 2019; Juliano et al., 2020). In wheat, the tissue-level production of flavonoids is linked to drought susceptibility. According to reports found, the wheat spike contains a higher concentration of flavonoid chemicals than the flag leaf. As a result, spikes exhibit greater resilience to drought stress (Li et al., 2020b).

Watermelon exhibits drought oxidative stress resistance when its citrulline level builds up (Akashi et al., 2001). The water deficit resistance of different plant species and tissues is greatly influenced by the amount and composition of lignin, which also undergoes differential changes under drought stress (Moura et al., 2010). Presoaking soybean seeds in SA have been shown to increase the amounts of lignin and pectin in the roots, hence mitigating the effects of drought (Al-Hakimi, 2007). In a similar vein, β-aminobutyric acid has the ability to increase maize resistance to drought stress (Shaw et al., 2016).

During water deficiency, antioxidants such as GSH and thionine help to detoxify ROS. Under water stress, it is discovered that the content of GLs is increased in plants such as Brassica species. By influencing ABA accumulation, Arabidopsis CML42 negatively regulates herbivory resistance and drought tolerance (Vadassery et al., 2012). Transcripts of glucosinolate production were observed to be increased in Chinese cabbage during drought stress, owing to stomatal closure of GLs (Eom et al., 2018). It is interesting to note that GLs have been demonstrated to communicate with phytohormones to enhance drought tolerance. Although metabolites containing nitrogen may contribute to drought tolerance, their unique existence in various plants places certain restrictions on their potential use. For example, GLs are found only in the Brassicaceae family.

Flooding stress for two days in four-day-old soybeans increased proteins and metabolites related to the TCA cycle and GABA shunt in the roots, including the hypocotyls (Komatsu et al., 2011). The study of Nakamura et al. (2012) depicted the time-dependent flooding stress in early-stage soybean roots, including the hypocotyl, altered metabolism related to the TCA cycle, alanine synthesis, and the GABA shunt. In response to excess water stress, the production of oxaloacetate in the TCA cycle, essential for phosphoenol synthesis, was activated in the root tip of soybeans. This activation, in turn, modulated the urea cycle in the root tip of four-day-old soybeans subjected to two days of flooding (Komatsu et al., 2014). Using ^{1}H NMR research, Coutinho et al. (2018) looked into the accumulation of carbohydrates between roots and leaves during flooding in soybean seedlings and found unique patterns. When cultivars are compared, the delicate cultivar BR4 has a greater concentration of sucrose in its roots and leaves.

9.4.2 METABOLITE-INDUCED TOLERANCE AGAINST ACIDIC STRESS

In addition to being a valuable range of natural products, plant SMs play a crucial part in the plant's defense mechanism against environmental stressors and pathogenic invasions. Plant SMs are being employed more and more as food additives and medication ingredients for medicinal, gastronomic, and aromatic purposes due to their amazing biological activity.

The global issue of aluminum (Al) toxicity is limiting crop output on acidic soils. In the crust of the planet, aluminum is the most prevalent but insoluble metal. On the other hand, the phytotoxic species Al^{3+} becomes solubilized to levels that significantly impede root growth in acidic soils with pH values below 5.5, which consequently impacts crop yield (Kochian, 1995). The suppression of root growth is the most obvious sign of Al poisoning. Al^{3+} exposure has been reported in both shoots and roots, in addition to the suppression of root growth that generates a wide array of symptoms. A portion of these modifications is exclusively attributable to the active Al^{3+} efflux across the plasma membrane and the release of ionic molecules by chelating chemical secretion.

Growing experimental data indicate that the most common tolerance mechanism in the kingdom of plants is the exclusion of aluminum from the root apices through the root exudation of organic acids. While the resistant response in wheat depends on the Al-inducible release of malate, the Al-tolerant phenotype in snapbean, maize, and *Cassia tora* is linked to citrate secretion (Yang et al., 2013).

According to Delhaize et al. (1993), organic acids aid in the internal and external detoxification of plants against aluminum toxicity. Certain organic acids have the ability to create stable complexes with aluminum that hinder its binding to cellular components. This leads to the detoxification of aluminum in plant species, as reported by Ma (2000), Raman and Selvaraj (2006), and Kalita et al. (2018). Citrate is said to have the highest aluminum binding activity of all the organic acids, followed by succinate and malate (Wang et al., 2007). According to Alotaibi et al. (2021), proline is a multifunctional stress-induced metabolite that builds up in mycorrhizal plants that are subjected to multiple abiotic stressors. The work elucidated the mechanism underlying the buildup of proline produced by AMF, scrutinizing the intricacies of proline production and degradation pathways in host plants subjected to Al toxicity.

The aluminum ion (Al^{3+}) interacts directly with the pre-existing proteins to quickly activate efflux. The phenomenon of a 12–24-h lag period between the injection of aluminum (Al) and the onset of citrate release is observed in various species, including soybean (*Glycine max* L.), rye (*Secale cereale*), and common bean (*Phaseolus vulgaris* L.) (Shen et al., 2002, Ma et al., 2001). Tolrà et al. (2009) found that maize's resistance to aluminum is influenced by increased concentrations of phenolic compounds with strong antioxidant and antiradical properties, in addition to aluminum being excluded from the roots. Liu et al. (2009) verified that root tip exudation of flavonoid-type phenolics in maize is triggered by aluminum. In acidic soils, however, phenolics are less recognized than organic acids because of their strong affinity for H^+.

The release of malate and citrate from roots has been associated with strong aluminum resistance in *Lespedeza bicolor*, while phosphorus deficiency downregulated the secretion of organic compounds in either of the two *Lespedeza* species (Dong et al., 2008). Research analysis reports that adding phosphorus (P) notably lowers the release of aluminum-triggered organic acids (like citrate and malate) in aluminum-tolerant species (Pradhan et al., 2023b). This decrease in organic acid release following phosphorus application is likely due to reduced aluminum effect in these resilient species. This finding aligns with Gaume et al. (2001), who found that phosphorus supplementation reduces organic acid exudation compared to situations where aluminum exposure occurs without phosphorus in maize. Additionally, Dong et al. (2004) reported the occurrence of interactions between Al and P in roots, showing that organic compound secretion is differently persuaded by Al toxicity and P deficiency in soybean plants. Root exudates containing SMs, such as citrate and malate, are the primary source of amelioration activity in response to Al toxicity in roots (Pradhan et al., 2023b; Kalita et al., 2018).

9.4.3 METABOLITE-INDUCED TOLERANCE AGAINST SALINITY STRESS

The excessive application of chemicals like fertilizers and pesticides, along with climate change, is contributing to the rise in soil salinity worldwide. Soils affected by salinity are characterized by a high electrical conductivity in the soil saturation extract (ECe) exceeding 4 dS/m and contain excessive amounts of salts, including bicarbonates, chlorides, and sulfates of sodium (Na^+), calcium (Ca^{2+}), and magnesium (Mg^{2+}), which negatively impact plant health (Rengasamy, 2006; Numan et al., 2018). Gengmao et al. (2015) found that NaCl concentrations below 100 mM did not hinder the growth of safflower seedlings in terms of plant height, root length, and dry weight (DW), nor the relative growth rate; however, it did enhance the medicinal flavonoid content in the leaves. Additionally, activities of superoxide dismutase (SOD), catalase (CAT), peroxidase (POD), and total antioxidant capacity (T-AOC) in safflower leaves were significantly increased under all NaCl treatments.

According to Teh et al. (2016), proline treatment during salt stress caused modifications in rice root surfaces, augmenting root proliferation per plant. L-Tryptophan (TRP) treatments and varying salinity levels were applied to *Capsicum annuum* that were cultivated in containers inside a wire-enclosed facility. Additional research on sugar beet (*Beta vulgaris*) revealed that, compared to plants exposed to salt stress without amino acid treatment, those receiving amino acid supplementation exhibited improved physiological and biochemical responses (Hozayn et al., 2020).

Administering GSH elevated the levels of chlorophyll (*a* and *b*) and relative water content (RWC) in *Vigna radiata* seedlings subjected to excess salt conditions in the laboratory setup, with improvements observed at 24 and 48 h after GSH treatment. Relative to the salt stress treatment alone, the levels of conventional stress markers, such as H_2O_2, MDA, and superoxide production level, were markedly reduced in the salt + GSH supplement. When analyzed with GSH therapy with only salt excess effect, the GSH/glutathione disulfide (GSSG) ratio was also drastically higher, suggesting a healthier redox state. When analyzed in comparison to control plants under excessive salt alone, plants treated to salt + GSH also exhibited higher ROS-scavenging system enzyme activity (SOD, glutathione S-transferase, ascorbate peroxidase [APX], glutathione peroxidase, glutathione reductase, dehydroascorbate reductase, and monodehydroascorbate reductase) (Nahar et al., 2015). Similarly, compared to the salt-treated controls, *Zea mays* grown in high-salinity soil with 100–200 ppm CA showed improved growth, increased chlorophyll *a* and *b* content, higher development, and better yield (El-Hawary and Nashed, 2019).

When mannitol (15 and 30 mM) is applied topically to maize, it has been shown to have protective benefits against saline stress (100 mM NaCl). Due to mannitol's role in cellular osmotic adjustment, the combination treatment enhanced total plant DW, RWC, and chlorophyll content (*a* and *b*). This showed clear symptoms of its impact in mitigating the adverse effects of salt stress and enhancing better yield (Kaya et al., 2013).

The optimal concentration of sorbitol, 5 mg/L, significantly improved plant height, fruit number, weight, and diameter, as well as increased levels of carotenoids and chlorophyll *a* and *b* in higher salt conditions as compared to that of plants without salt supplements (without sorbitol) (Issa et al., 2020). Additionally, both 50- and 100-mM sorbitol had a notably positive impact on plant growth, DW, proline amount, and level of photosynthetic ability in wheat. In rice, 10 mM sorbitol reduced H_2O_2 and MDA levels in plants supplemented with 170 mM NaCl, resulting in enhanced growth and DW (Theerakulpisut and Gunnula, 2012).

Relative to the control plants, exogenous administration of GB to lettuce (*Lactuca sativa*) cultivated in greenhouses under salt stress conditions boosted the plants' resistance to salinity, as seen by increased levels of DW, organic acid, and amino acids (Shams et al., 2016).

Semida et al. (2016) conducted a field experiment wherein they observed a noteworthy rise in growth metrics, including shoot length, number of leaves, leaf area, shoot DW, and FW, in onions (a susceptible species to salt) cultivated in a saline environment when α-Toc was applied. Lalarukh and Shahbaz (2018) observed that seeds of two types of sunflower (*Helianthus annuus*) treated with varying doses of α-Toc showed increases in total achene weight of up to 33% and hundred achene weights of up to 16 %, respectively (Lalarukh and Shahbaz, 2020).

When tomato plants underwent melatonin therapy in a growth chamber, they produced higher accumulations of chlorophyll *a* and *b*, enhanced carbonic anhydrase and rubisco activities, and improved growth characteristics (Siddiqui et al., 2019).

9.4.4 METABOLITE-INDUCED TOLERANCE AGAINST HEAVY METAL STRESS

Anthropogenic activities include over-fertilizer application and mineral extraction. Increases in heavy metal pollution, such as excess cadmium (Cd), hexa-chromium (Cr), copper (Cu), mercury (Hg), lead (Pb), and arsenic (As), are also linked to plants and pesticides. According to Manzoor et al. (2018), these pollutants have an adverse effect on photosynthesis, respiration, transpiration, and seed germination rates, which retard plant growth, yield, and quality. Keeping in view the detrimental effects of anthropogenic activity and climate change, these circumstances necessitate modifications to agricultural techniques. In the short term, applying artificial chemicals, SMs, sugars, or amino acids is known to be a successful method of increasing plant tolerance to stress. As per the report, the application of endogenous metabolites boosts the plant's ability to handle stress. These compounds are beneficial for aiding plants in managing different forms of stress (Godoy et al., 2021).

Proline's efficacy in reducing metal toxicity has also been investigated. Dry weight (DW) and root length increased with As^+ proline treatment, while arsenic availability in eggplant seedlings reduced. Additionally, H_2O_2 and MDA levels were reduced, and the activity of ROS-scavenging-related enzymes was elevated. Proline administration enhanced production (100-grain weight) in a container trial when *Triticum aestivum* was imposed with Cu stress; nevertheless, the cultivar utilized was a determining factor (Noreen et al., 2018).

TRP also increases crop output when crops are stressed by heavy metals. Compared to control plants, rice exposed to Cd stress exhibits low plant growth and yield, such as reduced tiller numbers, fertility, and harvest index. However, relative to plants exposed to Cd stress alone, the Cd^+ TRP treatment improved these parameters (Farooq et al., 2015), suggesting that TRP is useful in mitigating Cd-induced damage.

The low-mass molecule tripeptide GSH (γ-L-glutamyl-L-cysteinylglycine) is widely present and plays several important roles, including the detoxification of ROS, the creation of phytochelatins that bind heavy metals, leading to the purification of methylglyoxal, and its function as a reservoir of cysteine (Hasanuzzaman et al., 2017, 2018).

Khan et al. (2020) reported that upland cotton (*Gossypium hirsutum*) grown hydroponically, Pb toxicity elevated H_2O_2, MDA, and Pb levels while decreasing photosynthetic attributes like stomatal activity, CO_2 concentration in between the cells, photosynthetic capacity, transpiration rate, and pigment concentration. Furthermore, under Hg stress, GSH enhanced red pepper, tobacco (*Nicotiana tabacum*), Brassicaceae plant, and *Arabidopsis thaliana* as reported by Kim et al. (2017). Additionally, they showed that GSH has a greater affinity for binding Hg compared to that of the other toxic elements, such as Cd, Zn, or Cu, suggesting that GSH would be highly efficient for damage caused by Hg (Kim et al., 2017).

Foliar applications of citric acid (CA) to *Brassica* sp. increase the enzymatic antioxidant mechanism through enhanced CAT and SOD activities, thereby mitigating the detrimental level of Cd. In addition, compared to *Brassica* treated with cadmium (Cd), the plants showed enhanced photosynthetic efficiency (by 42%), growth (by 12%), and amount of chlorophyll (by 15%), all of which were accompanied by increases in the rate of stomatal conductance (Faraz et al., 2020). However, the use of CA can reverse the effects of lead (Pb) stress, including decreased growth and yield, hormone balance, germination, and mineral absorption. When castor beans (*Ricinus communis*) cultivated in soil containing 600 mg/kg Pb and 5 mM CA was applied, phenotypes that were parallel to control plants (not containing Pb or CA) were seen. Enhancements were observed in root and shoot lengths, as well as in leaf count and area, in contrast to those of plants cultivated in soil with just Pb. The roots and leaves of lead and CA affected plants, the regulation of antioxidants, such SOD, CAT, POX, and APX, were also increased (Mallhi et al., 2019). In high amounts, Cr can be hazardous to plants. When 100 μM CA was added to the irrigation solution of rice plant, the lengths of roots and shoot and DW were recovered. The ability of CA to bind metal ions was demonstrated by rice plants, which possess greater concentrations of Cr (by 50%).

Compared to plants without mannitol, those grown exposed to Cr stress (5 and 10 mg/kg soil) showed improvements in overall plant growth and yield when 50 and 100 mg/L mannitol was applied in a foliar pattern. The phenotypes of these plants were comparable to the control plants. Mannitol seems to provide protective benefits to maize under heavy metal stress, as indicated by low oxidation level and its content in the roots and leaves, comparably higher in tolerant cultivars. Additionally, it enhances the antioxidant enzyme efficacy (SOD, CAT, and APX) and increases levels of pigment concentrations (Habiba et al., 2019). Similar outcomes were observed in wheat seeds when treated with Pb (1.5 mM) and infused with 2 μM lipoic acid; the seeds produced longer roots and coleoptiles. The plants treated with LA were able to tolerate the potentially harmful effects of internal Pb, as indicated by the altered GSH/GSSG ratio (Turk et al., 2018).

9.4.5 METABOLITE-INDUCED TOLERANCE AGAINST NUTRIENT EXCESS AND DEFICIENCY STRESS

Plants tend to modify root morphology to enhance nutrient uptake efficiency. This includes changes in root length, density, and branching as well. It is found that rice (*Oryza sativa*), Brassica plants, sugar beet crop, and many other plants accumulate proline under high nitrogen conditions, which induces more GSH under high nitrate conditions (Monreal et al., 2007; Sánchez et al., 2001). Polyamines, such as putrescine, spermidine, and spermine, are involved in contributing to stress tolerance by stabilizing cellular structures and regulating gene expression in nitrogen-deficiency condition (Altman and Levin, 1993). Wheat (*Triticum aestivum*) is well-studied to accumulate phytate (phytate is a metabolite that is involved in responding to excess phosphorus) in response to high phosphorus availability in the soil. Phytate is stored within plant cells, which can develop its strategy to mitigate the toxicity induced from excess phosphorus availability (Elanchezhian et al., 2015). Organic acids, such as citrate and malate, are synthesized and released by plant roots to enhance the solubility of phosphorus in the rhizosphere, which makes it more available to uptake (Solangi et al., 2023). Phospholipids such as phosphatidylcholine may also induce under low phosphorus conditions. Malate, anthocyanin, and phytase are synthesized in tomato, rice, and barley plants individually to adapt phosphorus deficiency condition (Lipton et al., 1987). GB is one of the metabolites that can accumulate in some plant cells. It acts as an osmoprotectant that helps plants maintain osmotic balance and protect cellular structures from the high potassium stress. Several plant species respond to potassium deficiency by expressing genes related to organic acid synthesis as citrate and malate are among the organic acids released by the roots to improve potassium solubilization in the soil and enhance potassium uptake (Sung et al., 2015; Maruyama et al., 2019). Increased levels of oxalate have been reported in *A. thaliana* as a response to high calcium concentrations. Oxalate helps to detoxify calcium-induced toxicity and harmful effects on cellular processes (Zhang et al., 2019). In wheat plant (*Triticum aestivum*), organic acids help to solubilize calcium in the soil which makes the calcium more available for root absorption (Panchal et al., 2021). In rice plant (*Oryza sativa*) and certain fruiting plants, excessive magnesium availability has been found to induce extended proline accumulation (Kwon et al., 2019). Proline acts as a compatible solute, which helps in osmotic regulation and stress tolerance in toxic or high Mg availability (Lamichhane et al., 2023). Mulberry (*Morus alba*) is well known in proline and amino acid (citrate and malate) synthesis by its roots to enhance magnesium uptake and solubilization in the rhizosphere (Yang et al., 2013). Citrate improves the accessibility of Mg in the soil and improves magnesium uptake by plant roots (Jin et al., 2023; Hasanuzzaman et al., 2017). In *Arabidopsis* plant, it is found to accumulate GSH under excess sulfur conditions. GSH acts as an important antioxidant and contributes to sulfur storing and detoxification (Gutiérrez-Alcalá et al., 2000). Plants respond to low sulfur conditions by increasing the synthesis of metabolites, including jasmonic acid, SA, and abscisic acid, which regulates plants at its transcriptomic level (Forieri et al., 2017). Phenolic compounds, for example, catechins and organic acids, as citrate and malate, facilitate iron chelation and sequestration. In rice (*Oryza sativa*), accumulation of catechins and organic acids is found in excess availability of iron.

These metabolites help in iron detoxification, which prevents toxicity produced from high availability of iron (Anjitha et al., 2021). Barley (*Hordeum vulgare*) is a well-studied example of a plant responding to low iron stress by releasing phytosiderophores (Zhang et al., 2019). These molecules, such as mugineic acids, help to solubilize of iron in the rhizosphere for root uptake. Citrate, ascorbate, and flavonoids are also synthesized by some plants to adapt in iron-deficient condition (Erb and Kliebenstein, 2020). Caumarin synthesis is also another adaptive quality in Fe deficiency conditions (Perkowska et al., 2021).

Excessive levels of these micronutrients can be toxic, leading to impaired growth. Deficiencies of these micronutrients can cause specific metabolic disorders and impact enzyme activities critical for plant growth. *A. thaliana* is known to produce phytochelatins in high zinc availability (Kühnlenz et al., 2016). Barley (*Hordeum vulgare*) synthesizes nicotianamine to cope with low zinc availability and upregulates HvNAS genes to cope with low zinc availability (Clemens et al., 2013; Aleksza et al., 2024). Rice (*Oryza sativa*) produces metallothioneins in high copper availability conditions, which help to detoxify excessive metal availability and help to adapt in stressful situations (Zhou et al., 2006). Plants may produce metabolites, for example, phytoalexins like coumarins, which defend under copper-deficient conditions (Catapano et al., 2018). For example, in *A. thaliana*, the genes related to metallothionein synthesis (*AtMT1* and *AtMT2* genes) are induced under low copper conditions. These metallothioneins contribute to adapt in low copper availability by binding and storing copper ions, which reduces their toxic effects on cellular functioning (Llerena et al., 2021). Soybeans (*G. max*) are known to release citrate from their roots when plants are in excess manganese conditions. Citrate helps to chelate manganese in the rhizosphere, which makes it less available for uptake with the help of plant roots and reduces manganese toxicity (Li et al., 2019; Rosas et al., 2007). It has been found in wheat (*Triticum aestivum*) and some other plants that low manganese stress can alter their genetic regulation (Alejandro et al., 2020). Polyols, such as mannitol and sorbitol, can bind with boron in the cytoplasm, which helps to detoxify boron. In some plants, pectin tends to attach with boron, prevents its movement into the protoplast, and reduces toxic effects on cell structure (Reid et al., 2004). In *Brassica napus* plant, pectin responds to boron deficiency and gives protective mechanism. Plants sometimes release organic acids, for example, citrate, from their roots, which solubilize boron in the plants' rhizosphere, making this nutritient more available for uptake (Zhang et al., 2014). Barley (Hordeum vulgare) responds to boron deficiency by exuding citrate from its roots. The secretion of citrate helps to solubilize boron in the soil, facilitating its uptake by plant roots (Shireen et al., 2018).

9.4.6 METABOLITE-INDUCED AGAINST LOW TEMPERATURE STRESS

Soluble sugars, including sucrose and glucose, act as cryoprotectants by preventing crystal formation and can reduce freezing point inside the plant's cells (Jahed et al., 2023). In wheat plants, soluble sugars accumulate at the time of cold stress, enhancing their freezing tolerance (Raza et al., 2023a; Xu et al., 2024). In cold stress conditions, where freezing temperatures can lead to dehydration and cell damage, the accumulation of polyols serves as a protective mechanism (Longo et al., 2018). Apple trees, for example, are known to accumulate sorbitol in response to cold stress. This accumulation of polyols contributes to the overall cold tolerance of the plant, enabling it to survive and thrive in cold climates (Goswami et al., 2022). In freezing stress, proline can also have a cryoprotective effect by stabilizing proteins and cellular structures (Hossain et al., 2014). It can occasionally function as a molecular chaperone, aiding in the refolding of proteins and preventing denaturation that may occur under freezing conditions. Studies on *Arabidopsis* have given an insight that it can produce antioxidant protection mechanisms, by accumulating ascorbate (vitamin C) and GSH, to adapt itself to cold stress (Rajput et al., 2021; Rana et al., 2017). During cold stress, plants activate antioxidant-responsive genes, which help mitigate the excessive production of ROS (Hasanuzzaman et al., 2020).

9.4.7 METABOLITE-INDUCED FOR HIGH TEMPERATURE STRESS

Chaperone proteins have a crucial role in assisting the refolding of denatured proteins and preventing protein aggregation at the same time under heat stress conditions. Heat shock proteins (HSPs) are synthesized in response to different stressors, along with heat stress, in many plants and other organisms (Kumar et al., 2023; Saibil, 2013). *Arabidopsis* and other plants synthesize the expression of HSPs in response to elevated temperature (Yurina, 2023). Polyamines (spermidine and spermine) are organic cations that respond to heat stress and plays important roles in several physiological functions within plant cells. It acts as an antioxidant and helps mitigate the damaging effects of ROS. Synthesis of polyamine at the time of heat stress in tomato plants is an example of a stress response mechanism. This increased production of polyamines is considered a part of the plant's strategy that adapts in adverse situations of high temperatures and maintains cellular homeostasis (Pang et al., 2007). Proline and trehalose are examples of compatible solutes, also known as osmo protectants or osmolytes, that play crucial roles in maintaining cellular osmotic balance and stabilizing proteins and membranes under different adverse situations, including high-temperature stress (Raza et al., 2023b; Hassan et al., 2023). Accumulation of compatible solutes at the time of high-temperature stress is also found in maize (Shao et al., 2024). Phospholipids have a huge role in organizing and functioning of cell membrane proteins, which are engaged in stress response pathways. Heat-tolerant plants may increase the abundance of certain types of fatty acids, such as those with elongated carbon structure or higher degrees of saturation (Sharma et al., 2023). These changes in lipid composition help to maintain an optimal level of cell membrane fluidity and stability under elevated temperatures (Falcone et al., 2004).

HSPs represent a critical class of molecular chaperones that play a central role in sustaining cellular balance under both regular and challenging conditions. In the context of abiotic stresses, HSPs are particularly associated with heat stress, as revealed by omics studies conducted over the past decade (Yadav et al., 2020; Li et al., 2013). Their essential role in preserving protein availability and its assembly under increased temperature is well-documented (Khan and Shahwar, 2020). Plant species such as maize, wheat, and soybean exhibit acclimatization to heat stress by synthesizing HSPs when temperatures surpass

32–40°C (Yadav et al., 2020). Transgenic tobacco expressing *Brassica campestris* Hsp70 demonstrated enhanced heat tolerance, accompanied by increased superoxide dismutase and peroxidase activity, elevated soluble sugar content, and reduced conductance (Masand and Yadav, 2016). Similarly, the upregulation of Hsp90 increases significantly in response to high-temperature stress (Reddy et al., 2016). HSP70 has been implicated in imparting tolerance against low-temperature stress in various crop plants, including barley, tobacco, and wheat (Kosová et al., 2016). Recent findings by Elkelish et al. (2020) demonstrated the resistance of tomato plants to low-temperature stress through the upregulation of CAT and HSP genes. The upregulation of HSPs helped mitigate the severity of oxidative damage caused by cold stress. HSPs play a vital role in securing crops and averting damage caused by water deficiency (Ruibal et al., 2013). Huang et al. (2019) reported that both high temperature and water deficiency induced the expression of *CaHsp*16.4, a small *HSP* gene in *Capsicum annuum*.

9.5 IMPACT OF CLIMATE CHANGE ON METABOLIC REGULATIONS IN CROP PLANTS

The impact of climate change on metabolic regulations in crop plants is a topic of critical importance, with rising atmospheric CO_2 levels being a significant factor. Since the pre-industrial era, the concentration of atmospheric CO_2 has increased substantially and is projected to reach approximately 730–1000 ppm by the end of the 21st century. This rise in CO_2 levels has profound effects on the metabolic processes of crop plants. Notably, elevated CO_2 positively influences photosynthesis in C_3 plants, leading to enhanced plant growth. For instance, studies conducted on *Solanum lycopersicum* (tomato) revealed that exposure to atmospheric CO_2 at 800 μbar resulted in a relative carbon gain increase of about 12%, a photosynthesis boost of approximately 17%, and a reduction in photosynthetic induction loss by around 14% compared to levels at 400 μbar (Kaiser et al., 2017; Roy et al., 2024).

Additionally, elevated CO_2 concentrations have been found to influence various aspects of plant physiology and development. Lettuce, for example, exhibits increased net photosynthesis rates and enhanced growth when exposed to elevated CO_2, attributed to higher carbon availability, which promotes carbohydrate accumulation (Chen et al., 2023). Furthermore, the flowering-related events in plants are also affected by elevated CO_2 levels, impacting their fitness and altering the timing of flowering (Jagadish et al., 2016). The regulation of crucial flowering genes, such as MFT (MOTHER OF FT AND TFL1), is implicated in this process, with changes in sugar status potentially influencing signaling behavior and leading to delayed flowering (Jagadish et al., 2016).

In essence, as atmospheric CO_2 levels continue to rise due to climate change, the metabolic regulations of crop plants undergo significant alterations, impacting various physiological processes crucial for their growth, development, and reproductive success. Comprehending these metabolic responses is crucial for devising strategies to alleviate the negative impacts of climate change on agricultural productivity and food security (Kumar and Prasad, 2018; Rehman and Tanti, 2022).

9.6 FUTURE PERSPECTIVES

Plants have evolved intricate adaptive mechanisms in response to a myriad of abiotic and biotic stresses, altering their physiology and morphology to withstand environmental challenges. Despite these adaptations and internal defense systems, the impact of both biotic and abiotic stresses often leads to significant losses in crop performance and productivity. As plants encounter diverse stressors, there arises a critical need for effective approaches to elevate their durability and minimize yield losses.

Metabolomics and transcriptomics offer a deep understanding in the molecular aspects underlying plants' reactions or behavioral changes under absurd environmental conditions (Regon et al., 2022; Pradhan et al., 2019). By identifying and analyzing metabolites and gene expression patterns, these omics technologies serve as a promising bridge between genotype and morphological/physiological traits. The elucidation of metabolic and transcriptional signatures associated with stress tolerance provides researchers with essential tools for predicting and selecting phenotypic traits in crop-breeding programs. Integration of genome and metabolite data enables the development of predictive models to guide breeding efforts toward the cultivation of stress-tolerant crop varieties (Regon et al., 2021).

Moreover, SMs play a crucial role in sustaining the health and productivity of food crops, especially under stressful conditions. These compounds exhibit diverse biological activities, including antioxidant, antimicrobial, and allelopathic properties, which contribute to plant defense and stress tolerance. Harnessing the potential of SMs through targeted breeding or biotechnological approaches offers a promising avenue for enhancing crop resilience and productivity. By leveraging the natural chemical arsenal of plants, researchers can develop innovative strategies to minimize yield losses and ensure nutritional viability amid shifting environmental conditions (Rehman and Tanti, 2022; Shandilya and Tanti, 2020).

In conclusion, the integration of metabolomics, transcriptomics, genomics, and the study of SMs holds immense potential for advancing crop breeding practices and improving agricultural sustainability. By understanding and manipulating the intricate cellular dynamics underlying plant adaptations, researchers can develop resilient crop varieties capable of thriving in diverse environmental conditions, thereby ensuring global food security in the midst of ongoing climate change.

KEYWORDS

- **metabolomics**
- **climate change**
- **molecular crosstalk**
- **stress tolerance**

REFERENCES

Agarrwal, R., Bentur, J. S., Nair, S. Gas chromatography mass spectrometry based metabolic profiling reveals biomarkers involved in rice-gall midge interactions. *J. Integr. Plant Biol.*, 2014, 56(9), 837–848.

Akashi, K., Miyake, C., Yokota, A. Citrulline, a novel compatible solute in drought-tolerant wild watermelon leaves, is an efficient hydroxyl radical scavenger. *FEBS Lett.*, 2001, 508(3), 438–442.

Alejandro, S., Höller, S., Meier, B., Peiter, E. Manganese in plants: From acquisition to subcellular allocation. *Front. Plant Sci.*, 2020, 11, 300, https://doi.org/10.3389/fpls.2020.00300.

Aleksza, D., Spiridon, A., Tarkka, M., Hauser, M. T., Hann, S., Causon, T., Kratena, N., Stanetty, C., George, T. S., Russell, J., Oburger, E. Phytosiderophore pathway response in barley exposed to iron, zinc or copper starvation. *Plant Sci.*, 2024, 339, 111919.

Alexander, M. M., Cilia, M. A molecular tug-of-war: Global plant proteome changes during viral infection. *Curr. Plant Biol.*, 2016, 5, 13–24.

Al-Hakimi, A. M. A. Counteraction of drought stress on soybean plants by seed soaking in salicylic acid. *Int. J. Bot.*, 2007, 2(4), 421–426.

Alotaibi, M. O., Saleh, A. M., Sobrinho, R. L., Sheteiwy, M. S., El-Sawah, A. M., Mohammed, A. E., ad Elgawad, H. A. Arbuscular mycorrhizae mitigate aluminium toxicity and regulate proline metabolism in plants grown in acidic soil. *J. Fungi*, 2021, 7(7), 531.

Altman, A., Levin, N. Interactions of polyamines and nitrogen nutrition in plants. *Physiol. Plant.*, 1993, 89(3), 653–658.

Anjitha, K. S., Sameena, P. P., Puthur, J. T. Functional aspects of plant secondary metabolites in metal stress tolerance and their importance in pharmacology. *Plant Stress*, 2021, 2, 100038.

Ashraf, M. A., Iqbal, M., Rasheed, R., Hussain, I., Riaz, M., Arif, M. S. Environmental stress and secondary metabolites in plants: An overview. In: *Plant Metabolites and Regulation under Environmental Stress* (pp. 153–167). 2018.

Catapano, M. C., Karlíčková, J., Tvrdý, V., Sharma, S., Prasad, A. K., Saso, L., Mladěnka, P. Mono and dihydroxy coumarin derivatives: Copper chelation and reduction ability. *J. Trace Elem. Med. Biol.*, 2018, 46, 88–95.

Chamam, A., Sanguin, H., Bellvert, F., Meiffren, G., Comte, G., Wisniewski-Dyé, F., Prigent-Combaret, C. Plant secondary metabolite profiling evidences strain-dependent effect in the Azospirillum–Oryza sativa association. *Phytochemistry*, 2013, 87, 65–77.

Cheah, B. H., Lin, H. H., Chien, H. J., Liao, C. T., Liu, L. Y. D., Lai, C. C., Chuang, W. P. SWATH-MS-based quantitative proteomics reveals a uniquely intricate defense response in *Cnaphalocrocis medinalis*-resistant rice. *Sci. Rep.*, 2020, 10(1), 65–97.

Chen, W., Zhang, J., Zheng, S., Wang, Z., Xu, C., Zhang, Q., Wu, J., Lou, H. Metabolite profiling and transcriptome analyses reveal novel regulatory mechanisms of melatonin biosynthesis in hickory. *Horticult. Res.*, 2021, 8(1), 196–207.

Chen, D., Zhang, J., Zhang, Z., Lu, Y., Zhang, H., Hu, J. A high efficiency CO_2 concentration interval optimization method for lettuce growth. *Sci. Total Environ.*, 2023, 879, 162731.

Cheruiyot, E. K., Mumera, L. M., NG'ETICH, W. K., Hassanali, A., Wachira, F., Wanyoko, J. K. Shoot epicatechin and epigallocatechin contents respond to water stress in tea [*Camellia sinensis* (L.) O. Kuntze]. *Biosci. Biotechnol. Biochem.*, 2008, 72(5), 1219–1226.

Clemens, S., Deinlein, U., Ahmadi, H., Höreth, S., Uraguchi, S. Nicotianamine is a major player in plant Zn homeostasis. *Biometals*, 2013, 26, 623–632.

Coutinho, I. D., Henning, L. M. M., Döpp, S. A., Nepomuceno, A., Moraes, L. A. C., Marcolino-Gomes, J., Richter, C., Schwalbe, H., Colnago, L. A. Flooded soybean metabolomic analysis reveals important primary and secondary metabolites involved in the hypoxia stress response and tolerance. *Environ. Exp. Bot.*, 2018, 153, 176–187.

D'Amelia, V., Docimo, T., Crocoll, C., Rigano, M. M. Specialized metabolites and valuable molecules in crop and medicinal plants: The evolution of their use and strategies for their production. *Genes*, 2021, 12(6), 936.

Delfine, S., Loreto, F., Pinelli, P., Tognetti, R., Alvino, A. Isoprenoids content and photosynthetic limitations in rosemary and spearmint plants under water stress. *Agric. Ecosyst. Environ.*, 2005, 106(2–3), 243–252.

Delhaize, E., Ryan, P. R., Randall, P. J. Aluminium tolerance in wheat (*Triticum aestivum* L.) (II. Aluminium-stimulated excretion of malic acid from root apices). *Plant Physiol.*, 1993, 103(3), 695–702.

Dhal, S., Pal, H. Regulation of glycolysis and krebs cycle during biotic and abiotic stresses. In: *Photosynthesis and Respiratory Cycles during Environmental Stress Response in Plants* (pp. 263–308), Apple Academic Press, 2022.

Ding, Y., Gardiner, D. M., Powell, J. J., Colgrave, M. L., Park, R. F., Kazan, K. Adaptive defence and sensing responses of host plant roots to fungal pathogen attack revealed by transcriptome and metabolome analyses. *Plant, Cell Environ.*, 2021, 44(12), 3756–3774.

Dong, D., Peng, X., Yan, X. Organic acid exudation induced by phosphorus deficiency and/or aluminium toxicity in two contrasting soybean genotypes. *Physiol. Plant.*, 2004, 122(2), 190–199.

Dong, X. Y., Shen, R. F., Chen, R. F., Zhu, Z. L., Ma, J. F. Secretion of malate and citrate from roots is related to high Al-resistance in Lespedeza bicolour. *Plant Soil*, 2008, 306, 139–147.

Dong, Y., Fang, X., Yang, Y., Xue, G. P., Chen, X., Zhang, W., Chen, J. Comparative proteomic analysis of susceptible and resistant rice plants during early infestation by small brown planthopper. *Front. Plant Sci.*, 2017, 8, 1744–1760.

Elanchezhian, R., Krishnapriya, V., Pandey, R., Rao, A. S., Abrol, Y. P. Physiological and molecular approaches for improving phosphorus uptake efficiency of crops. *Curr. Sci.*, 2015, 10, 1271–1279.

El-Hawary, M. M., Nashed, M. E. Effect of foliar application by some antioxidants on growth and productivity of maize under saline soil conditions. *J. Plant Prod.*, 2019, 10(2), 93–99.

Elkelish, A., Qari, S. H., Mazrou, Y. S., Abdelaal, K. A., Hafez, Y. M., Abu-Elsaoud, A. M., El Nahhas, N. Exogenous ascorbic acid induced chilling tolerance in tomato plants through modulating metabolism, osmolytes, antioxidants, and transcriptional regulation of catalase and heat shock proteins. *Plants*, 2020, 9(4), 431–448.

Eom, S. H., Baek, S. A., Kim, J. K., Hyun, T. K. Transcriptome analysis in Chinese cabbage (*Brassica rapa* ssp. pekinensis) provides the role of glucosinolate metabolism in response to drought stress. *Molecules*, 2018, 23(5), 1186–1198.

Erb, M., Kliebenstein, D. J. Plant secondary metabolites as defenses, regulators, and primary metabolites: The blurred functional trichotomy. *Plant Physiol.*, 2020, 184(1), 39–52.

Fadiji, A. E., Babalola, O. O., Santoyo, G., Perazzolli, M. The potential role of microbial biostimulants in the amelioration of climate change-associated abiotic stresses on crops. *Front. Microbiol.*, 2022, 12, 829–842.

Falcone, D. L., Ogas, J. P., Somerville, C. R. Regulation of membrane fatty acid composition by temperature in mutants of Arabidopsis with alterations in membrane lipid composition. *BMC Plant Biol.*, 2004, 4, 1–15.

FAO. *Gene Editing and Agrifood Systems*. Rome, 2022. https://doi.org/10.4060/cc3579en

Faraz, A., Faizan, M., Sami, F., Siddiqui, H., Hayat, S. Supplementation of salicylic acid and citric acid for alleviation of cadmium toxicity to *Brassica juncea*. *J. Plant Growth Regul.*, 2020, 39, 641–655.

Farooq, H., Asghar, H. N., Khan, M. Y., Saleem, M., Zahir, Z. A. Auxin-mediated growth of rice in cadmium-contaminated soil. *Turk. J. Agric. For.*, 2015, 39(2), 272–276.

Forieri, I., Sticht, C., Reichelt, M., Gretz, N., Hawkesford, M. J., Malagoli, M., Hell, R. System analysis of metabolism and the transcriptome in Arabidopsis thaliana roots reveals differential co-regulation upon iron, sulfur and potassium deficiency. *Plant, Cell Environ.*, 2017, 40(1), 95–107.

Fukusaki, E., Kobayashi, A. Plant metabolomics: Potential for practical operation. *J. Biosci. Bioeng.*, 2005, 100(4), 347–354.

Gaume, A., Mächler, F., De León, C., Narro, L., Frossard, E. Low-P tolerance by maize (Zea mays L.) genotypes: Significance of root growth, and organic acids and acid phosphatase root exudation. *Plant Soil*, 2001, 228(2), 253–264.

Gengmao, Z., Yu, H., Xing, S., Shihui, L., Quanmei, S., Changhai, W. Salinity stress increases secondary metabolites and enzyme activity in safflower. *Ind. Crops Prod.*, 2015, 64, 175–181.

Godoy, F., Olivos-Hernández, K., Stange, C., Handford, M. Abiotic stress in crop species: Improving tolerance by applying plant metabolites. *Plants*, 2021, 10(2), 186.

Goswami, A. K., Maurya, N. K., Goswami, S., Bardhan, K., Singh, S. K., Prakash, J., Kumar, C. Physio-biochemical and molecular stress regulators and their crosstalk for low-temperature stress responses in fruit crops: A review. *Front. Plant Sci.*, 2022, 13, 1022167.

Gray, D. E., Pallardy, S. G., Garrett, H. E., Rottinghaus, G. E. Acute drought stress and plant age effects on alkamide and phenolic acid content in purple coneflower roots. *Planta Med.*, 2003, 69(01), 50–55.

Gutiérrez-Alcalá, G., Gotor, C., Meyer, A. J., Fricker, M., Vega, J. M., Romero, L. C. Glutathione biosynthesis in Arabidopsis trichome cells. *Proc. Natl. Acad. Sci.*, 2000, 97(20), 11108–11113.

Habiba, U., Ali, S., Rizwan, M., Ibrahim, M., Hussain, A., Shahid, M. R., Ahmad, P. Alleviative role of exogenously applied mannitol in maize cultivars differing in chromium stress tolerance. *Environ. Sci. Pollut. Res.*, 2019, 26, 5111–5121.

Hasanuzzaman, M., Bhuyan, M. B., Zulfiqar, F., Raza, A., Mohsin, S. M., Mahmud, J. A., Fotopoulos, V. Reactive oxygen species and antioxidant defense in plants under abiotic stress: Revisiting the crucial role of a universal defense regulator. *Antioxidants*, 2020, 9(8), 681–610

Hasanuzzaman, M., Nahar, K., Anee, T. I., Fujita, M. Glutathione in plants: Biosynthesis and physiological role in environmental stress tolerance. *Physiol. Mol. Biol. Plants*, 2017, 23, 249–268.

Hasanuzzaman, M., Nahar, K., Rahman, A., Mahmud, J. A., Alharby, H. F., Fujita, M. Exogenous glutathione attenuates lead-induced oxidative stress in wheat by improving antioxidant defense and physiological mechanisms. *J. Plant Interact.*, 2018, 13(1), 203–212.

Hassan, M. U., Nawaz, M., Shah, A. N., Raza, A., Barbanti, L., Skalicky, M., Qari, S. H. Trehalose: A key player in plant growth regulation and tolerance to abiotic stresses. *J. Plant Growth Regul.*, 2023, 42(8), 4935–4957.

Hossain, M. A., Hoque, M. A., Burritt, D. J., Fujita, M. Proline protects plants against abiotic oxidative stress: Biochemical and molecular mechanisms. In: *Oxidative Damage to Plants* (pp. 477–522), Academic Press, 2014.

Howlett, B. J. Secondary metabolite toxins and nutrition of plant pathogenic fungi. *Curr. Opin. Plant Biol.*, 2006, 9(4), 371–375.

Hozayn, M., Abd-Elmonem, A. A., Samaha, G. M. The physiological effect of pre-soaking with tryptophan on sugar beet (Beta vulgaris L.) productivity under different levels of salinity stresses. *Bull. Natl. Res. Centre*, 2020, 44, 1–9.

Huang, L. J., Cheng, G. X., Khan, A., Wei, A. M., Yu, Q. H., Yang, S. B., Gong, Z. H. CaHSP16. 4, a small heat shock protein gene in pepper, is involved in heat and drought tolerance. *Protoplasma*, 2019, 256, 39–51.

Iqbal, N., Ashraf, Y., Ashraf, M. Modulation of endogenous levels of some key organic metabolites by exogenous application of glycine betaine in drought stressed plants of sunflower (Helianthus annuus L.). *Plant Growth Regul.*, 2011, 63, 7–12.

Issa, D. B., Alturki, S. M., Sajyan, T. K., Sassine, Y. N. Sorbitol and lithovit-guano-25 mitigates the adverse effects of salinity on eggplant grown in pot experiment. *Agronomy Res.*, 2020, 18(1), 113–126.

Jagadish, S. K., Bahuguna, R. N., Djanaguiraman, M., Gamuyao, R., Prasad, P. V., Craufurd, P. Q. Implications of high temperature and elevated CO_2 on flowering time in plants. *Front. Plant Sci.*, 2016, 7, 913.

Jahed, K. R., Saini, A. K., Sherif, S. M. Coping with the cold: Unveiling cryoprotectants, molecular signaling pathways, and strategies for cold stress resilience. *Front. Plant Sci.*, 2023, 14, 1246093.

Jan, R., Asaf, S., Numan, M., Lubna, Kim, K. M. Plant secondary metabolite biosynthesis and transcriptional regulation in response to biotic and abiotic stress conditions. *Agronomy*, 2021, 11(5), 968.

Jin, X., Ackah, M., Wang, L., Amoako, F. K., Shi, Y., Essoh, L. G., Zhao, W. Magnesium nutrient application induces metabolomics and physiological responses in mulberry (*Morus alba*) plants. *Int. J. Mol. Sci.*, 2023, 24(11), 9650.

Jorge, T. F., Rodrigues, J. A., Caldana, C., Schmidt, R., van Dongen, J. T., Thomas-Oates, J., António, C. Mass spectrometry-based plant metabolomics: Metabolite responses to abiotic stress. *Mass Spectrom. Rev.*, 2016, 35(5), 620–649.

Juliano, F. F., Alvarenga, J. F. R., Lamuela-Raventos, R. M., Massarioli, A. P., Lima, L. M., Santos, R. C., Alencar, S. M. Polyphenol analysis using high-resolution mass spectrometry allows differentiation of drought tolerant peanut genotypes. *J. Sci. Food Agric.*, 2020, 100(2), 721–731.

Kaiser, E., Zhou, D., Heuvelink, E., Harbinson, J., Morales, A., Marcelis, L. F. Elevated CO_2 increases photosynthesis in fluctuating irradiance regardless of photosynthetic induction state. *J. Exp. Bot.*, 2017, 68(20), 5629–5640.

Kalita, J., Pradhan, A. K., Shandilya, Z. M., Tanti, B. Arsenic stress responses and tolerance in rice: Physiological, cellular and molecular approaches. *Rice Sci.*, 2018, 25(5), 235–249.

Kandaswamy, R., Ramasamy, M. K., Palanivel, R., Balasundaram, U. Impact of Pseudomonas putida RRF3 on the root transcriptome of rice plants: Insights into defense response, secondary metabolism and root exudation. *J. Biosci.*, 2019, 44, 1–13.

Kang, K., Yue, L., Xia, X., Liu, K., Zhang, W. Comparative metabolomics analysis of different resistant rice varieties in response to the brown plant hopper *Nilaparvata lugens* Hemiptera: Delphacidae. *Metabolomics*, 2019, 15, 1–13.

Kaya, C., Aydemir, S., Sonmez, O., Ashraf, M., Dikilitas, M. Regulation of growth and some key physiological processes in salt-stressed maize (*Zea mays* L.) plants by exogenous application of asparagine and glycerol. *Acta Botanica Croatica*, 2013, 72(1), 157–168.

Kermani, S. G., Saeidi, G., Sabzalian, M. R., Gianinetti, A. Drought stress influenced sesamin and sesamolin content and polyphenolic components in sesame (*Sesamum indicum* L.) populations with contrasting seed coat colors. *Food Chem.*, 2019, 289, 360–368.

Khan, N., Bano, A., Ali, S., Babar, M. A. Crosstalk amongst phytohormones from planta and PGPR under biotic and abiotic stresses. *Plant Growth Regul.*, 2020, 90, 189–203.

Khan, Z., Shahwar, D. Role of heat shock proteins (HSPs) and heat stress tolerance in crop plants. In: *Sustainable Agriculture in the Era of Climate Change* (pp. 211–234), 2020.

Kim, Y. O., Bae, H. J., Cho, E., Kang, H. Exogenous glutathione enhances mercury tolerance by inhibiting mercury entry into plant cells. *Front. Plant Sci.*, 2017, 8, 683.

Kochian, L. V. Cellular mechanisms of aluminium toxicity and resistance in plants. *Annu. Rev. Plant Biol.*, 1995, 46(1), 237–260.

Komatsu, S., Nakamura, T., Sugimoto, Y., Sakamoto, K. Proteomic and metabolomic analyses of soybean root tips under flooding stress. *Protein Pept. Lett.*, 2014, 21(9), 865–884.

Komatsu, S., Yamamoto, A., Nakamura, T., Nouri, M. Z., Nanjo, Y., Nishizawa, K., Furukawa, K. Comprehensive analysis of mitochondria in roots and hypocotyls of soybean under flooding stress using proteomics and metabolomics techniques. *J. Proteome Res.*, 2011, 10(9), 3993–4004.

Kopka, J., Fernie, A., Weckwerth, W., Gibon, Y., Stitt, M. Metabolite profiling in plant biology: Platforms and destinations. *Genome Biol.*, 2004, 5, 1–9.

Kosová, K., Urban, M. O., Vítámvás, P., Prášil, I. T. Drought stress response in common wheat, durum wheat, and barley: Transcriptomics, proteomics, metabolomics, physiology, and breeding for an enhanced drought tolerance. In: *Drought Stress Tolerance in Plants, Vol 2: Molecular and Genetic Perspectives* (pp. 277–314). 2016.

Kühnlenz, T., Hofmann, C., Uraguchi, S., Schmidt, H., Schempp, S., Weber, M., Clemens, S. Phytochelatin synthesis promotes leaf Zn accumulation of Arabidopsis thaliana plants grown in soil with adequate Zn supply and is essential for survival on Zn-contaminated soil. *Plant Cell Physiol.*, 2016, 57(11), 2342–2352.

Kumar, A., Prasad, M. N. V. Plant-lead interactions: Transport, toxicity, tolerance, and detoxification mechanisms. *Ecotoxicol. Environ. Saf.*, 2018, 166, 401–418.

Kumar, R. R., Dubey, K., Goswami, S., Rai, G. K., Rai, P. K., Salgotra, R. K., Chinnusamy, V. Transcriptional regulation of small heat shock protein 17 (sHSP-17) by Triticum aestivum HSFA2h transcription factor confers tolerance in arabidopsis under heat stress. *Plants*, 2023, 12(20), 3598.

Kwon, M. C., Kim, Y. X., Lee, S., Jung, E. S., Singh, D., Sung, J., Lee, C. H. Comparative metabolomics unravel the effect of magnesium oversupply on tomato fruit quality and associated plant metabolism. *Metabolites*, 2019, 9(10), 231–245.

Lalarukh, I. R. F. A. N. A., Shahbaz, M. Response of antioxidants and lipid peroxidation to exogenous application of alpha-tocopherol in sunflower (*Helianthus annuus* L.) under salt stress. *Pak. J. Bot.*, 2020, 52(1), 75–83.

Lalarukh, I., Shahbaz, M. Alpha-tocopherol induced modulations in morpho-physiological attributes of sunflower (*Helianthus annuus*) grown under saline environment. *Int. J. Agric. Biol.*, 2018, 20(03), 661–668.

Lamichhane, S., Tarpley, L., Dou, F. Impact of excess magnesium salt supply on rice yield, physiological response, and grain mineral content. *Sustainability*, 2023, 15(22), 15741.

Li, D., Wang, L., Teng, S., Zhang, G., Guo, L., Mao, Q., Chen, L. Proteomics analysis of rice proteins up-regulated in response to bacterial leaf streak disease. *J. Plant Biol.*, 2012, 55, 316–324.

Li, J., Jia, Y., Dong, R., Huang, R., Liu, P., Li, X., Chen, Z. Advances in the mechanisms of plant tolerance to manganese toxicity. *Int. J. Mol. Sci.*, 2019, 20(20), 5096.

Li, W., Wei, Z., Qiao, Z., Wu, Z., Cheng, L., Wang, Y. Proteomics analysis of alfalfa response to heat stress. *PLoS One*, 2013, 8(12), e82725.

Li, X., Zhang, X., Liu, G., Tang, Y., Zhou, C., Zhang, L., Lv, J. The spike plays important roles in the drought tolerance as compared to the flag leaf through the phenylpropanoid pathway in wheat. *Plant Physiol. Biochem.*, 2020, 152, 100–111.

Ling, Y., Ang, L., Weilin, Z. Current understanding of the molecular players involved in resistance to rice planthoppers. *Pest Manage. Sci.*, 2019, 75(10), 2566–2574.

Lipton, D. S., Blanchar, R. W., Blevins, D. G. Citrate, malate, and succinate concentration in exudates from P-sufficient and P-stressed *Medicago sativa L. seedlings. Plant Physiol.*, 1987, 85(2), 315–317.

Liu, C., Hao, F., Hu, J., Zhang, W., Wan, L., Zhu, L., He, G. Revealing different systems responses to brown planthopper infestation for pest susceptible and resistant rice plants with the combined metabonomic and gene-expression analysis. *J. Proteome Res.*, 2010, 9(12), 6774–6785.

Liu, F., Xu, G., Wu, X., Ding, Q., Zheng, J., Zhang, R., Gao, Y. Effect of drought stress and re-watering on emissions of volatile organic compounds from Rosmarinus officinalis. *J. Zhejiang A & F University*, 2014, 31(2), 264–271.

Liu, J., Magalhaes, J. V., Shaff, J., Kochian, L. V. Aluminium-activated citrate and malate transporters from the *MATE* and *ALMT* families function independently to confer Arabidopsis aluminium tolerance. *Plant J.*, 2009, 57(3), 389–399.

Liu, Q., Wang, X., Tzin, V., Romeis, J., Peng, Y., Li, Y. Combined transcriptome and metabolome analyses to understand the dynamic responses of rice plants to attack by the rice stem borer Chilo suppressalis (Lepidoptera: Crambidae). *BMC Plant Biol.*, 2016, 16, 1–17.

Llerena, J. P. P., Coasaca, R. L., Rodriguez, H. O. L., Llerena, S. Á. P., Valencia, Y. D., Mazzafera, P. Metallothionein production is a common tolerance mechanism in four species growing in polluted Cu mining areas in Peru. *Ecotoxicol. Environ. Saf.*, 2021, 212, 112009.

Longo, V., Janmohammadi, M., Zolla, L., Rinalducci, S. What can small molecules tell us about cold stress tolerance in plants? In: *Cold Tolerance in Plants: Physiological, Molecular and Genetic Perspectives* (pp. 127–157). 2018.

Loreto, F., Pollastri, S., Fineschi, S., Velikova, V. Volatile isoprenoids and their importance for protection against environmental constraints in the Mediterranean area. *Environ. Exp. Bot.*, 2014, 103, 99–106.

Lu, H. P., Luo, T., Fu, H. W., Wang, L., Tan, Y. Y., Huang, J. Z., Shu, Q. Y. Resistance of rice to insect pests mediated by suppression of serotonin biosynthesis. *Nat. Plants*, 2018, 4(6), 338–344.

Ludwig, N., Leidinger, P., Becker, K., Backes, C., Fehlmann, T., Pallasch, C., Keller, A. Distribution of miRNA expression across human tissues. *Nucleic Acids Res.*, 2016, 44(8), 3865–3877.

Ma, D., Sun, D., Wang, C., Li, Y., Guo, T. Expression of flavonoid biosynthesis genes and accumulation of flavonoid in wheat leaves in response to drought stress. *Plant Physiol. Biochem.*, 2014, 80, 60–66.

Ma, J. F. Role of organic acids in detoxification of aluminium in higher plants. *Plant Cell Physiol.*, 2000, 41(4), 383–390.

Ma, J. F., Ryan, P. R., Delhaize, E. Aluminium tolerance in plants and the complexing role of organic acids. *Trends Plant Sci.*, 2001, 6(6), 273–278.

Mahmood, T., Jan, A., Kakishima, M., Komatsu, S. Proteomic analysis of bacterial-blight defense-responsive proteins in rice leaf blades. *Proteomics*, 2006, 6(22), 6053–6065.

Mahmood, T., Kakishima, M., Komatsu, S. Proteome analysis of probenazole-effect in rice-bacterial blight interactions. *Protein Pept. Lett.*, 2009, 16(9), 1041–1052.

Mallhi, Z. I., Rizwan, M., Mansha, A., Ali, Q., Asim, S., Ali, S., Ahmad, P. Citric acid enhances plant growth, photosynthesis, and phytoextraction of lead by alleviating the oxidative stress in castor beans. *Plants*, 2019, 8(11), 525.

Manzoor, J., Sharma, M., Wani, K. A. Heavy metals in vegetables and their impact on the nutrient quality of vegetables: A review. *J. Plant Nutr.*, 2018, 41(13), 1744–1763.

Marone, D., Mastrangelo, A. M., Borrelli, G. M., Mores, A., Laidò, G., Russo, M. A., Ficco, D. B. M. Specialized metabolites: Physiological and biochemical role in stress resistance, strategies to improve their accumulation, and new applications in crop breeding and management. *Plant Physiol. Biochem.*, 2022, 172, 48–55.

Maruyama, H., Sasaki, T., Yamamoto, Y., Wasaki, J. AtALMT3 is involved in malate efflux induced by phosphorus deficiency in Arabidopsis thaliana root hairs. *Plant Cell Physiol.*, 2019, 60(1), 107–115.

Masand, S., Yadav, S. K. Overexpression of MuHSP70 gene from Macrotyloma uniflorum confers multiple abiotic stress tolerance in transgenic Arabidopsis thaliana. *Mol. Biol. Rep.*, 2016, 43, 53–64.

Mata-Pérez, C., Sánchez-Calvo, B., Padilla, M. N., Begara-Morales, J. C., Luque, F., Melguizo, M., Barroso, J. B. Nitro-fatty acids in plant signaling: Nitro-linolenic acid induces the molecular chaperone network in Arabidopsis. *Plant Physiol.*, 2016, 170(2), 686–701.

Monreal, J. A., Jimenez, E. T., Remesal, E., Morillo-Velarde, R., García-Mauriño, S., Echevarría, C. Proline content of sugar beet storage roots: Response to water deficit and nitrogen fertilization at field conditions. *Environ. Exp. Bot.*, 2007, 60(2), 257–267.

Moura, J. C. M. S., Bonine, C. A. V., de Oliveira Fernandes Viana, J., Dornelas, M. C., Mazzafera, P. Abiotic and biotic stresses and changes in the lignin content and composition in plants. *J. Integr. Plant Biol.*, 2010, 52(4), 360–376.

Munné-Bosch, S., Mueller, M., Schwarz, K., Alegre, L. Diterpenes and antioxidative protection in drought-stressed Salvia officinalis plants. *J. Plant Physiol.*, 2001, 158(11), 1431–1437.

Nahar, K., Hasanuzzaman, M., Alam, M. M., Fujita, M. Roles of exogenous glutathione in antioxidant defense system and methylglyoxal detoxification during salt stress in mung bean. *Biologia Plantarum*, 2015, 59, 745–756.

Nakamura, T., Yamamoto, R., Hiraga, S., Nakayama, N., Okazaki, K., Takahashi, H., Komatsu, S. Evaluation of metabolite alteration under flooding stress in soybeans. *Japan Agric. Res. Quart.: JARQ*, 2012, 46(3), 237–248.

Noreen, S., Akhter, M. S., Yaamin, T., Arfan, M. The ameliorative effects of exogenously applied proline on physiological and biochemical parameters of wheat (Triticum aestivum L.) crop under copper stress condition. *J. Plant Interact.*, 2018, 13(1), 221–230.

Nováková, S., Šubr, Z., Kováč, A., Fialová, I., Beke, G., Danchenko, M. Cucumber mosaic virus resistance: Comparative proteomics of contrasting Cucumis sativus cultivars after long-term infection. *J. Proteomics*, 2020, 214, 103626.

Numan, M., Bashir, S., Khan, Y., Mumtaz, R., Shinwari, Z. K., Khan, A. L., Ahmed, A. H. Plant growth promoting bacteria as an alternative strategy for salt tolerance in plants: A review. *Microbiol. Res.*, 2018, 209, 21–32.

Orimoloye, I. R. Agricultural drought and its potential impacts: Enabling decision-support for food security in vulnerable regions. *Front. Sustain. Food Syst.*, 2022, 6, 838824.

Ormeno, E., Mévy, J. P., Vila, B., Bousquet-Mélou, A., Greff, S., Bonin, G., Fernandez, C. Water deficit stress induces different monoterpene and sesquiterpene emission changes in Mediterranean species. Relationship between terpene emissions and plant water potential. *Chemosphere*, 2007, 67(2), 276–284.

Panchal, P., Miller, A. J., Giri, J. Organic acids: Versatile stress-response roles in plants. *J. Exp. Bot.*, 2021, 72(11), 4038–4052.

Pang, X. M., Zhang, Z. Y., Wen, X. P., Ban, Y., Moriguchi, T. Polyamines, all-purpose players in response to environment stresses in plants. *Plant Stress*, 2007, 1(2), 173–188.

Payton, P., Kottapalli, K. R., Kebede, H., Mahan, J. R., Wright, R. J., Allen, R. D. Examining the drought stress transcriptome in cotton leaf and root tissue. *Biotechnol. Lett.*, 2011, 33, 821–828.

Peng, L., Zhao, Y., Wang, H., Zhang, J., Song, C., Shangguan, X., He, G. Comparative metabolomics of the interaction between rice and the brown planthopper. *Metabolomics*, 2016, 12, 1–15.

Perkowska, I., Potrykus, M., Siwinska, J., Siudem, D., Lojkowska, E., Ihnatowicz, A. Interplay between coumarin accumulation, iron deficiency and plant resistance to Dickeya spp. *Int. J. Mol. Sci.*, 2021, 22(12), 6449.

Piasecka, A., Kachlicki, P., Stobiecki, M. Analytical methods for detection of plant metabolomes changes in response to biotic and abiotic stresses. *Int. J. Mol. Sci.*, 2019, 20(2), 379.

Pradhan, A. K., Shandilya, Z. M., Lahkar, L., Hasnu, S., Kalita, J., Borgohain, D., Tanti, B. Comparative metabolomics approach towards understanding chemical variation in rice under abiotic stress. In: *Advances in Rice Research for Abiotic Stress Tolerance* (pp. 537–550), Woodhead Publishing, 2019.

Pradhan, A. K., Shandilya, Z. M., Sarma, P., Bora, R. K., Regon, P., Vemireddy, L. N. R., Tanti, B. Concurrent effect of aluminum toxicity and phosphorus deficiency in the root growth of aluminum tolerant and sensitive rice cultivars. *Acta Physiologiae Plantarum*, 2023a, 45(2), 33.

Pradhan, A. K., Vemireddy, L. N. R., Tanti, B. Assessment of the genetic variability and population structure in boro rice cultivars of Assam, India using candidate gene based SSR markers. *Genet. Resour. Crop Evolut.*, 2023b, 70(6), 1747–1765.

Priya, and Singh, S. P. Factors influencing the adoption of sustainable agricultural practices: A systematic literature review and lesson learned for India. *Forum Soc. Econ.*, 2024, 53(1), 1–17.

Qu, X., Wang, H., Chen, M., Liao, J., Yuan, J., Niu, G. Drought stress-induced physiological and metabolic changes in leaves of two oil tea cultivars. *J. Am. Soc. Horticult. Sci.*, 2019, 144(6), 439–447.

Quan, N. T., Anh, L. H., Khang, D. T., Tuyen, P. T., Toan, N. P., Minh, T. N., Xuan, T. D. Involvement of secondary metabolites in response to drought stress of rice (*Oryza sativa* L.). *Agriculture*, 2016, 6(2), 23.

Rajput, V. D., Harish, Singh, R. K., Verma, K. K., Sharma, L., Quiroz-Figueroa, F. R., Mandzhieva, S. Recent developments in enzymatic antioxidant defence mechanism in plants with special reference to abiotic stress. *Biology*, 2021, 10(4), 267.

Raman, N., Selvaraj, T. Tripartite AMF, tripartite relationship and relationship host of rhizobium, promotion in growth AMF, of rhizobium, promotion. In: *Handbook of Microbial Biofertilizers* (p. 51), 2006.

Rana, S., Prakash, V., Sagar, A. Studies on analysis of antioxidant and enzyme inhibitory activities of Solanum indicum Linn. *IOSR J. Pharm. Biol. Sci.*, 2017, 12(4), 21–24.

Rani, A., Singh, K., Sood, P., Kumar, S., Ahuja, P. S. π-Coumarate: CoA ligase as a key gene in the yield of catechins in tea [*Camellia sinensis* (L.) O. Kuntze]. *Funct. Integr. Genomics*, 2009, 9, 271–275.

Raza, A., Charagh, S., Abbas, S., Hassan, M. U., Saeed, F., Haider, S., Varshney, R. K. Assessment of proline function in higher plants under extreme temperatures. *Plant Biol.*, 2023a, 25(3), 379–395.

Raza, A., Charagh, S., Najafi-Kakavand, S., Abbas, S., Shoaib, Y., Anwar, S., Siddique, K. H. Role of phytohormones in regulating cold stress tolerance: Physiological and molecular approaches for developing cold-smart crop plants. *Plant Stress*, 2023b, 8, 100–152.

Razavizadeh, R., Komatsu, S. Changes in essential oil and physiological parameters of callus and seedlings of *Carum copticum* L. under in vitro drought stress. *J. Food Meas. Charact.*, 2018, 12, 1581–1592.

Reddy, P. S., Chakradhar, T., Reddy, R. A., Nitnavare, R. B., Mahanty, S., Reddy, M. K. Role of heat shock proteins in improving heat stress tolerance in crop plants. In: *Heat Shock Proteins and Plants* (pp. 283–307). 2016.

Regon, P., Dey, S., Chowardhara, B., Saha, B., Kar, S., Tanti, B., Panda, S. K. Physio-biochemical and molecular assessment of Iron (Fe^{2+}) toxicity responses in contrasting indigenous aromatic *Joha Rice Cultivars of Assam, India. Protoplasma*, 2021, 258(2), 289–299.

Regon, P., Dey, S., Rehman, M., Pradhan, A. K., Chowra, U., Tanti, B., Talukdar, A. D., Panda, S. K. Transcriptomic analysis revealed reactive oxygen species scavenging mechanisms associated with ferrous iron toxicity in aromatic Keteki Joha rice. *Front. Plant Sci.*, 2022, 13, 798580.

Rehman, M., Tanti, B. Morpho-physiological responses in rice and cold stress induced acclimation involving biochemical and signaling pathways in Boro rice. *Vegetos*, 2022, 35(4), 859–868.

Reid, R. J., Hayes, J. E., Post, A., Stangoulis, J. C. R., Graham, R. D. A critical analysis of the causes of boron toxicity in plants. *Plant, Cell Environ.*, 2004, 27(11), 1405–1414.

Rengasamy, P. World salinization with emphasis on Australia. *J. Exp. Bot.*, 2006, 57(5), 1017–1023.

Rosas, A., Rengel, Z., de la Luz Mora, M. Manganese supply and pH influence growth, carboxylate exudation and peroxidase activity of ryegrass and white clover. *J. Plant Nutr.*, 2007, 30(2), 253–270.

Roy, S., Kapoor, R., Mathur, P. Revisiting changes in growth, physiology and stress responses of plants under the effect of enhanced CO_2 and temperature. *Plant Cell Physiol.*, 2024, 65(1), 4–19.

Ruibal, C., Castro, A., Carballo, V., Szabados, L., Vidal, S. Recovery from heat, salt and osmotic stress in Physcomitrella patens requires a functional small heat shock protein PpHsp16.4. *BMC Plant Biol.*, 2013, 13, 1–18.

Saibil, H. Chaperone machines for protein folding, unfolding and disaggregation. *Nat. Rev. Mol. Cell Biol.*, 2013, 14(10), 630–642.

Salem, M. A., Perez de Souza, L., Serag, A., Fernie, A. R., Farag, M. A., Ezzat, S. M., Alseekh, S. Metabolomics in the context of plant natural products research: From sample preparation to metabolite analysis. *Metabolites*, 2020, 10(1), 37–52.

Sana, T. R., Fischer, S., Wohlgemuth, G., Katrekar, A., Jung, K. H., Ronald, P. C., Fiehn, O. Metabolomic and transcriptomic analysis of the rice response to the bacterial blight pathogen *Xanthomonas oryzae* pv. oryzae. *Metabolomics*, 2010, 6, 451–465.

Sánchez, E., López-Lefebre, L. R., García, P. C., Rivero, R. M., Ruiz, J. M., Romero, L. Proline metabolism in response to highest nitrogen dosages in green bean plants (*Phaseolus vulgaris* L. cv. Strike). *J. Plant Physiol.*, 2001, 158(5), 593–598.

Sato, K., Kadota, Y., Shirasu, K. Plant immune responses to parasitic nematodes. *Front. Plant Sci.*, 2019, 10, 1165.

Savary, S., Willocquet, L., Pethybridge, S. J. The global burden of pathogens and pests on major food crops. *Nat. Ecol. Evol.* 2019, 3, 430–439.

Savoi, S., Wong, D. C., Arapitsas, P., Miculan, M., Bucchetti, B., Peterlunger, E., Castellarin, S. D. Transcriptome and metabolite profiling reveals that prolonged drought modulates the phenylpropanoid and terpenoid pathway in white grapes (*Vitis vinifera* L.). *BMC Plant Biol.*, 2016, 16, 1–17.

Semida, W. M., Abd El-Mageed, T. A., Howladar, S. M., Rady, M. M. Foliar-applied alpha-tocopherol enhances salt-tolerance in onion plants by improving antioxidant defence system. *Aust. J. Crop Sci.*, 2016, 10(7), 1030–1039.

Shams, M., Yildirim, E., Ekinci, M., Turan, M., Dursun, A., Parlakova, F., Kul, R. Exogenously applied glycine betaine regulates some chemical characteristics and antioxidative defence systems in lettuce under salt stress. *Horticult. Environ. Biotechnol.*, 2016, 57, 225–231.

Shandilya, Z. M., Tanti, B. Comparative biochemical, histochemical and expression analysis of SOD gene in a few traditional rice varieties under aluminium toxicity and phosphorous deficiency in Assam, *India. Vegetos*, 2020, 33(1), 145–157.

Shao, J., Wang, Q., Liu, P., Zhao, B., Han, W., Zhang, J., Ren, B. The complex stress of waterlogging and high temperature accelerated maize leaf senescence and decreased photosynthetic performance at different growth stages. *J. Agron. Crop Sci.*, 2024, 210(2), e12689.

Sharma, P., Lakra, N., Goyal, A., Ahlawat, Y. K., Zaid, A., Siddique, K. H. Drought and heat stress mediated activation of lipid signaling in plants: A critical review. *Front. Plant Sci.*, 2023, 14, 1216835.

Shaw, A. K., Bhardwaj, P. K., Ghosh, S., Roy, S., Saha, S., Sherpa, A. R., Hossain, Z. β-aminobutyric acid mediated drought stress alleviation in maize (Zea mays L.). *Environ. Sci. Pollut. Res.*, 2016, 23, 2437–2453.

Shen, H., Yan, X. L., Wang, X. R., Zheng, S. L. Exudation of citrate in common bean in response to aluminium stress. *J. Plant Nutr.*, 2002, 25(9), 1921–1932.

Shireen, F., Nawaz, M. A., Chen, C., Zhang, Q., Zheng, Z., Sohail, H., Bie, Z. Boron: Functions and approaches to enhance its availability in plants for sustainable agriculture. *Int. J. Mol. Sci.*, 2018, 19(7), 1856.

Siddiqui, M. H., Alamri, S., Al-Khaishany, M. Y., Khan, M. N., Al-Amri, A., Ali, H. M., Alsahli, A. A. Exogenous melatonin counteracts NaCl-induced damage by regulating the antioxidant system, proline and carbohydrates metabolism in tomato seedlings. *Int. J. Mol. Sci.*, 2019, 20(2), 353.

Slama, I., Abdelly, C., Bouchereau, A., Flowers, T., Savouré, A. Diversity, distribution and roles of osmoprotective compounds accumulated in halophytes under abiotic stress. *Ann. Bot.*, 2015, 115(3), 433–447.

Solangi, F., Zhu, X., Khan, S., Rais, N., Majeed, A., Sabir, M. A., Kayabasi, E. T. The global dilemma of soil legacy phosphorus and its improvement strategies under recent changes in agro-ecosystem sustainability. *ACS Omega*, 2023, 8(26), 23271–23282.

Sung, J., Lee, S., Lee, Y., Ha, S., Song, B., Kim, T., Krishnan, H. B. Metabolomic profiling from leaves and roots of tomato (*Solanum lycopersicum* L.) plants grown under nitrogen, phosphorus or potassium-deficient condition. *Plant Sci.*, 2015, 241, 55–64.

Swain, R., Sahoo, S., Rout, G. R. Instigating prevalent abiotic stress resilience in crop by exogenous application of phytohormones and nutrient. *Front. Plant Sci.*, 2023, 14, 1104874.

Tan, B. C., Lim, Y. S., Lau, S. E. Proteomics in commercial crops: An overview. *J. Proteomics*, 2017, 169, 176–188.

Teh, C. Y., Shaharuddin, N. A., Ho, C. L., Mahmood, M. Exogenous proline significantly affects the plant growth and nitrogen assimilation enzymes activities in rice (Oryza sativa) under salt stress. *Acta Physiol. Plant.*, 2016, 38, 1–10.

Theerakulpisut, P., Gunnula, W. Exogenous Sorbitol and Trehalose mitigated salt stress damage in salt-sensitive but not salt-tolerant rice seedlings. *Asian J. Crop Sci.*, 2012, 4(4), 165–170.

Tolrà, R., Barceló, J., Poschenrieder, C. Constitutive and aluminium-induced patterns of phenolic compounds in two maize varieties differing in aluminium tolerance. *J. Inorg. Biochem.*, 2009, 103(11), 1486–1490.

Turk, H., Erdal, S., Karayel, U., Dumlupinar, R. Attenuation of lead toxicity by promotion of tolerance mechanism in wheat roots by lipoic acid. *Cereal Res. Commun.*, 2018, 46, 424–435.

Turtola, S., Manninen, A. M., Rikala, R., Kainulainen, P. Drought stress alters the concentration of wood terpenoids in Scots pine and Norway spruce seedlings. *J. Chem. Ecol.*, 2003, 29, 1981–1995.

Uawisetwathana, U., Chevallier, O. P., Xu, Y., Kamolsukyeunyong, W., Nookaew, I., Somboon, T., Karoonuthaisiri, N. Global metabolite profiles of rice brown planthopper-resistant traits reveal potential secondary metabolites for both constitutive and inducible defenses. *Metabolomics*, 2019, 15, 1–15.

Vadassery, J., Reichelt, M., Hause, B., Gershenzon, J., Boland, W., Mithöfer, A. CML42-mediated calcium signaling coordinates responses to Spodoptera herbivory and abiotic stresses in Arabidopsis. *Plant Physiol.*, 2012, 159(3), 1159–1175.

Valentino, G., Graziani, V., D'Abrosca, B., Pacifico, S., Fiorentino, A., Scognamiglio, M. NMR-based plant metabolomics in nutraceutical research: An overview. *Molecules*, 2020, 25(6), 1444.

Valette, M., Rey, M., Gerin, F., Comte, G., Wisniewski-Dyé, F. A common metabolomic signature is observed upon inoculation of rice roots with various rhizobacteria. *J. Integr. Plant Biol.*, 2020, 62(2), 228–246.

Vallat, A., Gu, H., Dorn, S. How rainfall, relative humidity and temperature influence volatile emissions from apple trees in situ. *Phytochemistry*, 2005, 66(13), 1540–1550.

Vanderschuren, H., Lentz, E., Zainuddin, I., Gruissem, W. Proteomics of model and crop plant species: Status, current limitations and strategic advances for crop improvement. *J. Proteomics*, 2013, 93, 5–19.

Verma, N., Shukla, S. Impact of various factors responsible for fluctuation in plant secondary metabolites. *J. Appl. Res. Med. Aromatic Plants*, 2015, 2(4), 105–113.

Wang, J., Raman, H., Zhou, M., Ryan, P. R., Delhaize, E., Hebb, D. M., Mendham, N. High-resolution mapping of the Alp locus and identification of a candidate gene HvMATE controlling aluminium tolerance in barley (*Hordeum vulgare* L.). *Theor. Appl. Genet.*, 2007, 115, 265–276.

Wei, Z., Hu, W., Lin, Q., Cheng, X., Tong, M., Zhu, L., He, G. Understanding rice plant resistance to the Brown Planthopper (Nilaparvatalugens): A proteomic approach. *Proteomics*, 2009, 9(10), 2798–2808.

Xiang, C., Yang, X., Peng, D., Kang, H., Liu, M., Li, W., Liu, S. Proteome-wide analyses provide new insights into the compatible interaction of Rice with the root-knot nematode *Meloidogyne graminicola. Int. J. Mol. Sci.*, 2020, 21(16), 5640.

Xu, C., Wang, X., Wu, Y., Gao, J., Zhang, P., Zhao, Y., Huang, S. Molecular mechanisms underlying low temperature inhibition of grain filling in maize (*Zea mays* L.): Coordination of growth and cold responses. *Plant J.*, 2024, 119(2), 982–997. https://doi.org/10.1111/tpj.16811.

Yadav, A., Singh, J., Ranjan, K., Kumar, P., Khanna, S., Gupta, M., Sirohi, A. Heat shock proteins: Master players for heat-stress tolerance in plants during climate change. In: *Heat Stress Tolerance in Plants: Physiological, Molecular and Genetic Perspectives* (pp. 189–211). 2020.

Yang, L. T., Qi, Y. P., Jiang, H. X., Chen, L. S. Roles of organic acid anion secretion in aluminium tolerance of higher plants. *Biomed. Res. Int.*, 2013, 2013, 173682, https://doi.org/10.1155/2013/173682.

Yasmin, S., Hafeez, F. Y., Mirza, M. S., Rasul, M., Arshad, H. M., Zubair, M., Iqbal, M. Biocontrol of bacterial leaf blight of rice and profiling of secondary metabolites produced by rhizospheric *Pseudomonas aeruginosa* BRp3. *Front. Microbiol.*, 2017, 8, 287562.

Yurina, N. P. Heat shock proteins in plant protection from oxidative stress. *Mol. Biol.*, 2023, 57(6), 951–964.

Zha, W., You, A. Comparative iTRAQ proteomic profiling of proteins associated with the adaptation of brown planthopper to moderately resistant vs. susceptible rice varieties. *PLoS One*, 2020, 15(9), e0238549.

Zhang, J., Li, Y., Guo, J., Du, B., He, G., Zhang, Y., Li, J. Lipid profiles reveal different responses to brown planthopper infestation for pest susceptible and resistant rice plants. *Metabolomics*, 2018, 14, 1–13.

Zhang, N., Liu, B., Ma, C., Zhang, G., Chang, J., Si, H., Wang, D. Transcriptome characterization and sequencing-based identification of drought-responsive genes in potato. *Mol. Biol. Rep.*, 2014, 41, 505–517.

Zhang, X. M., Liu, L. X., Su, Z. M., Shen, Z. J., Gao, G. F., Yi, Y., Zheng, H. L. Transcriptome analysis of Medicago lupulina seedlings leaves treated by high calcium provides insights into calcium oxalate formation. *Plant Soil*, 2019, 444, 299–314.

Zhang, X., Zhang, D., Sun, W., Wang, T. The adaptive mechanism of plants to iron deficiency via iron uptake, transport, and homeostasis. *Int. J. Mol. Sci.*, 2019, 20(10), 2424.

Zhao, C., Liu, B., Piao, S., Wang, X., Lobell, D. B., Huang, Y., Huang, M., Yao, Y., Bassu, S., Ciais, P. Temperature increase reduces global yields of major crops in four independent estimates. *Proc. Natl. Acad. Sci. U.S.A.*, 2017, 114, 9326–9331.

Zhou, G., Xu, Y., Li, J., Yang, L., Liu, J. Y. Molecular analyses of the metallothionein gene family in rice (Oryza sativa L.). *BMB Rep.*, 2006, 39(5), 595–606.

Zogli, P., Pingault, L., Grover, S., Louis, J. Ento (o) mics: The intersection of 'omic' approaches to decipher plant defense against sap-sucking insect pests. *Curr. Opin. Plant Biol.*, 2020, 56, 153–161.

CHAPTER 10

Metabolomics for Heavy Metal Stress Responses and Tolerance in Plants: Cadmium and Arsenic

PROMA GHOSH, ANWESHA CHATTERJEE, and HARSHATA PAL*

Amity Institute of Biotechnology, Amity University Kolkata, Kolkata, West Bengal, India

*Corresponding authors

ABSTRACT

Accumulation of heavy metals (HM) like cadmium (Cd) and arsenic (As) poses a serious hazard to agricultural soils, plants, and ultimately human health. Human activities like mining, smelting, industrial waste, use of agrochemicals in excess, irrigation by sewage, and use of harmful fertilizers, pesticides, etc., are the prime sources of HM pollution. Among them, cadmium and arsenic are the most abundant metals. The presence of Pb (lead), Cd, As, and Ni (nickel) in agricultural lands can have detrimental effects on plant growth and development, reducing crop yields and jeopardizing food security. These substances can harm plants' general health and productivity by insterfering with vital physiological processes like photosynthesis, nitrogen uptake, and enzyme activity. HM causes significant changes in the metabolic activity of the plant tissue. Plant cells balance oxidants and antioxidants, controlling reactive oxygen species (ROS) accumulation. Excessive HM can unbalance this balance, damaging biological macromolecules. Plants counteract ROS by producing antioxidant molecules as secondary metabolites. Besides this, higher HM toxicity in plant cell affects different metabolic pathways that produce precursors for structural compounds and antioxidants. So a metabolomics study can give us a deeper insight into the mechanism of plant cells to tolerate and detoxify HM stress due to the accumulation of Cd and As. To better understand the mechanisms of stress resilience and detoxification, we will provide a thorough overview of the metabolomics investigation of the Cd- and As-stressed plant in this chapter.

10.1 INTRODUCTION

Plants are frequently subjected to a wide range of biotic and abiotic stressors in the environment, including disease and insect infestation, severe heat, drought, salt, pollution, heavy metals (HM), and dietary inadequacies. This results in harvest or quality losses, such as the

Exploring Plant Stress Tolerance: A Metabolomics Approach.
Shuvasish Choudhury, Debojyoti Moulick, and Mirza Hasanuzzaman (Eds.)

DOI: 10.1201/9781779641731-10

development of a strong, bitter off-flavor, and occasionally toxicological issues, along with significant costs to the world economy (Dawid and Hille, 2018). Soil pollution due to HM is increasing day by day because of anthropogenic activity like mining, industries, use of pesticides in agricultural fields, wrong disposal of electronic waste, etc., which is alarming for food safety and environmental stability (Sarwar et al., 2017). HM contamination in soil poses a significant abiotic stress in vegetable cultivation, impacting both crop growth and food safety. Metals such as lead, cadmium (Cd), and mercury, often originating from industrial activities, agricultural inputs, or atmospheric deposition, can accumulate in the soil and subsequently be taken up by plants. This accumulation not only hampers essential physiological processes in vegetables but also poses serious health risks to consumers. Studies have highlighted the adverse effects of HM on plant growth, nutrient uptake, and overall crop productivity. Mitigation strategies, including soil amendments and phytoremediation techniques, are essential to address this abiotic stress and ensure the sustainability of vegetable cultivation in contaminated environments (Khan et al., 2008; Gupta, 2021).

Cadmium and arsenic (As) pollution in agricultural soil and vegetables represents a critical environmental and health challenge, as substantiated by scientific literature. Numerous studies, including research published in the *Environmental Pollution* journal (Li et al., 2022a; Chen and Costa, 2021), emphasize the pervasive presence of these toxic HM in agricultural soils. These pollutants often result from industrial activities, mining operations, and the use of certain agrochemicals. The uptake of cadmium and arsenic by vegetables is well-documented in publications such as *Journal of Agricultural and Food Chemistry* (Signes-Pastor et al., 2008), heightening concerns about potential health risks for consumers. The documented adverse health effects, such as organ damage and increased cancer risks, underscore the urgency of implementing effective mitigation strategies. Scientific evidence serves as a crucial foundation for policy interventions aimed at minimizing the severity of cadmium and arsenic pollution in agricultural systems, ensuring food safety and protecting public health.

Cd is a HM that poses a significant threat to both plants and humans due to its widespread environmental presence, primarily through industrial activities, mining, and agricultural practices. In plants, cadmium stress disrupts various physiological and biochemical processes, leading to adverse effects on growth, development, and overall productivity. It interferes with nutrient uptake, photosynthesis, and enzyme activities, causing oxidative stress and ultimately hampering plant fitness (Cuypers et al., 2010; Clemens and Simm, 2003). The repercussions of cadmium exposure extend beyond the plant kingdom, as this toxic metal can accumulate in crops, entering the human food chain and posing potential health risks. Chronic exposure to cadmium has been associated with various human health issues, including renal dysfunction, osteoporosis, and an increased risk of certain cancers (Satarug and Moore, 2004; Jarup et al., 1998). The mechanisms of cadmium toxicity in humans involve its accumulation in vital organs, such as the kidneys, and interference with essential cellular functions.

Efforts to mitigate the impact of cadmium stress on both plants and humans require interdisciplinary research and effective environmental management strategies. Understanding the intricate molecular and cellular responses to cadmium stress in plants and unraveling the pathways of cadmium toxicity in humans are crucial steps toward developing sustainable solutions to minimize the adverse effects of this HM on ecosystems and human health (Figure 10.1).

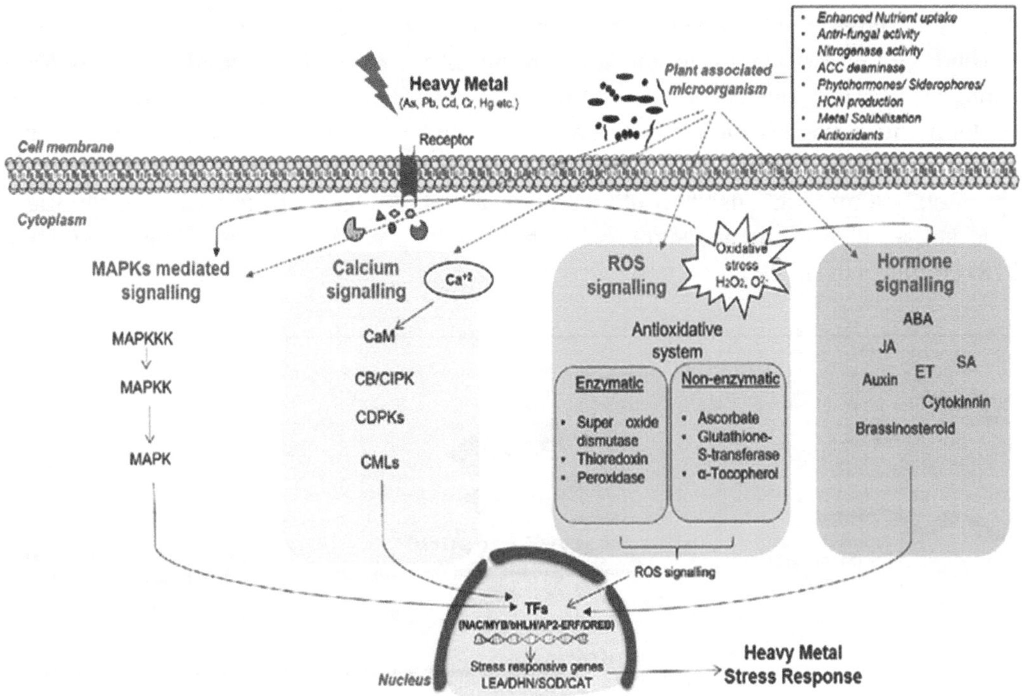

FIGURE 10.1 Heavy metal stress signaling in a plant cell.

Source: Reprinted from Tiwari and Lata, 2018. Copyright © 2018 Tiwari and Lata. https://creativecommons.org/licenses/by/4.0/

As is a metalloid widely distributed in the environment, and its presence poses significant threats to both plants and humans. In plants, arsenic stress induces various physiological and biochemical alterations that negatively impact growth, development, and overall plant health. Arsenic interferes with essential processes such as photosynthesis, nutrient uptake, and enzymatic activities, leading to oxidative stress and impaired plant performance (Rahman et al., 2009; Su et al., 2010; Zhao et al., 2010).

The consequences of arsenic exposure extend to human health, as chronic ingestion of arsenic-contaminated water and food can result in severe health problems. Arsenic is classified as a human carcinogen, and long-term exposure has been linked to skin lesions, cardiovascular diseases, neurotoxicity, and an increased risk of various cancers, including skin, lung, and bladder cancers. The mechanisms of arsenic toxicity in humans involve its ability to disrupt cellular functions and DNA integrity, leading to detrimental health outcomes. Addressing the impact of arsenic stress on both plants and humans requires comprehensive research and effective environmental management strategies. Understanding the molecular and cellular responses of plants to arsenic stress and elucidating the pathways of arsenic toxicity in humans are crucial steps toward developing strategies to mitigate the adverse effects of arsenic contamination on ecosystems and human health.

The majority of studies using metabolomics to investigate metal toxicity in any organism have only recently been published, making this an innovative field of research. Environmental toxicometabolomics is a subdiscipline of environmental metabolomics, which is a relatively new area that has just recently been studied (Booth et al., 2011).

The breeding of stress-tolerant plants requires a new understanding of metabolomics, genomics, transcriptomics, and proteomics datasets. This knowledge optimizes crop trait development, enhances yields and quality, and ensures nutritional and health factors for functional food or feedstuffs. Techniques like sensomics, lipidomics, hormonomics, and phytometabolomics are used to characterize metabolome alterations during abiotic and biotic stress, identifying resistant food crops with preferred quality (Dawid and Hille, 2018) (Figure 10.2).

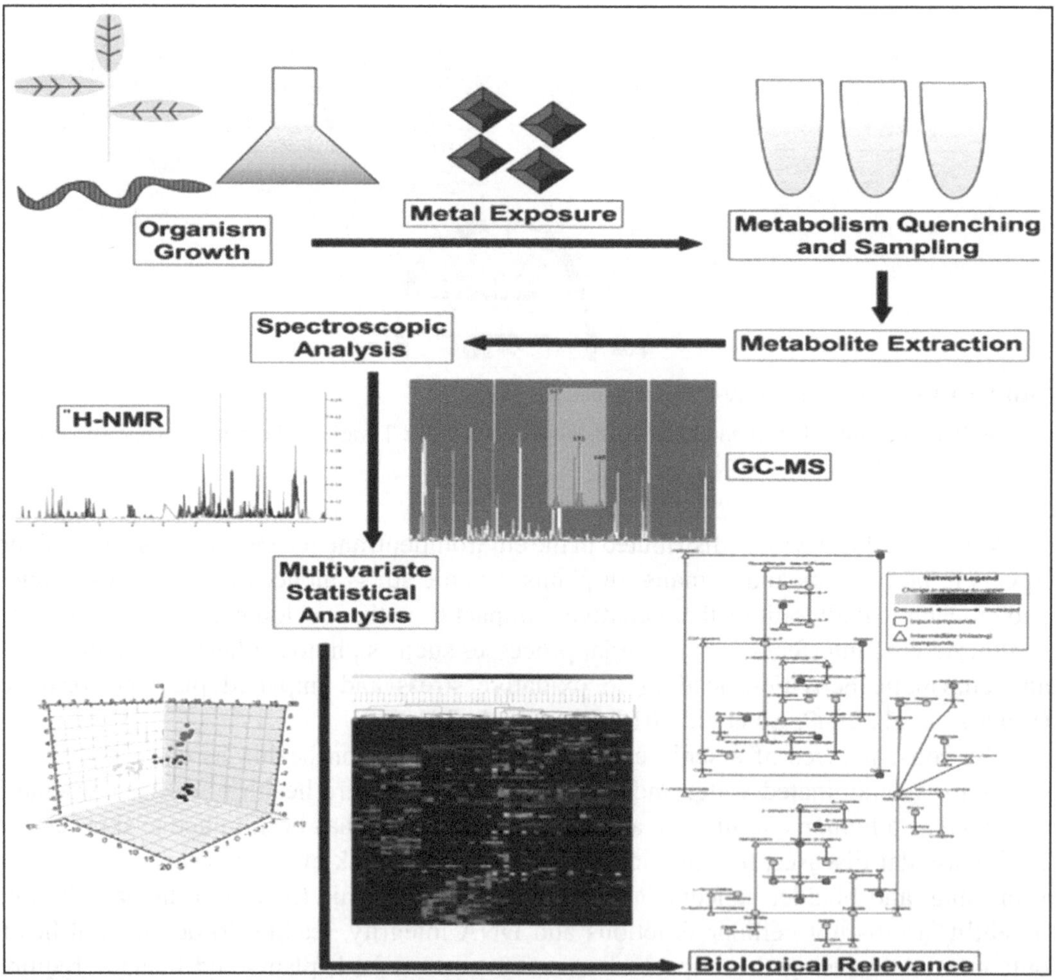

FIGURE 10.2 Process of metabolomics study in organisms.

Source: Reprinted with permission from Booth et al. (2011). Copyright © 2011, © The Royal Society of Chemistry 2011.

Metabolites play a crucial role in conferring HM stress tolerance to plants, representing a dynamic aspect of the plant's adaptive response (AR) to environmental challenges. The synthesis and accumulation of specific metabolites serve as key mechanisms for detoxification, chelation, and sequestration of HM. Notably, phytochelatins, organic acids, and antioxidants have been identified as essential metabolites involved in mitigating HM stress in plants (Gill and Tuteja, 2011; Hasanuzzaman et al., 2017). These metabolites act as chelators, binding HM and facilitating their compartmentalization in vacuoles, thus preventing their translocation to vital cellular components. The upregulation of antioxidant metabolites, such as glutathione (GSH) and proline, helps in neutralizing reactive oxygen species (ROS) generated under HM stress conditions. Understanding the importance of metabolites in HM stress tolerance provides valuable insights for developing strategies to enhance plant resilience in contaminated environments, contributing to sustainable agriculture and environmental management.

10.2 CADMIUM STRESS TOLERANCE STUDY THROUGH METABOLOMICS

Metabolomics is a data-generating technique that helps to provide comprehensive explanations for a living organism's metabolic condition. The most popular methods for plant metabolomics are chromatography-based methods like liquid chromatography (LC) and gas chromatography (GC), which have a high separation efficiency for metabolites with varying degrees of volatility and polarity. Prior research has shown that the uptake and accumulation of Cd in plants can cause oxidative stress, disrupt photosynthetic and respiratory processes, lower the activity of enzymes and other physiological metabolites, and disrupt the structure and function of cell membranes. All of these effects result in a decline in biomass, which has a major impact on plant growth and development (Chen et al., 2013).

Studies on metal toxicity in plants, particularly cadmium, have been conducted on crop plants like tomatoes and rice and model systems like *Arabidopsis thaliana* (Hédiji et al., 2010).

The metabolites in the response of cadmium-stressed plants play various roles in the plants' tolerance mechanisms. For instance, in radish roots responding to HM (Pb or Cd) stress, metabolites such as amino acids, organic acids, and sugars were identified, which are involved in energy metabolism, osmotic regulation, and the maintenance of cellular homeostasis (Wang et al., 2015). Additionally, in the metabonomic analysis of *Brassica napus* by liquid chromatography-mass spectrometry (LC-MS), the production of reducing agents such as monoterpenoids and carotenoids was found to play a role in the plant's response to cadmium stress (Huo et al., 2023). Furthermore, the study of *Gynostemma pentaphyllum*'s response to cadmium stress highlighted the important role of metabolites such as galactose, lipid (glycerophosphatide metabolism), and amino acids in the plant's antioxidant system and the impairment of amino acid biosynthesis, enzyme activity, and oxidative stress response (Al-Khayri et al., 2023).

In general, the identified metabolites in cadmium-stressed plants are involved in various physiological and biochemical processes, including energy metabolism, osmotic regulation, antioxidant defense, and the maintenance of cellular homeostasis. These metabolites play a crucial role in the plants' AR to cadmium stress, helping them to mitigate the toxic effects of cadmium and maintain their physiological functions.

High-performance liquid chromatography (HPLC) has proven to be an indispensable tool in elucidating the intricate biochemical responses of plants to cadmium stress. HPLC facilitates the precise analysis of various key metabolites involved in detoxification processes, such as phytochelatins and metallothioneins. Researchers have employed HPLC techniques to quantitatively measure the changes in these metabolites, providing valuable insights into the molecular mechanisms that underlie cadmium stress tolerance in plants. For example, a study utilized HPLC to assess the accumulation of phytochelatins in [*Brassica napus*], revealing significant alterations in the metabolite profile as a response to cadmium-induced stress (Mendoza-Cózatl et al., 2008). Such studies contribute to our understanding of plant adaptive strategies and provide a foundation for developing effective approaches to mitigate the impact of cadmium stress on plant growth and agricultural productivity.

An investigation looked at the function of key metabolic pathways and epigenetic changes in the Cd-induced AR in plants. AR was hindered by metabolic blockers of de novo translation, oxidative stress, DNA repair, and protein kinase cascades; autophagy and postreplication repair inhibitors were unsuccessful. Another finding that suggests DNA methylation plays a role in AR is that cells treated with 5-aza-2′deoxycytidine no longer exhibit AR. This is the first research linking the AR generated by Cd in plants under genotoxic stress to DNA methylation (Chakrabarti and Mukherjee, 2021).

According to a research, cadmium enhanced the overall antioxidant capacity as well as the amounts of phenolic and flavonoid components. In the presence of cadmium, rutin, epicatechin, and derivatives of cinnamic acid rose at concentrations that did not change the ratio of chlorophylls (Márquez-García et al., 2012)

Metabolomics studies using HPLC have provided valuable insights into the impact of cadmium on plant metabolism and the strategies employed by plants to alleviate this stress. It has been instrumental in unraveling the metabolic responses of plants to cadmium stress, offering valuable insights into the molecular mechanisms underlying cadmium tolerance and providing a foundation for the development of strategies to mitigate the impact of cadmium on plant metabolism. These studies contribute to the understanding of plant responses to cadmium stress and the development of potential approaches to enhance cadmium tolerance in plants.

10.2.1 ULTRA-HIGH-PERFORMANCE LIQUID CHROMATOGRAPHY (UHPLC)

Because to its affordability and ease of use, high-performance thin-layer chromatography has been employed extensively. Because of its speed, efficiency, sensitivity, and selectivity, it can determine complicated samples. It can enhance UHPLC high resolution, increase peak efficiency, quick separations, and reduce solvent usage compared to standard HPLC.

The effect of cadmium stress on plants has been a subject of extensive study, particularly in the field of UHPLC. A study on sassafras seedlings demonstrated that cadmium stress in soil negatively influenced plant growth, reducing net growth in plant height and leaf area. It was found that cadmium stress can affect the synthesis of normal proteins and the production of stress proteins, leading to a decrease in biomass and a weakening of photosynthetic production capacity. Additionally, cadmium toxicity in medicinal plants has been recognized as having a deleterious impact on plant development, and various biotechnological and omics approaches have been explored to alleviate metal stress. These approaches include genetic engineering and CRISPR Cas 9 gene editing to enhance the ability of plants to survive cadmium stress. Furthermore, a study on fountain grass revealed that the influence of cadmium on plant performance correlates closely with metabolite profiles, providing insights into the metabolic plasticity and cadmium tolerance mechanisms of plants (Al-Khayri et al., 2023; Zhao et al., 2021).

The research on the effect of cadmium stress on plants has provided valuable insights into the physiological and metabolic responses of plants to cadmium toxicity. These studies have not only enhanced our understanding of the negative impact of cadmium on plant growth and development but have also explored potential strategies to alleviate cadmium stress in plants, which is crucial for ensuring the sustainable growth of medicinal plants and other plant species in cadmium-contaminated environments.

10.2.2 NMR

NMR is a nondisintegrating fingerprint technology that identifies metabolites based on size, charge, volatility, or stability. It provides structural information for identification. However, due to limited sensitivity, it only detects highly abundant metabolites and does not detect valuable metabolites in reduced concentrations, hindering plant metabolomics.

Metabolomics studies on the tolerance mechanisms of plants under cadmium stress have provided valuable insights into the impact of cadmium on plant metabolism and the strategies employed by plants to alleviate this stress. Several research articles have investigated the metabolome and transcriptome responses of various plant species to cadmium stress, shedding light on the key factors and metabolic pathways involved in the tolerance mechanisms.

Two studies on cadmium's effects on tomato plants using ^{1}H-NMR metabolomics found that metabolite responses depend on tissue type, age, exposure concentration, and similarities and differences between acute and chronic exposure (Hédiji et al., 2010; Djebali et al., 2011). Citrate was found to be increased in both roots and leaves, possibly due to photosynthetic activity and chelating agents. A decrease in carotenoid metabolites was observed in mature leaves, while chronic exposure increased sucrose but decreased fructose and glucose levels in leaves, suggesting a more long-term metabolic change.

Exogenous GSH in Maize: A study by Wang et al. (2021) revealed that exogenous GSH significantly alleviated cadmium-induced growth issues in maize seedlings. The researchers identified 145 and 133 metabolites in the leaves and roots, respectively, and found that GSH-mediated ROS and Mg metabolism were involved in plant HM tolerance.

Melatonin in Cotton: Another study by Li et al. (2022b) investigated the molecular mechanisms of melatonin-mediated cadmium detoxification in cotton seedlings. The researchers found that melatonin significantly induced the expression of cadmium-responsive genes and helped cotton seedlings tolerate cadmium stress.

Buckwheat and *Gynostemma pentaphyllum*: Research by Huo et al. (2023) analyzed the key metabolic pathways in response to cadmium stress in novel buckwheat and cultivated species. The study found that polysaccharides and proteins in the cell wall could bind with cadmium to prevent cadmium from entering the cell, and chelators such as glutathione, plant chelating peptides, and metallothionein in plant cells could bind to cadmium to prevent cadmium from binding to proteins with important physiological functions.

The integration of metabolomics and transcriptomics has provided comprehensive insights into the molecular mechanisms underlying cadmium tolerance in plants, offering a valuable resource for further research in this field. For example, Li et al. (2022b) combined transcriptome and metabolome analysis to reveal the regulatory network of melatonin-mediated cadmium stress in cotton.

In conclusion, NMR-based metabolomics studies on cadmium-stressed plants and their tolerance mechanisms have revealed significant changes in metabolites and metabolic processes. These findings contribute to the development of strategies for reducing cadmium accumulation in plants and ensuring safe food production. Further research integrating metabolomics and transcriptomics is expected to provide even more comprehensive insights into the molecular mechanisms underlying cadmium tolerance in plants.

10.2.3 LC-MS/MS

Metabolomics studies using LC-MS/MS have provided valuable insights into the impact of cadmium on plant metabolism and the strategies employed by plants to alleviate this stress. Several research articles have investigated the metabolome and transcriptome responses of various plant species to cadmium stress, shedding light on the key factors and metabolic pathways involved in the tolerance mechanisms.

A study by Yuan et al. (2022) investigated the responses of the edible and medicinal plant *Salvia miltiorrhiza* to different levels of cadmium stress. The study revealed that cadmium stress inhibited growth, accumulated cadmium residue, and affected secondary metabolites of the plant's roots. The untargeted LC–MS/MS-based metabolomic profiling provided insights into the metabolic mechanisms of *S. miltiorrhiza*'s resistance to cadmium stress. The study revealed that cadmium stress inhibited growth, accumulated cadmium residue, and affected secondary metabolites of *S. miltiorrhiza* roots, shedding light on the plant's response to cadmium stress (Yuan et al., 2022).

Furthermore, an integrated transcriptome and metabonomic analysis of key metabolic pathways in response to cadmium stress in novel buckwheat and cultivated species provided insights into the metabolic responses of these plants to cadmium stress. The study highlighted the significant changes in cell composition and expression regulation during cadmium treatment, offering valuable information on the metabolic pathways involved in the plants' tolerance mechanisms.

The hyperaccumulator *Erigeron annuus*'s tolerance and detoxifying processes to cadmium stress were the subject of research. The study shed insight into the unique molecular reactions to cadmium stress (Fourier Transform Ion Cyclotron Resonance) by demonstrating that the plant's detoxification and tolerance mechanisms to the stress included the creation of molecules in root exudate (Zhang et al., 2021).

Integrated transcriptome and metabonomic analysis of key metabolic pathways in response to cadmium stress in novel buckwheat and cultivated species 3 provided insights into the metabolic responses of these plants to cadmium stress, highlighting the key metabolic pathways involved in their tolerance mechanisms.

LC-MS/MS-based metabolomics studies on cadmium-stressed plants have revealed significant changes in metabolites and metabolic processes, contributing to the development of strategies for reducing cadmium accumulation in plants and ensuring safe food production (Huo et al., 2023). Further research integrating metabolomics and transcriptomics is expected to provide even more comprehensive insights into the molecular mechanisms underlying cadmium tolerance in plants.

In conclusion, LC-MS/MS-based metabolomics studies have been instrumental in unraveling the metabolic responses of plants to cadmium stress, offering valuable insights into the molecular mechanisms underlying cadmium tolerance and providing a foundation for the development of strategies to mitigate the impact of cadmium on plant metabolism.

10.2.4 GAS CHROMATOGRAPHY-MASS SPECTROMETRY (GC-MS)

GC-MS is widely used for plant metabolomics due to its efficient separation of metabolites, high resolution, sensitivity, and reproducibility. It is particularly useful for analyzing volatile and thermostable compounds, which provide information about plant physiological states. Gas chromatography-time of flight-mass spectrometry is used for metabolic analysis of forest species under abiotic and biotic stress, and for estimating metabolite fluxes.

Metabolomics studies using GC-MS have provided valuable insights into the impact of cadmium on plant metabolism and the strategies employed by plants to alleviate this stress. Several research articles have investigated the metabolome and transcriptome responses of various plant species to cadmium stress, shedding light on the key factors and metabolic pathways involved in the tolerance mechanisms.

Metabolomics has been used to study model systems (*Arabidopsis thaliana*) and agricultural plants (such as rice and tomatoes) (Dubey et al., 2010; Hédiji et al., 2010; Djebali et al., 2011). GC-MS metabolomics was utilized in the most current investigation of the effects of cadmium on *A. thaliana* entire plants to find some intriguing metabolite alterations (Djebali et al., 2011). Antioxidant molecules like campesterol and isoflavone were observed to rise together with a number of chemicals with suitable solute characteristics, including alanine, proline, glycerol, and trehalose in order to mitigate the osmotic and oxidative stress caused by cadmium exposure, respectively (Booth et al., 2011).

A study analyzed the metabolic changes in radish roots responding to HM (Pb or Cd) stress using a GC-MS-based metabolomic approach. The study revealed potential metabolites and biological pathways involved in the plant's response to cadmium stress (Wang et al., 2015).

Another study by Hédiji et al. (2010) investigated the effects of long-term cadmium exposure on the growth and metabolomic profile of tomato plants. The researchers identified significant changes in the metabolites involved in energy metabolism, amino acid metabolism, and secondary metabolism, highlighting the metabolic pathways involved in the plant's response to cadmium stress (Hédiji et al., 2010).

Research analyzed the responses of *G. pentaphyllum* to cadmium stress at both the physiological and molecular levels. The study provided insights into the plant's Cd-resistance mechanisms based on metabolomics, highlighting the key metabolic pathways involved in the plant's tolerance mechanisms (Zhou et al., 2023).

GC-MS-based metabolomics studies on cadmium-stressed plants have revealed significant changes in metabolites and metabolic processes, contributing to the development of strategies for reducing cadmium accumulation in plants and ensuring safe food production (Paul and Das, 2023). Further research integrating metabolomics and transcriptomics is expected to provide even more comprehensive insights into the molecular mechanisms underlying cadmium tolerance in plants (Huo et al., 2023).

A research investigates the metabolic changes in radish roots in response to lead (Pb) and Cd stress using GC-MS. The study found that Pb and Cd stress led to significant alterations in the levels of various metabolites, including sugars, amino acids, and organic acids. The authors also identified several biological pathways that were affected by Pb and Cd stress, such as carbohydrate metabolism, energy metabolism, and glutathione metabolism. The study provides valuable insights into the metabolic responses of radish roots to HM stress, highlighting the potential of GC-MS-based metabolomic approaches in examining the effects of HM stress on plant metabolism. The findings contribute to a better understanding of the mechanisms of HM stress in plants and the effective management of HM contamination in vegetable crops (Wang et al., 2015).

It is obvious that GC-MS-based metabolomics studies have been instrumental in unraveling the metabolic responses of plants to cadmium stress, offering valuable insights into the molecular mechanisms underlying cadmium tolerance and providing a foundation for the development of strategies to mitigate the impact of cadmium on plant metabolism (Wang et al., 2015).

Cd stress can negatively impact plant growth, leading to reduced shoot length, shoot fresh and dry weight, and altered biochemical processes. However, the use of plant growth-promoting microbes (PGP) and genetic analysis techniques such as GC-MS can help alleviate Cd stress in plants.

One study found that inoculation of rice plants with the bacterial strain S6D1–105 under cadmium stress conditions led to increased shoot length and biomass compared to uninoculated plants (Wang et al., 2015). This suggests that PGP microbes can help mitigate the negative effects of Cd stress on plant growth.

Another study investigated the use of heavy-metal-tolerant plant growth-promoting (HMT-PGP) microbes in the remediation of contaminated soil (Jabeen et al., 2022). These microbes can accumulate, transform, or detoxify HM, thereby improving plant growth and overall health. The interaction between HMT-PGP microbes and plants has been shown to enhance growth under heavy-metal stress or increase the plant's ability to tolerate such stress conditions (Jabeen et al., 2022).

In addition to PGP microbes, endophytic bacteria have been found to alleviate Cd stress in plants through their antioxidant potential, including sodium oxide dismutase (SOD) synthesis (Ullah et al., 2019). This indicates that the use of endophytic bacteria could be a promising approach for mitigating Cd stress in plants. Furthermore, metabolomic analysis using GC-MS has been employed to investigate the metabolic changes in response to Cd stress in radish roots. This approach has helped identify potential metabolites and biological pathways involved in the Cd stress response, providing valuable insights into the plant's defense mechanisms and adaptations to heavy-metal stress.

So the use of PGP microbes, such as PGP and HMT-PGP microbes, and genetic analysis techniques like GC-MS can help alleviate Cd stress in plants, improve their growth, and enhance their ability to tolerate heavy-metal stress conditions. These approaches offer promising potential for the remediation of contaminated soils and the sustainable production of crops in environments affected by heavy-metal pollution.

10.3 ARSENIC STRESS TOLERANCE STUDY THROUGH METABOLOMICS

Arsenic is a toxic metalloid that poses a significant threat to human health and the environment. Arsenic contamination in soil and water is a global problem, affecting millions of people worldwide. Plants are the primary source of arsenic in the food chain, and their ability to tolerate arsenic stress is crucial for ensuring food safety. As can exist in both inorganic and organic forms, with highly oxidized Arsenate (AsV) in aerobic environments and reduced Arsenite (AsIII) in anaerobic environments (Zhao et al., 2010). Microbes can biotransform inorganic As into organic forms, impacting plant metabolism. The plant's response to As exposure impacts carbon, nitrogen, and sulfur metabolism (Finnegan and Chen, 2012). As affects plant carbon metabolism by stimulating the accumulation of ascorbate to protect against ROS. It can inhibit net photosynthesis and decrease chlorophyll content, potentially reducing ATP and NADPH production needed for carbon fixation. The effects of As on gene expression related to carbon metabolism are largely unaffected, suggesting that plant metabolism has sufficient plasticity to maintain adequate carbon flow without the need to adjust enzyme amounts in central pathways (Finnegan and Chen, 2012).

Arsenic exposure affects the movement of nitrogen-containing compounds through plant tissues (Foyer et al., 2011). It can decrease cellular Cys pools and deplete GSH pools, impacting the antioxidant capacity of the cell (Zhou et al., 2011; Pavlík et al., 2010). As a consequence of As exposure, metabolic acclimations are required, including the diversion of nitrogen from normal growth and development to support the biosynthesis of GSH and PC.

The role of metabolites in arsenic stress tolerance and detoxification in plants is crucial for maintaining the plant's physiological functions and preventing oxidative stress. Arsenic exposure can lead to the accumulation of ROS, which can cause oxidative stress in plants (Nahar et al., 2022). Metabolites play a significant role in combating this oxidative stress and helping plants adapt to arsenic stress.

Key roles of metabolites in arsenic stress tolerance and detoxification include the following.

10.3.1 KEY ROLES OF METABOLITES IN ARSENIC STRESS TOLERANCE AND DETOXIFICATION

10.3.1.1 OSMOREGULATION

Arsenic exposure can lead to impaired nutrient absorption and disruption of the photosynthetic apparatus (Abbas et al., 2018).

10.3.1.2 PROLINE METABOLISM

Metabolites such as proline, which is a compatible solute, can help regulate osmotic balance and maintain cellular homeostasis during arsenic stress (Abbas et al., 2018).

10.3.1.3 ANTIOXIDANT DEFENSE

Arsenic-induced oxidative stress can lead to the production of ROS, which can cause oxidative stress in plants (Nahar et al., 2022). Metabolites such as ascorbic acid, glutathione, and tocopherol can help scavenge ROS and maintain redox homeostasis in plants exposed to arsenic stress (Nahar et al., 2022).

10.3.1.4 THIOL LIGAND COMPOUNDS

Arsenic exposure can lead to the accumulation of thiol ligand compounds, which can help detoxify arsenic and reduce its toxicity in plants (Martínez-Castillo et al., 2022).

10.3.1.5 PHYTOCHEMICALS

The production of phytochemicals, such as phenolic compounds and flavonoids, can be induced by arsenic stress (Martínez-Castillo et al., 2022). These compounds can help scavenge ROS and provide additional antioxidant defense in plants exposed to arsenic stress.

10.3.1.6 GLUTATHIONE S-TRANSFERASES (GSTS)

GSTs play a crucial role in arsenic detoxification in plants (Kumar and Trivedi, 2018). These enzymes can conjugate arsenic with glutathione, a process that can help reduce the toxicity of arsenic and enhance its excretion from the plant.

In conclusion, metabolites play a significant role in arsenic stress tolerance and detoxification in plants. These metabolites are involved in various physiological and biochemical processes, including osmoregulation, antioxidant defense, and the maintenance of cellular homeostasis. By understanding the role of these metabolites, researchers can develop strategies to enhance arsenic tolerance in plants and mitigate the toxic effects of arsenic on plant metabolism.

Metabolomics studies have emerged as a powerful tool for investigating the metabolic responses of plants to arsenic stress. Metabolomics involves the comprehensive analysis of small molecules, or metabolites, in biological systems, providing a snapshot of the metabolic state of the organism. Several research articles have investigated the metabolome and transcriptome responses of various plant species to arsenic stress, shedding light on the key factors and metabolic pathways involved in the tolerance mechanisms. For example, a study by Martínez-Castillo et al. (2022) aimed to conduct a comprehensive biochemical profiling and metabolomics analysis to investigate the effects of arsenic-induced metabolic disorders. The study revealed potential metabolites and biological pathways involved in the plant's response to arsenic stress. Another study investigated the effects of long-term arsenic exposure on the growth and metabolomic profile of tomato plants. The researchers identified significant changes in the metabolites involved in energy metabolism, amino acid metabolism, and secondary metabolism, highlighting the metabolic pathways involved in the plant's response to arsenic stress.

The significant metabolites in response to arsenic stress differ from those in nonstressed plants. Arsenic exposure can lead to significant changes in primary and secondary metabolism in plants (Martínez-Castillo et al., 2022). A review article by Abbas et al. (2018) states that arsenic exposure can impede nutrient absorption, negatively affect photosynthetic apparatus, disrupt plant water status, and interact with the functional groups of enzymes, leading to changes in the plant's metabolic pathways.

Research focused on the uptake and metabolism of arsenic in various plant species, including hair grass, Norway spruce, wild strawberry, bilberry, raspberry, and yarrow. The authors identify and quantify various arsenic species in these plants, including AsV, arsenite, monomethyl-arsenic acid, dimethylarsinic acid, and tetramethyl-arsenium ion. The study highlights the role of intracellular compartmentalization in adaptation to arsenic stress and the importance of maintaining low cytoplasmic concentrations of arsenic to maintain cellular function. The authors also discuss the relationship between AsV sensitivity and phosphate nutrition, with increased phosphate status leading to reduced AsV uptake. The chapter provides valuable insights into the mechanisms of arsenic resistance in plants and highlights the need for further research to improve our understanding of the physiological basis of metal(loid) resistances (Meharg and Hartley-Whitaker, 2002).

Metabolomic studies have identified specific metabolites that are induced or altered in response to arsenic stress. For example, a study by Finnegan and Chen (2012) investigated the effects of arsenic-induced metabolic disorders on biochemical profiling and metabolomics analysis. The study revealed significant changes in the proteome and metabolome of both humans and animal models exposed to chronic arsenic exposure.

Furthermore, a review article by Martínez-Castillo et al. (2022) provides a metabolomic perspective on arsenic stress in plants. The article highlights the significant changes in primary metabolites in response to arsenic stress, including amino acids, organic acids, and sugars. The article also notes that arsenic accumulation in plant tissue generates oxidative stress, mainly during detoxification into less toxic metabolites and disruption of electron transport.

Plant hormones and signaling molecules play a significant role in arsenic stress tolerance and detoxification (Nahar et al., 2022). For example, the use of plant nutrients, phytohormones, signaling molecules, and chelating agents can help enhance oxidative stress tolerance in plants exposed to arsenic. Additionally, the integration of ROS and hormone signaling during abiotic stress has been studied to better understand the cellular metabolic processes that accompany arsenic stress (Abbas et al., 2018).

In conclusion, the identified metabolites in response to arsenic stress differ from those in nonstressed plants. Arsenic exposure can lead to significant changes in primary and secondary metabolism in plants, including amino acids, organic acids, and sugars. These changes are involved in various physiological and biochemical processes, including osmoregulation, antioxidant defense, and the maintenance of cellular homeostasis. By understanding the role of these metabolites, researchers can develop strategies to enhance arsenic tolerance in plants and mitigate the toxic effects of arsenic on plant metabolism.

So, metabolomics studies have been instrumental in unraveling the metabolic responses of plants to arsenic stress, offering valuable insights into the molecular mechanisms underlying arsenic tolerance and providing a foundation for the development of strategies to mitigate the impact of arsenic on plant metabolism.

10.3.2 HPLC

Arsenic stress has significant impacts on plant metabolism and the environment due to its highly toxic and carcinogenic nature (Nahar et al., 2022). Understanding the role of metabolites in arsenic stress tolerance and detoxification is crucial for developing strategies to enhance plant resistance to arsenic exposure and mitigate its toxic effects on plant metabolism. Arsenic exposure can lead to the accumulation of various forms of arsenic, such as AsV and arsenite (AsIII). The uptake and accumulation of arsenic in plants can cause biochemical and metabolic changes, including impeded nutrient absorption, negative effects on photosynthetic apparatus, disruption of plant water status, and interaction with the functional groups of enzymes.

A study by Abbas et al. (2018) investigated the effects of arsenic stress on plant metabolism using HPLC-based approaches. The study found that arsenic exposure led to significant changes in the levels of several important amino acids, including a remarkable fall in the levels of few important amino acids. The study also reported that arsenic accumulation in plant tissue generated oxidative stress, mainly during detoxification into less toxic metabolites and disruption of electron transport. The study highlighted the role of metabolites in mitigating oxidative stress and the plant's antioxidant defense mechanisms in response to arsenic stress. The study by Abbas et al. (2018) provides valuable insights into the metabolic changes and tolerance mechanisms in plants exposed to arsenic stress using HPLC-based approaches. The study highlights the role of metabolites in mitigating oxidative stress and the plant's antioxidant defense mechanisms in response to arsenic stress. Further research in this area is essential for developing effective strategies to enhance arsenic tolerance in plants and mitigate the toxic effects of arsenic on plant metabolism.

A study by Malik et al. (2011) investigated the effect of arsenic stress on the growth and development of plants using HPLC-based approaches. The study found that arsenic exposure led to a significant decrease in the levels of several important amino acids, including a remarkable fall in the levels of few important amino acids. The study also reported that arsenic accumulation in plant tissue generated oxidative stress, mainly during detoxification into less toxic metabolites and disruption of electron transport. The study highlighted the role of metabolites in mitigating oxidative stress and the plant's antioxidant defense mechanisms in response to arsenic stress. The study provides valuable insights into the metabolic changes and tolerance mechanisms in plants exposed to arsenic stress using HPLC-based approaches. The study highlights the role of metabolites in mitigating oxidative stress and the plant's antioxidant defense mechanisms in response to arsenic stress. Further research in this area is essential for developing effective strategies to enhance arsenic tolerance in plants and mitigate the toxic effects of arsenic on plant metabolism.

The role of metabolites in arsenic stress tolerance and detoxification in plants is crucial for understanding the AR of plants to arsenic stress and developing strategies to enhance their resistance. The identified metabolites are involved in various physiological and biochemical processes, including osmoregulation, antioxidant defense, and the maintenance of cellular homeostasis. Further research on the role of metabolites in arsenic stress tolerance and detoxification can provide valuable insights into the mechanisms of plant adaptation to arsenic stress and help in developing strategies to mitigate the toxic effects of arsenic on plant metabolism.

10.3.3 NMR

NMR spectroscopy has emerged as a powerful tool in the study of metabolites, providing valuable insights into the biochemical processes occurring within living organisms. In the context of arsenic-stressed plants, NMR offers a noninvasive and highly sensitive approach to unravel the metabolic responses triggered by arsenic exposure.

Arsenic, a toxic metalloid, poses a significant threat to plants as it can accumulate in various plant tissues, affecting growth, development, and overall health. Understanding the metabolic adjustments plants undergo in response to arsenic stress is crucial for developing strategies to mitigate its impact on both plant and human health. NMR enables researchers to analyze a wide range of metabolites, including sugars, organic acids, amino acids, and secondary metabolites, without the need for extensive sample preparation. By studying the NMR spectra of plant extracts, scientists can identify and quantify metabolites, providing a comprehensive view of the metabolic landscape under arsenic stress.

This approach facilitates the identification of key metabolites associated with stress response mechanisms, metabolic pathways altered by arsenic exposure, and potential biomarkers indicative of plant health under such conditions. The NMR study of metabolites in arsenic-stressed plants not only enhances our understanding of plant physiology but also contributes to the development of strategies for sustainable agriculture and environmental management.

A study by Tripathi et al. (2019) investigated the effect of arsenic stress on the growth and development of plants using NMR-based metabolomic approaches. The study found that arsenic exposure led to significant changes in the levels of several important metabolites, including amino acids, sugars, organic acids, phosphagens, and osmolytes. The study highlighted the role of metabolites in mitigating oxidative stress and the plant's antioxidant defense mechanisms in response to arsenic stress.

In another study, researchers used a green microalga, IITRIND2, to investigate the differential cellular responses during mitigation of arsenic(III, V) using NMR spectroscopy. The study identified and quantified approximately 45 metabolites, including amino acids, sugars, organic acids, phosphagens, and osmolytes. The results showed that the two different arsenic states distinctly influenced the microalgal cellular mechanisms.

These studies demonstrate the potential of NMR-based metabolomic approaches in examining the effects of arsenic stress on plant growth and development, as well as the role of metabolites in the plant's response to arsenic-induced oxidative stress and detoxification. Further research in this area is essential for developing effective strategies to enhance arsenic tolerance in plants and mitigate the toxic effects of arsenic on plant metabolism.

A study identified and quantified approximately 45 metabolites, including amino acids, sugars, organic acids, phosphagens, and osmolytes. The NMR metabolomic experiments were complemented with various biophysical techniques to establish the mechanisms of plant uptake and metabolism of arsenic, as well as mitigation strategies. Impact of arsenic stress on growth, oxidative stress, and the antioxidant system: The study investigates the mechanisms adopted by microalgae to counteract the detrimental effects of high arsenic concentrations and on. The two different arsenic states distinctly influenced the microalgal cellular mechanisms due to applied organic metallic chemistry (Naseem Malik, 2010).

In conclusion, the chapter provides valuable insights into the mechanisms of arsenic uptake and metabolism in plants, as well as the potential strategies for mitigating the toxic effects of arsenic on plant growth and development. The study emphasizes the role of *GSTs* in arsenic detoxification and highlights the need for further research to improve agricultural productivity and food safety in the context of arsenic stress.

10.3.4 LC-MS/MS

LC-MS has emerged as a versatile and powerful technique for the comprehensive analysis of metabolites in complex biological samples. In the context of arsenic-stressed plants, LC-MS plays a crucial role in unraveling the dynamic metabolic changes induced by arsenic exposure.

Arsenic, a pervasive environmental contaminant, poses significant challenges to plants by influencing various metabolic pathways and cellular processes. Studying the metabolomic profile of plants under arsenic stress using LC-MS allows researchers to explore a broad spectrum of metabolites, including lipids, polar and nonpolar compounds, amino acids, and other secondary metabolites.

LC-MS involves the separation of complex mixtures of metabolites using LC, followed by their identification and quantification based on mass spectrometry. This dual analytical approach provides high sensitivity, resolution, and accuracy in detecting and characterizing diverse metabolites in plant samples.

By examining LC-MS spectra, researchers can identify specific metabolites that are modulated in response to arsenic stress. This information is pivotal for understanding the metabolic pathways affected by arsenic exposure and for identifying potential biomarkers indicative of plant stress. LC-MS studies contribute valuable insights into the adaptive strategies employed by plants to cope with environmental stressors, aiding the development of strategies for enhancing plant resilience in arsenic-contaminated environments.

In summary, LC-MS-based metabolomics is a robust tool for investigating the metabolite composition of arsenic-stressed plants, providing a deeper understanding of the plant's biochemical responses to stress and offering opportunities for the development of sustainable agricultural practices and environmental remediation strategies.

10.3.5 GC-MS

GC-MS has emerged as a pivotal analytical technique in the study of metabolites, providing a robust platform for the comprehensive analysis of complex biological samples. In the specific context of arsenic-stressed plants, GC-MS offers a high-resolution and sensitive approach to unravel the intricate metabolic changes induced by arsenic exposure.

Arsenic, a pervasive environmental contaminant, poses significant challenges to plants by disrupting cellular processes and metabolic pathways. Studying the metabolomic profile of plants under arsenic stress using GC-MS allows researchers to identify and quantify a diverse range of metabolites, including sugars, amino acids, organic acids, and other secondary metabolites. The GC-MS technique involves two primary steps: GC to separate individual components within a complex mixture and mass spectrometry for precise identification based on their mass-to-charge ratio. This dual approach provides detailed information about the composition of metabolites present in plant samples, offering insights into the biochemical adaptations triggered by arsenic exposure.

By examining the GC-MS spectra of plant extracts, researchers can discern alterations in metabolite abundance and composition, enabling the identification of potential biomarkers indicative of arsenic stress. This information is critical for understanding how plants respond to environmental challenges and for devising strategies to enhance their resilience.

In conclusion, GC-MS-based metabolomics is a powerful tool for unraveling the metabolic intricacies of arsenic-stressed plants, contributing to our broader understanding of plant stress responses and facilitating the development of strategies for sustainable agriculture and environmental management.

A study that uses GC-MS to identify potential metabolites and biological pathways involved in the response of radish roots to Pb and Cd stress. The study found that Pb and Cd stress led to significant changes in the levels of several important metabolites, including amino acids, organic acids, and sugars. The authors also identified several biological pathways that were affected by Pb and Cd stress, including glycolysis, the tricarboxylic acid cycle, and the pentose phosphate pathway. The study highlights the potential of GC-MS-based metabolomic approaches in examining the effects of HM stress on plant growth and development, as well as the role of metabolites in the plant's response to HM-induced oxidative stress and detoxification. The paper provides valuable insights into the mechanisms of HM stress in plants and highlights the need for further research to improve our understanding of the physiological basis of HM resistances (Wang et al., 2015).

The investigation on the effects of As on the accumulation of As in plant tissues, enzymatic activities, and the content of total phenols and proline was studied using different As concentrations (0, 50, 100, and 150 μM) and treatment durations (1, 3, and 5 days) on *Artemisia annua* plants. Plants accumulated As in both roots and shoots in a dose and time-dependent manner. The maximum accumulation of As was found after 7 days of treatment, and the results were statistically similar to the amount of As accumulated after 5 days. Therefore, further experiments were confined up to 5 days. It was found that As stress affected the activities of various enzymes in leaves and roots. For example, SOD and ascorbate peroxidase (APX) activities increased under As stress, while catalase (CAT) and peroxidase (POX) activities were affected differently in leaves and roots. The content of GSH in As-treated leaves and roots decreased compared to the control in a dose-dependent manner. However, the total thiol content increased in response to 100 μM As exposure, with a higher induction in leaves than in roots. The study analyzed the essential oil components in leaves and roots using GC-MS. The results showed that some terpenes were slightly reduced under As stress, while others, such as β-bergamotene, β-caryophellene, α and β-isocomene, modephene, and A. kumarsii, were not significantly affected (Kumari et al., 2017). The study provides insights into the effects of arsenic on *A. annua*, including As accumulation, enzymatic activities, and the content of total phenols and proline. The results can be useful for understanding the mechanisms of As toxicity in plants and developing strategies for phytoremediation and bioremediation applications.

Arsenic stress can negatively impact plant growth and development, leading to reduced biomass production and impaired metabolic processes. However, the use of PGP microbes and genetic analysis techniques such as GC-MS can help alleviate arsenic stress in plants.

One study found that inoculation of rice plants with the bacterial strain RRF3 resulted in improved growth and reduced arsenic accumulation in the roots (Ngalimat et al., 2021). This suggests that PGP microbes can help mitigate the negative effects of arsenic stress on plant growth. Another study demonstrated that a microbial consortium containing *Pseudomonas pseudoalcaligenes* and *Chlorella vulgaris* could ameliorate arsenic toxicity in rice, leading to improved growth and reduced arsenic accumulation in the plants (Mishra et al., 2017). The positive responses were attributable to a significant decline in arsenic accumulation in the root and shoot tissues. In addition to PGP microbes, mycorrhizal fungi have been shown to play a role in arsenic stress tolerance in plants. For example, arbuscular mycorrhiza has been found to augment arsenic tolerance in wheat by strengthening the plant's antioxidant defense mechanisms.

Furthermore, transcriptomic analysis of rice plant roots has shown that inoculation with the bacterium *Pseudomonas putida* strain RRF3 reorganizes the root transcriptome and protects the plant from arsenic stress (Kandaswamy et al., 2019). This indicates that PGP bacteria can help plants cope with arsenic stress by altering their gene expression patterns.

So, the use of PGP microbes, such as PGP bacteria and mycorrhizal fungi, and genetic analysis techniques like GC-MS can help alleviate arsenic stress in plants, improve their growth, and enhance their ability to tolerate heavy-metal stress conditions. These approaches offer promising potential for the remediation of contaminated soils and the sustainable production of crops in environments affected by arsenic pollution.

10.4 CONCLUSION

Metabolomics has emerged as a powerful tool to understand the stress responses and tolerance mechanisms in plants exposed to HM such as cadmium and arsenic. By employing a combination of transcriptomic, metabolomic, and phenomic approaches, researchers have gained insights into the key metabolic pathways, gene expression patterns, and biochemical changes that enable plants to cope with heavy-metal stress (Figure 10.3).

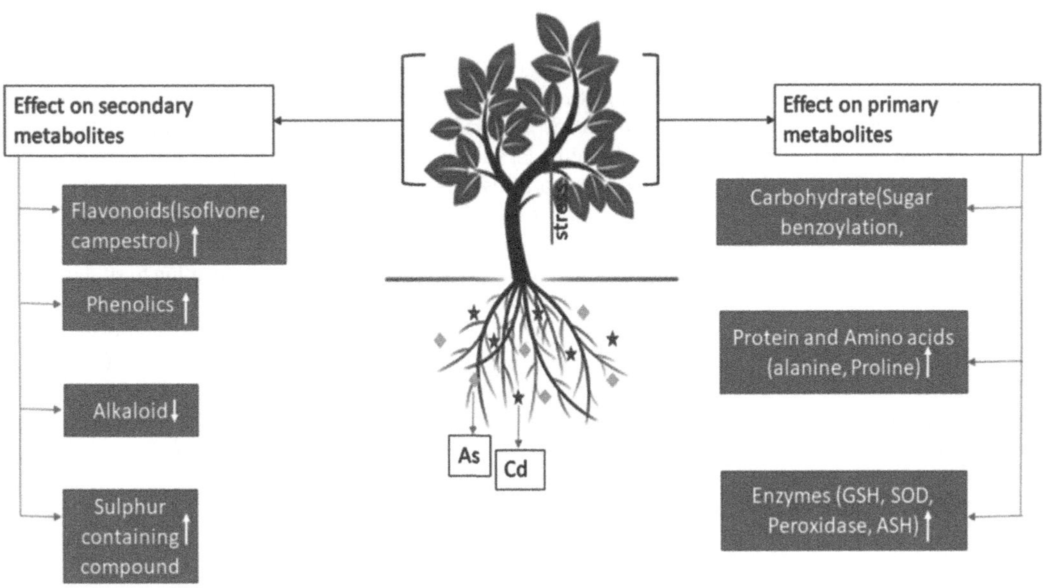

FIGURE 10.3 Effects of arsenic and cadmium stress on plant metabolites.

Key findings from the studies discussed include:

Cadmium and arsenic stress lead to changes in ROS and the chlorophyll system, affecting plant growth and biomass production. PGP microbes can help alleviate heavy-metal stress by altering gene expression patterns and improving plant growth. Mycorrhizal fungi play a role in heavy-metal stress tolerance in plants, enhancing their ability to cope with arsenic stress. Metabolomic analysis has been used to identify metabolites and biological pathways involved in the stress response, providing valuable insights into plant defense mechanisms and adaptations to heavy-metal stress conditions.

These findings highlight the potential of metabolomics as a valuable tool for studying heavy-metal stress responses and tolerance in plants, ultimately contributing to the development of strategies for the sustainable production of crops in contaminated soils and the remediation of heavy-metal pollution.

KEYWORDS

- **metabolomics**
- **heavy metal toxicity**
- **arsenic stress**
- **cadmium stress**

REFERENCES

Abbas, G., Murtaza, B., Bibi, I., Shahid, M., Niazi, N. K., Khan, M. I., Amjad, M., Hussain, M., Natasha. Arsenic uptake, toxicity, detoxification, and speciation in plants: Physiological, biochemical, and molecular aspects. *Int. J. Environ. Res. Public Health*, 2018, 15(1), 59. https://doi.org/10.3390/ijerph15010059

Al-Khayri, J. M., Banadka, A., Rashmi, R., Nagella, P., Alessa, F. M., Almaghasla, M. I. Cadmium toxicity in medicinal plants: An overview of the tolerance strategies, biotechnological and omics approaches to alleviate metal stress. In: *Frontiers in Plant Science (Vol. 13)*. Frontiers Media S.A., 2023, https://doi.org/10.3389/fpls.2022.1047410

Booth, S. C., Workentine, M. L., Weljie, A. M., Turner, R. J. Metabolomics and its application to studying metal toxicity. *Metallomics*, 2011, 3(11), 1142–1152. https://doi.org/10.1039/c1mt00070e

Chakrabarti, M., Mukherjee, A. Investigating the underlying mechanism of cadmium-induced plant adaptive response to genotoxic stress. *Ecotoxicol. Environ. Saf.*, 2021, 209, 111817. https://doi.org/10.1016/j.ecoenv.2020.111817

Chen, Q. Y., Costa, M. Annual review of pharmacology and toxicology arsenic: A global environmental challenge. The annual review of pharmacology and toxicology is Annu. *Rev. Pharmacol. Toxicol.*, 2021, 61, 47–63. https://doi.org/10.1146/annurev-pharmtox-030220

Chen, W., Gong, L., Guo, Z., Wang, W., Zhang, H., Liu, X., Yu, S., Xiong, L., Luoa, J. A novel integrated method for large-scale detection, identification, and quantification of widely targeted metabolites: Application in the study of rice metabolomics. *Mol. Plant*, 2013, 6(6), 1769–1780. https://doi.org/10.1093/mp/sst080

Clemens, S., Simm, C. Schizosaccharomyces pombe as a model for metal homeostasis in plant cells: The phytochelatin-dependent pathway is the main cadmium detoxification mechanism. *New Phytol.*, 2003, 159(2), 323–330. https://doi.org/10.1046/j.1469-8137.2003.00811.x

Cuypers, A., Plusquin, M., Remans, T., Jozefczak, M., Keunen, E., Gielen, H., Opdenakker, K., Nair, A. R., Munters, E., Artois, T. J., Nawrot, T., Vangronsveld, J., Smeets, K. Cadmium stress: An oxidative challenge. *BioMetals*, 2010, 23(5), 927–940. https://doi.org/10.1007/s10534-010-9329-x

Dawid, C., Hille, K. Functional metabolomics-a useful tool to characterize stress-induced metabolome alterations opening new avenues towards tailoring food crop quality. In: *Agronomy* (Vol. 8, Issue 8). MDPI, 2018. https://doi.org/10.3390/agronomy8080138

Djebali, W., Abbes, Z., Maucourt, M. Metabolite modifications in Solanum lycopersicum roots and leaves under cadmium stress. *Afr. J. Biotechnol.*, 2011, 10(4), 567–579. http://www.academicjournals.org/AJB

Dubey, S., Misra, P., Dwivedi, S., Chatterjee, S., Bag, S. K., Mantri, S., Asif, M. H., Rai, A., Kumar, S., Shri, M., Tripathi, P., Tripathi, R. D., Trivedi, P. K., Chakrabarty, D., Tuli, R. Transcriptomic and metabolomic shifts in rice roots in response to Cr (VI) stress. *BMC Genomics*, 2010, 11(1), 648. https://doi.org/10.1186/1471-2164-11-648

Finnegan, P. M., Chen, W. Arsenic toxicity: The effects on plant metabolism. *Front. Physiol.*, 2012, 3, 182. https://doi.org/10.3389/fphys.2012.00182

Foyer, C. H., Noctor, G., Hodges, M. Respiration and nitrogen assimilation: Targeting mitochondria-associated metabolism as a means to enhance nitrogen use efficiency. *J. Exp. Bot.*, 2011, 62(4), 1467–1482. https://doi.org/10.1093/jxb/erq453

Gill, S. S., Tuteja, N. Cadmium stress tolerance in crop plants: Probing the role of sulfur. *Plant Signal. Behav.*, 2011, 6(2), 215–222. https://doi.org/10.4161/psb.6.2.14880

Gupta, A. *Heavy Metal and Metalloid Contamination of Surface and Underground Water*. Taylor and Francis Group, LLC, 2021. https://doi.org/10.1016/j.fct.2017.06.013

Hasanuzzaman, M., Nahar, K., Anee, T. I., Fujita, M. Glutathione in plants: Biosynthesis and physiological role in environmental stress tolerance. *Physiol. Mol. Biol. Plants*, 2017, 23(2), 249–268. https://doi.org/10.1007/s12298-017-0422-2

Hédiji, H., Djebali, W., Cabasson, C., Maucourt, M., Baldet, P., Bertrand, A., Boulila Zoghlami, L., Deborde, C., Moing, A., Brouquisse, R., Chaïbi, W., Gallusci, P. Effects of long-term cadmium exposure on growth and metabolomic profile of tomato plants. *Ecotoxicol. Environ. Saf.*, 2010, 73(8), 1965–1974. https://doi.org/10.1016/j.ecoenv.2010.08.014

Huo, D., Hao, Y., Zou, J., Qin, L., Wang, C., Du, D. Integrated transcriptome and metabonomic analysis of key metabolic pathways in response to cadmium stress in novel buckwheat and cultivated species. *Front. Plant Sci.*, 2023, 14. https://doi.org/10.3389/fpls.2023.1142814

Jabeen, Z., Irshad, F., Habib, A., Hussain, N., Sajjad, M., Mumtaz, S., Rehman, S., Haider, W., Hassan, M. N. Alleviation of cadmium stress in rice by inoculation of Bacillus cereus. *PeerJ*, 2022, 10. https://doi.org/10.7717/peerj.13131

Jarup, L., Berglund, M., Elinder, C. G., Nordberg, G., Vahter, M. Health effects of cadmium exposure—A review of the literature and a risk estimate. *Scand. J. Work Environ. Health*, 1998, 24, 1–51.

Kandaswamy, R., Ramasamy, M. K., Palanivel, R., Balasundaram, U. Impact of Pseudomonas putida RRF3 on the root transcriptome of rice plants: Insights into defense response, secondary metabolism and root exudation. *J. Biosci.*, 2019, 44(4), 98. https://doi.org/10.1007/s12038-019-9922-2

Khan, S., Cao, Q., Zheng, Y. M., Huang, Y. Z., Zhu, Y. G. Health risks of heavy metals in contaminated soils and food crops irrigated with wastewater in Beijing, China. *Environ. Pollut.*, 2008, 152(3), 686–692. https://doi.org/10.1016/j.envpol.2007.06.056

Kumar, S., Trivedi, P. K. Glutathione S-transferases: Role in combating abiotic stresses including arsenic detoxification in plants. In: *Frontiers in Plant Science (Vol. 9)*. Frontiers Media S.A., 2018. https://doi.org/10.3389/fpls.2018.00751

Kumari, A., Pandey, N., Pandey-Rai, S. Protection of Artemisia annua roots and leaves against oxidative stress induced by arsenic. *Biologia Plantarum*, 2017, 61(2), 367–377. https://doi.org/10.1007/s10535-016-0686-3

Li, C., Cao, Y., Li, T., Guo, M., Ma, X., Zhu, Y., Fan, J. Changes in antioxidant system and sucrose metabolism in maize varieties exposed to Cd. *Environ. Sci. Pollut. Res.*, 2022a, 29(43), 64999–65011. https://doi.org/10.1007/s11356-022-20422-8

Li, L., Yan, X., Li, J., Wu, X., Wang, X. Metabolome and transcriptome association analysis revealed key factors involved in melatonin mediated cadmium-stress tolerance in cotton. *Front. Plant Sci.*, 2022b, 13. https://doi.org/10.3389/fpls.2022.995205

Malik, J. A., Goel, S., Sandhir, R., Nayyar, H. Uptake and distribution of arsenic in chickpea: Effects on seed yield and seed composition. *Commun. Soil Sci. Plant Anal.*, 2011, 42(14), 1728–1738. https://doi.org/10.1080/00103624.2011.584593

Márquez-García, B., Fernández-Recamales, M. Á., Córdoba, F. Effects of cadmium on phenolic composition and antioxidant activities of *Erica andevalensis*. *J. Bot.*, 2012, 2012, 1–6. https://doi.org/10.1155/2012/936950

Martínez-Castillo, J. I., Saldaña-Robles, A., Ozuna, C. Arsenic stress in plants: A metabolomic perspective. In: *Plant Stress (Vol. 3)*. Elsevier B.V., 2022. https://doi.org/10.1016/j.stress.2022.100055

Meharg, A. A., Hartley-Whitaker, J. Arsenic uptake and metabolism in arsenic resistant and nonresistant plant species. *New Phytologist*, 2002, 154(1), 29–43. https://doi.org/10.1046/j.1469-8137.2002.00363.x

Mendoza-Cózatl, D. G., Butko, E., Springer, F., Torpey, J. W., Komives, E. A., Kehr, J., Schroeder, J. I. Identification of high levels of phytochelatins, glutathione and cadmium in the phloem sap of Brassica napus. A role for thiol-peptides in the long-distance transport of cadmium and the effect of cadmium on iron translocation. *Plant J.*, 2008, 54(2), 249–259. https://doi.org/10.1111/j.1365-313X.2008.03410.x

Mishra, J., Singh, R., Arora, N. K. Alleviation of heavy metal stress in plants and remediation of soil by rhizosphere microorganisms. *Front. Microbiol.*, 2017, 8(SEP), 1706. https://doi.org/10.3389/fmicb.2017.01706

Nahar, K., Rhaman, M. S., Parvin, K., Bardhan, K., Marques, D. N., García-Caparrós, P., Hasanuzzaman, M. Arsenic-induced oxidative stress and antioxidant defense in plants. *Stresses*, 2022, 2(2), 179–209. https://doi.org/10.3390/stresses2020013

Naseem Malik, R. Heavy metal contamination and accumulation in soil and wild plant species from industrial area of Islamabad, Pakistan. 2010, https://www.researchgate.net/publication/228117034

Ngalimat, M. S., Hata, E. M., Zulperi, D., Ismail, S. I., Ismail, M. R., Zainudin, N. A. I. M., Saidi, N. B., Yusof, M. T. Plant growth-promoting bacteria as an emerging tool to manage bacterial rice pathogens. *Microorganisms*, 2021, 9(4), 682. https://doi.org/10.3390/microorganisms9040682

Paul, A., Das, S. Gas Chromatography Mass Spectrometry based metabolomic investigation on five different rice cutivars (*Oryza sativa* L.) under different induced Cadmium stress. *Food Chem. Adv.*, 2023, 2, 100175. https://doi.org/10.1016/j.focha.2022.100175

Pavlík, M., Pavlíková, D., Staszková, L., Neuberg, M., Kaliszová, R., Száková, J., Tlustoš, P. The effect of arsenic contamination on amino acids metabolism in *Spinacia oleracea* L. *Ecotoxicol. Environ. Saf.*, 2010, 73(6), 1309–1313. https://doi.org/10.1016/j.ecoenv.2010.07.008

Rahman, M. M., Owens, G., Naidu, R. Arsenic levels in rice grain and assessment of daily dietary intake of arsenic from rice in arsenic-contaminated regions of Bangladesh - Implications to groundwater irrigation. *Environ. Geochem. Health*, 2009, 31(SUPPL. 1), 179–187. https://doi.org/10.1007/s10653-008-9238-x

Sarwar, N., Imran, M., Shaheen, M. R., Ishaque, W., Kamran, M. A., Matloob, A., Rehim, A., Hussain, S. Phytoremediation strategies for soils contaminated with heavy metals: Modifications and future perspectives. *Chemosphere*, 2017, 171, 710–721. https://doi.org/10.1016/j.chemosphere.2016.12.116

Satarug, S., Moore, M. R. Adverse health effects of chronic exposure to low-level cadmium in foodstuffs and cigarette smoke. *Environ. Health Perspect.*, 2004, 112(10), 1099–1103. https://doi.org/10.1289/ehp.6751

Signes-Pastor, A. J., Mitra, K., Sarkhel, S., Hobbes, M., Burló, F., De Groot, W. T., Carbonell-Barrachina, A. A. Arsenic speciation in food and estimation of the dietary intake of inorganic arsenic in a rural village of West Bengal, India. *J. Agric. Food Chem.*, 2008, 56(20), 9469–9474. https://doi.org/10.1021/jf801600j

Su, Y. H., McGrath, S. P., Zhao, F. J. Rice is more efficient in arsenite uptake and translocation than wheat and barley. *Plant Soil*, 2010, 328(1), 27–34. https://doi.org/10.1007/s11104-009-0074-2

Tripathi, P., Tripathi, R.D. Metabolome modulation during arsenic stress in plants. In Srivastava, S., Srivastava, A., Suprasanna, P. (Eds.), *Plant–Metal Interactions,* Springer: Singapore, 2019, pp. 119–140

Ullah, I., Al-Johny, B. O., AL-Ghamdi, K. M. S., Al-Zahrani, H. A. A., Anwar, Y., Firoz, A., AL-Kenani, N., Almatry, M. A. A. Endophytic bacteria isolated from Solanum nigrum L., alleviate cadmium (Cd) stress response by their antioxidant potentials, including SOD synthesis by sodA gene. *Ecotoxicol. Environ. Saf.*, 2019, 174, 197–207. https://doi.org/10.1016/j.ecoenv.2019.02.074

Wang, R., Lin, K., Chen, H., Qi, Z., Liu, B., Cao, F., Chen, H., Wu, F. Metabolome analysis revealed the mechanism of exogenous glutathione to alleviate cadmium stress in maize (*Zea mays* L.) seedlings. *Plants*, 2021, *10*(1), 105, https://doi.org/10.3390/plants

Wang, Y., Xu, L., Shen, H., Wang, J., Liu, W., Zhu, X., Wang, R., Sun, X., Liu, L. Metabolomic analysis with GC-MS to reveal potential metabolites and biological pathways involved in Pb and Cd stress response of radish roots. *Sci. Rep.*, 2015, 5, 18296. https://doi.org/10.1038/srep18296

Yuan, J., Liu, R., Sheng, S., Fu, H., Wang, X. Untargeted LC–MS/MS-based metabolomic profiling for the edible and medicinal plant salvia miltiorrhiza under different levels of cadmium stress. *Front. Plant Sci.*, 2022, 13, 889370. https://doi.org/10.3389/fpls.2022.889370

Zhang, H., Heal, K., Zhu, X., Tigabu, M., Xue, Y., Zhou, C. Tolerance and detoxification mechanisms to cadmium stress by hyperaccumulator Erigeron annuus include molecule synthesis in root exudate. *Ecotoxicol. Environ. Saf.*, 2021, 219, 112359. https://doi.org/10.1016/j.ecoenv.2021.112359

Zhao, F. J., McGrath, S. P., Meharg, A. A. Arsenic as a food chain contaminant: Mechanisms of plant uptake and metabolism and mitigation strategies. *Annu. Rev. Plant Biol.*, 2010, 61, 535–559. https://doi.org/10.1146/annurev-arplant-042809-112152

Zhao, H., Guan, J., Liang, Q., Zhang, X., Hu, H., Zhang, J. Effects of cadmium stress on growth and physiological characteristics of sassafras seedlings. *Sci. Rep.*, 2021, 11(1), 9913. https://doi.org/10.1038/s41598-021-89322-0

Zhou, X., Sun, X., Cooper, K. L., Wang, F., Liu, K. J., Hudson, L. G. Arsenite interacts selectively with zinc finger proteins containing C3H1 or C4 motifs. *J. Biol. Chem.*, 2011, 286(26), 22855–22863. https://doi.org/10.1074/jbc.M111.232926

Zhou, Y., Yao, L., Huang, X., Li, Y., Wang, C., Huang, Q., Yu, L., Pan, C. Transcriptomics and metabolomics association analysis revealed the responses of *Gynostemma pentaphyllum* to cadmium. *Front. Plant Sci.*, 2023, 14. https://doi.org/10.3389/fpls.2023.1265971.

CHAPTER 11

Metabolomics for High- and Low-Temperature Stress Responses and Tolerance in Plants

SHAIMAA I. M. ELSAYED[1*], FARID M. ROHIM[2], MAGDI T. ABDELHAMID[3,4], and ELSAYED A. OMER[5]

[1]*Horticultural Department, National Research Centre, Giza, Egypt*

[2]*Horticultural Crops Technology Department, National Research Centre, Giza, Egypt*

[3]*Botany Department, National Research Centre, Dokki, Cairo, Egypt*

[4]*Department of Research Centers, Montana State University, Bozeman, MT, USA*

[5]*Medicinal and Aromatic Plants Research Department, National Research Centre, Giza, Egypt*

*Corresponding authors

ABSTRACT

High temperature or heat stress (HS) and low temperature or cold stress (CS) are critical environmental elements affecting plant growth and development, and the continuous seasonal and daily temperature changes affect plants and have different effects depending on whether they stay within a natural temperature range for each species or whether the range's extremes are reached. However, "cold acclimation" or "acquired thermotolerance," adaptive processes by which plants increase their chilling or heat tolerance in response to a previous low nonfreezing or high temperature exposure, respectively, can extend the temperature range of survival in some species. The substantial role of central amino acids, carbohydrates, nitrogen metabolisms, and metabolites complicated in biosynthesis and catalyzing actions play a role in HS and CS response and help plants cope with them. These metabolites appeared to be an integral component of reprogramming the metabolome under temperature stress, which is one aspect of metabolism that is constant across all of the temperature–stress metabolomic studies conducted to date. This chapter emphasizes how crops under HS and CS may be improved using a cutting-edge plant metabolomics method. Developing plants with specific characteristics using integrated omics (metabolites and transcriptomes) and metabolomics-assisted breeding are new directions for future research to increase plants' tolerance to HS and CS to achieve abundance in their productivity and overcome world hunger problems.

Exploring Plant Stress Tolerance: A Metabolomics Approach.
Shuvasish Choudhury, Debojyoti Moulick, and Mirza Hasanuzzaman (Eds.)

DOI: 10.1201/9781779641731-11

11.1 INTRODUCTION

Global food production systems are threatened by population growth and climate change. Because there is a limited supply of new arable land, greater productivity must come from using existing land. This calls for effective harvesting and storage methods and minimizing losses from biotic and abiotic pressures. Over the previous century, surface temperatures have risen by around 0.6°C worldwide. Global mean temperatures are predicted by climate models to increase (over 1990 levels) by 0.4–1.1°C by 2025, 0.8–2.6°C by 2050, and 1.4–5.8°C by 2100. Furthermore, projections show a rise in the frequency of brief hot, cold, and frosty days. Strategies for adaptation are required to counteract the adverse consequences of climate change and variability (Maduraimuthu and Vara-Prasad, 2014).

The distribution of a plant species is influenced by climatic circumstances, especially temperature, which might be affected directly through physiological restrictions (growth and reproduction) or indirectly through ecological considerations (competition for resources). Climate changes during the past century, particularly those related to temperature, have had a substantial impact on the physiology, phenology, abundance, and distribution of a wide variety of agricultural species. Studies using modeling techniques suggest that a significant shift in the geographic range and species' survival might result from climate change (Maduraimuthu and Vara-Prasad, 2014).

Nearly two decades ago, NASA scientist James Hansen warned about the obvious effects of global warming and recommended lowering CO_2 levels from the present 385.0 mg L^{-1} to a maximum of 350.0 mg L^{-1} to preserve human civilization (Hansen et al., 2008). Unpredictable temperature increases and rainfall patterns in particular pose a major danger to the productive production of crops needed to feed a growing population (Hassan et al., 2022).

Due to its extensive history of traditional agriculture and reliance on the Nile River as its primary water source, Egypt is particularly susceptible to the effects of climate change. In addition, erosion and increased development are already problems for the Nile's shoreline (Rosenzweig et al., 2002). Climate change may affect Egypt's agricultural industry and the stability of food security in several ways. Drought, extreme temperatures, and future temperature rise will all have a detrimental impact on crop productivity. In addition, floods brought on by sea level rise, seawater intrusion, and secondary salinization will alter the way land is used. Reduced precipitation and global warming may have an impact on water resources. Furthermore, an increase in agricultural water demand is anticipated. The primary difficulty for researchers is the confusing influence of all these elements (Mahmoud, 2017).

Crop production is reduced by several environmental stressors, both biotic and abiotic, that negatively impact crop growth and development. On the other hand, abiotic variables, such as soil pH, drought, salt, cold, heat, ultraviolet (UV) radiation, reactive oxygen species (ROS), and trace metals, are very damaging and reduce crop output globally. Abiotic stressors are anticipated to account for over 50% of agricultural output losses. Furthermore, these elements are in charge of the physiological and biochemical alterations that occur in plants. The way that various plant species react to these kinds of pressures is a complicated process that varies depending on the species (Salam et al., 2023).

11.2 TEMPERATURE STRESS IN PLANTS

The heat stress results in different negative impacts for the different plant tissues that translated into reduction of the yield of different crops, as shown in Table 11.1. Raising the temperature at night impacts rice yield since studies have shown that every 1°C increase in temperature results in a nearly 5% yield loss (Tao et al., 2006). Over the past 25 years, night temperature augmentation has resulted in a significant yield decline in Indian rice cultivars in the Philippines (Banerjee et al., 2022; Sarkar et al., 2022). It is interesting to note that rising global temperatures are expected to drastically lower rice production shortly (Lobell and Gourdji, 2012). Rice ultimate output is impacted by crop phenophase lengthening, which occurs when summer temperatures rise by an average of 1–4°C (Hatfield and Prueger, 2015). According to global climate models, humid and subtropical regions may have to pay a price for impending heat waves (Battisti and Naylor, 2009). Global warming negatively impacts two significant yield-related characteristics, such as the number of grains per m^2, average grain weight, and grain quality. When rice is exposed to high temperatures, its blooming patterns vary, its grain weight decreases, and its seed set is reduced (Qasim et al., 2020). The relationship between high temperatures, pollen shedding, and seed setting is inverse in rice plants. The high daytime temperature in rice was found to cause longer seed setting times (Matsui et al., 2001).

TABLE 11.1 Some Negative Impacts as a Result of Heat Stress for Some Crops

Crops	Heat Stress	Impact	References
Rice	Increasing temperature at night (1°C)	5% loss of yield	Tao et al., (2006)
Indian rice cultivars	Night temperature augmentation	Decline in yield	Banerjee et al. (2022); Sarkar et al. (2022)
Rice	Summer temperatures rise by an average of 1–4 °C	Ultimate output	Hatfield and Prueger (2015)
Rice	High temperatures	Its blooming patterns vary, its grain weight decreases, and its seed set is reduced	Qasim et al. (2020)
Rice	The high daytime	Longer seed setting times	Matsui et al. (2001)
Maize	heat stress (30 °C)	Less anthesis-silking intermissions	Edreira et al. (2011)
Maize	High temperatures over 35°C	Nonviability of pollen, reduced pollination potential, spikelet dropping, and aberrant kernel number	Vega et al. (2001)
Betel vine	Excessive fog and chilling damage	Browning leaves	Chaudhary et al. (2011)
Lemon	Excessive fog and chilling damage	Squeezing of leaves	Pautasso et al. (2012)
Pineapple	Excessive fog and chilling damage	Rot of Roots	Vedwan and Rhoades (2001)

The production of maize, another vital food source for the world, is also in danger because of temperature enhancement. The lower number of grains generated in maize might be attributed to reproductive processes and poor photosynthetic accumulation. According to Edreira et al. (2011), maize exposed to heat stress (30°C) appears to have less anthesis-silking intermissions throughout the blooming phase. According to Vega et al. (2001), high temperatures over 35°C are the primary cause of the nonviability of pollen, reduced pollination potential, spikelet dropping, and aberrant kernel number, all of which are associated with the active reproductive period.

Examples research on weather extremes brought on by global warming, Chaudhary et al. (2011), on the shedding and browning of betel vine leaves, the squeezing of lemon leaves, Pautasso et al. (2012), Vedwan and Rhoades (Vedwan and Rhoades, 2001) on rot of pineapple roots resulting from excessive fog and chilling damage.

Numerous studies looked at how climate change will affect Egypt's crop output, and the results generally showed that most crops would be less productive. Between 10% and 60% of agricultural operations are expected to drop due to rising temperatures (Smith et al., 2013). Increased temperatures brought on by predicted climate change would directly impact agricultural yields, perhaps resulting in a decrease of 10–30% for produces; for example, barley, soybeans, sunflower, rice, maize, and wheat (Abou Hadid, 2006; El-Ramady et al., 2013). Temperature increase of 2°C by 2050 is predicted to result in 15% lower wheat yields (Medany and Hassanein, 2006), 11% lower rice output (Eid and El-Marsafawy, 2002), (Hassan, 2013), 19% lower maize yields (Eid et al., 1997), 28% lower soybean yields (Eid and El-Marsafawy, 2002), and 11% lower barley yields.

Egyptian clover is in decline because producing food for cattle would be a significant difficulty for farmers (Banerjee et al., 2022). When the temperature rises by 3.5°C, the expected output of maize, barley, wheat, soybean, rice, sorghum, sunflower, and sugarcane will decline by 18%, 18%, 18%, 28%, 11%, 19%, 27%, and 25%, individually. Temperature rise of 1.5°C and 3.5°C, individually, can result in a 14% and 50% drop in tomato output (Hassan, 2013). However, if the temperature increased by 2°C and 4°C, individually, cotton crop would increase by 17–31% (Eid et al., 1997; Hassan, 2013). Potatoes and cotton are predicted to increase by 15–40%, mainly if they are grown during the major growing seasons (Abou Hadid, 2006; El-Ramady et al., 2013).

According to six and five emission situations, the estimated wheat production in Egypt for 2020 and 2050 was expected to exceed the baseline of 1970–2000. However, wheat yield in 2080 is anticipated to be lower than the baseline level under all emission scenarios. A decline in rice yield was simulated from 2020 to 2080. Under all emission scenarios, maize yield is predicted to decline with climate change (Fawaz and Soliman, 2016).

11.3 METABOLOMICS

11.3.1 AN OVERVIEW

Stresses impact plants' morphological and physiological features, including growth and metabolic processes, as well as the inhibition of functional groups of essential molecules, such as enzymes and polynucleotides, and transport mechanisms for large ions and nutrients (Hayat et al., 2021; Ullah et al., 2020). Plants, however, employ a variety of strategies, either singly or in combination, to deal with these challenges, such as metabolomics, transcriptomics, proteomics, and genomics. Plant cells undergo a series of chemical changes known as metabolism necessary for survival. Primary metabolism, on one hand, refers to all metabolic pathways vital to plant's life and produces substances known as metabolites that are directly related to the organism's growth and development. Conversely, secondary metabolism does not directly contribute to growth, reproduction, or development; instead, they are generated close to the stationary period of growth and a vast array of specialized molecules that are necessary for the plant to thrive in its surroundings but are not vital to the plant's operation. Long-term secondary plant metabolites are beneficial, frequently for defenses; they also give plants their color and are involved in signaling, communication, and controlling primary metabolic pathways (e.g., phytohormones) (http://www.usadellab.org/cms/index.php?page=metabolism).

Any alteration to the growth and developmental settings that interfere with metabolic homeostasis and necessitate the modification of metabolic pathways through a process commonly referred to as adaptation is considered plant stress (Shulaev et al., 2008). Metabolomics has advanced quickly over the past 10 years and is now acknowledged as the standard method for analyzing or clarifying test phenotypes in various living systems and adapting to changing climate circumstances (Guy et al., 2008). Significantly, identifying different molecules, such as by-products of stress metabolism and compounds of stress signal transduction and connected to the plant acclimation responses, may aid in studying stress biology in plants. In addition to its usefulness in other domains, metabolomics may be widely employed in selecting plants that are resilient to changing environmental conditions. Various studies have demonstrated that severe temperature stress may lead to notable instability in the plant metabolome pattern (Obata and Fernie, 2012). Metabolomics denotes the highest genetic level inside a biological system or indicates changes in an organism's characteristics or function.

11.3.2 METABOLIC CHANGES UNDER HIGH-TEMPERATURE STRESS

Higher temperatures can harm vital physiological functions such as the proper balance between primary and secondary metabolites and hormones or the regularity of water and membranes, respiration, and photosynthesis (Hemantaranjan, 2014). To survive in high temperatures, higher plants are subjected to excess heat of at least 5°C over their ideal growth conditions and display a distinct set of cellular and metabolic responses (Guy, 1999). In addition, Bray et al. (2000) pointed to a decrease in the synthesis of standard proteins and an increase in the dictation and translation of heat shock proteins (HSPs). These effects include also changes in the association of cellular structures and membrane functions. Additionally, the construction of abscisic acid (ABA) as phytohormones, antioxidants, and other protective molecules is also impacted (Maestri et al., 2002).

Figure 11.1 summarizes the various effects of heat stress on the vital processes and plant yield.

Plants use a complex network of sensors in different cellular compartments to perceive changes in the surrounding temperature. Enhanced Ca^{2+} inflow, cytoskeletal restructuring,

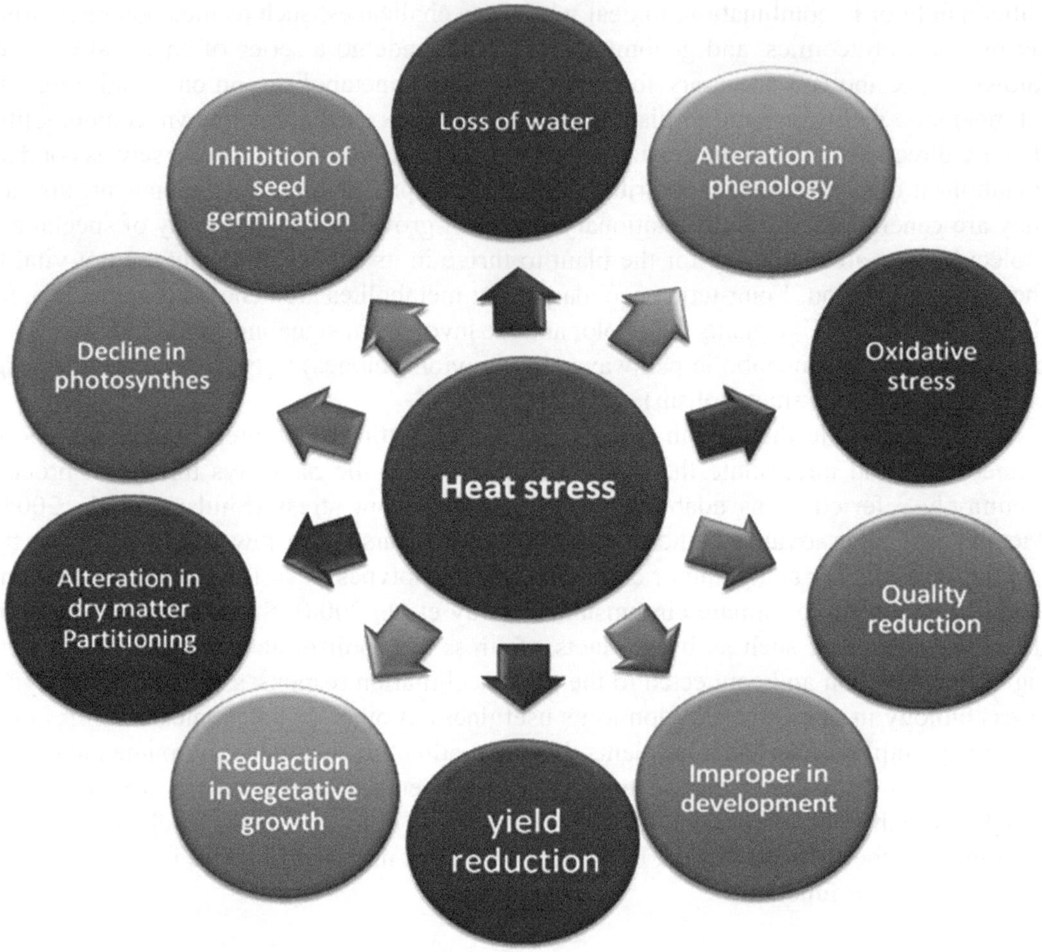

FIGURE 11.1 Different impacts of heat stress on physiology, growth, behavior, and yield.

and the activation of lipid-based signaling cascades are all caused by the membrane's improved fluidity. When heat stress occurs, signaling across these pathways causes the body to produce antioxidants and osmolytes. For instance, one of the main thermosensors in land plant cells is the membrane cyclic nucleotide-gated Ca^{2+} channel, which is encoded by the *Arabidopsis CNGC*$_2$ gene. According to Saidi et al. (2009), these plasma membrane channels react to temperature changes in the surrounding environment by starting an ideal heat shock response. The information highlights the critical function of lipid membranes in the response to heat stress in plants (Horváth et al., 2012). It has recently been demonstrated that vegetative and reproductive tissues may differ in the tissue-specific activation of different signaling pathways (Mittler et al., 2011). High temperatures uniquely impact photosynthetic membranes, causing grana stacks to expand and stack abnormally. Ion leaks from heat-exposed leaf cells and modifications in the energy budget allocated to the photosystems coexist with such structural alterations (Wahid and Shabbir, 2005). Thus, a sustained capacity for photosynthetic and respiratory processes depends on preserving cellular membrane occupation below high-temperature stress (Chen et al., 2010). The generation of harmful ROS is also linked to the adverse effects of heat on chlorophyll and the photosynthetic machinery (Camejo et al., 2006; Guo et al., 2007). Heat stress eventually lowers the photosynthetic and respiratory activities of plants by raising chlorophyllase activity and reducing the concentration of photosynthetic pigments (Talanova et al., 2003; Sharkey and Zhang, 2010). In terms of reproductive success, a decrease in photosynthesis would ultimately lead to a drop in energy stores that causes plant famine, limiting the resources available for imitation in paternal and gametophytic tissues (Young et al., 2004; Sumesh et al., 2008).

In plant tissues exposed to high temperatures, homeostasis is often disrupted, affecting biosynthesis and metabolite compartmentalization (Maestri et al., 2002). By downregulating specific genes entangled in metabolism of carbohydrate, high temperatures alter the activity of enzymes entangled in carbon metabolism, sucrose production, and starch accumulation (Ruan et al., 2010). Proline, betaine, glycine, and resolvable sugars are some of main metabolites that build up as reaction to high temperature (Wahid, 2007). Numerous plant species also collect additional osmolytes, such as quaternary and tertiary ammonium compounds or sugar alcohols (polyols) (Sairam and Tyagi, 2004). Under heat stress, osmolyte synthesis is expected to improve protein stability and maintain the membrane bilayer's structure (Mirzaei et al., 2012).

Under heat stress, the levels of several vital phytohormones also rise, such as ABA, salicylic acid, and ethylene (ET), whereas the other phytohormones, such as cytokinin, auxin (AUX), and gibberellic acids, fall. These fluctuations finally result in premature plant senescence (Larkindale et al., 2005). For illustration, it is known that elevated ABA and ET levels and decreased AUX levels and transport contribute to the abscission of reproductive organs, a significant heat stress consequence (Binder and Patterson, 2009). Similarly, pollen sterility was proposed to be associated with changed AUX biosynthesis in developing anthers (Sakata et al., 2010). Furthermore, it was discovered that a similar change in the creatine kinase (CK) concentration was the reason for the decreased grain filling in cereals (Banowetz et al., 1999).

Plants' responses to high temperature are also greatly influenced by secondary metabolites, such as flavonoids, steroids, and other compounds in plant. These compounds, likewise, often contribute to abiotic stress responses linked to heat tolerance (Wahid, 2007). For instance, as tomato plants acclimate to heat, thermal stress results in the buildup of soluble phenolics, rise phenylalanine ammonia–lyase action, and reduced polyphenol oxidase and peroxidase activity (Rivero et al., 2001). In the chamomile plant, there was an increase in the production of volatile oil, and the percentage of the main component of the volatile oil was due to heat stress (Elsayed et al., 2021).

11.3.3 METABOLIC CHANGES UNDER LOW-TEMPERATURE STRESS

Low temperature (LT) is one of the abiotic stressors, as mentioned by Shi et al. (2012), that plants experience LT stress. This stress affects photosynthesis, cell survival, cell division, water transport, and metabolism, which has a detrimental impact on plant development and ultimately limits agricultural yield (Sanghera et al., 2011; Jena et al., 2012). Figure 11.2 shows the variety traditions for plants respond to cold temperatures.

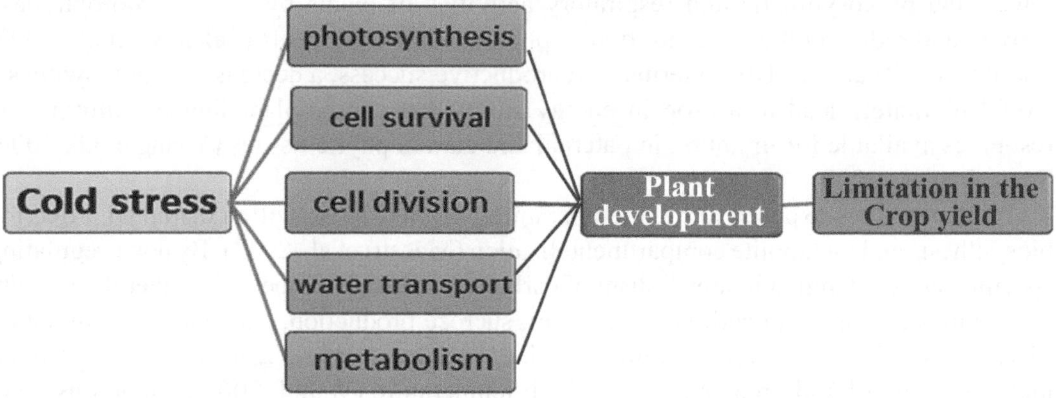

FIGURE 11.2 Imaginary planning of the effect of abiotic stress resulting from cold on the vital processes of the plant and its final yield.

LT stress is divided into two categories: (1) freezing stress (<0°C), which results in ice formation inside plant tissues and (2) chilling stress, which occurs at LTs (0–15°C) and causes harm without ice crystal construction in plant cells. We refer to both types of stress as cold or LT stress (Singh et al., 2017). Exposure to LT hindered various metabolic pathways, preventing the plant from expressing its full genetic potential, as manifested by a range of phenotypic symptoms (Chinnusamy et al., 2007). LTs hinder the metabolism of carbohydrates, slow down electron transport, cause partial stomatal closure, and interfere with phloem loading, all of which reduce photosynthesis (Ruelland and Zachowski, 2010).

Anjum et al. (2012) reported that plant oxidative stress, which arises from an imbalance between the production of ROS and their metabolism through enzymatic and nonenzymatic antioxidants, is finally brought on by chilling stress (Gill, 2010). LT stress causes a variety of physiological and metabolic alterations in plants. Under LT stress, a variety of ROS is produced, including superoxide radical (O_2•⁻), singlet oxygen ($^1O_2$), hydrogen peroxide (H_2O_2), and hydroxyl radical (OH•) (Miller et al., 2010). Extreme cold caused the plant's tolerance limit of ROS accumulation to exceed because antioxidant enzymes, which are in charge of detoxifying ROS, were not as active. Increased ROS levels result in oxidative stress, which can harm proteins, carbohydrates, DNA, and membrane lipids, among other things (Gill, 2010; Mo et al., 2010; Zhu et al., 2007). Additionally, they change the activities of enzymes, biochemical reactions, and plant functions, including respiration and photosynthesis, all of which have a detrimental effect on the plant's survival rate (Solanke and Sharma, 2008). On the other hand, plants generate a variety of antioxidants. The amount of harm caused by ROS is determined by equilibrium among the buildup of ROS products and the antioxidant scavenging system's ability to detoxify them (Azooz et al., 2009). The plant's susceptibility to cold is mainly determined by how well its antioxidant defense system scavenges ROS (Anjum et al., 2012; Gill, 2010; Zhao et al., 2009).

Different organelles' ascorbate peroxidase (APX), superoxide dismutase (SOD), and catalase (CAT) detoxify a greater quantity of H_2O_2 generated under LT stress (Das and Roychoudhury, 2014). H_2O_2 is transformed into O_2 and water by catalase.

11.4 METABOLOMICS-DRIVEN INSIGHTS INTO TEMPERATURE TOLERANCE MECHANISMS

Plants have created a variety of adaptation mechanisms to endure abiotic impacts, such as fluctuations in metabolism across various directions, to indicate their survival under harsh environmental conditions (Ashapkin et al., 2020). Metabolomics may help clear up the processes behind these responses and help plants develop resistance mechanisms (Carrera et al., 2021). Dimethyl sulfoniopropionate (DMSP) and glycine betaine are two examples of these metabolites; other examples include sugars such as sucrose and fructan and amino acids such as proline and ectoine, in addition to specific metabolites of polyols such as sorbitol and mannitol (Dikilitas et al., 2020; Kebert and Vuksanović, 2022).

Recent years have seen several studies on the metabolomic responses to cold stress in several crops, including tomatoes, wheat, maize, and *Miscanthus* (Paupière et al., 2020; Paupière et al., 2017; Qi et al., 2017; Sun et al., 2016; Le Gall et al., 2017), as well as some medicinal plants. Ghassemi et al. (2021) said that the secondary metabolites (SMs) play a part in defensive mechanisms that adapt to different types of stress. Terpenoids, chemicals containing nitrogen and sulfur, and phenolic substances make up plant SMs. The modulation of SMs and cryoprotective chemicals in plants to improve their resistance to cold is greatly influenced by LTs. LT stress has been shown in many researches to be an efficient means of promoting the accumulation of SMs. Rani et al. (2017), Upadhyaya (2012), and Wallaart et al. (2000) showed that how cold stress affects the synthesis of bioactive chemicals in a few aromatic and medicinal plants.

Arabidopsis thaliana is the species that has historically been the subject of the most research in this area. Most of this grass's metabolites, as well as proline, sugars, and tricarboxylic acid (TCA) cycle intermediates, rise in response to cold stress (Cook et al., 2004). One promising prospective biomarker for *Arabidopsis* cold tolerance is raffinose, as postulated by Korn et al. (2010) and Wienkoop et al. (2008), albeit the overall response varies by ecotype. Most of this species' reactions to heat shock were similar to those brought on by cold, including increase in the TCA cycle and amino acids derived from pyruvate (Kaplan et al., 2004).

Due to increasing global temperatures, heat is a significant abiotic stress that crops in several countries are constantly subjected to. Plants under heat stress produce excessive amounts of flavonoids, phenylpropanoids, and phenolic metabolites (Carrera et al., 2021). The heat stress factor B1 (HsfB1) was both overexpressed and suppressed in tomato (*Solanum lycopersicum* L.) leaves, and each had a distinct effect on the plant's ability to tolerate cold. Numerous isomers of caffeoylquinic acid were among the products of the phenylpropanoid and flavonoid pathways that accumulated due to HsfB1 overexpression (Paupière et al., 2020). However, putrescine, a polyamine, and the sugars such as glucose and sucrose accumulated in tomato leaves with reduced HsfB1. As a result, heat tolerance did not depend on the metabolite group. Thus, metabolomics and transcriptomics were utilized to identify changes in respiratory pathways at the substrate level, providing unique insights into the responses of *A. thaliana* plant to heat stress, resulting from either prolonged warming or heat shocks. Long-term warming increased the rate of glycolysis but inhibited the TCA cycle; in contrast, a heat shock significantly reduced the amount of pyruvate converted to acetyl coenzyme A. Wang et al. (2020) reported that thereby impairing the TCA cycle as well as glycolysis but not both in this measure. Furthermore, after 10 days of anthesis, the flag leaves of wheat (*Triticum aestivum* L.) exposed to high temperatures exhibited a general increase in pipecolate and L-tryptophan; however, anthranilate and drummondol dropped. The aminoacyl-tRNA pathway and the production of secondary metabolites were the metabolic pathways most affected by high temperature stress (Thomason et al., 2018).

11.5 METABOLOMICS AND BREEDING FOR TEMPERATURE STRESS RESILIENCE

In the past few years, significant advancements in software and technology have been made in metabolomics, offering a great chance to investigate the whole metabolome profiling of many plant species. Numerous research fields, especially biotechnology, which includes plant breeding and plant functional genomics, have benefited from metabolic applications (Rai and Saito, 2016; Christ et al., 2018). Furthermore, applications related to metabolomes provide new avenues for translational metabolome research concerning plant breeding. Recent advances in postgenomic methods have improved the analysis approach. Genomic methods have improved the analysis approach. It will take less time to produce crops with greater stress tolerance when plant metabolomics is combined with additional high-throughput platforms (HTPs) (Christ et al., 2018). Also, according to Christ et al. (2018), metabolomics has the tremendous potential to thoroughly examine several metabolic and phenotyping analyses of plants under stressful conditions. Due to the availability of comprehensive datasets for multiple omics approaches, the focus of recent studies has been to map for quantitative characteristics and separate genetic variations at the mRNA, protein, and metabolic stages by integrating metabolomics with other omics mechanisms, such as protein quantitative trait locus (pQTL), metabolic QTL (mQTL), and epigenomic QTL (eQTL) (Raza, 2022).

Interestingly, metabolomics, metabolome-wide study of association, mQTLs, and genome–phenome-wide association studies, in addition to genome-wide association studies (GWAS), are potent tools for delving into genetic variations linked with metabolic traits of plants (Fang et al., 2019; Templer et al., 2017). Comprehending the metabolic systems that operate a complex mechanism of metabolomics is crucial for metabolomics-assisted breeding (MAB), which aims to create enhanced cultivars resilient to various stressors.

Additionally, quantitative genetics knowledge is expanded using data obtained from mQTL investigations (Acuña-Galindo et al., 2015). By reducing the distance between genotype and phenotype and opening up new avenues for metabolic dissection, metabolomics begins with identifying potential genes using either QTL mapping analysis or recording single nucleotide polymorphism (SNP) markers. Moreover, metabolic markers are essential for identifying agronomic traits and investigating the genetic processes associated by plant phenotype (Sweetlove et al., 2014).

Templer et al. (2017) used the combination for drought, and HS was used to profile the metabolomes of 81 barley (*Hordeum vulgare* L.) accessions to identify QTLs associated with stress tolerance. Under HS, it was discovered that 57 metabolites were connected to the metabolism of antioxidant defense and found QTLs related to the pathways that produce antioxidant enzymatic metabolites in response to stress (succinate, glutathione, and γ-tocopherol). Barley's antioxidant defense based on these mQTLs helps it adapt to a harsh environment (Templer et al., 2017). Wheat mQTL analysis was previously done in drought and high temperatures. According to the findings, there were 234 QTLs associated with HS and 66 million QTLs spread across the wheat genome. Furthermore, both stressors and 43 mQTL showed a correlation, whereas just two were HS specific. Fifty SNPs within mQTL were identified by a mixture of 137 SNP markers for HG-related applicant genes; these genes are involved in sugar metabolism, ROS scavenging, and ABA-induced stomatal opening and closure. Accuña-Galindo et al. (2015) suggested that future research and genetic development of wheat under HS may consider the identified mQTL and genes. After completing the rice mQTL analysis, several mQTLs have been identified in developing seeds and flag leaves across 12 chromosomes (Gong et al., 2013). The metabolic and genetic analysis reliant on glucosinolate production in rapeseed (Brassica napus L.) was previously carried out by Feng et al. (2012). Noteworthy, 105 mQTLs linked to the production of glucosinolate in rapeseed leaves and seeds have been found.

Muthuramalingam et al. (2018) used computational metabolomics GWAS to conduct a worldwide inquiry on the genes controlling rice's threonine (Thr) metabolism under various abiotic stimuli, including HS. The results demonstrated that 16 abiotic stress-associated Thr metabolite producing genes (Thr MPG) altered the levels of metabolites and significantly influenced the physiological and nutritional conditions of rice plants. A total of 1373 and 1028 SNPs group were linked to genetic variations and complicated traits. Concerning stress-Thr MPG mapping, chromosomal collinearity with C4 grasses was revealed. Furthermore, these genes' computational appearance design predicted a distinct expression describing several progressive issues. The results of the protein interaction analysis indicate that Thr MPG is a multigenic abiotic stressor.

11.6 FUTURE DIRECTIONS AND CHALLENGES

Future climate scenarios may include crops' immediate or progressive exposure to high temperatures. It makes the study of stress effect temperature a problematic challenge. In order to lessen the negative effects of climate change and global warming on agricultural yield, more research on the identification of putative metabolites functioning under temperature stress may be essential.

The creation of metabolomic quantitative trait loci maps, for instance, is the result of merging libraries from various omics to create systematic backdrops. The subject of metabolomics-assisted biological engineering has great promise in achieving plant stress tolerances. More specifically, mass spectrometry, a potent instrument that helps to explain crucial biological processes connected to the plant kingdom, has greatly advanced the field of plant metabolomics research during the past 20 years. The future prospects of metabolomics include the creation of giant libraries of metabolites to determine and comprehend chemical configurations, as well as to establish and adjust the spectrum singularities of the metabolites assessed with cured spectra of observation patterns.

Creating databases that enable functional comparisons and integrating metabolites from higher plant species is another move researchers should take in this same direction to increase agricultural yield and environmental adaptation. The creation of bioinformatic software to purify the metabolites to validate the compounds' identity in the current databases is one of the most urgent metabolomics possibilities. Finally, but just as importantly, scientists working with this omic technology must help the research teams studying plant metabolites by being more meticulous and stringent throughout the whole process of identifying and analyzing plant metabolites.

11.7 CONCLUSION

In conclusion, several physiological functions, including secondary and primary metabolism, lipid or photosynthesis, and hormone signaling, are adversely impacted by high or LTs. Heat stress damages several proteins, membranes, and cytoskeleton structures adversely influencing plant growth and development. The process of pollen grain development is impacted, and the most affected stage is reproductive growth. Heat produces an imbalance in metabolism and the buildup of harmful by-products such as ROS, which eventually impacts plant vegetative and reproductive growth and has an adverse effect on fruit set and yield quality. It also stimulates the buildup of HSPs, which impede protein breakdown.

Plants display a plethora of responses to LT stress, such as transcriptomic reprogramming and membrane modification. Protective proteins such as cold-regulated (COR) proteins (COR polypeptides), antifreeze proteins (AFPs), and some of them even shield lipids are produced by LT-responsive genes. The ability of cells to withstand repeated cycles of dehydration and rehydration is a critical factor in determining their survival during cold stress. The amount of damage in the plasma membrane is determined by the lipid composition and the presence or absence of certain cryoprotectants.

Creating databases, which enable functional comparisons and integrating metabolites from higher plant species, is another move researchers should take in this same direction to increase agricultural yield and environmental adaptation. One of the most critical metabolomics opportunities is the development of bioinformatic tools to purify the metabolites in order to verify the compounds' identification in the existing databases. Last, but no less crucially, scientists employing this omic technology need to support the research groups examining plant metabolites by being more exacting and rigorous during the entire process of finding and evaluating plant metabolites.

KEYWORDS

- **heat stress**
- **cold stress**
- **metabolites**
- **metabolomics-assisted breeding**
- **systems biology**
- **extreme temperature**

REFERENCES

Abou Hadid, A. F. A. *Assessment of Impacts, Adaptation, and Vulnerability to Climate Change in North Africa: Food Production and Water Resources* (p. 148). AIACC, AF 90P: Washington, DC, 2006.

Acuña-Galindo, M. A., Mason, R. E., Subramanian, N. K., Hays, D. B. Meta-analysis of wheat QTL regions associated with adaptation to drought and heat stress. *Crop Sci.*, 2015, 55, 477–492.

Anjum, N. A., Ahamd, I., Mohmood, I., Pacheco, M., Duarte, A. C., Pereira, E. Modulation of glutathione and its related enzymes in plants responses to toxic metals and metalloids-a review. *Environ. Exp. Bot.*, 2012, 75, 307–324.

Ashapkin, V. V., Kutueva, L. I., Aleksandrushkina, N. I., Vanyushin, B. F. Epigenetic mechanisms of plant adaptation to biotic and abiotic stresses. *Int. J. Mol. Sci.*, 2020, 21, 7457.

Azooz, M. M., Ismail, A. M., Elhamd, M. F. Growth, lipid peroxidation and antioxidant enzyme activities as a selection criterion for salt tolerance of maize cultivars grown under salinity stress. *Int. J. Agric. Biol. Eng.*, 2009, 11, 21–26.

Banerjee, H., Sarkar, S., Dutta, S. K., Garai, S., Ray, K., Zingore, S., Goswami, R., Majumdar K. Nitrogen management trade-offs in hybrid rice for agronomy, carbon, and energy efficiency. *Nutr. Cycl. Agroecosystems*, 2022, 123, 5–18.

Banowetz, G., Ammar, K., Chen, D. Temperature effects on cytokinin accumulation and kernel mass in a dwarf wheat. *Ann. Bot.* 1999, 83, 303–307.

Battisti, D. S., Naylor, R. L. Historical warnings of future food insecurity with unprecedented seasonal heat. *Science*, 2009, 323, 240–244

Binder, B. M., Patterson, S. E. Ethylene-dependent and -independent regulation of abscission. *Stewart Postharvest Rev.*, 2009, 5, 1–10.

Bray, E. A., Bailey-Serres, J., Weretilnyk, E. Responses to abiotic stresses. In: Gruissem, W., Buchannan, B. B., Jones, R. L. (Eds.), *Biochemistry and Molecular Biology of Plants* (pp. 1158–1203). American Society of Plant Physiologists: Rockville, 2000.

Camejo, D., Jiménez, A., Alarcón, J. J., Torres, W., Gómez, J. M., Sevilla, F. Changes in photosynthetic parameters and antioxidant activities following heat shock treatment in tomato plants. *Funct. Plant. Biol.*, 2006, 33, 177–187.

Carrera, F. P., Noceda, C., Maridueña-Zavala, M. G., Cevallos-Cevallos, J. M. Metabolomics, a powerful tool for understanding plant abiotic stress. *Agronomy*, 2021, 11, 824.

Chaudhary, P., Rai, S., Wangdi, S. Mao, A., Rehman, N., Chettri, S., Bawa, K. S. Consistency of local perceptions of climate change in the Kangchenjunga Himalaya landscape. *Curr. Sci.*, 2011,101, 504–513.

Chen, J., Wang, P., Mi, H. L., Chen, G. Y., Xu, D. Q. Reversible association of ribulose-1,5-bisphosphatecarboxy-lase oxygenase active as with the thylakoid membrane depends up on the ATP level and pH in rice without heat stress. *J. Exp. Bot.*, 2010, 61, 2939–2950.

Chinnusamy, V., Zhu, J., Zhu, J. K. Cold stress regulation of gene expression in plants. *Trends. Plant. Sci.*, 2007, 12, 444–451.

Christ, B., Pluskal, T., Aubry, S., Weng, J. K. Contribution of untar-geted metabolomics for future assessment of biotech crops. *Trends Plant Sci.*, 2018, 23, 1047–1056.

Cook, D., Fowler, S., Fiehn, O., Thomashow, M. F. From the cover: A prominent role for the CBF cold response pathway in configuring the low-temperature metabolome of Arabidopsis. *Proc. Natl. Acad. Sci.*, 2004, 101, 15243–15248.

Das, K., Roychoudhury, A. Reactive oxygen species (ROS) and response of antioxidants as ROS-scavengers during environmental stress in plants. *Front. Environ. Sci.*, 2014, 2, 1–13.

Dikilitas, M., Simsek, E., Roychoudhury, A. Role of proline and glycine betaine in overcoming abiotic stresses. In: *Protective Chemical Agents in the Amelioration of Plant Abiotic Stress: Biochemical and Molecular Perspectives* (pp. 1–23). Wiley: Hoboken, NJ, USA, 2020.

Edreira, J. I. R., Budakli, C. E., Sammarro, D., Otegui, M. E. Heat stress effects around flowering on kernel set of temperate and tropical maize hybrids. *Field Crops Res.*, 2011, 123, 62–73.

Eid, H. M., El-Marsafawy, S. M. In: *3rd International Symposium on Sustainable Agro-environmental Systems: New Technologies and Applications Adaptation to Climate Change in Egyptian Agriculture and Water Resources* (pp. 26–29). Cairo, 2002.

Eid, H. M., El-Marsafawy, S. M., Salib, A. Y., Ali, M. A. In: *Meteorology and Environmental Cases Conference, Vulnerability of Egyptian Cotton Productivity to Climate Change* (pp. 2–6). Cairo, 1997.

El-Ramady, H. R., El-Marsafawy, S. M., Lewis, L. N. Sustainable agriculture and climate changes in Egypt. In: Lichtfouse, E., (Ed.), *Sustainable Agriculture Reviews* (pp. 41–95). Springer: Dordrecht, 2013.

Elsayed, S., El-Gohary, E. A., Hendawy, F. S., Abd El-Naby, K. S. Mitigation of heat stress effects on chamomile and its essential oil using melatonin or gibberellic acid and some agricultural treatments. *Egypt. J. Chem.*, 2021, 64, 6017–6028.

Fang, C., Fernie, A. R., Luo, J. Exploring the diversity of plant metabolism. *Trends Plant Sci.*, 2019, 24, 83–98.

Fawaz, M. M., Soliman, S. A. The potential scenarios of the impacts of climate change on Egyptian resources and agricultural plant production. *Open. J. Appl. Sci.*, 2016, 6, 270–286.

Feng, J., Long, Y., Shi, L., Shi, J., Barker, G., Meng, J. Characterization of metabolite quantitative trait loci and metabolic networks that control glucosinolate concentration in the seeds and leaves of *Brassica napus*. *New Phytol.*, 2012, 193, 96–108.

Ghassemi, S., Delangiz, N., Lajayer, B. A., Saghafi, D., Maggi, F. Review and future prospects on the mechanisms related to cold stress resistance and tolerance in medicinal plants. *Acta Ecol. Sin.*, 2021, 41, 120–129.

Gill, T. Reactive oxygen species and antioxidant machinery in abiotic stress tolerance in crop plants. *Plant Physiol. Biochem.*, 2010, 48, 909–930.

Gong, L., Chen, W., Gao, Y., Liu, X., Zhang, H., Xu, C., Yu, S., Zhang, Q., Luo, J. Genetic analysis of the metabolome exemplified using a rice population. *Proceed. Nati. Acad. Sci.*, 2013, 110, 20320–20325.

Guo, T. R., Zhang, G. P., Zhang, Y. H. Physiological changes in barley plants under combined toxicity of aluminum, copper and cadmium. *Colloids. Surf. B*, 2007, 57, 182–188.

Guy, C. Molecular responses of plants to cold shock and cold acclimation. *J. Mol. Microbiol. Biotechnol.*, 1999, 1, 231–242.

Guy, C., Kaplan, F., Kopka, J., Selbig, J., Hincha, D. K. Metabolomics of temperature stress. *Physiol. Plant.* 2008, 132, 220–235.

Hansen, J., Sato, M., Kharecha, P., Beerling D., Berner, R., Masson-Delmotte V., Pagani M., Raymo, M., Royer L. D., Zachos, J. C. Target atmospheric CO_2: Where should humanity aim? *Open Atoms Sci J.*, 2008, 2, 217–231.

Hassan, K. E. In: *XXVII IUSSP International Population Conference Impacts of Future Climate Change on Egyptian Population* (pp. 1–15). Busan, Republic of Korea, 2013.

Hassan, M. U., Ghareeb, R. Y., Nawaz, M., Mahmood, A., Shah, A. N., Abdel-Megeed, A., Abdelsalam, N. R., Hashem, M., Alamri, S., Thabit, M. A., Qari, H. S. Melatonin: A vital pro-tectant for crops against heat stress: Mechanisms and prospects. *Agronomy*, 2022, 12, 1116.

Hatfield, J. L., Prueger, J. H. Temperature extremes: Effect on plant growth and development. *Weather. Clim. Extremes*, 2015, 10, 4–10.

Hayat, K., Khan, A., Bibi, F., Murad, W., Fu, Y., Batiha, G. E.-S., Alqarni, M., Khan, A., Al-Harrasi, A. Effect of cadmium and copper exposure on growth, physio-chemicals and medicinal properties of *Cajanus cajan* L. (Pigeon Pea). *Metabolites*, 2021, 11, 769.
Hemantaranjan, A. Heat stress responses and thermotolerance. *Adv. Plants Agric. Res.*, 2014, 1, 1–10.
Horváth, I., Glatz, A., Nakamoto, H., Mishkind, M. L., Munnik, T., Saidi, Y., Goloubinoff, P., Harwood, J. L., Vigh, L. Heat shock response in photosynthetic organisms: Membrane and lipid connections. *Prog. Lipid Res.* 2012, 51, 208–220. http://www.usadellab.org/cms/index.php?page=metabolism
Jena, K. K., Kim, S. M., Suh, J. P., Yang, C. I., Kim, Y. J. Identification of cold-tolerant breeding lines by quantitative trait loci associated with cold tolerance in rice. *Crop Sci.*, 2012, 51, 517–523.
Kaplan, F., Kopka, J., Haskell, D. W., Zhao, W., Schiller, K. C., Gatzke, N., Guy, C. L. Exploring the temperature-stress metabolome of arabidopsis. *Plant Physiol.*, 2004, 136, 4.
Kebert, M., Vuksanović, V. Species-level differences in osmoprotectants and antioxidants contribute to stress tolerance of quercus robur L., and Q. cerris L. seedlings under water deficit and high temperatures. *Plants*, 2022, 11, 1744.
Korn, M., Gärtner, T., Erban, A., Kopka, J., Selbig, J., Hincha, D. K. Predicting arabidopsis freezing tolerance and heterosis in freezing tolerance from metabolite composition. *Mol. Plant.*, 2010, 3, 224–235.
Larkindale, J., Hall, J. D., Knight, M. R., Vierling, E. Heat stress phenotypes of Arabidopsis mutants implicate multiple signaling pathways in the acquisition of thermotolerance. *Plant Physiol.* 2005, 138, 882–897.
Le Gall, H., Fontaine, J.-X., Molinié, R., Pelloux, J., Mesnard, F., Gillet, F., Fliniaux, O. NMR-based metabolomics to study the cold-acclimation strategy of two miscanthus genotypes. *Phytochem. Anal.*, 2017, 28, 58–67.
Lobell, D. B., Gourdji, S. M. The influence of climate change on global crop productivity. *Plant Physiol.*, 2012, 160, 1686–1697.
Maduraimuthu, D., Vara-Prasad, P. V. High temperature stress. In: Jackson, M., Ford-Lloyd, B. V., Parry, M. (Eds.), *Plant Genetic Resources and Climate Change* (p. 291). CABI, 2014.
Maestri, E., Klueva, N., Perrotta, C., Gulli, M., Nguyen, H. T., Marmiroli, N. Molecular genetics of heat tolerance and heat shock proteins in cereals. *Plant Mol. Biol.*, 2002, 48, 667–681.
Mahmoud, M. A. Impact of climate change on the agricultural sector in Egypt. In: Negm, A. M. (Eds.), *Environmental Chemistry* (vol. 74, pp. 213–227), Springer: Cham, 2017.
Matsui, T., Omasa, K., Horie, T. The difference in sterility due to high temperatures during the flowering period among japonica rice varieties. *Plant Prod. Sci.*, 2001, 4, 90–93.
Medany, M. A., Hassanein, M. K. Assessment of the impact of climate change and adaptation on potato production. *Egypt. J. Appl. Sci.*, 2006, 21, 623–638.
Miller, G., Suzuki, N., Ciftci-Yilmaz, S., Mittler, R. Reactive oxygen species homeostasis and signalling during drought and salinity stresses. *Plant Cell Environ.*, 2010, 33, 453–467.
Mirzaei, M., Pascovici, D., Atwell, B. J., Haynes, P. A. Differential regulation of aquaporins, small GTPases and V-ATPases proteins in rice leaves subjected to drought stress and recovery. *Proteomics*, 2012, 12, 864–877.
Mittler, R., Vanderauwera, S., Suzuki, N., Miller, G., Tognetti, V. B., Vandepoele, K., Gollery, M., Shulaev, V., Breusegem, V. F. ROS signaling: The new wave? *Trends Plant Sci.* 2011, 16, 300–309.
Mo, Y., Liang, G., Shi, W., Xie, J. Metabolic responses of alfalfa *(Medicago Sativa* L.) leaves to low and high temperature induced stresses. *Afr. J. Biotechnol.*, 2010, 10, 1117–1124.
Muthuramalingam, P., Krishnan, S. R., Pandian, S., Mareeswaran, N., Aruni, W. Pandian, S. K., Ramesh, M. Global analysis of threonine metabolism genes unravel key players in rice to improve the abiotic stress tolerance. *Sci. Rep.*, 2018, 8, 1–14.
Obata, T., Fernie, A. R. The use of metabolomics to dissect plant responses to abiotic stresses. *Cell. Mol. Life Sci.*, 2012, 69, 3225–3243.
Paupière, M. J., Müller, F., Li, H., Rieu, I., Tikunov, Y. M., Visser, R. G. F., Bovy, A. G. Untargeted metabolomic analysis of tomato pollen development and heat stress response. *Plant Reprod.*, 2017, 30, 81–94.
Paupière, M. J., Tikunov, Y., Schleiff, E., Bovy, A., Fragkostefanakis, S. Reprogramming of tomato leaf metabolome by the activity of heat stress transcription factor HsfB. *Front. Plant Sci.*, 2020, 11, 1–10.
Pautasso, M., Döring, T. F., Garbelotto, M., Pellis, L., Jeger, M. J. Impacts of climate change on plant diseases—Opinions and trends. *Eur. J. Plant. Pathol.*, 2012,133, 295–313.

Qasim, M. Z., Hammad, H. M., Abbas, F. Saeed, S., Bakhat, H. F., Nasim, W., Farhad, W., Rabbani, F., Fahad, S. The potential applications of picotechnology in biomedical and environmental sciences. *Environ. Sci. Pollut. Res. Int.*, 2020, 27, 133–142.

Qi, X., Xu, W., Zhang, J., Guo, R., Zhao, M., Hu, L., Wang, H., Dong, H., Li, Y. Physiological characteristics and metabolomics of transgenic wheat containing the maize C4 phosphoenolpyruvate carboxylase (PEPC) gene under high temperature stress. *Protoplasma*, 2017, 254, 1017–1030.

Rai, A., Saito, K. Omics data input for metabolic modeling. *Curr. Opin. Biotechnol.*, 2016, 37, 127–134.

Rani, R., Khan, M. A., Kayani, W. K., Ullah, S., Naeem, I., Mirza, B. Metabolic signatures altered by in vitro temperature stress in *Ajuga bracteosa* Wall. ex. Benth. *Acta Physiol. Plant.*, 2017, 39, 97

Raza, A. Metabolomics: A systems biology approach for enhancing heat stress tolerance in plants. *Plant Cell Rep.*, 2022, 41, 741–763.

Rivero, R. M., Ruiz, J. M., Garcia, P. C., López-Lefebre, L. R., Sánchez, E., Romero, L. Resistance to cold and heat stress: Accumulation of phenolic compounds in tomato and watermelon plants. *Plant Sci.*, 2001, 160, 315–321.

Rosenzweig C., Tubiello, F. N., Goldberg, R., Mills, E., Bloomfield, J. Increased crop damage in the US from excess precipitation under climate change. *Glob. Environ. Chang.*, 2002, 12, 197–202.

Ruan, Y. L., Jin, Y., Yang, Y. J., Li, G. J., Boyer, J. S. Sugar input, metabolism, and signaling mediated by invertase: Roles in development, yield potential, and response to drought and heat. *Mol. Plant*, 2010, 3, 942–955.

Ruelland, E., Zachowski, A. How plants sense temperature. *Environ. Exp. Bot.*, 2010, 69, 225–232.

Saidi, Y., Finka, A., Muriset, M., Bromberg, Z., Weiss, Y. G., Maathuis, F. J. M., Goloubinoff, P. The heat shock response in moss plants is regulated by specific calcium-permeable channels in the plasma membrane. *Plant Cell.*, 2009, 21, 2829–2843.

Sairam, R., Tyagi, A. Physiology and molecular biology of salinity stress tolerance in plants. *Curr. Sci.*, 2004, 86, 407–421.

Sakata, T., Oshino, T., Miura, S., Tomabechi, M., Tsunaga, Y., Higashitani, N., Miyazawa, y., Takahashi, H., Watanabe, M., Higashitani, A. Auxins reverse plant male sterility caused by high temperatures. *PNAS*, 2010, 107, 8569–8574

Salam, U., Ullah, S., Tang, Z.-H., Elateeq, A. A., Khan, Y., Khan, J., Khan, A., Ali, S. Plant metabolomics: An overview of the role of primary and secondary metabolites against different environmental stress factors. *Life*, 2023, *13*, 706.

Sanghera, G. S., Wani, S. H., Hussain, W., Singh, N. B., Engineering cold stress tolerance in crop plants *Curr. Genomics*, 2011, 12, 30–43.

Sarkar, S., Gaydon, D. S., Brahmachari, K., Poulton, P. L., Chaki, A. K., Ray, K., Ghosh, A., Manoj K. N., Mainuddin M. Testing APSIM in a complex saline coastal cropping environment. *Environ. Model. Softw.*, 2022,147, 105239.

Sharkey, T. D., Zhang, R. High temperature effects on electron and proton circuits of photosynthesis. *J. Integr. Plant. Biol.*, 2010, 52, 712–722.

Shi, Y., Tian, S., Hou, L., Huang, X., Zhang, X., Guo, H. Ethylene signalling negatively regulates freezing tolerance by repressing expression of CBF and type-A ARR genes in Arabidopsis. *Plant Cell*, 2012, 24, 2578–2595.

Shulaev, V., Cortes, D., Miller, G., Mittler, R. Metabolomics for plant stress response. *Physiol. Plant.*, 2008, 132, 199–208.

Singh, B. K., Sutradhar, M., Singh, A. K., Mandal, N. Cold stress in rice at early growth stage-An overview. *IJPAB*, 2017, 5, 407–419.

Smith, J., Deck, L., McCarl, B., Kirshen, P., Malley, J., Abdrabo, M. *Potential Impacts of Climate Change on the Egyptian Economy* (p. 143). UNDP: Cairo, 2013.

Solanke, A. U., Sharma, A. K. Signal transduction during cold stress in plants. *Physiol. Mol. Biol. Plants.*, 2008, 14, 69–79.

Sumesh, K., Sharma-Natu, P., Ghildiyal, M. Starch synthase activity and heat shock protein in relation to thermal tolerance of developing wheat grains. *Biol. Plant.* 2008, 52, 749–753.

Sun, C. X., Gao, X. X., Li, M. Q., Fu, J. Q., Zhang, Y. L. Plastic responses in the metabolome and functional traits of maize plants to temperature variations. *Plant Biol.*, 2016, 18, 249–261.

Sweetlove, L. J., Obata, T., Fernie, A. R. Systems analysis of metabolic phenotypes: What have we learnt? *Trends Plant Sci.*, 2014, 19, 222–230.

Talanova, V., Akimova, T., Titov, A. Effect of whole plant and local heating on the ABA content in cucumber seedling leaves and roots and on their heat tolerance. *Russ. J. Plant Physiol.*, 2003, 50, 90–94.

Tao, F., Yokozawa, M., Xu, Y., Hayashi, Y., Zhang, Z. Climate changes and trends in phenology and yields of field crops in China, 1981–2000. *Agric. For. Meteorol.*, 2006, 138, 82–92.

Templer, S. E., Ammon, A., Pscheidt, D., Ciobotea, O., Schuy, C., McCollum, C., Sonnewald, U., Hanemann, A., Förster, J., Ordon, F. Metabolite profiling of barley flag leaves under drought and combined heat and drought stress reveals metabolic QTLs for metabolites associated with antioxidant defense. *J. Exp. Bot.*, 2017, 68, 1697–1713.

Thomason, K., Babar, A., Erickson, J. E., Mulvaney, M., Beecher, C., Macdonald, G. Comparative physiological and metabolomics analysis of wheat (*Triticum aestivum* L.) following post-anthesis heat stress. *PLoS One*, 2018, 13, e0197919.

Ullah, S., Khan, J., Hayat, K., Abdelfattah Elateeq, A., Salam, U., Yu, B., Ma, Y., Wang, H., Tang, Z.-H. Comparative study of growth, cadmium accumulation and tolerance of three chickpea (*Cicer arietinum* L.) cultivars. *Plants*, 2020, 9, 310.

Upadhyaya, H. Changes in antioxidative responses to low temperature in tea [*Camellia sinensis* (L) *O. Kuntze*] cultivars. *Int. J. Modern Bot.*, 2012, 2, 83–87.

Vedwan, N., Rhoades, R. E. Climate change in the Western Himalayas of India: A study of local perception and response. *Clim. Res.*, 2001, 19, 109–117.

Vega, C. R. C., Andrade, F. H., Sadras, V. O., Uhart, S. A., Valentinuz, O. R. Seed number as a function of growth. A comparative study in soybean, sunflower, and maize. *Crop Sci.*, 2001, 41, 748–754.

Wahid, A. Physiological implications of metabolite biosynthesis for net assimilation and heat-stress tolerance of sugarcane (*Saccharum officinarum*) sprouts. *J. Plant Res.*, 2007, 120, 219–228.

Wahid, A., Shabbir, A. Induction of heat stress tolerance in barley seedling sbypre-sowing seed treatment with glycine betaine. *Plant Growth Regul.*, 2005, 46, 133–141.

Wallaart, T. E., Pras, N., Beekman, A. C., Quax, W. J. Seasonal variation of artemisinin and its biosynthetic precursors in plants ofArtemisia annua of different geographical origin: Proof for the existence of chemotypes. *Planta Med.*, 2000, 66, 57–62.

Wang, L., Ma, K. B., Lu, Z. G., Ren, S. X., Jiang, H. R., Cui, J. W., Jin, B. Differential physiological, transcriptomic and metabolomic responses of Arabidopsis leaves under prolonged warming and heat shock. *BMC Plant Biol.*, 2020, 20, 1–86.

Wienkoop, S., Morgenthal, K., Wolschin, F., Scholz, M., Selbig, J., Weckwerth, W. Integration of metabolomic and proteomic phenotypes. *Mol. Cell. Proteom.*, 2008, 7, 1725–1736.

Young, L. W., Wilen, R. W., Bonham-Smith, P. C. High temperature stress of Brassica napus during flowering reduces micro-and megagametophyte fertility, induces fruit abortion, and disrupts seed production. *J. Exp. Bot.*, 2004, 55, 485–495.

Zhao, D. Y., Shen, L., Yu, M. M., Zheng, Y., Sheng, J. P. Relationship between activities of antioxidant enzymes and cold tolerance of postharvest tomato fruits. *Food Sci.*, 2009, 14, 309–313.

Zhu, J. H., Dong, C. H., Zhu, J. K. Interplay between cold-responsive gene regulation, metabolism and RNA processing during plant cold acclimation. *Curr. Opin. Plant Biol.*, 2007, 10, 290–295.

[illegible], K., [illegible], M., [illegible], A. [illegible] responses to the [illegible] and functional traits of [illegible] plants to temperature variation. *Plant Biol.* [illegible]
[illegible], [illegible]. [illegible] in [illegible]. *[illegible]* [illegible], 13, 226–240.
Ramírez, M., [illegible], A. Effect of whole plant and local heating on the [illegible] content in [illegible] leaves and corn and cucumber plants under heat stress. *Russ. J. Plant Physiol.* 2011, 58, [illegible].
Tao, F., Yokozawa, M., [illegible], Y., Zhang, Z. Climate change, plant traits and crop yields on [illegible] field crops in China 1981–2000. *Agric. For. Meteorol.* 2008, 148, 82–92.
Templer, S. E., Ammon, A., Riede, D., Grosser, O., [illegible], C., [illegible], S., [illegible], [illegible], A., [illegible], J., [illegible], T. Metabolite profiling of barley flag leaves under drought and combined heat and drought stress reveals metabolic QTLs for metabolites associated with antioxidant defense. *J. Exp. Bot.* 2017, 68, [illegible].
Thomason, K., Babar, M. A., [illegible], J. E., [illegible], A., [illegible], [illegible], C. Comparative physiological and metabolomics analysis of wheat (*Triticum aestivum* L.) following post-anthesis heat stress. *PLoS One* 2018, 13, e0197919.
[illegible], [illegible], [illegible], [illegible], [illegible] [illegible], Z., [illegible], [illegible], [illegible] Comparative study of the [illegible] accumulation and [illegible] of [illegible] *J. [illegible]* 2021, [illegible].
[illegible], [illegible]. Changes in antioxidative response to low temperature in tea (*Camellia sinensis* (L.) Kuntze) cultivars. *[illegible]* 2011, [illegible].
[illegible], [illegible], [illegible]. Climate change, the [illegible] and [illegible] of [illegible] and responses. *[illegible]* [illegible], 19, [illegible].
[illegible], [illegible], C., Andrade, F. H., Sadras, V. O., [illegible], [illegible]. Kernel number as a function of growth: A comparative study in maize, sunflower, and [illegible]. *[illegible]* 2017, 47, [illegible].
Wahid, A. [illegible] of [illegible] [illegible] in [illegible] and heat stress tolerance of [illegible]. *[illegible]* [illegible].
Wahid, A., [illegible] tolerance: [illegible] in [illegible] sugarcane [illegible]. *J. Plant Physiol.* 2007, 164, 283–294.
[illegible], [illegible], [illegible], Z., Zhou, [illegible], [illegible], [illegible], W. Seasonal variation of [illegible] and [illegible] in [illegible] of different geographical origin. [illegible] an evaluation of [illegible]. [illegible] 2020, [illegible].
[illegible], [illegible], [illegible], [illegible], [illegible], [illegible], [illegible]. [illegible] [illegible] and [illegible] in *Arabidopsis* leaves under [illegible] and warming. [illegible] 2021, [illegible].
[illegible], [illegible], [illegible], [illegible], [illegible], [illegible]. [illegible] and [illegible]. *[illegible]* 2005, [illegible].
Young, [illegible], [illegible], [illegible], [illegible]. High temperature [illegible] during [illegible] and [illegible]. [illegible] 2004, 55, 485–495.
Zhang, [illegible], [illegible], [illegible], [illegible], [illegible]. Relationship between [illegible] and [illegible] in [illegible] of [illegible] tobacco. *[illegible]* 2006, 14, 309–315.
Zhu, [illegible], [illegible], [illegible]. The [illegible] between [illegible] and [illegible] during [illegible]. *[illegible]* 20[illegible], 10, [illegible].

CHAPTER 12

Metabolomics for Heavy Metal Stress Responses and Tolerance in Plants

NIKHA BORAH[1], ANKITA GOGOI[1], KABYASHREE BURAGOHAIN[1], SUKANYA SONOWAL[1], RATUL NATH[1*], L. R. SAIKIA[2], BHABEN CHOWARDHARA[3], and AKBAR HOSSAIN[4]

[1]*Microbiology Laboratory, Department of Life Sciences, Dibrugarh University, Dibruagarh, Assam, India*

[2]*Department of Life Sciences, Dibrugarh University, Dibrugarh, Assam, India*

[3]*Department of Botany, Arunachal University of Studies, Namsai, Arunachal Pradesh, India*

[4]*Division of Soil Science, Bangladesh Wheat and Maize Research Institute (BWMRI), Nashipur, Dinajpur, Bangladesh*

*Corresponding authors

ABSTRACT

Heavy metal stress significantly impacts plants, microorganisms, and animals of different ecosystems. Modulating metabolic activities due to heavy metal stress poses remarkable changes in the production and regulation of metabolites in plants, microorganisms, and animals. Plants responses to heavy metal stress induce changes in certain primary and secondary metabolites that provide baseline defense to alleviate the stress condition. Metabolomics, a valuable tool for studying plant responses to heavy metal stress, allows for the comprehensive analysis of plant metabolite profiles, providing insights into stress-induced metabolic changes. Further, it can provide information on the metabolic changes that plants perform to preserve energy, deal with oxidative stress, and maintain cellular homeostasis under heavy metal stress situations. The present chapter highlights the pivotal role of key metabolites such as phytochelatins, melatonins, antioxidants, and osmoprotectants in heavy metal stress responses involved in metal detoxification, ion sequestration, and reactive oxygen species scavenging with an emphasis on the tolerance mechanisms incorporated by plants.

12.1 INTRODUCTION

The harmonious maintenance of their internal environment by plants and other organisms is known as "homeostasis." Any alteration in their surroundings that disrupt this equilibrium is

Exploring Plant Stress Tolerance: A Metabolomics Approach.
Shuvasish Choudhury, Debojyoti Moulick, and Mirza Hasanuzzaman (Eds.)

DOI: 10.1201/9781779641731-12

recognized as a “plant stress condition.” These stress conditions can be broadly categorized into two distinct types: (1) abiotic and (2) biotic stresses. Abiotic stresses arise from nonliving factors such as temperature fluctuations, variations in light exposure, or exposure to chemicals such as salinity and heavy metals (HMs). These environmental challenges can have negative consequences for plants, as they adapt rapidly from an ideal habitat to an unsettled state. Biotic stresses, in contrast, originate from the adverse impact of living organisms such as pathogens and pests, which can pose significant threats to plant well-being (Swamy et al., 2019). The global concern for HM pollution in ecosystems has intensified in recent years (Feng et al., 2021). While certain HMs such as zinc, manganese, copper, iron, and nickel serve as enzyme cofactors crucial for physiological functions, most of HMs disrupt normal cellular metabolism, leading to improper growth and development (Singh et al., 2016; Banach et al., 2020; Goyal et al., 2020).

In plants, HM toxicity interferes with growth rate, photosynthesis, respiration, and ion efflux such as vital physiological processes through alterations in key enzyme activities, such as phosphoenolpyruvate carboxylase (PEPC) and ribulose-1,5-bisphosphate carboxylase/oxygenase (RuBisCO). Further, HMs disturb the delicate balance of redox homeostasis, promoting the excessive production of reactive oxygen species (ROS) such as singlet oxygen ($^1O_2$), hydroxyl radicals (OH^-), superoxide (O_2^-), and hydrogen peroxide (H_2O_2) (Phurailatpam et al., 2022). Rhizosphere microorganisms associated with plant growth and development also face HM toxicity, causing a decrease in the bioavailability of plants' essential nutrients. Similar to the HM stress in plants, HM toxicity in microorganisms also disrupts essential enzymatic functions, generates ROS, disturbs ion regulation, and directly affects DNA and proteins. For instance, chromium [Cr (III)] modifies enzyme activity and interacts with DNA, while copper [Cu (I) and Cu (II)] catalyzes ROS production. This toxicity damages cell membranes, disrupts nucleic acid structures, inhibits enzymes, and adversely affects microbial growth and metabolism, ultimately jeopardizing the viability and functionality of microorganisms (Igiri et al., 2018).

The accumulation of excessive HMs in plants not only hampers their growth and development but also poses a threat to animals and humans through the food chain (Järup and Åkesson, 2009; Anjum et al., 2016; Feng et al., 2021). Consumption of food and water contaminated with HMs leads to severe health complications, including kidney damage, cardiovascular diseases, neurological disorders, lung damage, and gastrointestinal problems (Järup, 2003; Johri et al., 2010; Phurailatpam et al., 2022). To combat such adverse environmental conditions, plants have evolved various mechanisms for acclimation and adaptation. These mechanisms involve the activation of specific genes and regulatory pathways in response to stress. For instance, plants employ harmonious solutes and organic compounds to combat drought conditions, preventing protein and membrane denaturation. They accumulate amino acids, amides, and imino acids to thrive under abiotic stresses such as drought and salinity. Additionally, the hormone abscisic acid (ABA) plays a pivotal role in stimulating adaptive responses in plants under stress conditions (Xiong et al., 2002; Koyro et al., 2012; Zhu, 2016). Microorganisms also play a vital role in inducing stress-tolerant responses in plants, particularly beneficial soil microorganisms such as plant growth-promoting rhizobacteria (PGPR) and arbuscular mycorrhizal fungi (AMF). These microorganisms, closely associated with plant roots, engage in various mechanisms to boost plant resilience to stress. Some PGPR secrete phytohormones such as indole-3-acetic acid (IAA), while others produce plant-specific metabolites and volatile compounds, promoting nutrient uptake, antioxidant mechanisms, and maintaining homeostasis (Mishra et al., 2017).

In the context of this rapidly advancing era in plant biology research, the advent of new technologies has opened revolutionary avenues for understanding cellular components' functional properties. Among these pioneering approaches, "metabolomics" stands out prominently. Metabolomics enables us to qualitatively and quantitatively analyze plant metabolites before and after exposure to external stressors, providing profound insights into plant metabolic networks and related gene functions (Choudhury et al., 2021; Feng et al., 2021). This chapter focuses on metabolites involved in plant stress responses against HMs and the mechanisms adopted by plants to combat HM stress situations.

12.2 HEAVY METAL TOXICITY IN PLANTS, MICROBES, AND HUMAN

As environmental pollution has increased, HM toxicity has drawn a lot of attention. If the plant's internal metal concentration exceeds normal, HMs have direct effects such as cytoplasmic enzyme inhibition and oxidative stress-induced cell assembly disintegration. HM can attach to sulfhydryl groups in proteins causing structural disruption and can cause deficient consequences by displacing necessary elements (Hall, 2002). In its elemental or chemically mixed forms, Hg, Pb, Cd, Cr, As, etc., are poisonous to plants, and their reactions range widely from tolerance to toxicity. For instance, high amounts of these HMs may replace necessary elements for pigmentation and enzymes, impairing their functionality and resulting in oxidative stress. This is commonly associated with the advancement of senescence (Shah et al., 2010). High Cd exposure reduces plant's ability to photosynthesis, absorb water, and absorb nutrients. Plants that are cultivated in high Cd soil exhibit obvious signs of damage, such as chlorosis, retardation of growth, browning of the tips of their roots, and ultimately death of the plant (Yadav, 2010). Elevated soil Zn levels inhibit plant metabolic processes, leading to stunted growth and aging. Excessive Zn restricts roots and shoots development, causing chlorosis in young leaves exposed to it. Moreover, high Zn concentrations can cause deficiencies in Mn and Cu in plant shoots, as it hinders the transfer of these micronutrients from roots to shoots. This issue is exacerbated by higher concentrations of Fe and Mn in plant roots when grown in Zn-rich soil (Ebbs and Kochian, 1997; Yadav, 2010). In addition to altering the germination process, Cr can have toxic effects on the growth of roots, stems, and leaves in plants. These effects can impact the yield and overall output of dry matter. Plants have also been shown to undergo metabolic changes as a result of exposure to Cr, either directly affecting enzymes or other metabolites or indirectly through the production of ROS, which can lead to oxidative stress (Shanker et al., 2005). Elevated levels of Hg^{2+} disturb mitochondrial function and cause oxidative stress by releasing ROS, which in turn affects plant membrane lipids and cellular metabolism. Exposure to elevated Hg concentrations affects the functions of multiple antioxidant enzymes and the metal-chelating substance pool (Zhou et al., 2007). Plants growing in elevated Ni^{2+} soil had impaired nutritional balance, which led to cell membrane dysfunction. Water content in dicot and monocot plant species decreased due to high Ni^{2+} uptake. Hence, the reduction in water uptake is used to track the Ni^{2+} toxicity in plants (Pandey and Sharma, 2002; Yadav, 2010).

HM toxicity on microorganisms is regulated by a variety of parameters, including pH, chelating substances concentration, speciation, and organic matter. Changes in biomass, microbial diversity, and element cycling can all be consequences of those elements being in the environment for ecosystems (Roane and Pepper, 1999; Lima de Silva et al., 2012). Increases in metal content often have a negative impact on the respiration rate and enzyme activity of soil microorganisms, which seem to be an indicator of soil contamination. Microorganisms, such as yeast, fungus, bacteria, and algae, are excellent models for examining the detrimental effects of metals on cells (Ashraf and Ali, 2007). Prolonged exposure to heavy metals in soils has a negative impact on microbial activity, particularly microbial respiration. Numerous studies have demonstrated significant decrease in microbial activity as a result of brief exposure to harmful metals, in addition to long-term metal-mediated alterations in soil enzyme activities (Sobolev and Begonia, 2008). HMs disrupt functions, denaturize proteins, or compromise the integrity of cell membranes, all of which have an impact on the development, morphology, and metabolism of soil microorganisms. Various investigations have demonstrated that changes in microbial communities were caused by HM pollution (Xie et al., 2016).

HMs pose the greatest risk to human health when exposed beyond the level. Consumption of food crops cultivated in areas with dangerous levels of HMs, as well as consumption of live cattle fed, produce from these areas, may result in HM poisoning in humans. Excess HMs concentrate mostly in the blood, liver, and kidneys. The formation of ROS and nitrogen species, such as superoxides, hydroperoxides, and nitric oxide, in these tissues is a key feature of HM toxicities that result in carcinogenicity (Gupta and Gupta, 2012). Even at extremely low levels of exposure, Pb is a hazardous HM in humans. Depending on the amount taken and exposure situation, the impact on the body of an individual can be both acute and long lasting. Because of its systemic toxicity, Pb affects various organs in the body, causing neurological, cardiovascular, renal, gastrointestinal, hematological, and reproductive consequences. Human Pb exposure is often assessed by blood samples. Lead accumulated in the bones might manifest as a remobilized type of lead exposure later in the individuals life (Mahurpawar, 2015). Inorganic As is a recognized carcinogen that can cause skin, lung, liver, and bladder cancer (Table 12.1). Lower level exposure can produce nausea and vomiting, reduced red and white blood cell production, irregular heart rhythm, blood vessel damage, and a sensation of "pins and needles" in the hands and feet (Martin and Griswold, 2009).

12.3 METABOLITES PRODUCED BY PLANTS AND ASSOCIATED MICROORGANISMS IN RESPONSE TO HEAVY METAL STRESS

Singh et al. (2016) conducted a comprehensive assessment of plants resilience to elemental metal-induced stress, utilizing a multidisciplinary approach encompassing ionomics, metabolomics, proteomics, and transcriptomics. Their research highlighted the severe

TABLE 12.1 Toxic Effect of Heavy Metals on Selected Crops

Sr. No.	Heavy Metals	Plants	Toxicity in Plants	References
1	Cd	*Oryza sativa*	Cadmium toxicity has been shown to reduce seed germination, growth, mineral nutrition, photosynthesis, and grain production. In rice, it also induces oxidative stress and genotoxicity.	(Rizwan et al., 2016)
		Zea mays	Plants exposed to Cd stress had significantly higher Cd concentrations in all plant sections, notably the roots. It demonstrates an increase in ROS production, which is the end outcome of oxidative stress.	(Anjum et al., 2015)
		Triticum aestivum L.	Cd reduced wheat germination, growth, mineral nutrition, photosynthesis, and grain production, and plant response to the toxic effects of Cd varies with cultivars, growth circumstances, and stress duration. Cadmium caused wheat plants to experience oxidative stress and genotoxicity.	(Rizwan et al., 2016)
2	Pb	*Triticum aestivum* L.	Pb exposure causes pro-oxidative effects in plants and alters a variety of biochemical and physiological processes. Pb, not being a vital element for plant development and growth, reduces the germination of seeds and seedling growth, destroys pigments, and can even kill plants.	(Sharma and Dubey, 2005; Kaur et al., 2013)
		Hordeum vulgare L.	Significant reduction in growth as a result of lead poisoning; reduction in biomass of roots rather than shoots.	(Aery and Jagetiya, 1997)
3	Cu/Cd	*Oryza sativa*	Rapid declination in yield, root, and shoot dry weight in rice.	(Xu et al., 2006)
		Solanum lycopersicum	The development of the tomato plants was impacted by the presence of either Cu or Cd in the culture media. The primary visible indications of poisoning are browning and shortened length of the roots, chlorosis, and necrosis of the leaves.	(Mediouni et al., 2006)
4	Zn	*Zea mays*	Fresh biomass from roots and shoots was decreased as a result of the harmful effects of a high zinc content.	(Bokor et al., 2014)
		Solanum nigrum	Decreased root and shoot growth and reduced number of leaves.	(Al Khateeb and Al-Qwasemeh, 2014)
5	Hg	*Triticum aestivum* L.	Lipid peroxidation and H_2O_2 levels increased in wheat leaves exposed to mercury poisoning. Changes in proline concentration, oxidative and antioxidative status, and gene expression patterns were among the stress responses brought on by mercury exposure.	(İşkil et al., 2022)
		Vigna radiata	Decreased *Chl* content. The vegetative development was suppressed by a higher concentration of Hg salt solution. It also considerably reduced the nodulation of *V. radiata* seedlings. Nodule size and quantity have been gradually declining as mercury salt solution concentrations have increased.	(Mondal et al., 2015)

repercussions of metal toxicity on both soil and water, profoundly impacting plant productivity and overall production. The deleterious effects of metals in these environments have compounded the challenges faced by plants, affecting their tissues, cellular structures, physiological functions, biochemical processes, and molecular activities. They underscored that the capacity of plants to withstand elemental metal stress within soil ecosystems can be significantly improved through the application of omics techniques, focusing on stressed metabolites and inducible proteins with transcriptional factors. They also emphasized the potential role of metallophytes, plants that accumulate metals, in mitigating the adverse impacts of metal incursion on various biological processes. Furthermore, Singh et al. (2016) advocated for the reduction of human activities that negatively influence ecosystems, as this reduction could enhance global plant production and overall productivity.

A study carried out by Vinocur and Altman (2005) explored the latest advancements in enhancing plants' ability to tolerate nonliving ecological stressors. They investigated how biotic stress factors could be regulated through the engineering of monogenic traits in plants. These traits act as components of tolerance that signal pathways in the complex, multigenic nature of plants. They proposed that strategies for engineering plants to enhance their resilience to heavy metals in the soil should encompass the synthesis of structural and functional metabolites and proteins that confer tolerance to these metals. Additionally, they noted that genes in plants responsive to metals exhibit untapped potential that can be harnessed in next-generation studies, employing proteomics and metabolomics. These studies offer promising avenues to combat the challenges posed by HMs and ultimately contribute to a more sustainable and productive crop production system (Anani et al., 2022).

Analyzing plant responses to Pb stress has revealed alterations in organic acids, amino acids, and coenzymes. Notably, under Pb stress, Vetiver plants show increased levels of asparagine, proline, histidine, tryptophan, valine, isoleucine, threonine, nicotinamide, and methionine. However, the response can vary among different plant species, as radish roots exposed to high Pb levels exhibited suppressed amino acid synthesis. The analysis of halophyte *Suaeda salsa* roots under Pb stress revealed increases in fructose, glucose, and isocaproate, coupled with decreased levels of sucrose, citrate, and malate. These responses highlight the importance of plants synthesizing more amino acids and organic acids under heavy metal stress, contributing to their tolerance. Pb stress also affects key enzymes involved in various metabolic pathways, triggering defense responses (glutaredoxin, aldose 1-epimerase, malate dehydrogenase, ascorbate–glutathione cycle, and thioredoxin) to mitigate damage. This response includes the upregulation of genes involved in amino acid metabolism, signal transduction, membrane transport, and secondary metabolite and xenobiotic biosynthesis (Wu et al., 2012; Wang et al., 2015, 2016; Pidatala et al., 2018).

Research on plant responses to Cd stress revealed the release of root exudates rich in organic acids and amino acids, including tartaric acid and histidine. In response to Cd stress, plants increase the production of antioxidant enzymes such as ascorbate peroxidase (APX), catalase (CAT), superoxide dismutase (SOD), and peroxidase (POD). Mwamba et al. (2020) investigated the metabolomic responses in plant genotypes with varying levels of cumulative cadmium exposure. Their research revealed that Cd-tolerant, Cd-accumulating genotypes exhibited an enhanced overall antioxidant capacity. This improvement was characterized by higher levels of substances such as ascorbic acid (AsA), serotonin, phytosteroids, monoterpenes, carotenoids, and lignans. In a study by Navarro-Reig et al. (2017), it was observed that Cd(II) had a significant impact on secondary metabolism, affecting purine, amino acid, glycerolipid, and carbon metabolism in *Oryza sativa* L. Similarly, Xie et al. (2019) reported substantial increase in the content of phytochelatins (PCs) and noteworthy alterations in metabolites associated with antioxidation and osmotic balance in *Amaranthus hypochondriacus* when exposed to Cd stress (Irfan et al., 2014; Gupta et al., 2017; Navarro-Reig et al., 2017; Çatav et al., 2020; Fu et al., 2018; Xie et al., 2019).

Hg stress leads to changes in enzyme activities and the depletion of glutathione. Lipid peroxidation is increased, affecting the cellular ultrastructure of plant seedlings. Proteomics analysis suggests that ABA plays a protective role in mitigating Hg stress. The impact of inorganic Hg on soluble compartments differs from that of methyl-Hg, which primarily impacts nonsoluble compartments. Plants respond to Hg stress by increasing the content of phenylalanine, threonine, and tryptophan, which contribute to their defense mechanisms. A deeper understanding of the effects of Hg on crop metabolism enhances our knowledge of the molecular and physiological responses of crops to Hg contamination (Chen et al., 2012; Kang et al., 2015; Cosio and Renault, 2020).

Plants respond to Cr stress by increasing the contents of free proline, total phenols, and soluble sugars. The expression of soluble proteins decreases, while the activities of antioxidant enzymes such as SOD, APX, POD, CAT, glutathione reductase (GR), and glutathione peroxidase (GPX) are enhanced in response to Cr stress. Changes in lipids and their oxidation products, which serve as signaling molecules, play a role in plant resistance to stress. In addition, Cr(VI) directly participates in the nonenzymatic oxidation of fatty acids. The content of amino acids, including phenylalanine, threonine, and tryptophan, increases in plants exposed to Cr stress. Understanding these responses at the metabolomics level contributes to a more comprehensive understanding of how plants react to Cr contamination (Dubey et al., 2010; Anjum et al., 2017; Gonzalez Ibarra et al., 2019; Banerjee and Roychoudhury, 2020).

Cu contributes to multiple metabolic processes, including photosynthesis, and is an integral component of essential antioxidant enzymes such as SOD and ascorbate oxidase (Nagajyoti et al., 2010). However, the release of harmful copper-containing dust from industrial production and vehicle emissions, coupled with the excessive use of copper-based fungicides and feed additives in agriculture, has significantly elevated copper levels in agricultural soils. Studies show that castor plant biomass decreases significantly when copper exceeds 250 mmol L^{-1}, with increased tartaric, malic, and succinic acid levels. Copper exposure in cucumbers can lead to the accumulation of threonic acid, tyrosine, phenylalanine, and polyphenols. Certain organic acids and amino acids can help detoxify excess copper in *Celosia argentea* (Chaoui et al., 2004; Ou et al., 2012; Huang et al., 2016; Branco-Neves et al., 2017; Zhao et al., 2018; Wang et al., 2021b).

Excessive absorption of ferrous ions by plants leads to the production of highly oxidizing hydroxyl radicals, causing damage to cell membrane systems and, in severe cases, plant cell death. Iron stress can inhibit the growth of plant roots and shoots, resulting in leaf bronzing. Linolenic acid, linoleic acid, and elaidic acid are among the marker metabolites that respond to iron stress in rice (Turhadi et al., 2019; Feng et al., 2021).

Elevated intracellular levels of zinc ions, typically exceeding 5 mg L^{-1}, have detrimental effects, including protein degradation, inhibition of new protein synthesis, and disruption in cell membrane and chloroplast structures (Wang et al., 2009; Subba et al., 2014; Rouphael et al., 2016). Recent research by Zhang et al. (2017) has shown that an excess of zinc can impact flavonoid metabolism, while zinc deficiency can affect carbohydrate metabolism. Notably, in tea leaves under zinc stress, the content of flavonoid metabolites, such as catechin gallate, epicatechin, and epigallocatechin, was significantly increased.

High concentrations of HMs in ecosystems, such as the case of Quanzhou Bay, which is contaminated with HMs and has an invasion of *Spartina alterniflora* (smooth cordgrass), can have harmful effects on the environment. Researchers used advanced sequencing techniques (pyrosequencing) to study the bacterial communities associated with *S. alterniflora* and found that these microbes had ecological functions, such as producing a substance called 1-aminocyclopropane-1-carboxylic acid (ACC) deaminase that helps plants grow and cope with HMs (Yang et al., 2020).

In recent decades, scientific studies have underscored the pivotal role of the microbiome in shaping plant health and their responses, especially during stress conditions (Phurailatpam and Mishra, 2020; Phurailatpam et al., 2022). These microscopic organisms hold great promise as valuable resources for uncovering solutions to alleviate HM stress in crop plants. Research aimed at deciphering the microbiome's function in enhancing HM tolerance is gaining momentum. Plant-associated microorganisms play a crucial role in regulating various aspects of their host plants' biology. They employ a range of detoxification mechanisms, which include binding to the fungal cell wall, scavenging HM elements, facilitating precipitation, engaging in extracellular scavenging, promoting vaporization, and aiding in compartmentalization, among other mechanisms. Understanding the ways in which these microorganisms operate holds the key to addressing HM-related stress in plants.

Metabolomics research has been instrumental in identifying microbes that can tolerate HMs and aid in the cleanup of contaminated soils. One study suggested that when certain plants produce specific metabolites such as amino acids, phenols, and organic acids, it acts as a nutrient source for helpful microbes, enhancing the removal of pollutants such as benzopyrene and pyrene from the environment (Toyama et al., 2011). Additionally, when examining the metabolites of wetland grass (*Phragmites australis*) in different root areas (rhizosphere vs. endosphere), researchers found variations in the secretion of particular metabolites due to the presence of dissolved substances, various metals, and changes in pH. In another study, scientists aimed to understand how plants tolerate heavy metals such as Zn, Cr, Fe, and Cu by conducting metabolomic studies on plants either inoculated with or without certain microbes. They discovered that the presence of specific microbes increased the levels of proline and antioxidant activity, demonstrating the plant's enhanced tolerance to HM (Zhu, 2016).

Metabolomics studies have played a crucial role in elucidating the mechanisms underlying HM tolerance in plants. A notable example is the research on *Salix integra*, where targeted metabolomics were employed to investigate its Pb accumulation ability following inoculation with Pb-resistant *Bacillus* sp. and *Aspergillus niger*. Inoculation with these microbes led to increased proline levels, a crucial compound for stress adaptation, and heightened activity of antioxidants, SOD, and CAT. This bolstered the plant's defense against oxidative stress induced by HMs. The study also identified 410 metabolites, primarily consisting of organic acids, amino acids, and carbohydrates. This diverse array of metabolites underscores the intricate nature of plant–microbe interactions in the context of HM tolerance. In yet another study, a combination of proteomics and metabolomics was used to uncover how *Triticum aestivum* became more tolerant to Pb and Cd stress when inoculated with the *Enterobacter bugandensis* TJ6 strain. The research revealed that multiple mechanisms, such as extracellular absorption, reduced metal accumulation, and the secretion of certain substances, contributed to the plant's improved metal tolerance and stress mitigation (Han et al., 2021; Tiwari and Bae, 2023).

12.4 OMICS APPROACHES FOR PLANT STRESS RESPONSE

As immobile organisms, plants cannot evade undesired environmental shifts. Exposure to heavy metals induces a broad spectrum of physiological and biochemical adjustments, necessitating the development or adoption of strategies to confront the adverse effects of HM toxicity. Plants react to external stimuli, including HM toxicity, through a series of mechanisms. These encompass (1) detecting external stress signals; (2) transmitting these signals into the cell through signal transduction; and (3) initiating appropriate responses to counteract the detrimental impacts of stress signals by modulating the cellular physiological, biochemical, and molecular state.

At the whole-plant level, assessing the detection and alterations in signal transduction caused by exposure to HMs can be challenging. However, monitoring early responses, such as oxidative stress, changes in transcripts and proteins, or the accumulation of metabolites, can serve as valuable indicators for investigating how plants sense and process these signals following exposure to stress (Singh et al., 2016).

In recent years, there has been a notable shift in the realm of crop improvement under toxic HM stress, with a predominant focus on unraveling the intricate molecular regulatory networks governing plant responses to such challenging conditions. As a result, "omics" approaches have gained substantial significance in identifying candidate genes, regulatory elements, differential expression analysis, molecular targets, and signaling networks in plants grappling with HM stress. The integration of high-throughput techniques in sequencing, microarrays, chromatography, mass spectroscopy, and computational biology has ignited a paradigm shift in the field of plant stress biology. Well-established omics technologies such as genomics, transcriptomics, and metabolomics, as well as more modern approaches such as RNAomics and multiomics, play a vital role in elucidating plant stress responses and improving crops. These cutting-edge technologies have not only revolutionized the study of plant responses to HM stress but have also opened up new horizons for crop enhancement (Jamla ct al., 2021).

Genomics stands at the forefront of this revolution, acting as a cornerstone for understanding how plants exhibit resilience against toxic HMs. This essential omics approach involves the comprehensive study of genes and their intricate interplay within specific physiological states. By utilizing genomics, researchers have the power to sequence and annotate genes, which is instrumental in improving crop varieties. It also enables us to unveil the roles of genes in metal stress responses at a molecular level. Gene sequencing, structural–functional characterization, and annotation are pivotal techniques within genomics. Moreover, computational tools play a pivotal role in the analysis of the colossal amount of sequencing data generated from diverse plant cultivars. By leveraging databases and advanced software, scientists can explore gene structure, gene–gene interactions, and the identification of cis-regulatory elements (Bohnert et al., 2006; Khalid et al., 2019; Sahoo et al., 2020).

Building upon the foundation of genomics, transcriptomics emerges as the next crucial layer in the omics framework. While genomics provides insights into the presence or absence of specific genes, transcriptomics delves deeper, examining the functional role of these genes and their expression patterns under specific physiological states or stress conditions. Transcriptomic profiles furnish us with invaluable insights into responses of plants to HM stress. Through techniques such as next-generation sequencing, RNA sequencing, microarrays, and reverse transcription polymerase chain reaction (RT-PCR), researchers can pinpoint known and novel differentially expressed genes (DEGs). The intricate web of gene expression alterations in response to HM stress becomes more comprehensible through this approach (Khalid et al., 2019; Mehta et al., 2019; Jamla et al., 2021; Zhou et al., 2022).

Metabolomics adds another layer of insight by focusing on the comprehensive profiling of plant metabolites, offering a holistic view of the metabolic changes that occur under stress conditions. Metabolites, encompassing primary and secondary metabolites, are essential players in plant responses to stress. Proline, putrescine, carotenoids, flavonoids, and phenols are examples of metabolites that play pivotal roles in stress responses, detoxification, and protection of the photosynthetic machinery. Metabolomics unveils the dynamic nature of plant responses to stress, a valuable resource for crop improvement, transgenic plant development, phenotyping, and more. Advanced techniques such as gas chromatography-mass spectrometry (GC-MS) and liquid chromatography-mass spectrometry (LC-MS) broaden the scope of metabolomics in unraveling plant metabolomes. Data analysis and interpretation in metabolomics require extensive libraries, databases, and computational tools for characterizing metabolites (Sharma and Dietz, 2006; Wang et al., 2015; Khalid et al., 2019; Mehta et al., 2019; Razzaq et al., 2019).

miRNAomics further enriches our understanding of plant responses to HM stress. Under abiotic stress conditions, including HM stress, plants modulate gene expression through miRNAs, which are small noncoding RNA molecules that act as posttranscriptional regulators. Various miRNAs have been identified, such as miR156, miR171, miR396, and miR390, which actively participate in regulating HM stress tolerance in plants. Computational tools and databases play a crucial role in the identification and characterization of these HM-responsive miRNAs and their targets. Understanding these miRNAs and their functions is instrumental in the development of plants with enhanced tolerance to HM stress (Shriram et al., 2016; Khalid et al., 2019; Wani et al., 2020).

Finally, the emerging field of metallomics sheds light on the study of the metallome, encompassing metal-containing biomolecules within plant cells. This field explores the intricate world of metalloproteins, metalloenzymes, and metal metabolites. HM stress can significantly influence the structure and function of these metalloproteins, reshaping the metallome of plant cells. Techniques such as MS, synchrotron radiation X-ray absorption spectroscopy (XAS), and X-ray fluorescence (XRF) allow for in-depth exploration of metalloproteins and their interactions with HMs. Complementing these experimental methods, computational approaches are harnessed to analyze and interpret metallomics data. While metallomics is a promising and evolving field, our understanding of plant metallomes is still in its infancy, with ongoing research poised to unlock the intricate interactions within this complex domain (Shi and Chance, 2008; Jamla et al., 2021).

Collectively, these omics approaches are not isolated but interconnected, providing a comprehensive framework for the scientific community to comprehend and ultimately enhance plant responses to HM stress, thereby contributing to the development of robust and resilient crop varieties.

12.5 PLANT METABOLITES INVOLVED IN HEAVY METAL STRESS TOLERANCE

HMs not only accumulate in plants but also disrupt their water relations and lead to osmotic stress. When soils are contaminated with HMs, the osmotic potential in the soil becomes lower than in plant roots, limiting water uptake and causing osmotic disturbances. This disruption of water relations can affect hormone balance, nutrient levels, photosynthesis, and more in plants. The combination of osmotic stress and HM exposure can severely impact plant growth, reducing water content, root length, and plant height. Osmotic stress exacerbates other stressors on plants, leading to oxidative damage and the overproduction of harmful ROS; HMs themselves can also lead to oxidative stress. The persistence of HMs in plant cells further compounds the damage they cause. Overall, the effects of HM and osmotic stress on plants are significant and should be further studied and understood (Mashabela et al., 2023).

To counteract and regulate the detrimental effects of osmotic stress and HM toxicity, plants have evolved adaptive methods. This response involves a systematic approach, including the observation of environmental signals, intracellular signal transmission, and the activation of genes and metabolic processes to counteract stressors. The initial metal toxicity symptoms resemble responses to other stressors, such as oxidative stress, nutrient imbalances, drought, and changes in photosynthesis and development. These stress responses involve interconnected regulatory networks that can be considered a standardized plant reaction to various environmental stimuli. In the case of osmotic stress and HM toxicity, genes involved in metal chelation and ion transport become active through different pathways. While research has primarily focused on mechanisms related to common stresses such as drought and salinity, the response mechanisms for HM poisoning and osmotic stress are closely linked, with HM toxicity contributing to osmotic imbalances in plants (Rucińska-Sobkowiak, 2016; Singh et al., 2016; Jalmi et al., 2018).

Mitogen-activated protein kinases (MAPKs) are a group of regulatory enzymes that play a pivotal role in steering a variety of cellular responses by transmitting signals from outside the cell to the inside. They regulate processes such as cell growth, differentiation, movement, and survival and are highly conserved molecules involved in plant responses and development (Cargnello and Roux, 2011). In the context of HM toxicity and osmotic stress, MAPKs are activated by specific metal molecules, and ROS is generated during metal stress. These signals set off a cascade of events, involving three types of kinases (MAPKs, MAPKKs, and MAPKKKs) that ultimately lead to the phosphorylation of downstream targets, including transcription factors (Cargnello and Roux, 2011; Singh et al., 2016; Jalmi et al., 2018). Numerous studies have examined how MAPKs function in the response of plants to HM toxicity and osmotic stress, with a focus on model plants such as *Arabidopsis*. For example, in *Arabidopsis*, exposure to Cd activated MPK6, which in turn triggered processes resembling programmed cell death as a defense mechanism against stress. The role of MPK3 in responding to metals was also demonstrated when exposed to Cu and Cd (Ye et al., 2013; Amaral Dos Reis et al., 2018).

In a study involving naked oat seedlings facing drought stress, the expression of MAPKs (Asmap1 and Aspk11), along with transcription factors (WRKY1, DREB2, and MYB), increased (Gao et al., 2018). This led to the regulation of genes related to antioxidant capacity and drought stress response, ultimately enhancing the plant's ability to tolerate stress. While there have been relatively few studies on the MAPK-mediated responses of plants to metal toxicity and osmotic stress in recent years, research in other organisms, such as fish, fungi, yeast, and erythrocytes, has shown that MAPKs are activated in response to HM and osmotic stress. This suggests that MAPK cascades are indeed involved in stress responses across various biological systems (Zhou et al., 2010; Gatidis et al., 2011; Esquivel-Naranjo et al., 2016; Jing et al., 2020).

Plants have evolved response mechanisms to adapt to HM and osmotic stress conditions, one of which is calcium signaling. When plants encounter these stressors, they swiftly increase cytosolic calcium ions ([Ca^{2+}] cyt) from low to high concentrations, initiating complex signaling and metabolic pathways. This rise in [Ca^{2+}] cyt activates calcium sensors such as calcium-dependent protein kinases (CDPKs), which relay information and trigger downstream responses via signal transduction pathways (Steinhorst and Kudla, 2014; Lephatsi et al., 2021). In the case of HM exposure, Ca plays a crucial role in reducing toxicity, particularly in the case of Cd due to similarities in physicochemical properties. Plants use Ca sensors to elevate intracellular Ca^{2+} levels as an adaptive response to Cd, enhancing tolerance and reducing toxicity (Jalmi et al., 2018). Exogenous Ca^{2+} treatment has been shown to improve plant tolerance to Cd stress, and crosstalk between signaling pathways has been observed in responses to Cd (Ahmad et al., 2015; Zhao et al., 2018).

Plants also employ calcium signaling in response to osmotic stress, where it leads to a rapid increase in [Ca^{2+}] cyt, signaling the involvement of cytosolic calcium in the osmotic stress response. In addition to this, plants activate various mechanisms when facing osmotic stress, such as accumulating ABA, reprogramming gene expression, altering growth, and improving stress tolerance (Chen et al., 2020; Huang et al., 2017). Calcium signaling helps plants adapt to diverse environmental challenges, including heavy metals and osmotic stress.

Plant hormones, particularly ABA, are crucial in orchestrating a plant's response to environmental stresses such as osmotic drought, HM toxicity, and salinity. ABA triggers rapid protective measures such as stomatal closure for short-term responses and regulates stress-responsive genes for long-term adaptation. Genes responsible for ABA biosynthesis, including zeaxanthin epoxidase (ZEP), abscisic aldehyde oxidase 3

(AAO3), 9-*cis*-epoxycarotenoid dioxygenase 3 (NCED3), and molybdenum cofactor sulfurase, become upregulated during osmotic stress, increasing ABA levels and initiating ABA-dependent gene expression (Yoshida et al., 2014; Verma et al., 2016).

Other plant hormones such as auxins, jasmonic acid (JA), cytokinins, and ethylene also play significant roles in response to osmotic stress. For example, osmotic stress stimuli regulate auxin responses by influencing auxin biosynthesis (YUC, TAA1), transport (PIN), and perception (TIR/AFB, Aux/IAA) (Naser and Shani, 2016). In contrast, JA are known to enhance resistance to osmotic stress in rice, as evidenced by elevated JA content in the wild type comparison to a mutant with disrupted JA biosynthesis (Tang et al., 2020). Plants employ various mechanisms to mitigate the effects of heavy metal toxicity and prevent osmotic stress. These include regulating ROS production and activating antioxidant defenses. These phytohormones also play a role in modulating lateral root formation and patterning, suggesting their involvement in the remodeling of root systems and architecture in response to HM stress (Jalmi et al., 2018). In addition to limiting the detrimental effects of HM, plants deploy various mechanisms to shield themselves from exposure to osmotic stress. One such mechanism involves preventing the excessive production of ROS, ensuring they do not surpass toxic thresholds, thereby aiding the plant in acclimatizing to environmental stress conditions. Plants employ antioxidant defense systems to manage ROS production, translocation, and activity through scavenging, well-documented classical plant defense mechanisms. Notably, melatonin, identified as a pleiotropic signaling molecule, has gained significant attention as a universal regulator central to the plant's defense response strategies discussed above (Li et al., 2016).

Numerous studies have highlighted the accumulation of metallothionein (MT) in plants responding to HM toxicity. MT has been shown to effectively reduce Cd uptake and mitigate toxicity in plants. For instance, the supplementation of MT significantly enhanced Cd tolerance in tomato plants, leading to reduced Cd content in leaves. This was evident through decreased growth inhibition, photoinhibition, and electrolyte leakage. MT has also proven valuable in conferring Cd stress tolerance in transgenic *Arabidopsis* plants by upregulating the alfalfa serotonin *N*-acetyltransferase (SNAT) gene. In these plants, MT treatment showed the upregulation of *PDR8* and *HMA4* genes, ultimately reducing Cd accumulation and reestablishing microRNA-mediated redox homeostasis (Li et al., 2016; Gu et al., 2021). Similar observations have been made in studies involving various plant species, including *Hordeum vulgare* (barley), *Lupinus albus* (lupin), *Nicotiana tabacum* (tobacco), and *Solanum lycopersicum* (tomato). These studies, as reviewed by Hoque et al. (2021), have highlighted MT biosynthesis as a key regulatory mechanism, acting as a biostimulator to alleviate HM toxicity. MT achieves this by regulating the expression of target genes and modulating HM transporters.

Furthermore, Gu et al. (2021) have proposed that MT functions as an elicitor of miRNA-mediated ROS homeostasis in plants reacting to metal stress. These findings underscore the critical functions of MT in plant responses to HM toxicity and its potential as a regulator of ROS balance in such stressful conditions. Plants respond to both osmotic stress and HM toxicity inducers such as salinity and drought by enhancing MT biosynthesis. MT fosters HM tolerance by synthesizing stress-related genes, enhancing ROS scavenging, and improving various plant features that support growth.

MT's role in managing osmotic stress is gaining attention. It has been reported to protect the antioxidant properties of plants under osmotic stress, enhancing stress tolerance and promoting growth in various plant species. While the exact biochemical pathways through which MT enhances osmotic stress tolerance are not fully understood, recent research has focused on uncovering these mechanisms. For instance, the receptor for phytomelatonin CAND2/PMTR1 has been found to play a vital role in MT-mediated plant responses to osmotic stress (Wang et al., 2021a).

Glutathione (GSH) is a vital nonenzymatic antioxidant produced in plant cells. It contributes in mitigating osmotic stress, countering oxidative damage, and maintaining cellular redox balance via the 1-ascorbate-glutathione (GSH-AsA) cycle. GSH is also associated with observed tolerance in plants exposed to heavy metals, such as Cd. Studies have shown that MT priming induces the accumulation of GSH and AsA in plants reacting to drought-induced stress. This accumulation, in turn, enhances enzymatic antioxidant activity, aids in osmotic adjustment, and boosts the production of osmoprotectants. Additionally, MT enhances the nonenzymatic GSH-AsA cycle in plants and activates various antioxidative enzymes, including peroxide dismutase, CAT, SOD, APX, POD, GSH reductase, and dehydroascorbate reductase (DHAR) during osmotic stress (Forman et al., 2009; Qin et al., 2018; Hasanuzzaman et al., 2019; Khan et al., 2019; Sadak and Bakry, 2020; Alharby and Fahad, 2020; Zeng et al., 2022).

MT plays a pivotal role in plant responses to both HM toxicity and osmotic stress. It influences the primary regulatory mechanisms in plants' reactions to environmental stress, making it a promising avenue for enhancing stress tolerance. For instance, various studies have reported that MT impacts MAPKs, key regulatory proteins that direct cellular responses to diverse stimuli, affecting cell proliferation, differentiation, motility, and survival. These findings suggest a new dimension to MT's involvement in plant stress responses. While the primary regulatory mechanisms for plant stress responses have been extensively studied, the exact role of MT in these processes remains partly understood. This knowledge gap presents an opportunity for advanced research into the application of MT to elevate tolerance to HM toxicity and osmotic stress in plants (Figure 12.1). Monitoring metabolic pathways and metabolite changes during abiotic stress can provide valuable insights into maintaining metabolic balance and promoting the production of defense-related compounds that aid in stress tolerance (Jalmi et al., 2018; Lephatsi et al., 2021; Mashabela et al., 2023).

Numerous reports have highlighted the pivotal role of signaling molecules in conferring abiotic stress tolerance to plants through interactions with beneficial microbes. For instance, studies have revealed that the expression of MAPK5 was altered in rice roots treated with *Bacillus amyloliquefaciens*, a PGPR, under salt stress. This observation indicates the activation of MAPK signaling in the presence of PGPR, illustrating the significance of microbial involvement in stress responses in plants (Nautiyal et al., 2013). Similarly, changes in the expression of At3g57530, a gene associated with calcium and calcium-dependent protein kinase activity, have been documented in *Arabidopsis* when subjected to treatments with *Pseudomonas putida* and *Pseudomonas fluorescens* (Wang et al., 2005; Srivastava et al., 2012). Additionally, the application of PGPR has been found to modulate

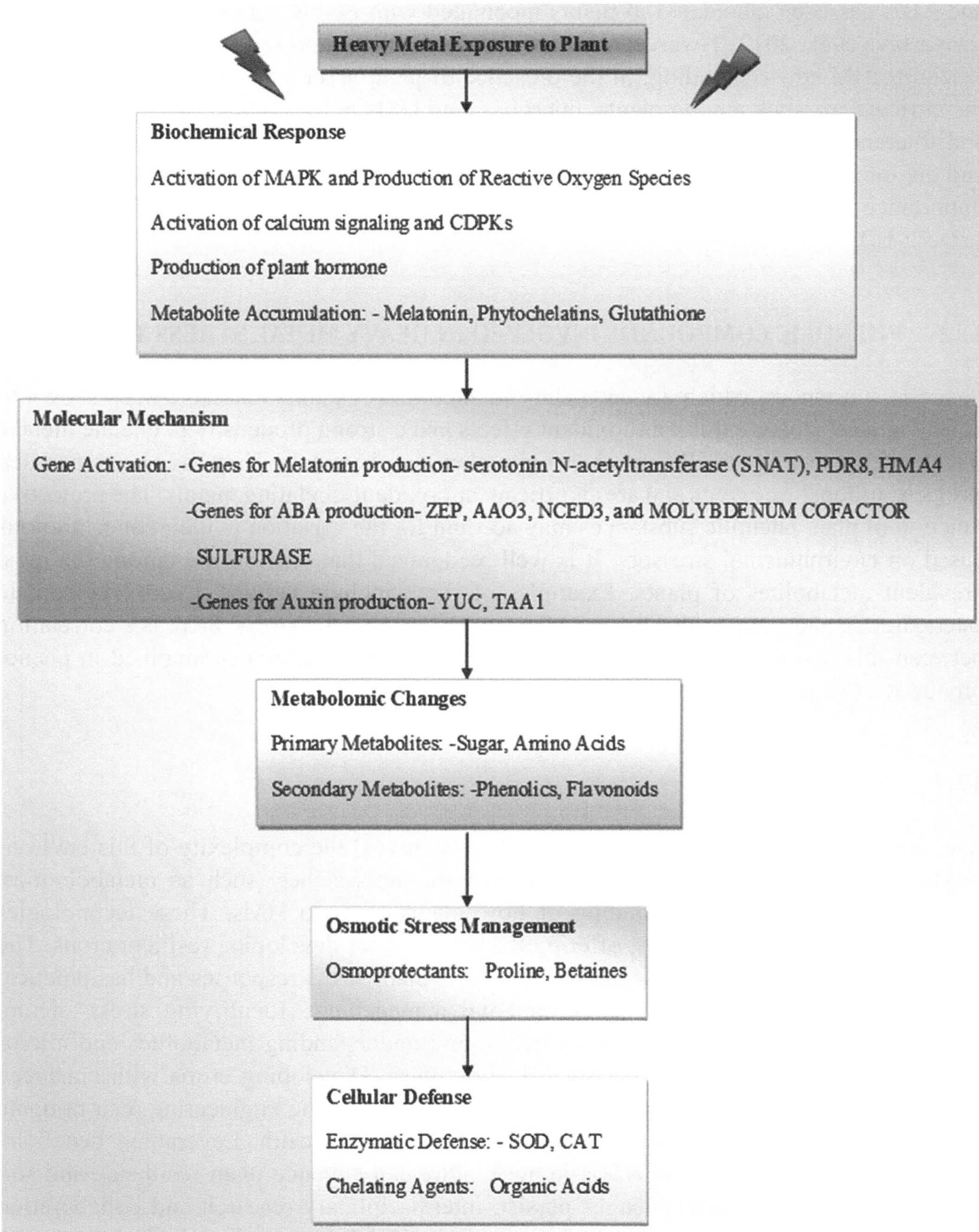

FIGURE 12.1 Plant metabolites involved in HM tolerance.

the expression of downstream stress-responsive transcription factors such as *Myb*, *Nac*, and b*ZIP* in various plants, including *Oryza sativa*, *Cicer arietinum*, and *Arabidopsis*. Moreover, the involvement of phytohormones such as salicylic acid (SA), ethylene, JA, and ABA has been elucidated in plants inoculated with PGPR under stressful conditions (Srivastava et al., 2012; Tiwari et al., 2016, 2017). The activation of these genes, which are crucial to HM stress signaling, in the presence of plant-associated microbes underscores the intricate crosstalk among plants, microbes, and HMs in the context of stress responses and tolerance. Therefore, comprehending the complex metal stress signaling pathways and the interconnected networks involving plant–microbe–metal interactions is of utmost importance for unraveling the intricate stress-responsive networks in plants (Tiwari and Lata, 2018).

12.6 PHENOLIC COMPOUNDS INVOLVED IN HEAVY METAL STRESS TOLERANCE

Phenolic substances, which are secondary metabolites of plants and have been shown to defend against stress, exhibit antioxidant effects and a strong propensity to chelate metals. Plants that are exposed to HMs produce more phenolic chemicals. They have been regarded as electron-donating agents and are significant antioxidant-chelating metals. The protective function of plant phenolic substances may account for the variation in their concentrations based on environmental stressors. It is well recognized that phenols are among the most prevalent metabolites of plants. Examples of these include tannins, hydroxycinnamate esters, lignin, and flavonoids (Table 12.2). Under heavy metal stress, there is a correlation between this rise in phenolics and increased activity of the enzymes involved in phenol production (Singh et al., 2016; Yadav et al., 2021).

12.7 CONCLUSION

Intricate responses of plants to heavy metal stress reveal the complexity of this environmental challenge. Recent advancements in omics approaches, such as metabolomics, have transformed our understanding of how plants react to HMs. These technologies provide comprehensive insights, offering the potential for developing resilient crops. The knowledge gained enhances our understanding of plant stress responses and has practical applications in agriculture and environmental management. Identifying stress-tolerant genotypes, exploring stress-responsive genes, and understanding metabolites and microbial interactions contribute to sustainable agriculture. Developing crops with enhanced metal tolerance, using both conventional breeding and genetic engineering, can mitigate metal contamination's impact on food chains and human health. Leveraging beneficial microbes, such as PGPRs and AMF, in agriculture can enhance plant resilience and soil health. As environmental challenges persist, interdisciplinary research and collaboration are crucial. Combining molecular biology, omics technologies, and ecological principles can lead to transformative strategies for global food security and environmental sustainability. Ongoing knowledge pursuit in this field opens doors to innovative solutions in agriculture and environmental science.

TABLE 12.2 Changes in the Production of Phenolic Compounds to Combat HM Stress

Heavy Metals	Plants	Effect on Different Phenolic Compounds	References
Cd, Cu, Pb	*Zea mays*	Increased level of chlorogenic acid and rutin in leaves, whereas decreased level of ferulic acid and caffeic acid in all HM exposure.	(Kısa et al., 2016)
Cd	*Phaseolus vulgaris*	Both soluble and insoluble phenols were accumulated in the leaves of this plant under Cd stress.	(Fuhrer, 1982)
Cu, Pb	*Triticum aestivum* L.	There was a noticeable rise in the production of flavonoids and other phenolic substances.	(Jańczak-Pieniążek et al., 2022)
Cu, Cd, Pb	*Solanum lycopersicum*	Decrease the contents of caffeic acid and chlorogenic acid in all applications of HMs; nonetheless, there has been a notable variation in the applied dosages of heavy metals about their contents. All of the Cu and Cd treatments result in a considerable reduction in the quantity of vanillic acid.	(Kısa et al., 2019)
Cd	*Triticum aestivum*	Increases the salicylic level in the leaf of the plant, which triggers the antioxidant and metal detoxification systems.	(Kovács et al., 2014)
Cd	*Brassica juncea*	Variation in the level of phenolic compounds was found. Due to higher accumulation of rutin, oxidative damage caused by HM was prevented.	(Kapoor et al., 2016)
Cd	*Lactuca sativa* L.	Phenolic compound, lignin, increases in the roots when the Zn concentration increases, which inhibits or lessens the amount of Cd that may enter plant tissues.	(Gao et al., 2022)
Cu and Zn	*S. lycopersicum*	Increase in the concentration of polyphenols and flavonoids in roots and leaves, and they are involved in plants' defense mechanisms quenching ROS and functioning as metal chelators; these chemicals offer protection against oxidative stress.	(Badiaa et al., 2018)
Al, Cd	*Vaccinium corymbosum* L.	The exposure to Al and Cd, the phenolic compound especially chlorogenic and ellagic acids significantly increases in the plantlet, which plays a crucial role in ROS removal.	(Manquián-Cerda et al., 2018)
Al	*Zea mays* L.	The flavonoid-type phenolics (quercetin and catechin) and catechol were the most noticeable phenolics found in significant amounts in the exudates of the experimental plants utilized in this study.	(Kidd et al., 2001)
Pb	*Lupinus luteus* L.	Increased flavonoid content in both cotyledons exhibited enhanced tolerance to the heavy metal.	(Izbiańska, Arasimowicz-Jelonek, and Deckert, 2014)

KEYWORDS

- **heavy metal stress**
- **plant responses**
- **tolerance mechanisms**
- **metabolomics**
- **metabolites**

REFERENCES

Aery, N. C., Jagetiya, B. L. Relative toxicity of cadmium, lead, and zinc on barley. *Commun. Soil Sci. Plant Anal.*, 1997, 28(11–12), 949–960.

Ahmad, P., Sarwat, M., Bhat, N. A., Wani, M. R., Kazi, A. G., Tran, L.-S. P. Alleviation of cadmium toxicity in *Brassica juncea* L. (Czern. & Coss.) by calcium application involves various physiological and biochemical strategies. *PLoS One*, 2015, 10(1), e0114571.

Al Khateeb, W., Al-Qwasemeh, H. Cadmium, copper and zinc toxicity effects on growth, proline content and genetic stability of *Solanum nigrum* L., a crop wild relative for tomato; comparative study. *Physiol. Mol. Biol. Plants*, 2014, 20(1), 31–39.

Alharby, H. F., Fahad, S. Melatonin application enhances biochar efficiency for drought tolerance in maize varieties: Modifications in physio-biochemical machinery. *Agron. J.*, 2020, 112(4), 2826–2847.

Amaral Dos Reis, R., Keunen, E., Mourato, M. P., Martins, L. L., Vangronsveld, J., Cuypers, A. Accession-specific life strategies affect responses in leaves of arabidopsis thaliana plants exposed to excess Cu and Cd. *J. Plant Physiol.*, 2018, 223, 37–46.

Anani, O. A., Abel, I., Olomukoro, J. O., Onyeachu, I. B. Insights to proteomics and metabolomics metal chelation in food crops. *J. Proteins Proteomics*, 2022, 13(3), 159–173.

Anjum, S. A., Ashraf, U., Khan, I., Tanveer, M., Shahid, M., Shakoor, A., Wang, L. Phyto-toxicity of chromium in maize: Oxidative damage, osmolyte accumulation, anti-oxidative defense and chromium uptake. *Pedosphere*, 2017, 27(2), 262–273.

Anjum, S. A., Tanveer, M., Hussain, S., Bao, M., Wang, L., Khan, I., Ullah, E., Tung, S. A., Samad, R. A., Shahzad, B. Cadmium toxicity in maize (*Zea mays* L.): Consequences on antioxidative systems, reactive oxygen species and cadmium accumulation. *Environ. Sci. Pollut. Res.*, 2015, 22(21), 17022–17030.

Anjum, S. A., Tanveer, M., Hussain, S., Ullah, E., Wang, L., Khan, I., Samad, R. A., Tung, S. A., Anam, M., Shahzad, B. Morpho-physiological growth and yield responses of two contrasting maize cultivars to cadmium exposure. *CLEAN—Soil Air Water*, 2016, 44(1), 29–36.

Ashraf, R., Ali, T. Effect of heavy metals on soil microbial community and mung seed germination. *Pak. J. Bot.*, 2007, 39.

Badiaa, O., Yssaad, H. A. R., Topcuoglu, B. Effect of heavy metals (copper and zinc) on proline, polyphenols and flavonoids content of tomato (*Lycopersicon esculentum* Mill.). In: *PlantArch* (pp. 1229–1238), 2018.

Banach, A. M., Kuźniar, A., Grządziel, J., Wolińska, A. Azolla filiculoides L. as a source of metal-tolerant microorganisms. *PLoS One*, 2020, 15(5), e0232699.

Banerjee, A., Roychoudhury, A. Oxylipins and strobilurins as protective chemical agents to generate abiotic stress tolerance in plants. In: Roychoudhury, A., Tripathi, D. K. (Eds.), *Protective Chemical Agents in the Amelioration of Plant Abiotic Stress* (pp 483–490). Wiley, 2020.

Bohnert, H. J., Gong, Q., Li, P., Ma, S. Unraveling abiotic stress tolerance mechanisms – Getting genomics going. *Curr. Opin. Plant Biol.*, 2006, 9(2), 180–188.

Bokor, B., Vaculík, M., Slováková, Ľ., Masarovič, D., Lux, A. Silicon does not always mitigate zinc toxicity in maize. *Acta Physiol. Plant.*, 2014, 36(3), 733–743.

Branco-Neves, S., Soares, C., De Sousa, A., Martins, V., Azenha, M., Gerós, H., Fidalgo, F. An efficient antioxidant system and heavy metal exclusion from leaves make *Solanum Cheesmaniae* more tolerant to Cu than its cultivated counterpart. *Food Energy Secur.*, 2017, 6(3), 123–133.

Cargnello, M., Roux, P. P. Activation and function of the MAPKs and their substrates, the MAPK-activated protein kinases. *Microbiol. Mol. Biol. Rev.*, 2011, 75(1), 50–83.

Çatav, Ş. S., Genç, T. O., Oktay, M. K., Küçükakyüz, K. Cadmium toxicity in wheat: Impacts on element contents, antioxidant enzyme activities, oxidative stress, and genotoxicity. *Bull. Environ. Contam. Toxicol.*, 2020, 104(1), 71–77.

Chaoui, A., Jarrar, B., El Ferjani, E. Effects of cadmium and copper on peroxidase, NADH oxidase and IAA oxidase activities in cell wall, soluble and microsomal membrane fractions of pea roots. *J. Plant Physiol.*, 2004, 161(11), 1225–1234.

Chen, K., Li, G., Bressan, R. A., Song, C., Zhu, J., Zhao, Y. Abscisic acid dynamics, signaling, and functions in plants. *J. Integr. Plant Biol.*, 2020, 62(1), 25–54.

Chen, Y.-A., Chi, W.-C., Huang, T.-L., Lin, C.-Y., Quynh Nguyeh, T. T., Hsiung, Y.-C., Chia, L.-C., Huang, H.-J. Mercury-induced biochemical and proteomic changes in rice roots. *Plant Physiol. Biochem.*, 2012, 55, 23–32.

Choudhury, S., Sharma, P., Moulick, D., Mazumder, M. K. Unrevealing metabolomics for abiotic stress adaptation and tolerance in plants. *J. Crop Sci. Biotechnol.*, 2021, 24(5), 479–493.

Cosio, C., Renault, D. Effects of cadmium, inorganic mercury and methyl-mercury on the physiology and metabolomic profiles of shoots of the macrophyte elodea nuttallii. *Environ. Pollut.*, 2020, 257, 113557.

Dubey, S., Misra, P., Dwivedi, S., Chatterjee, S., Bag, S. K., Mantri, S., Asif, M. H., Rai, A., Kumar, S., Shri, M., Tripathi, P., Tripathi, R. D., Trivedi, P. K., Chakrabarty, D., Tuli, R. Transcriptomic and metabolomic shifts in rice roots in response to Cr (VI) stress. *BMC Genom.*, 2010, 11(1), 648.

Ebbs, S. D., Kochian, L. V. Toxicity of zinc and copper to brassica species: Implications for phytoremediation. *J. Environ. Qual.*, 1997, 26(3), 776–781.

Esquivel-Naranjo, E. U., García-Esquivel, M., Medina-Castellanos, E., Correa-Pérez, V. A., Parra-Arriaga, J. L., Landeros-Jaime, F., Cervantes-Chávez, J. A., Herrera-Estrella, A. A *Trichoderma atroviride* stress-activated MAPK pathway integrates stress and light signals. *Mol. Microbiol.*, 2016, 100(5), 860–876.

Feng, Z., Ji, S., Ping, J., Cui, D. Recent advances in metabolomics for studying heavy metal stress in plants. *TrAC Trends Anal. Chem.*, 2021, 143, 116402.

Forman, H. J., Zhang, H., Rinna, A. Glutathione: Overview of its protective roles, measurement, and biosynthesis. *Mol. Aspects Med.*, 2009, 30(1–2), 1–12.

Fu, H., Yu, H., Li, T., Zhang, X. Influence of cadmium stress on root exudates of high cadmium accumulating rice line (*Oryza sativa* L.). *Ecotoxicol. Environ. Saf.*, 2018, 150, 168–175.

Fuhrer, J. Early effects of excess cadmium uptake in phaseolus vulgaris. *Plant Cell Environ.*, 1982, 5(4), 263–270.

Gao, F., Zhang, X., Zhang, J., Li, J., Niu, T., Tang, C., Wang, C., Xie, J. Zinc Oxide Nanoparticles improve lettuce (*Lactuca sativa* L.) plant tolerance to cadmium by stimulating antioxidant defense, enhancing lignin content and reducing the metal accumulation and translocation. *Front. Plant Sci.*, 2022, 13, 1015745.

Gao, W., Zhang, Y., Feng, Z., Bai, Q., He, J., Wang, Y. Effects of melatonin on antioxidant capacity in naked oat seedlings under drought stress. *Molecules*, 2018, 23(7), 1580.

Gatidis, S., Zelenak, C., Fajol, A., Lang, E., Jilani, K., Michael, D., Qadri, S. M., Lang, F. P38 MAPK activation and function following osmotic shock of erythrocytes. *Cell. Physiol. Biochem.*, 2011, 28(6), 1279–1286.

Gonzalez Ibarra, A. A., Wrobel, K., Yanez Barrientos, E., Corrales Escobosa, A. R., Gutierrez Corona, J. F., Enciso Donis, I., Wrobel, K. Impact of Cr(VI) on the oxidation of polyunsaturated fatty acids in helianthus annuus roots studied by metabolomic tools. *Chemosphere*, 2019, 220, 442–451.

Goyal, D., Yadav, A., Prasad, M., Singh, T. B., Shrivastav, P., Ali, A., Dantu, P. K., Mishra, S. Effect of heavy metals on plant growth: An overview. In: Naeem, M., Ansari, A. A., Gill, S. S. (Eds.), *Contaminants in Agriculture* (pp 79–101). Springer International Publishing: Cham, 2020.

Gu, Q., Wang, C., Xiao, Q., Chen, Z., Han, Y. Melatonin confers plant cadmium tolerance: An update. *Int. J. Mol. Sci.*, 2021, 22(21), 11704.

Gupta, U. C., Gupta, S. C. Heavy metal toxicity in humans and its preventive and control measures. *Curr. Nutr. Food Sci.*, 2012, 7(4), 221–231.

Gupta, D. K., Pena, L. B., Romero-Puertas, M. C., Hernández, A., Inouhe, M., Sandalio, L. M. NADPH oxidases differentially regulate ROS metabolism and nutrient uptake under cadmium toxicity. *Plant Cell Environ.*, 2017, 40(4), 509–526

Hall, J. L. Cellular mechanisms for heavy metal detoxification and tolerance. *J. Exp. Bot.*, 2002, 53(366), 1–11.

Han, H., Zhang, H., Qin, S., Zhang, J., Yao, L., Chen, Z., Yang, J. Mechanisms of *Enterobacter bugandensis* TJ6 immobilization of heavy metals and inhibition of Cd and Pb uptake by wheat based on metabolomics and proteomics. *Chemosphere*, 2021, 276, 130157.

Hasanuzzaman, M., Bhuyan, M. H. M. B., Anee, T. I., Parvin, K., Nahar, K., Mahmud, J. A., Fujita, M. Regulation of ascorbate-glutathione pathway in mitigating oxidative damage in plants under abiotic stress. *Antioxidants*, 2019, 8(9), 384.

Hoque, Md. N., Tahjib-Ul-Arif, Md., Hannan, A., Sultana, N., Akhter, S., Hasanuzzaman, Md., Akter, F., Hossain, Md. S., Sayed, Md. A., Hasan, Md. T., Skalicky, M., Li, X., Brestič, M. Melatonin modulates plant tolerance to heavy metal stress: Morphological responses to molecular mechanisms. *Int. J. Mol. Sci.*, 2021, 22(21), 11445.

Huang, F., Luo, J., Ning, T., Cao, W., Jin, X., Zhao, H., Wang, Y., Han, S. Cytosolic and nucleosolic calcium signaling in response to osmotic and salt stresses are independent of each other in roots of arabidopsis seedlings. *Front. Plant Sci.*, 2017, 8, 1648.

Huang, G., Guo, G., Yao, S., Zhang, N., Hu, H. Organic acids, amino acids compositions in the root exudates and Cu-accumulation in castor (*Ricinus communis* L.) under Cu stress. *Int. J. Phytoremediation*, 2016, 18(1), 33–40.

Igiri, B. E., Okoduwa, S. I. R., Idoko, G. O., Akabuogu, E. P., Adeyi, A. O., Ejiogu, I. K. Toxicity and bioremediation of heavy metals contaminated ecosystem from tannery wastewater: A review. *J. Toxicol.*, 2018, 2018, 1–16.

Irfan, M., Ahmad, A., Hayat, S. Effect of cadmium on the growth and antioxidant enzymes in two varieties of brassica Juncea. *Saudi J. Biol. Sci.*, 2014, 21(2), 125–131.

İşkil, R., Surgun-Acar, Y., Çatav, Ş. S., Zemheri-Navruz, F., Erden, Y. Mercury toxicity affects oxidative metabolism and induces stress responsive mechanisms in wheat (*Triticum aestivum* L.). *Physiol. Mol. Biol. Plants*, 2022, 28(4), 911–920.

Izbiańska, K., Arasimowicz-Jelonek, M., Deckert, J. Phenylpropanoid pathway metabolites promote tolerance response of lupine roots to lead stress. *Ecotoxicol. Environ. Saf.*, 2014, 110, 61–67.

Jalmi, S. K., Bhagat, P. K., Verma, D., Noryang, S., Tayyeba, S., Singh, K., Sharma, D., Sinha, A. K. Traversing the links between heavy metal stress and plant signaling. *Front. Plant Sci.*, 2018, 9, 12.

Jamla, M., Khare, T., Joshi, S., Patil, S., Penna, S., Kumar, V. Omics approaches for understanding heavy metal responses and tolerance in plants. *Curr. Plant Biol.*, 2021, 27, 100213.

Jańczak-Pieniążek, M., Cichoński, J., Michalik, P., Chrzanowski, G. Effect of heavy metal stress on phenolic compounds accumulation in winter wheat plants. *Molecules*, 2022, 28(1), 241.

Järup, L. Hazards of heavy metal contamination. *Br. Med. Bull.*, 2003, 68(1), 167–182.

Järup, L., Åkesson, A. Current status of cadmium as an environmental health problem. *Toxicol. Appl. Pharmacol.*, 2009, 238(3), 201–208.

Jing, H., Zhang, Q., Li, S., Gao, X. Pb Exposure triggers MAPK-dependent inflammation by activating oxidative stress and miRNA-155 expression in carp head kidney. *Fish Shellfish Immunol.*, 2020, 106, 219–227.

Johri, N., Jacquillet, G., Unwin, R. Heavy metal poisoning: The effects of cadmium on the kidney. *BioMetals*, 2010, 23(5), 783–792.

Kang, G., Li, G., Wang, L., Wei, L., Yang, Y., Wang, P., Yang, Y., Wang, Y., Feng, W., Wang, C., Guo, T. Hg-responsive proteins identified in wheat seedlings using iTRAQ analysis and the role of ABA in Hg stress. *J. Proteome Res.*, 2015, 14(1), 249–267.

Kapoor, D., Rattan, A., Bhardwaj, R., Kaur, S., Gupta, A., Manoj. Antioxidative defense responses and activation of phenolic compounds in *Brassica juncea* plants exposed to cadmium stress. *Int. J. Green Pharm.*, 2016, 10, 228–234.

Kaur, G., Singh, H. P., Batish, D. R., Kohli, R. K. Lead (Pb)-induced biochemical and ultrastructural changes in wheat (*Triticum aestivum*) roots. *Protoplasma*, 2013, 250(1), 53–62.

Khalid, N., Aqeel, M., Noman, A. System biology of metal tolerance in plants: An integrated view of genomics, transcriptomics, metabolomics, and phenomics. In: Sablok, G. (Ed.), *Plant Metallomics and Functional Omics* (pp. 107–144). Springer International Publishing: Cham, 2019.

Khan, M. N., Zhang, J., Luo, T., Liu, J., Rizwan, M., Fahad, S., Xu, Z., Hu, L. Seed priming with melatonin coping drought stress in rapeseed by regulating reactive oxygen species detoxification: Antioxidant defense system, osmotic adjustment, stomatal traits and chloroplast ultrastructure perseveration. *Ind. Crops Prod.*, 2019, 140, 111597.

Kidd, P. S., Llugany, M., Poschenrieder, C., Gunsé, B., Barceló, J. The role of root exudates in aluminium resistance and silicon-induced amelioration of aluminium toxicity in three varieties of maize (*Zea mays* L.). *J. Exp. Bot.*, 2001, 52(359), 1339–1352.

Kısa, D., Elmastaş, M., Öztürk, L., Kayır, Ö. Responses of the phenolic compounds of zea mays under heavy metal stress. *Appl. Biol. Chem.*, 2016, 59(6), 813–820.

Kisa, D., Kayir, O., Saglam, N., şahin, S., Öztürk, L., Elmastaş, M. Changes of phenolic compounds in tomato associated with the heavy metal stress. *JONAS*, 2019, 2, 35–43.

Kovács, V., Gondor, O. K., Szalai, G., Darkó, É., Majláth, I., Janda, T., Pál, M. Synthesis and role of salicylic acid in wheat varieties with different levels of cadmium tolerance. *J. Hazard. Mater.*, 2014, 280, 12–19.

Koyro, H.-W., Ahmad, P., Geissler, N. Abiotic stress responses in plants: An overview. In: Ahmad, P., Prasad, M. N. V. (Eds.), *Environmental Adaptations and Stress Tolerance of Plants in the Era of Climate Change* (pp 1–28)Springer New York: New York, NY, 2012.

Lephatsi, M. M., Meyer, V., Piater, L. A., Dubery, I. A., Tugizimana, F. Plant responses to abiotic stresses and rhizobacterial biostimulants: Metabolomics and epigenetics perspectives. *Metabolites*, 2021, 11(7), 457.

Li, M., Hasan, Md. K., Li, C., Ahammed, G. J., Xia, X., Shi, K., Zhou, Y., Reiter, R. J., Yu, J., Xu, M., Zhou, J. Melatonin mediates selenium-induced tolerance to cadmium stress in tomato plants. *J. Pineal Res.*, 2016, 61(3), 291–302.

Lima de Silva, A. A., de Carvalho, M. A. R., de Souza, S. A. L., Dias, P. M. T., da Silva Filho, R. G., de Meirelles Saramago, C. S., de Melo Bento, C. A., Hofer, E. Heavy metal tolerance (Cr, Ag AND Hg) in bacteria isolated from sewage. *Braz. J. Microbiol. Publ. Braz. Soc. Microbiol.*, 2012, 43(4), 1620–1631.

Mahurpawar, M. Effects of heavy metals on human health. *Int J Res Granthaalayah*, 2015, 3, 1–7.

Manquián-Cerda, K., Cruces, E., Escudey, M., Zúñiga, G., Calderón, R. Interactive effects of aluminum and cadmium on phenolic compounds, antioxidant enzyme activity and oxidative stress in blueberry (*Vaccinium corymbosum* L.) plantlets cultivated in vitro. *Ecotoxicol. Environ. Saf.*, 2018, 150, 320–326.

Martin, S., Griswold, W. Human health effects of heavy metals. *Environ. Sci. Technol. Briefs Citizens*, 2009, 15, 1–6.

Mashabela, M. D., Masamba, P., Kappo, A. P. Applications of metabolomics for the elucidation of abiotic stress tolerance in plants: A special focus on osmotic stress and heavy metal toxicity. *Plants*, 2023, 12(2), 269.

Mediouni, C., Benzarti, O., Tray, B., Ghorbel, M. H., Jemal, F. Cadmium and copper toxicity for tomato seedlings. *Agron. Sustain. Dev.*, 2006, 26(4), 227–232.

Mehta, S., James, D., Reddy, M. K. Omics technologies for abiotic stress tolerance in plants: Current status and prospects. In: Wani, S. H. (Ed.), *Recent Approaches in Omics for Plant Resilience to Climate Change* (pp. 1–34). Springer International Publishing: Cham, 2019.

Mishra, J., Singh, R., Arora, N. K. Alleviation of heavy metal stress in plants and remediation of soil by rhizosphere microorganisms. *Front. Microbiol.*, 2017, 8, 1706.

Mondal, N. K., Das, C., Datta, J. K. Effect of mercury on seedling growth, nodulation and ultrastructural deformation of *Vigna radiata* (L) wilczek. *Environ. Monit. Assess.*, 2015, 187(5), 241.

Mwamba, T. M., Islam, F., Ali, B., Lwalaba, J. L. W., Gill, R. A., Zhang, F., Farooq, M. A., Ali, S., Ulhassan, Z., Huang, Q., Zhou, W., Wang, J. (2020). Comparative metabolomic responses of low-and high-cadmium accumulating genotypes reveal the cadmium adaptive mechanism in Brassica napus. *Chemosphere*, 250, 126308.

Nagajyoti, P. C., Lee, K. D., Sreekanth, T. V. M. Heavy metals, occurrence and toxicity for plants: A review. *Environ. Chem. Lett.*, 2010, 8(3), 199–216.

Naser, V., Shani, E. Auxin response under osmotic stress. *Plant Mol. Biol.*, 2016, 91(6), 661–672.

Nautiyal, C. S., Srivastava, S., Chauhan, P. S., Seem, K., Mishra, A., Sopory, S. K. Plant growth-promoting bacteria bacillus amyloliquefaciens NBRISN13 modulates gene expression profile of leaf and rhizosphere community in rice during salt stress. *Plant Physiol. Biochem.*, 2013, 66, 1–9.

Navarro-Reig, M., Jaumot, J., Piña, B., Moyano, E., Galceran, M. T., Tauler, R. Metabolomic analysis of the effects of cadmium and copper treatment in oryza sativa L. Using untargeted liquid chromatography coupled to high resolution mass spectrometry and all-Ion fragmentation. *Metallomics*, 2017, 9(6), 660–675.

Ou, X., Zhang, Y., Xu, C., Lin, X., Zang, Q., Zhuang, T., Jiang, L., Von Wettstein, D., Liu, B. Transgenerational inheritance of modified DNA methylation patterns and enhanced tolerance induced by heavy metal stress in rice (*Oryza sativa* L.). *PLoS One*, 2012, 7(9), e41143.

Pandey, N., Sharma, C. P. Effect of heavy metals Co^{2+}, Ni^{2+} and Cd^{2+} on growth and metabolism of cabbage. *Plant Sci.*, 2002, 163(4), 753–758.

Phurailatpam, L., Dalal, V. K., Singh, N., Mishra, S. Heavy metal stress alleviation through omics analysis of soil and plant microbiome. *Front. Sustain. Food Syst.*, 2022, 5, 817932.

Phurailatpam, L., Mishra, S. Role of plant endophytes in conferring abiotic stress tolerance. In: Hasanuzzaman, M. (Ed.), *Plant Ecophysiology and Adaptation under Climate Change: Mechanisms and Perspectives II* (pp. 603–628). Springer Singapore: Singapore, 2020.

Pidatala, V. R., Li, K., Sarkar, D., Wusirika, R., Datta, R. Comparative metabolic profiling of vetiver (*Chrysopogon zizanioides*) and maize (*Zea mays*) under lead stress. *Chemosphere*, 2018, 193, 903–911.

Qin, S., Liu, H., Nie, Z., Gao, W., Li, C., Lin, Y., Zhao, P. AsA–GSH cycle and antioxidant enzymes play important roles in Cd tolerance of wheat. *Bull. Environ. Contam. Toxicol.*, 2018, 101(5), 684–690.

Razzaq, A., Sadia, B., Raza, A., Khalid Hameed, M., Saleem, F. Metabolomics: A way forward for crop improvement. *Metabolites*, 2019, 9(12), 303.

Rizwan, M., Ali, S., Abbas, T., Zia-ur-Rehman, M., Hannan, F., Keller, C., Al-Wabel, M. I., Ok, Y. S. Cadmium minimization in wheat: A critical review. *Ecotoxicol. Environ. Saf.*, 2016, 130, 43–53.

Rizwan, M., Ali, S., Adrees, M., Rizvi, H., Zia-ur-Rehman, M., Hannan, F., Qayyum, M. F., Hafeez, F., Ok, Y. S. Cadmium stress in rice: Toxic effects, tolerance mechanisms, and management: A critical review. *Environ. Sci. Pollut. Res.*, 2016, 23(18), 17859–17879.

Roane, T. M., Pepper, I. L. Microbial responses to environmentally toxic cadmium. *Microb. Ecol.*, 1999, 38(4), 358–364.

Rouphael, Y., Colla, G., Bernardo, L., Kane, D., Trevisan, M., Lucini, L. Zinc excess triggered polyamines accumulation in lettuce root metabolome, as compared to osmotic stress under high salinity. *Front. Plant Sci.*, 2016, 7.

Rucińska-Sobkowiak, R. Water relations in plants subjected to heavy metal stresses. *Acta Physiol. Plant.*, 2016, 38(11), 257.

Sadak, M. S., Bakry, B. A. Alleviation of drought stress by melatonin foliar treatment on two flax varieties under sandy soil. *Physiol. Mol. Biol. Plants*, 2020, 26(5), 907–919.

Sahoo, J. P., Behera, L., Sharma, S. S., Praveena, J., Nayak, S. K., Samal, K. C. Omics studies and systems biology perspective towards abiotic stress response in plants. *Am. J. Plant Sci.*, 2020, 11(12), 2172–2194.

Shah, F. U. R., Ahmad, N., Masood, K. R., Peralta-Videa, J. R., Ahmad, F. U. D. Heavy metal toxicity in plants. In: Ashraf, M., Ozturk, M., Ahmad, M. S. A. (Eds.), *Plant Adaptation and Phytoremediation* (pp. 71–97). Springer Netherlands: Dordrecht, 2010.

Shanker, A., Cervantes, C., Lozatavera, H., Avudainayagam, S. Chromium toxicity in plants. *Environ. Int.*, 2005, 31(5), 739–753.

Sharma, P., Dubey, R. S. Lead toxicity in plants. *Braz. J. Plant Physiol.*, 2005, 17(1), 35–52.

Sharma, S. S., Dietz, K.-J. The significance of amino acids and amino acid-derived molecules in plant responses and adaptation to heavy metal stress. *J. Exp. Bot.*, 2006, 57(4), 711–726.

Shi, W., Chance, M. R. Metallomics and metalloproteomics. *Cell. Mol. Life Sci.*, 2008, 65(19), 3040–3048.

Shriram, V., Kumar, V., Devarumath, R. M., Khare, T. S., Wani, S. H. MicroRNAs as potential targets for abiotic stress tolerance in plants. *Front. Plant Sci.*, 2016, 7.

Singh, S., Parihar, P., Singh, R., Singh, V. P., Prasad, S. M. Heavy metal tolerance in plants: Role of transcriptomics, proteomics, metabolomics, and ionomics. *Front. Plant Sci.*, 2016, 6, 1143.

Sobolev, D., Begonia, M. Effects of heavy metal contamination upon soil microbes: Lead-induced changes in general and denitrifying microbial communities as evidenced by molecular markers. *Int. J. Environ. Res. Public. Health*, 2008, 5(5), 450–456.

Srivastava, S., Chaudhry, V., Mishra, A., Chauhan, P. S., Rehman, A., Yadav, A., Tuteja, N., Nautiyal, C. S. Gene expression profiling through microarray analysis in *Arabidopsis Thaliana* Colonized by *Pseudomonas Putida* MTCC5279, a plant growth promoting rhizobacterium. *Plant Signal. Behav.*, 2012, 7(2), 235–245.

Steinhorst, L., Kudla, J. Signaling in cells and organisms — Calcium holds the line. *Curr. Opin. Plant Biol.*, 2014, 22, 14–21.

Subba, P., Mukhopadhyay, M., Mahato, S. K., Bhutia, K. D., Mondal, T. K., Ghosh, S. K. Zinc stress induces physiological, ultra-structural and biochemical changes in mandarin orange (*Citrus reticulata* Blanco) seedlings. *Physiol. Mol. Biol. Plants*, 2014, 20(4), 461–473.

Swamy, M. K., Nalina, N., Nalina, D., Akhtar, M. S., Purushotham, B. Heavy metal stress and tolerance in plants mediated by rhizospheric microbes. In: Akhtar, M. S. (Ed.), *Salt Stress, Microbes, and Plant Interactions: Causes and Solution* (pp. 181–198). Springer Singapore: Singapore, 2019.

Tang, G., Ma, J., Hause, B., Nick, P., Riemann, M. Jasmonate is required for the response to osmotic stress in rice. *Environ. Exp. Bot.*, 2020, 175, 104047.

Tiwari, P., Bae, H. Trends in harnessing plant endophytic microbiome for heavy metal mitigation in plants: A perspective. *Plants*, 2023, 12(7), 1515.

Tiwari, S., Lata, C. Heavy metal stress, signaling, and tolerance due to plant-associated microbes: An overview. *Front. Plant Sci.*, 2018, 9, 452.

Tiwari, S., Lata, C., Chauhan, P. S., Nautiyal, C. S. Pseudomonas putida attunes morphophysiological, biochemical and molecular responses in *Cicer arietinum* L. during drought stress and recovery. *Plant Physiol. Biochem.*, 2016, 99, 108–117.

Tiwari, S., Prasad, V., Chauhan, P. S., Lata, C. Bacillus amyloliquefaciens confers tolerance to various abiotic stresses and modulates plant response to phytohormones through osmoprotection and gene expression regulation in rice. *Front. Plant Sci.*, 2017, 8, 1510.

Toyama, T., Furukawa, T., Maeda, N., Inoue, D., Sei, K., Mori, K., Kikuchi, S., Ike, M. Accelerated biodegradation of pyrene and Benzo[a]Pyrene in the phragmites australis rhizosphere by bacteria–root exudate interactions. *Water Res.*, 2011, 45(4), 1629–1638.

Turhadi, T., Hamim, H., Ghulamahdi, M., Miftahudin, M. Iron toxicity-induced physiological and metabolite profile variations among tolerant and sensitive rice varieties. *Plant Signal. Behav.*, 2019, 14(12), 1682829.

Verma, V., Ravindran, P., Kumar, P. P. Plant hormone-mediated regulation of stress responses. *BMC Plant Biol.*, 2016, 16(1), 86. https://doi.org/10.1186/s12870-016-0771-y.

Vinocur, B., Altman, A. Recent advances in engineering plant tolerance to abiotic stress: Achievements and limitations. *Curr. Opin. Biotechnol.*, 2005, 16(2), 123–132.

Wang, C., Zhang, S. H., Wang, P. F., Qian, J., Hou, J., Zhang, W. J., Lu, J. Excess Zn alters the nutrient uptake and induces the antioxidative responses in submerged plant hydrilla verticillata (L.f.) royle. *Chemosphere*, 2009, 76(7), 938–945.

Wang, L.-F., Li, T.-T., Zhang, Y., Guo, J.-X., Lu, K.-K., Liu, W.-C. CAND2/PMTR1 is required for melatonin-conferred osmotic stress tolerance in arabidopsis. *Int. J. Mol. Sci.*, 2021a, 22(8), 4014.

Wang, S., He, T., Xu, F., Li, X., Yuan, L., Wang, Q., Liu, H. Analysis of physiological and metabolite response of *Celosia Argentea* to Copper Stress. *Plant Biol.*, 2021b, 23(2), 391–399.

Wang, Y., Ohara, Y., Nakayashiki, H., Tosa, Y., Mayama, S. Microarray analysis of the gene expression profile induced by the endophytic plant growth-promoting rhizobacteria, *Pseudomonas Fluorescens* FPT9601-T5 in *Arabidopsis*. *Mol. Plant-Microbe Interactions®*, 2005, 18(5), 385–396.

Wang, Y., Xu, L., Shen, H., Wang, J., Liu, W., Zhu, X., Wang, R., Sun, X., Liu, L. Metabolomic analysis with GC-MS to reveal potential metabolites and biological pathways involved in Pb & Cd stress response of radish roots. *Sci. Rep.*, 2015, 5(1), 18296.

Wang, Y., Xu, L., Tang, M., Jiang, H., Chen, W., Zhang, W., Wang, R., Liu, L. Functional and integrative analysis of the proteomic profile of radish root under Pb exposure. *Front. Plant Sci.*, 2016, 7, 1871.

Wani, S. H., Kumar, V., Khare, T., Tripathi, P., Shah, T., Ramakrishna, C., Aglawe, S., Mangrauthia, S. K. miRNA applications for engineering abiotic stress tolerance in plants. *Biologia (Bratisl.)*, 2020, 75(7), 1063–1081.

Wu, H., Liu, X., Zhao, J., Yu, J., Pang, Q., Feng, J. Toxicological effects of environmentally relevant lead and zinc in halophyte Suaeda salsa by NMR-based metabolomics. *Ecotoxicology*, 2012, 21(8), 2363–2371.

Xie, M., Chen, W., Lai, X., Dai, H., Sun, H., Zhou, X., Chen, T. Metabolic responses and their correlations with phytochelatins in amaranthus hypochondriacus under cadmium stress. *Environ. Pollut.*, 2019, 252, 1791–1800.

Xie, Y., Fan, J., Zhu, W., Amombo, E., Lou, Y., Chen, L., Fu, J. Effect of heavy metals pollution on soil microbial diversity and bermudagrass genetic variation. *Front. Plant Sci.*, 2016, 7, 755.

Xiong, L., Schumaker, K. S., Zhu, J.-K. Cell signaling during cold, drought, and salt stress. *Plant Cell*, 2002, 14(suppl 1), S165–S183.

Xu, J., Yang, L., Wang, Z., Dong, G., Huang, J., Wang, Y. Toxicity of copper on rice growth and accumulation of copper in rice grain in copper contaminated soil. *Chemosphere*, 2006, 62(4), 602–607.

Yadav, B., Chhaya; Dubey, R., Gnanasekaran, P., Narayan, O. P. OMICS approaches towards understanding Plant's responses to counterattack heavy metal stress: An insight into molecular mechanisms of plant defense. *Plant Gene*, 2021, 28, 100333.

Yadav, S. K. Heavy metals toxicity in plants: An overview on the role of glutathione and phytochelatins in heavy metal stress tolerance of plants. *South Afr. J. Bot.*, 2010, 76(2), 167–179.

Yang, Y., Ding, J., Chi, Y., Yuan, J. Characterization of bacterial communities associated with the exotic and heavy metal tolerant wetland plant spartina alterniflora. *Sci. Rep.*, 2020, 10(1), 17985.

Ye, Y., Li, Z., Xing, D. Nitric oxide promotes MPK6-mediated Caspase-3-like activation in Cadmium-induced *Arabidopsis Thaliana* programmed cell death. *Plant Cell Environ.*, 2013, 36(1), 1–15.

Yoshida, T., Mogami, J., Yamaguchi-Shinozaki, K. ABA-dependent and ABA-independent signaling in response to osmotic stress in plants. *Curr. Opin. Plant Biol.*, 2014, 21, 133–139.

Zeng, W., Mostafa, S., Lu, Z., Jin, B. Melatonin-mediated abiotic stress tolerance in plants. *Front. Plant Sci.*, 2022, 13, 847175.

Zhang, Y., Wang, Y., Ding, Z., Wang, H., Song, L., Jia, S., Ma, D. Zinc stress affects ionome and metabolome in tea plants. *Plant Physiol. Biochem.*, 2017, 111, 318–328.

Zhao, L., Huang, Y., Paglia, K., Vaniya, A., Wancewicz, B., Keller, A. A. Metabolomics reveals the molecular mechanisms of copper induced cucumber leaf (*Cucumis Sativus*) Senescence. *Environ. Sci. Technol.*, 2018, 52(12), 7092–7100.

Zhou, X., Joshi, S., Patil, S., Khare, T., Kumar, V. Reactive oxygen, nitrogen, carbonyl and sulfur species and their roles in plant abiotic stress responses and tolerance. *J. Plant Growth Regul.*, 2022, 41(1), 119–142.

Zhou, X., Ma, Y., Sugiura, R., Kobayashi, D., Suzuki, M., Deng, L., Kuno, T. MAP Kinase (MAPKKK)-Dependent and -independent activation of Sty1 stress MAPK in fission yeast*. *J. Biol. Chem.*, 2010, 285(43), 32818–32823.

Zhou, Z. S., Huang, S. Q., Guo, K., Mehta, S. K., Zhang, P. C., Yang, Z. M. Metabolic adaptations to mercury-induced oxidative stress in roots of *Medicago sativa* L. *J. Inorg. Biochem.*, 2007, 101(1), 1–9.

Zhu, J.-K. Abiotic stress signaling and responses in plants. *Cell*, 2016, 167(2), 313–324.

CHAPTER 13

Analytical Methods for Metabolomics: Mass Spectrometry

AHMED H. EL-DESOKY[1*], AHMED M. TAWEELA[1], KAMEL MAHMOUD[2], and ELSAYED A. OMER[3]

[1]Pharmacognosy Department, Institute of Pharmaceutical and Drug Industries Research, National Research Centre (NRC), Dokki, Giza, Egypt

[2]Pharmacognosy Department, College of Pharmaceutical Sciences and Drug Manufacturing, Misr University for Sciences and Technology (MUST), City, Egypt

[3]Medicinal and Aromatic Plants Research Department, Pharmaceutical and Drug Industries Institute, National Research Centre (NRC), Dokki, Giza, Egypt

*Corresponding authors

ABSTRACT

Metabolomics is one of omics aspects that can give a comprehensive description of metabolites in biological samples, including plants, plant extracts, fractions, headspace, exudates, or volatile oils. It is a powerful tool to describe variable states of plant cells or products under different conditions, such as normal and stress conditions. One of the most powerful tools for metabolomics profiling is mass spectroscopy, either being directly implemented to the plant sample or being coupled with chromatographic tools such as gas (gas chromatography-mass spectrometry) or liquid (liquid chromatography-mass spectrometry) chromatography. Metabolomic analyses allow the identification of a wide diversity of chemical scaffolds, from inorganic compounds to lipoidal matter, phenolic compounds, organic and amino acids, alkaloids, carbohydrates, glycosides, and a range of other small molecules such as terpenes. Plant samples are, in most cases, complex matrices of compounds with variable physicochemical properties regarding their polarity, boiling points, and molecular weights. This complexity creates enormous challenges to the analytical methodologies used in current plant metabolomics. Thus, reliable analytical tools for the separation and identification of such complex mixtures are needed. Analytical platforms using mass spectrometry (MS) provide such tools to profile stress-responsive metabolites to facilitate the output of biotechnological procedures used in developing stress-tolerant plants. In this chapter, we discussed the main techniques using MS in

Exploring Plant Stress Tolerance: A Metabolomics Approach.
Shuvasish Choudhury, Debojyoti Moulick, and Mirza Hasanuzzaman (Eds.)

DOI: 10.1201/9781779641731-13

the workflow of plant metabolomics under stress conditions, to spotlight guidelines for metabolomic analysis workflows.

13.1 GENERAL INTRODUCTION

13.1.1 METABOLOMICS AND MASS SPECTROMETRY (MS)

Metabolomics is a rapidly evolving field within the broader discipline of systems biology, focusing on the comprehensive study of small molecules, known as secondary metabolites, present in biological systems. These metabolites serve as crucial indicators of cellular activities, reflecting the dynamic interplay of biochemical processes within living organisms. Analyzing the metabolome, the complete set of metabolites in a biological sample, provides valuable insights into the physiological and pathological states of cells, tissues, and organisms (Schmitt-Kopplin, 2021).

One of the key analytical techniques at the forefront of metabolomics research is MS. MS enables the identification and quantification of metabolites by measuring the mass-to-charge (*m*/*z*) ratio of ions generated from a sample. This powerful analytical tool has gained prominence in metabolomics due to its high sensitivity, specificity, and capability to analyze a diverse range of metabolites in complex biological matrices (Noureddine Benkeblia, 2023).

13.1.2 COUPLING MASS SPECTROSCOPY WITH CHROMATOGRAPHIC SEPARATION

The application of MS in metabolomics involves several distinct steps, each contributing to the comprehensive characterization of the metabolome. Sample preparation is a critical stage where biological samples are extracted, purified, and often derivatized to enhance the detection of specific metabolites. Following this, the mass spectrometer is employed to ionize the metabolites, generating ions that are subsequently separated based on their *m*/*z* ratios. Detection and quantification are achieved through the measurement of ion intensities, and the resulting data are processed and analyzed using sophisticated computational tools (Salem et al., 2020).

MS techniques commonly employed in metabolomics include gas chromatography-mass spectrometry (GC-MS) and liquid chromatography-mass spectrometry (LC-MS). GC-MS is particularly well-suited for volatile and thermally stable metabolites (Fiehn, 2016), while LC-MS accommodates a broader range of metabolite classes, including polar and nonpolar compounds (Gowda and Djukovic, 2014). Additionally, advancements in high-resolution MS have enhanced the precision and accuracy of metabolite identification, allowing researchers to unravel the intricacies of the metabolome with unprecedented detail (Ren et al., 2018).

The integration of MS with other analytical technologies, such as nuclear magnetic resonance (NMR) spectroscopy, chromatography, and various data analysis approaches, further enhances the depth and breadth of metabolomics studies (Danzi et al., 2023). The continuous refinement of MS methodologies, coupled with advancements in data processing and interpretation, continues to drive innovation in metabolomics research, offering new opportunities for understanding biological systems at the molecular level (Salem et al., 2020).

Therefore, MS has emerged as a cornerstone in the field of metabolomics, enabling researchers to explore and decipher the complex landscape of metabolites in biological samples. The ongoing development of analytical methods and instrumentation holds promise for expanding our understanding of metabolic processes.

13.1.3 SIGNIFICANCE OF HIGH-RESOLUTION MS

Metabolic investigations using high-resolution MS, regardless of the chromatographic technique used, yield a significant amount of data in a single run. The most difficult part of metabolomic processes is processing and analyzing these massive datasets, since it requires both peak identification and the identification of the quantitative relationships between these compounds. Metabolomics experiment designs emphasize either qualitative or quantitative information, or both, depending on the goal of the investigations. Authentic or isotope-labeled standards are not always available for optimizing MS parameters. For example, traditional targeted metabolomics using ion transition monitoring (i.e., selective reaction monitoring experiments) focused on the quantization of metabolites. Absolute quantification is typically performed by isotope-labeled internal standards or a similar molecule in terms of physicochemical properties. The boundaries between qualitative and quantitative analyses are becoming less clear in the new high-resolution MS paradigm due to advancements in mass spectrometer devices and data processing algorithms. High-quality quantitative and qualitative data can be simultaneously collected using the many data gathering methods that high-resolution mass spectroscopy offers (full scan, data-independent acquisition, and data-dependent acquisition, for example). To achieve experiment goals like biomarker identification, metabolite profiling, or diagnostics, the data treatment and processing will be classified between quantitative and qualitative determination (i.e., relative, absolute, or semiquantitative analysis) (Wang et al., 2019).

Both the methods for the investigation of plant metabolites and the applications of plant metabolomics are extremely varied. Metabolomics studies have made use of hyphenated techniques like capillary electrophoresis mass spectrometry, direct injection mass spectrometry, NMR, high-performance liquid chromatography (HPLC) with photodiode array detection, and thin-layer chromatography with ultraviolet (UV) detection, among others. Among these approaches, GC-MS and LC-MS are the most frequently used ones, followed by matrix-assisted laser desorption/ionization-mass spectrometry (MALDI-MS). GC-MS is a conventional analytical technique used for the study of polar, volatile, thermally stable, low-molecular-weight chemicals. This is mainly due to the predictability and reproducibility of fragmentation patterns and the availability of databases. Here, we would like to highlight some of the most recent developments in plant metabolomics research, including GC-MS, LC-MS, and MALDI-MS, which are current methods for the investigation of both polar and nonpolar molecules. Given that electro spray ionization is the most often used ionization source in these techniques, particular attention will be paid to the constraints (Ernst et al., 2014).

13.2 MASS SPECTROSCOPY

13.2.1 MODES OF IONIZATION IN MASS SPECTROSCOPY

MS is an analysis technique that separates ions depending on their *m/z* ratio. This analysis has two steps: creating ions in the gas state and then separating and analyzing them. Converting analytes to ions in the gas state is performed by several techniques like electron ionization (EI), chemical ionization (CI), electrospray ionization (ESI), photoionization, plasma, and matrix-assisted laser desorption. The second step of analyzing them is usually done by several different types of mass analyzers such as magnetic sectors, quadrupoles, time of flight (TOF), ion traps, Orbitrap, and ion cyclotron resonance.

Two of these mass analyzers, whether identical or different, could be combined in a series to perform tandem mass spectroscopy (MS/MS) (Thomas et al., 2022).

13.2.2 TANDEM MASS SPECTROSCOPY

MS/MS could be identical or hybrid. It could be composed of two or more identical mass analyzers like a triple quadrupole analyzer. While a hybrid type consists of two or more different types of mass analyzers like a quadrupole combined to TOF, or a quadrupole combined with a linear ion trap to yield Q-Trap (Thomas et al., 2022).

In case of LC-MS, analytes eluted by chromatographic mobile phase are carried to mass spectrometer's ion source, producing gas phase ions. Such ions are first filtered by the first mass analyzer, and only precursor ions are allowed to proceed into the collision cell. Then, fragmentation of precursor ions happens and product ions are finally analyzed in the tandem mass spectrometer (Thomas et al., 2022).

13.2.3 IONIZATION TECHNIQUES

Converting analytes to ions in the gas state is performed by several techniques. The categorization of these techniques depends on the method of transferring analytes to the gas state, ranging from evaporation to nebulization to laser desorption. The chemical and physical nature of the targeted analytes and their matrices significantly determine the technique of their vaporization and ionization (Bhardwaj and Hanley, 2014).

13.2.3.1 ELECTRON IMPACT (EI)

This type of ionization technique is the most widely used in organic MS. Analytes that best ionize by this type are usually volatile and thermally stable, as these analytes are heated mildly and evaporated to form gaseous neutrals, which are then ionized by collisions with 70 eV kinetic energy electrons to form radical cations, Mc^{+}.

Due to its highly reproducible results even in different instruments, it participated significantly in creating mass spectral libraries of high quality that are employed extensively in the identification of known or even new unknown compounds (Bhardwaj and Hanley, 2014).

13.2.3.2 CHEMICAL IONIZATION

The CI technique depends on using reagent gases as intermediates in the ionization process, unlike EI, where the analytes are bombarded by accelerated electrons. Examples of reagent gases widely employed in CI are ammonia, methane, water, and isobutene (Mellon, 2003; Saitman, 2019).

In this technique, reagent gas ions collide with the analytes to produce ions characteristic of the analyte. This indirect ionization provides lower energy impact than direct ionization, hence it is called "soft ionization"; in which product ions can be produced either positively or negatively charged.

Reagent gases like ammonia generate NH_{4+} ions. These ions generally react with analyte molecules (M) by protonation to give positively charged ions of the analytes.

$$4M + NH_{4+} \rightarrow MH+ + NH_3.$$

While reagent gas of water generates OH^- ions and reacts by proton abstraction to give negatively charged ions of analytes.

$$5M + OH- \rightarrow M-H-+H_2O.$$

13.2.3.3 ELECTROSPRAY IONIZATION

ESI is a soft ionization technique like CI that involves ionization from liquid droplets rather than in the gas phase (Bhardwaj and Hanley, 2014).

In ESI, the sample solution passes through a capillary tube under a strong electric field. At the end of the capillary, the solution is released as a micron or submicron charged droplets that pass through a curtain of heated inert gas to cause collisions resulting in the formation of ions (Bhardwaj and Hanley, 2014).

ESI can produce both protonated molecules in positive mode and deprotonated molecules in negative mode. It can also form multiply charged ions, enabling the study of large molecules at lower *m/z* values, making it easier for their mass analysis (Bhardwaj and Hanley, 2014).

ESI is one of the softest ionization methods available nowadays, as its resulting fragmentation is minimal, which makes it ideal for highly polar, least volatile, and thermally unstable compounds. It is very simple to determine the molecular mass of compounds by it, as fragmentation of most compounds stops at protonation or deprotonation of compounds without subsequent fragmentation to further ions.

The three main techniques using MS to study MS-based metabolomics in plants under stress conditions will be discussed in detail in the following parts.

13.3 MATRIX-ASSISTED LASER DESORPTION/IONIZATION-TIME OF FLIGHT (MALDI-TOF) MS IMAGING

13.3.1 INTRODUCTION

This imaging technique has emerged as a powerful tool in the field of plant science (Etalo et al., 2015). It has proven to be a versatile tool across various plant species, offering valuable insights into metabolite distribution (Sturtevant et al., 2016). While specific findings vary between species, the technique consistently contributes to a better understanding of plant biology and holds promise for further advancements in the field (Sturtevant et al., 2016). Moreover, MALDI-TOF MS imaging is nondestructive, preserving the sample for additional analyses. This is advantageous when limited amounts of plant material are available (Kaspar et al., 2011). At its essence, MALDI-TOF MS is rooted in the principles of MS (Dreisewerd, 2003). The process begins with the application of a matrix material onto the sample of interest, which can range from biological tissues to thin sections of plant specimens. This matrix plays a pivotal role in the ionization process during laser irradiation. Upon exposure to a laser beam [typical N_2 and Nd:YAG laser with UV wavelengths at 337, 355, or 266 nm and/or infrared laser at 2.94 μm or 10.6 μm], the matrix absorbs energy and facilitates the desorption and ionization of analytes present in the sample (Cha and Yeung, 2007). These ionized molecules are then accelerated through a flight tube by an electric field (Stoeckli et al., 2001). The time taken by ions to travel through this tube is directly proportional to their *m/z* ratio, forming the basis for the TOF measurement (Dreisewerd, 2003). The instrumental setup includes a two-dimensional stage that allows the systematic rostering of the laser beam across the sample (Jurchen et al., 2005). As the laser pulses hit different positions, mass spectra are generated, capturing the molecular composition at each specific location (Jurchen et al., 2005). The culmination of these individual spectra results in a comprehensive image (Jurchen et al., 2005). This allows the visualization of gradients across the tissue for individual analytes using specialized computer programs (Cha and Yeung, 2007). The versatility of MALDI-TOF MS imaging extends its applications across various scientific disciplines (Lanekoff et al., 2012). In the realm of biology and medicine, it enables the visualization of biomolecules in tissues, providing insights into disease pathology, drug distribution, and cellular processes (Seeley et al., 2008). Moreover, the application of MALDI-TOF MS imaging across diverse plant species has significantly expanded our understanding of plant metabolism, specialized metabolite distribution, and ecological interactions (Cha and Yeung, 2007).

13.3.2 PROCEDURE OVERVIEW

13.3.2.1 SAMPLE PREPARATION

Preparing samples presents a specific challenge due to the fact that optimal ionization, especially for visualizing certain compound classes, depends on specific conditions. While performing tissue sectioning, the goal is to slice through most cells, revealing their contents for either direct ionization or facilitating cocrystallization with matrices that aid in ionization (Goodwin, 2012). The sample typically consists of microtome tissue sections prepared in a cryostat-microtome, affixed to the target plate and covered with a matrix compatible with the molecular species targeted for detection, usually proteins/peptides or metabolites (Caprioli et al., 1997). The target plate, moved within the MALDI-TOF instrument source, is irradiated by a fixed laser beam, generating a mass spectrum specific to the laser-irradiated region (Cornett et al., 2007). The dataset can be presented as an average spectrum, and images of individual ions or multiple ions' *m/z* values can be extracted, illustrating their spatial distribution based on signal intensities (intensity map) (Cha, and Yeung, 2007; Stoeckli et al., 2001). Diligent sample treatment is crucial to ensure signal quality and prevent lateral displacement of analytes. Artifacts can potentially occur at any point between sample collection and mass spectrometry imaging (MSI) analysis (Schwartz et al., 2003; Chaurand et al., 2006). The optimization of sample preparation tends to be somewhat dependent on the sample type, given inherent differences in the composition of plant tissues. Contamination with substances that can interfere with MS analysis poses a significant risk during the preparation of tissue sections (Goodwin, 2012). Although optimal cutting temperature (OCT) media is often employed to facilitate cryo-sectioning of fragile and small tissues, they cannot be used for matrix-assisted laser desorption/ionization-mass spectrometry imaging (MALDI-MSI) due to their ready ionization, acting as ion suppressors (Cha and Yeung, 2007). Therefore, it is preferable to treat only the base of the tissue with a minimal amount of OCT media, carefully avoiding contamination of the sectioned region. Sectioning poses challenges for leaves, pollen, and grains, and efforts are made to optimize sample preparation. Kawamoto introduced an adhesive tape method to address these challenges, allowing the preparation of sections from various samples with reduced distortion and dislocation, as seen in the analysis of rice grains (Kawamoto, 2003). Besides that, the thickness of the section can impact data quality. While early MALDI-MSI users worked with relatively thick sections (50–100 μm), recent applications have reduced this to less than 10 μm, which appears to optimize the signal-to-noise ratio (Caprioli et al., 1997; Stoeckli et al., 2001; Sugiura et al., 2006). Sections can be attached to the target plate using adhesive tape or thaw-mounting, the latter requiring a brief warming of the target plate's back to create a localized warm path (Chaurand et al., 2004).

13.3.2.2 POSTSECTIONING SAMPLE HANDLING

13.3.2.2.1 Washing and Drying

After sections have been cut and placed on slides, they can undergo immediate further processing or be stored in the freezer. However, long-term storage on slides may not be advisable. When preparing to resume the analysis of a sample, there are several on-target processing options available that can significantly enhance the obtained results. Certain processing methods involve eliminating ionization-suppressing species (such as salts and lipids) from the samples, while other options involve adding compounds to modify the sample (such as enzymes or derivatizing agents) or, most importantly, aid ionization (such as matrix application). Upon retrieval of a sample from the freezer or transfer on dry ice from the cryostat microtome, researchers typically address sample instability by employing various drying methods. Commonly used techniques include freeze-drying and vacuum desiccation, dehydration by solvent washing, and air-drying under a stream of nitrogen (Chaurand et al., 2008; Goodwin et al., 2008, 2011, Goodwin and Pitt, 2010a). Tissue washing is a crucial processing stage that requires careful consideration, as it has the potential to significantly impact the analyte composition within the sample. In addition to dehydrating the tissue, washing with organic solvents serves as a means of removing ionization-suppressing small molecules and lipids (Seeley et al., 2008). The choice of a washing protocol is determined by the target analyte and should be avoided if small-molecule drugs or metabolites are the intended target unless their specific solubility in different solvents is known (Seeley et al., 2008). It is recommended that any sample washing be performed with warm solutions. Researchers should be aware that when washing tissues, there is a risk of residue streaking on the target, which may also adhere to adjacent samples, so precautions should be taken to limit such a risk (Angel et al., 2012).

13.3.2.2.2 Enzymatic Digestion or Chemical Derivatization

Sensitivity becomes a challenge when attempting to directly detect intact proteins from tissue sections. Currently, the upper limit for reasonably sensitive detection is approximately 25–30 kDa. Nonetheless, researchers can indirectly obtain information about the abundance and distribution of larger proteins by treating tissues with enzymes, typically trypsin. This enzymatic treatment cleaves the proteins into peptides that can be easily ionized, detected, and subsequently identified (Caprioli et al., 1997).

Chemical derivatization facilitates the enhanced detection of compounds, ranging from small molecules to glycopeptides, which may be challenging to ionize or detect due to the presence of endogenous compounds of similar mass (Liu et al., 2010; Ullmer et al., 2006). Methods involving on-tissue sample derivatization have also been employed to improve protein identification through MALDI MSI (Franck et al., 2009). However, like any sample processing method, there is a potential risk that the derivatization solution might lead to analyte suppression or delocalization. A recent instance of on-tissue derivatization involves the use of reagent precoated targets for the detection of antituberculosis drugs via MALDI MSI (Manier et al., 2011).

13.3.2.2.3 Application of the Matrix

In matrix application, the fundamental approach involves applying the matrix using a rapidly evaporating solvent solution (including 0.1–1% trifluoroacetic acid, TFA for positive ion analysis (Schwartz et al., 2003). Extracting the maximum amount of analyte from the sample achieves proper crystallization of the matrix and minimizes analyte diffusion. The selection of the matrix depends on each experiment's mass range. The ideal matrix should possess several characteristics, including the formation of micron-sized crystals, uniform coating across the sample, minimal volatility under high vacuum conditions (approximately 10^{-7} mbar), strong chemical stability, a low generation of matrix-related ions, efficient matrix utilization, a low threshold of laser desorption/ionization power, and enhanced safety during handling (Wang et al., 2013). However, there are no fixed guidelines determining which matrix will yield the most precise results for a given analyte. Consequently, it is recommended to assess all matrices, considering both positive and negative ionization modes. Various solvent combinations involving acetonitrile, ethanol, methanol, and isopropanol in different proportions with water have been explored in solution. Nevertheless, a 50:50 ratio with 0.1% TFA is commonly considered a suitable starting point for experimentation and optimization concerning target analytes (Schwartz et al., 2003).

The exploration and identification of novel MALDI matrices constitute a crucial dimension in the realm of mass spectrometric imaging. Currently, compounds like 9-aminoacridine (9-AA), paranitroaniline (pNA), and 2,5-dihydroxyacetophenone stand as popular choices for MALDI imaging of low-molecular-weight compounds in the negative ion MS detection mode (Shroff et al., 2015; Steven et al., 2013; Tarzi et al., 2013). Specifically, in the positive ion mode, α-cyano-4-hydroxycinnamic acid (CHCA) has gained widespread usage for lipid detection (Cillero-Pastor and Heeren, 2014). However, its applicability in tissue imaging is constrained by the prevalent background signals associated with the matrix. Another matrix, 2,5-dihydroxybenzoic acid (DHB), finds extensive application in imaging for both positive and negative ion modes (Goodwin et al., 2008; Goodwin et al., 2010b). Nevertheless, challenges arise from nonhomogeneous crystallization and the large crystal size of DHB, leading to molecular delocalization and suboptimal spot-to-spot reproducibility. Through spray matrix deposition, 2-mercaptobenzothiazole (2-MBT) and dithranol have emerged as promising matrices for MALDI imaging (Zemski Berry et al., 2019). pNA has demonstrated effectiveness in MALDI tissue imaging of endogenous lipids in the positive ion mode, particularly on intermediate pressure MS systems (Steven et al., 2013). However, its volatility poses limitations when applied to high-vacuum MALDI sources, which are customary for Fourier-transform ion cyclotron resonance MS instruments. Recent studies have demonstrated the efficacy of colloidal graphite and colloidal silver for the detection of small molecular weight compounds in plants (Zhang et al., 2007; Jun et al., 2010). Furthermore, the characteristics of hydroxyflavone matrices, including quercetin and morin, establish them as a valuable, plant-derived family of matrices for metabolomics investigations, offering hypotoxic attributes (Wang et al., 2013).

13.3.2.2.4 Data handling and processing

Advancements in imaging technology highlight the necessity for improved computational and statistical tools to analyze data. Several MALDI imaging software packages have been developed, some of which are open-source such as BioMap and MMSITE and commercial packages like Flex Imaging (Bruker, Bremen, Germany) and Tissue View (Applied Biosystems, Darmstadt, Germany). The commercial packages are designed to process data from specific MS devices, ensuring better performance, but they may not be compatible with data from nonrecommended MS instruments. A standardized data format within the

COMPUTIS consortium, called imzML, facilitates the exchange of imaging MS data between different instruments and analysis software. Raw or processed spectra, or lists of peaks, can be loaded and read. Initial data inspection involves examining an overall average spectrum of the complete data set, and specific mass signals are selected for intensity distribution analysis. Statistical comparisons between regions of interest, individual samples, or sample groups can then be performed (Kaspar et al., 2011).

13.3.2.3 MALDI-TOF MS IMAGING OF PLANT METABOLITES

The extensive specialization in the metabolism of plant tissues gives rise to highly varied distributions of metabolites. Examples include the distinctive pigmentation patterns of flowers that attract pollinators and the accumulation of defense compounds in vulnerable tissues (Martins et al., 2013; Halkier and Møller, 1989). The exploration of these metabolites has relied for a long time on extracting the entire plant organ, chromatographic separation, and chemical analyses. Mostly, these studies have been guided by specific molecular characteristics of these metabolites, where the details of the metabolite distributions are lost (Buschmann, 2000). On the other hand, imaging techniques that depend on functional group histochemistry or specific staining lack specificity for individual compounds may cause dislocation or distortions of localization within tissues (Satyan et al., 2010; Sánchez-Pérez, et al., 2009).

At the cellular level, some compounds such as terpenoids and alkaloids are known to accumulate in single or few cells or specialized cells interspersed among others in a given tissue. Hydroxynitrile glucosides, defense compounds present in barley, are exclusively found in the single cell layer of the epidermis. Similarly, in various plants, hydroxynitrile glucosides (commonly known as cyanogenic glucosides) are stored in the vacuoles of cells (Nielsen et al., 2002; Gruhnert et al., 1994). Obtaining such results has involved time-consuming procedures to isolate specific tissues, cells, or compartments for subsequent chemical analysis. With MSI, a single analysis can provide a comprehensive image of the metabolic status in the investigated tissue, offering information on various compounds.

Furthermore, the utilization of MS imaging across a variety of plant species has greatly enhanced our comprehension of plant metabolism (Cha and Yeung, 2007). The biosynthetic pathway of monoterpenoid indole alkaloids in Madagascar periwinkle (*Catharanthus roseus*) is distributed among a minimum of three distinct cell types and five subcellular compartments. These findings predominantly stem from investigations into the localization and *in vitro* activities of the relevant enzymes (Facchini and St-Pierre, 2005). As a result, gaining deeper insights into the specifics of plant metabolism centers on understanding of the subcellular localization of metabolites, the substrates, and products of enzymes existing in diverse cell compartments.

Secondary metabolites, vital for plant health, are distributed throughout the plant, and their spatial distribution patterns, along with primary metabolites, have not been extensively documented using MSI. While numerous metabolites were effectively identified, the study faced a common challenge of being unable to attribute a mass to them. Given the belief that plants produce over 200,000 metabolites (Wink, 2003; Fiehn, 2002), and considering that

MALDI-MSI can only resolve a few hundred masses, the straightforward assignment of mass proves insufficient for unambiguous identification. Additionally, different components can be distinguished using reverse ion reflector modes, either in the positive or negative mode. For example, the negative ion reflector mode is effective for identifying phenolic acids and flavonoids, while the positive mode is well-suited for detecting alkaloids and amino acids (Liu et al., 2012; Sun et al., 2020). Here, we highlight some of the applications of MALDI-MS imaging in plant tissues (Table 13.1) for secondary metabolites localization.

13.3.2.4 CHALLENGE AND FUTURE

As with any cutting-edge technology, MALDI-TOF MS imaging comes with both opportunities and challenges. Advances in matrix development, laser technology, and data analysis methods continue to enhance the capabilities of the technique. Challenges, such as matrix interference and the complexity of data interpretation, drive ongoing research to refine and optimize MALDI-TOF MS imaging protocols. Looking ahead, the future of MALDI-TOF MS imaging holds promise for even greater integration with complementary techniques such as confocal microscopy, to push the boundaries of spatially resolved secondary metabolites analysis.

13.4 LIQUID CHROMATOGRAPHY-MASS SPECTROMETRY

Liquid chromatography is a technique for the separation of analytes depending on the differential equilibrium of substances between the mobile phase and the stationary phase. This differential equilibrium resulted from interactions of analytes with both phases; these interactions are based on multiple mechanisms like polarity, ionic interaction, and size exclusion (Thomas et al., 2022).

13.4.1 MODES OF LIQUID CHROMATOGRAPHY

Liquid chromatography could be normal-phase or reversed-phase chromatography, depending on the differential polarity of both phases. The mobile phase is less polar than the stationary phase in normal-phase chromatography; this allows more polar analytes to attach to the stationary phase while the nonpolar analytes elute first. While in reversed phase chromatography, which is the most commonly used type, the opposite situation applies, so the nonpolar analytes attach to the stationary phase until the more polar analytes pass through (López-Fernández et al., 2020).

13.4.1.1 HPLC

HPLC is the most reliable and adaptable separation technique for the routine analysis of simple and complex samples of pharmaceuticals, biological, clinical, and natural products.

TABLE 13.1 Examples of the Utilization of MALDI-MSI in Examining Plant Metabolites

Plant Name	Organ	Matrix Type	Compounds Investigated	References
Accent Grape	Fruit	1,5-Diaminonaphthalene (DAN)	Anthocyanins	Berisha et al. (2014)
Allium macrostemon Bge. and *Allium chinense* G. Don	Fresh bulbs	DHB	Saponins	Duan et al. (2022)
Arabidopsis thaliana	Flowers	DHB	Various flavonoids (e.g., quercetin, kaempferol rhamnoside, isorhamnetin)	Hölscher et al. (2009)
	Leaves	9-AA	Glucosinolates	Shroff et al. (2015)
	Flowers, Siliques	9-AA	Glucosinolates	Sarsby et al. (2012)
Arachis hypogaea (peanut)	Skin	matrix-free	Aflatoxins and stilbenoids	de Oliveira et al. (2014)
Arctium lappa	Root	DHB	caffeoylquinic acids and flavonoids	Li et al. (2022)
Atropa belladonna (Deadly Nightshade)	Berries	4-maleicanhydridoproton sponge	Tropane alkaloids (e.g., Atropine, scopolamine)	Corinti et al. (2018)
Braeburn apple	Hypanthium, skin	Norharmane	Dihexose, Glycoside quercitin pentoside	Soltwisch et al. (2015)
Catharanthus roseus	Dark pink and white petals	1:1 mixture of DHB and α-cyano-4-hydroxycinnamic acid	Vinca alkaloids	Dutkiewicz et al. (2021)
Citrus sinensis (Orange)	Leaves and stem	DHB and α-cyano-4-hydroxycinnamic acid	Hesperixdin, rutin	Soares et al. (2015)
Cryptocarya pulchrinervia	Leaves	CHCA	cryptobrachytone C	Ratnasari et al. (2021)
Eucalyptus globulus, E. grandis	Stem	Silica	Lignin	Araújo et al. (2014)
Fragaria ananassa (Strawberry)	Berries	DHB	Citric acid, sugars, and anthocyanins	Wang et al. (2021)
Ginkgo biloba (Ginkgo)	Leaves	CHCA	Cyclodimeric flavonoids (biginkgoside) Besides, ginkgolides, saccharides, various phospholipids, and chlorophyll a	Li et al. (2018)

TABLE 13.1 *(Continued)*

Plant Name	Organ	Matrix Type	Compounds Investigated	References
Gliricidia sepium	Leaves	α-cyano-4-hydroxycinnamic acid, DHB, and 2-mercaptobenzothiazole	Kaempferol, isorhamnetin, and some fatty acids	Hertel Pereira et al. (2022)
Glycyrhiza glabra	Rizhome	DHB	Flavonoids and saponins	Li et al. (2014a)
Gossypium hirsutum (cotton)	Seed	DHB	Polyphenols	Horn and Chapman (2014)
Golden Delicious Apple	Hypanthium, peel, pericarp	CHCA	Flavonols, Dihydrochalcones	Franceschi et al. (2012)
Helianthus annuus (Sunflower)	Leaf trichomes	DHB and CHCA (mixture 1:1)	Sesquiterpenes and polymethoxylated flavone	Brentan Silva et al. (2017)
Hordeum vulgare (Barley)	Seeds	DHB	Hordatines	Gorzolka et al. (2014)
Hypericum perforatum (St. John's Wort)	Hypericum dark glands	9-aminacridine	Naphthodianthrones (e.g., hypericin, pseudohypericin) and biflavonoids	Hölscher et al. (2009)
Hypericum perforatum, H. olympucum and *H. patulum*	Leaves	CHCA	Naphtodianthrone	Kusari et al. (2015)
Linum usitatissimum (Flax)	Intact flax (capsule and Seed)	DHB	Cyanogenic glycosides, e.g., (Linamarin, linustatin) and lignans	Dalisay et al. (2015)
Lychnophora species	Leaves	Matrix-free	Flavonoids	Silva et al. (2014)
Medicago truncatula (symbiosis with *Sinorhizobium meliloti*)	Root nodules	DHB	Amino acids and others	Gemperline et al. (2014)
Musa acuminata	Fruit (epidermis)	Matrix-free	Phenylphenalenones	Hölscher et al. (2015)
Oryza sativa (Rice)	Seeds	DHB	Lysophosphatidylcholine	Zaima et al. (2014)
	Seeds	DHB	Anthocyanins	Yoshimura et. al. (2012a)
Panax ginseng	Root	DHB and CHCA	Ginsenosides	Lee et al. (2017)
Pisum sativum (Pea)	Pea pods	DHB	Phenylpropanoids, lignans, and pterocarpans	Seneviratne et al. (2015)
Putterlickia verrucosa *Putterlickia retrospinosa*	Root	DHB	Maytansine	Kusari et al. (2014)

TABLE 13.1 *(Continued)*

Plant Name	Organ	Matrix Type	Compounds Investigated	References
Podophyllum hexandrum and *P. peltatum*	Rhizomes	CHCA, 4- chloro-α-cyano-cinnamic acid, or (E)-2-cyano-3(napthalen2-yl) acrylic acid	Alkaloid (magnoflorine)	Marques et al. (2014)
Raphanus sativus	Bulbs and leaves	CHCA and DHB	Anthocyanin and others	Seaman et al. (2014)
Salvia miltiorrhiza (Danshen)	Roots	1,5-naphthalenediamine (1,5-DAN) and 1,1′-binaphthyl-2,2′-diamine (BNDM)	Salvianolic acids, phenolic acids, amino acids, polyamines, and oligosaccharides	Sun et al. (2022)
Scutellaria baicalensis Georgi	Root	N-(1-naphthyl)-ethylenediamine dihydrochloride, 1,1′-binaphthyl-2,2′-diamine (BNDM), and 1,5-diaminonaphthalene (DAN) and 9-AA	Flavones, flavones glycosides, and phenolic acids	Sun et al. (2020)
Solanum habrochaites	Leaves	Matrix-free (carbon)	Flavonoids, sugars, and glycoalkaloids	Li et al. (2014b)
Soybean	Leaves	2D graphene	Flavonoids and others	Friesen et al. (2015)
Toxicodendron radicans(Poison ivy)	Stem	DHB	Urushiols	Aziz et al. (2017
Vaccinium ashei (Blueberry)	Berries	DHB	Anthocyanins	Yoshimura et al. (2012b)
Vitis vinifera (Grapes)	Leaves	DHB	Resveratrol, pterostilbene and Viniferins	Becker et al. (2014)
Zea mays (Corn)	Leaves	1,5-Diaminonaphthalene (DAN)	Amino acids, organic acids, glycerolipids, flavone aglycones	Korte et al. (2015)

It has been extensively employed successfully for years in the identification, quantification, and purification of large and small molecules.

The advantage of HPLC is that it offers more efficient analyte separation than conventional column chromatography as it employs smaller particle-sized stationary phases applied in small diameter columns; hence, it requires employing high-pressure pumps to inject mobile phases; hence, it is also known as high-pressure liquid chromatography and very high-pressure liquid chromatography (Durai Ananda Kumar et al., 2020).

Increasing the flow rate by pressure and decreasing the particle size provided better separation. However, the employment of pressure on the mobile phase creates back-pressure, which renders further increase of the flow rate. To overcome this limitation, ultraperformance liquid chromatography (UPLC) was developed with a shorter stationary phase column, smaller particle-sized stationary phase, and higher flow rate (Durai Ananda Kumar et al., 2020).

13.4.1.2 HPLC VERSUS UPLC

UPLC is a rapid and cost-effective liquid chromatography. It provides a flow rate ten times faster than HPLC, which results in three times better separation.

The UPLC is based on the principle of decreasing particle size of the stationary phase to less than 2 μm in comparison with 3–5 μm particle size in HPLC columns. This particle size reduction results in an increase in the surface area to a larger capacity. This allows decreasing the solvent consumed by 83% and the sample consumed by 93%. Thus, it provides higher sensitivity and faster analysis (Shalini et al., 2014).

13.4.1.3 FAST PROTEIN LIQUID CHROMATOGRAPHY (FPLC) VERSUS HPLC

The chromatography of biomolecules such as lipids, nucleic acids, and proteins is very critical and sensitive, as they cannot endure high temperatures, high pressures, or the solvents that are usually used in HPLC. For these reasons, an alternative approach to HPLC was developed: the FPLC.

While the principal aim of HPLC is the identification and qualification of small compounds ranging in size from a few atoms up to approximately 3000 Da, the principal aim of FPLC is to purify large biomolecules of several kilodaltons (kDa).

For this purpose, several changes have been made in FPLC like using buffer as eluent instead of solvents in HPLC. Also, different methods of separation have been incorporated like ion exchange, size exclusion, and affinity, instead of the reversed phase method in HPLC.

In all liquid chromatography techniques, eluted analytes have different masses; these can be converted to ions that can be separated and analyzed depending on their different m/z ratio; here comes the role of MS coupled to the liquid chromatography (Thomas et al., 2022).

13.4.2 SAMPLE PREPARATION

The extraction and characterization of different compounds in plant matrices are complex, as the compounds may be in simple form or polymerized or complexed with various other plant-matrix components, such with sugar moieties, for example. Consequently, Different methods of extraction and different solvents with gradient polarities are employed to recover them. In addition, the compounds in the plant matrix have different characteristics; some of them are photosensitive. Thus, quick extraction methods are necessary for them to avoid their degradation by light. The most widely used methods to make a successive and selective extraction of compounds from plant matrix are liquid–liquid extraction (LLE) or solid–liquid extraction (SLE) followed by selective retention of analytes (López-Fernández et al., 2020).

The plant material must be dried and grounded before extraction. For complete extraction, stirring devices are usually employed such as vortexes, orbital shakers, automatic shakers, or ultrasonic baths. The successful extraction of plant active constituents is significantly affected by multiple factors like gradient solvent polarity, contact time, pH, and temperature. These factors are constantly modified for fulfilling complete extraction according to the nature of plant material (seed, leaf, bark, root, flower, etc.) and the nature of targeted plant constituents (phenols, alkaloids, tannins, volatile oils, resins, etc.) (López-Fernández et al., 2020).

However, all extraction methods include the coextraction of excess unwanted substances, such as fats, waxes, sugars, and pigments. These impurities frequently interfere with targeted substances. Thus, for their removal, an additional step of clean-up in advance of liquid chromatography analysis is necessary. Again, the LLE and SLE are the most common clean-up techniques. The most common clean-up procedure in plant extraction is the removal of fats and waxes by using nonpolar solvents like hexane or petroleum ether (López-Fernández et al., 2020).

13.4.3 STRESS IN PLANTS

Generally, plants are susceptible to various environmental stress factors that negatively affect their growth, development, and crop productivity. These stress factors trigger the formation of excess reactive oxygen species (ROS) and nitrogen species through alteration of their metabolism. Stress factors could be biotic factors like fungi, bacteria, and animals, or abiotic factors like temperature, salinity, humidity, light, minerals, and pH (Chaki et al., 2020).

Naturally, the antioxidant defense mechanism of enzymatic and nonenzymatic antioxidants defends against these reactive species. When reactive species concentration exceeds the ability of antioxidant enzymes, it leads to oxidative stress that causes damage to the macromolecules in plants, like nucleic acids, membrane lipids, and proteins, followed by nitro-oxidative stress and plant cell death (Chaki et al., 2020).

Stress factors sometimes trigger plant defenses through enhancing the production of secondary metabolic compounds. Some researchers used this mechanism to increase the levels of production of valuable and favorable secondary metabolites. For example, some researchers succeeded in increasing levels of production of polyphenols, anthocyanins, and tannins in Tinto Cão berries, and polyphenols and tannins in Touriga Franca berries through the application of a small concentration of chitosan (Chaki et al., 2020).

13.5 GAS CHROMATOGRAPHY-MASS SPECTROMETRY

13.5.1 BACKGROUND

The advent of advanced analytical techniques has revolutionized the field of plant biology, providing scientists with powerful tools to monitor and understand various physiological processes in plants. GC-MS has emerged as a cornerstone in this regard, offering unparalleled capabilities in analyzing the chemical composition of plant tissues. The principles of GC-MS and its diverse applications in monitoring plant stress will be discussed here, by providing an overview of the principles of GC-MS, discussing the physiological responses of plants to stress, discussing the applications of GC-MS in monitoring plant stress, and highlighting case studies demonstrating the effectiveness of GC-MS in plant stress analysis.

13.5.1.1 GAS CHROMATOGRAPHY

GC separates complex mixtures of compounds based on their volatility and chemical properties. The vaporized sample is injected into a chromatographic column, where it travels through a stationary phase, separating into individual components.

13.5.1.1.1 Components of Gas Chromatograph (GC)

As GC is an analytical instrument used for separating and analyzing volatile compounds in a gas phase, it consists of several key components that work together to achieve accurate and efficient chromatographic separations. The main components of a GC include the following.

1. Injector

Injectors introduce the sample onto the chromatograph. It vaporizes the sample and injects it into the chromatographic column. There are several types of injectors, whose selection depends on the application. Common types include split/splitless injectors, which are suitable for most applications, large volume injector (LVI), on-column injectors, cool on-column injector, solvent vent injector, and programmable temperature vaporizers (Houck and Siegel, 2015).

In a split injector, only a portion of the sample is directed into the column, while the remaining portion is vented out. This is useful when the sample concentration is too high for the column or when the goal is to focus on specific components. It is suitable for applications where samples have a wide concentration range, where high sensitivity for lower concentrations is desired.

A splitless injector allows the entire sample to enter the column initially, holding the split vent closed for a specified time before opening it. This enhances the sensitivity for low-concentration components. It is ideal for trace analysis or when dealing with low-concentration samples.

On-column injectors involve directly injecting the sample onto the head of the chromatographic column, bypassing the need for a separate injection port. This is common in capillary column systems. It is used in applications where minimizing sample discrimination or achieving sharp peaks is critical.

Programmed temperature vaporizing injectors allow for precise control over the sample vaporization temperature, injection volume, and injection time. This flexibility can enhance the separation of complex samples with a variety of analytes and when optimization of injection conditions is crucial (Breuninger et al., 2023).

LVIs enable the injection of larger sample volumes onto the column, improving detection limits for trace-level analysis or to improve the detection of low concentration samples (Breuninger et al., 2023).

Cool on-column injector allows the sample to be introduced at a lower temperature than the column's initial temperature, reducing the potential for sample dissemination during injection. It is beneficial when analyzing thermally labile compounds.

Solvent vent injectors allow the injection of large volumes of sample by venting the solvent away from the column after injection. This helps to minimize the impact of high-boiling solvents on column performance. It is useful for samples dissolved in high-boiling-point solvents.

The choice of injector depends on the specific requirements of the analysis, and it is often influenced by the characteristics of the sample and the chromatographic column. Experimentation and optimization are typically required to determine the most suitable injector for a given application.

2 Column

It is the heart of the chromatograph, as it plays a crucial role in the separation of sample components. The choice of column type depends on the specific requirements of the analysis.

There are several types of GC columns including:

Capillary columns that are widely used due to their high resolution and efficiency compared to packed columns. They are long, narrow tubes coated with a stationary phase. Their high resolution qualifies them for analysis of complex mixtures, rendering them popular in a wide range of applications.

The high resolution of capillary columns can be explained in terms of the Van Deemter equation. It is a fundamental formula in GC that relates height equivalent to theoretical plates (HETP) of a chromatographic column to different factors affecting the separation efficiency. The equation is expressed as:

$$\mathrm{HETP} = A + B/u + Cu$$

where A is the Eddy diffusion term, B is the longitudinal diffusion term, u is the linear velocity of the mobile phase, and C is the mass transfer term.

Eddy diffusion is caused by the irregularities in the flow of the mobile phase through the column. Since capillary columns have a very small internal diameter, this reduces the impact of eddy diffusion. The long, narrow path helps maintain a more uniform flow, minimizing the contribution of A to plate height.

Longitudinal diffusion results from the tendency of molecules to spread out along the column axis. Thus, the small diameter of capillary columns reduces the distance over which longitudinal diffusion occurs, minimizing the impact of B on plate height.

Finally, mass transfer occurs when analyte molecules move between the mobile and stationary phases, impacting the separation efficiency. Therefore, a high surface area-to-volume ratio in capillary columns minimizes mass transfer effects, allowing efficient analyte transfer between the phases, reducing the contribution of C to plate height.

Linear velocity represents the speed at which the mobile phase travels through the column. Capillary columns often operate at lower linear velocities than packed columns. This is advantageous as it helps to reduce the impact of longitudinal diffusion, contributing to the high resolution.

Packed columns that consist of a solid support material coated with a liquid stationary phase, and are used for analyses where high resolution is not critical, and faster analysis times are desired.

Similar to capillary columns, open tubular and wall-coated open tubular are open tubular columns coated with a stationary phase on the interior wall. They are excellent for separating compounds with high efficiency and are commonly used in modern GC systems (Mikaliunaite et al., 2023).

Support-coated open tubular columns have a solid support coated with a stationary phase as a thin layer. They combine the benefits of both packed and capillary columns, providing high resolution with improved sample capacity (Mikaliunaite et al., 2023).

Porous layer open tubular columns have a porous layer that provides increased surface area for improved separation. They are ideal for analyzing gases and volatile compounds due to their high permeability (Mikaliunaite et al., 2023).

Chiral columns contain a chiral stationary phase designed to separate enantiomers (stereoisomers). They are used for analyzing chiral compounds, where the stereochemistry is crucial (Wang et al., 2023).

Wide-bore columns are columns with a larger internal diameter, allowing for the analysis of higher-molecular-weight compounds and suited for the analysis of heavy hydrocarbons and larger molecules (Corbally et al., 2023).

On the other hand, specialty columns include columns designed for specific applications, such as GC-MS interface columns or columns tailored for particular compound classes.

The selection of the appropriate GC column is crucial for achieving the desired separation and analysis goals. It depends on factors such as the type of compounds being analyzed, the sample matrix, and the analytical objectives.

3 Oven

The oven provides a controlled temperature environment for the chromatographic column. Temperature programming can be employed to optimize separation. Isothermal ovens maintain a constant temperature, while temperature-programmed ovens allow varying temperatures during the analysis.

4 Carrier Gas Supply

The carrier gas is the mobile phase that transports the sample through the column. Common gases include helium, nitrogen, and hydrogen. The choice of carrier gas depends on factors such as detector compatibility, column type, and analysis requirements. GC-MS analyses usually utilize helium or hydrogen.

5 Detector

The detector monitors the effluent from the column and produces a signal that is proportional to the concentration of each compound.

Common detectors include flame ionization detector, thermal conductivity detector, electron capture detector, mass spectrometer (MS), and others (Stein and Scott, 1994).

6 Data System

The data system records and processes the detector signals, generating a chromatogram that represents the separation of compounds in the sample. It includes data acquisition software, integrators, and data analysis tools.

7 Injector Port

It houses the injector and ensures a controlled environment for sample introduction. It may have features like a septum purge to prevent sample contamination.

8 Column Oven Controller

It regulates the temperature of the column oven. It allows for precise temperature programming during the analysis.

9 Column Inlet

That connects the injector to the chromatographic column, ensuring a smooth flow of the sample into the column. Temperature control of the column inlet is important to prevent the condensation of the sample or the diffusion of the sample, resulting in wide or tailed peaks.

10 Column Outlet

It connects the column to the detector, allowing the separated components to enter the detector for analysis. Its temperature is also crucial to ensure the acceleration of separated analytes to the detector to avoid peak widening or postcolumn analyte condensation.

11 Flow Control Devices

These devices, such as pressure regulators and flow controllers, ensure a constant and controlled flow of carrier gas through the system.

Pressure control devices, including pressure regulators, maintain optimal pressures in different parts of the chromatograph for efficient operation.

Understanding the functions and interactions of these components is essential for optimizing gas chromatographic analyses and obtaining accurate results. The choice of specific components depends on the nature of the samples being analyzed and the analytical requirements of the experiment.

13.5.1.2 MASS SPECTROMETRY

MS identifies and quantifies compounds based on their m/z ratio. The ionized molecules from the GC column enter the mass spectrometer, producing a mass spectrum that represents the unique fingerprint of each compound (Fang et al., 2015).

The coupling of GC and MS enhances the analytical power of both techniques. GC separates compounds, and MS provides precise identification and quantification, allowing for the detection of a wide range of chemical compounds in complex samples (Sparkman et al., 2011).

13.5.1.2.1 Varieties of Detectors in MS

The primary mass spectrometer commonly used with GCs is the quadrupole MS, also known as the "Mass Selective Detector" (Holmes and Morrell, 1957). Another frequently encountered one is the ion trap MS. While magnetic sector mass spectrometers are an option, they are typically costly and cumbersome, making them less prevalent in high-throughput service laboratories. Other detectors that one might come across include TOF, tandem quadrupoles, or, in the instance of an ion trap MS^n, where "*n*" denotes the number of MS stages (Gohlke, 1959; Sparkman et al., 2011).

13.5.1.2.2 Gas Chromatography–Tandem Mass Spectrometry

The addition of a second phase of mass fragmentation, such as incorporating a second quadrupole into a quadrupole MS, is termed tandem MS (MS/MS). MS/MS is occasionally employed to quantify low concentrations of target compounds within a high sample matrix background (Hübschmann, 2015).

The initial quadrupole (Q1) is linked to a collision cell (Q2) and a second quadrupole (Q3). Either of the two quadrupoles can operate in scanning or static mode, according to the nature and purpose of the MS/MS analysis. Various analyses can be performed, including precursor ion scan, product ion scan, selected reaction monitoring (SRM, also known as multiple reaction monitoring), and neutral loss scan. For instance, when Q1 is in static mode and Q3 is scanning (focused on a single mass, as in selected ion monitoring), a product ion spectrum, also known as a "daughter spectrum," is obtained. From this spectrum, a significant product ion can be selected, serving as the selected precursor ion's product ion. This pair is termed a "transition" and forms the foundation for SRM. SRM is remarkably specific and effectively diminishes background interference from the matrix (Hübschmann, 2015).

13.5.1.2.3 Ionization Modes Common in GC-MS Analyses

Upon completing their journey through the column, traversing the transfer line, and entering the mass spectrometer, molecules undergo ionization through various methods, typically employing only one method at a time. Following fragmentation, the sample is usually detected by an electron multiplier, converting the ionized fragment into a detectable signal.

1 Electron Impact

The predominant mode of ionization is EI. In this process, molecules enter the mass spectrometer and encounter free electrons emitted from a filament. The bombardment of electrons causes characteristic and reproducible fragmentation of the molecules. The utilization of molecular electron bombardment is expressed as a hard ionization technique that yields several fragments of lower *m/z* ratio and sometimes, molecules with molecular mass unit, while in "soft ionization," molecules collide with an introduced gas. The fragmentation pattern depends on the applied collision energy, typically set at 70 eV (electron volts) (McLafferty et al., 1974). The use of 70 eV renders the comparison of obtained spectra with spectra in libraries and depositories easy, using software supplied by the manufacturer or developed by the National Institute of Standards (NIST-USA). Library searches use matching algorithms such as Probability Based Matching and dot-product matching, utilized by various method standardization agencies. Libraries from sources like NIST, Wiley, the AAFS, and instrument manufacturers are available (Stein and Scott, 1994).

2 Cold Electron Ionization (Cold-EI)

The "hard ionization" of EI can be mitigated by cooling the molecules before ionization, resulting in more informative mass spectra. In cold-EI, molecules exit the GC column, mix with added helium makeup gas, and expand into a vacuum *via* a specific supersonic nozzle, forming a supersonic molecular beam. Then, collisions with the gas during the expansion of the supersonic jet reduce the internal energy (vibrational and rotational) of the analyzed molecules, minimizing the fragmentation caused by electrons during ionization. Cold-EI mass spectra feature an abundant molecular ion while retaining the usual fragmentation pattern, making them compatible with library search identification techniques (Amirav et al., 2008). The enhanced molecular ions enhance identification of known and unknown compounds, by studying the isotope pattern and linking it to isotope abundance via amplifying isomer mass spectral effects (Stein and Scott, 1994).

13.5.2 PHYSIOLOGICAL RESPONSES OF PLANTS TO STRESS

13.5.2.1 ABIOTIC STRESS

Plants respond to various environmental stresses, including drought, salinity, and temperature fluctuations. These stressors trigger physiological responses, such as the accumulation of ROS and changes in metabolic pathways (Sachdev et al., 2021).

13.5.2.2 BIOTIC STRESS

Pathogen attacks and insect infestations induce biotic stress in plants, leading to alterations in defense mechanisms and the production of secondary metabolites (Al-Khayri et al., 2023).

Applications of GC-MS in Monitoring Plant Stress are the following.

13.5.3 METABOLITE PROFILING

GC-MS enables comprehensive metabolite profiling, allowing researchers to identify and quantify a wide array of metabolites. This approach aids in understanding how plants reprogram their metabolic pathways in response to stress (Wang et al., 2015).

13.5.3.1 VOLATILE ORGANIC COMPOUNDS (VOCS)

VOCs play a crucial role in plant stress signaling. GC-MS allows for the identification of specific VOCs associated with stress conditions, providing insights into the plant's response mechanisms.

Common VOCs associated with plant stress can include green leaf volatiles that are released in response to mechanical damage or herbivore feeding and are involved in plant defense (Pierik et al., 2014), terpenes that are compounds having diverse roles, including antimicrobial properties and serving as precursors for the synthesis of other defense-related compounds (Dong et al., 2011), methyl salicylate (MeSA) which is involved in systemic acquired resistance against pathogens (Shirasu et al., 1997), and methyl jasmonate (MeJA) a signaling molecule associated with induced systemic resistance (Karban et al., 2013).

13.5.3.2 HORMONAL ANALYSIS

GC-MS facilitates the analysis of plant hormones, including abscisic acid, ethylene, and jasmonic acid. Monitoring changes in hormone levels helps unravel the intricate signaling networks involved in stress responses (Engelberth and Engelberth, 2009).

13.5.3.3 CASE STUDIES

Illustrative case studies demonstrate the efficacy of GC-MS in monitoring plant stress. Examples include the analysis of drought-induced metabolomic changes and the identification of stress-specific VOCs in response to pathogen attacks (Kishimoto et al., 2005; Piesik et al., 2011, 2012).

13.5.4 FUTURE PROSPECTS

Continued advancements in GC-MS technology, such as high-resolution MS, hold promise for enhanced sensitivity and specificity in plant stress analysis.

13.5.5 CHALLENGES

Challenges, including data standardization and the integration of multiomics approaches, must be addressed to maximize the potential of GC-MS in unraveling the complexities of plant stress responses.

In conclusion, GC-MS stands as a powerful tool in monitoring plant stress, offering unparalleled insights into the biochemical and molecular changes associated with various stressors. As technology advances and methodologies become more sophisticated, the application of GC-MS in plant stress research will undoubtedly continue to flourish, contributing to our understanding of plant stress physiology and facilitating the development of stress-tolerant crops for sustainable agriculture.

KEYWORDS

- **metabolomics**
- **metabolic profiling**
- **analytical techniques**
- **gas chromatography-mass spectrometry**
- **liquid chromatography-mass spectrometry**

REFERENCES

Al-Khayri, J. M., Rashmi, R., Toppo, V., Chole, P. B., Banadka, A., Sudheer, W. N., Nagella, P., Shehata, W. F., Al-Mssallem, M. Q., Alessa, F. M., Almaghasla, M. I., Rezk, A. A.-S. Plant secondary metabolites: The weapons for biotic stress management. *Metabolites*, 2023, 13(6), 716. https://doi.org/10.3390/metabo13060716

Amirav, A., Gordin, A., Poliak, M., Fialkov, A. B. Gas chromatography-mass spectrometry with supersonic molecular beams. *J. Mass Spectrom.*, 2008, 43(2), 141–163. https://doi.org/10.1002/jms.1380

Angel, P. M., Spraggins, J. M., Baldwin, H. S., Caprioli, R. Enhanced sensitivity for high spatial resolution lipid analysis by negative ion mode matrix assisted laser desorption ionization imaging mass spectrometry. *Anal. Chem.*, 2012, 84(3), 1557–1564. https://doi.org/10.1021/ac202383m

Araújo, P., Ferreira, M. S., de Oliveira, D. N., Pereira, L., Sawaya, A. C., Catharino, R. R., Mazzafera, P. Mass spectrometry imaging: An expeditious and powerful technique for fast *in situ* lignin assessment in eucalyptus. *Anal. Chem.*, 2014, 86(7), 3415–3419. https://doi.org/10.1021/ac500220r

Aziz, M., Sturtevant, D., Winston, J., Collakova, E., Jelesko, J., Chapman, K. MALDI-MS imaging of Urushiols in poison ivy stem. *Molecules*, 2017, 22(5), 711. https://doi.org/10.3390/molecules22050711

Becker, L., Carré, V., Poutaraud, A., Merdinoglu, D., Chaimbault, P. MALDI mass spectrometry imaging for the simultaneous location of Resveratrol, Pterostilbene and viniferins on grapevine leaves. *Molecules*, 2014, 19(7), 10587–10600. https://doi.org/10.3390/molecules190710587

Benkeblia, N. Gas chromatography-mass spectrometry and liquid chromatography-mass spectrometry metabolomics platforms: Tools for plant oligosaccharides analysis. *Carbohydr. Polym. Technol. Appl.*, 2023, 5, 100304. https://doi.org/10.1016/j.carpta.2023.100304

Berisha, A., Dold, S., Guenther, S., Desbenoit, N., Takats, Z., Spengler, B., Römpp, A. A comprehensive high-resolution mass spectrometry approach for characterization of metabolites by combination of ambient ionization, chromatography and imaging methods. *Rapid Commun. Mass Spectrom.*, 2014, 28(16), 1779–1791. https://doi.org/10.1002/rcm.6960

Bhardwaj, C., Hanley, L. Ion sources for mass spectrometric identification and imaging of molecular species. *Nat. Prod. Rep.*, 2014, 31(6), 756–767. https://doi.org/10.1039/c3np70094a

Brentan Silva, D., Aschenbrenner, A.-K., Lopes, N., Spring, O. Direct analyses of secondary metabolites by mass spectrometry imaging (MSI) from sunflower (helianthus annuus L.) trichomes. *Molecules*, 2017, 22(5), 774. https://doi.org/10.3390/molecules22050774

Buschmann, H. Accumulation of hydroxycoumarins during post-harvest deterioration of tuberous roots of cassava (manihot esculenta Crantz). *Ann. Bot.*, 2000, 86(6), 1153–1160. https://doi.org/10.1006/anbo.2000.1285

Caprioli, R. M., Farmer, T. B., Gile, J. Molecular imaging of biological samples: Localization of peptides and proteins using MALDI-TOF MS. *Anal. Chem.*, 1997, 69(23), 4751–4760. https://doi.org/10.1021/ac970888i

Cha, S., Yeung, E. S. Colloidal graphite-assisted laser desorption/ionization mass spectrometry and MS*n* of small molecules. 1. imaging of cerebrosides directly from rat brain tissue. *Anal. Chem.*, 2007, 79(6), 2373–2385. https://doi.org/10.1021/ac062251h

Chaki, M., Begara-Morales, J. C., Barroso, J. B. Oxidative stress in plants. *Antioxidants*, 2020, 9(6), 481. https://doi.org/10.3390/antiox9060481

Chaurand, P., Norris, J. L., Cornett, D. S., Mobley, J. A., Caprioli, R. M. New developments in profiling and imaging of proteins from tissue sections by MALDI mass spectrometry. *J. Proteome Res.*, 2006, 5(11), 2889–2900. https://doi.org/10.1021/pr060346u

Chaurand, P., Rahman, M. A., Hunt, T., Mobley, J. A., Gu, G., Latham, J. C., Caprioli, R. M., Kasper, S. Monitoring Mouse Prostate Development by profiling and imaging mass spectrometry. *Mol. Cell. Proteomics*, 2008, 7(2), 411–423. https://doi.org/10.1074/mcp.m700190-mcp200

Chaurand, P., Sanders, M. E., Jensen, R. A., Caprioli, R. M. Proteomics in diagnostic pathology. *Am. J. Pathol.*, 2004, 165(4), 1057–1068. https://doi.org/10.1016/s0002-9440(10)63367-6

Cillero-Pastor, B., Heeren, R. M. Applications of α-cyano-4-hydroxycinnamic acid in tissue imaging: Challenges and solutions. *Anal. Chem.*, 2014, 86(9), 4284–4290.

Corinti, D., Crestoni, M. E., Fornarini, S., Pieper, M., Niehaus, K., Giampà, M. An integrated approach to study novel properties of a MALDI matrix (4-Maleicanhydridoproton Sponge) for MS imaging analyses. *Anal. Bioanal.Chem.*, 2018, 411(4), 953–964. https://doi.org/10.1007/s00216-018-1531-7

Cornett, D. S., Reyzer, M. L., Chaurand, P., Caprioli, R. M. MALDI imaging mass spectrometry: Molecular snapshots of biochemical systems. *Nat. Methods*, 2007, 4(10), 828–833. https://doi.org/10.1038/nmeth1094

Dalisay, D. S., Kim, K. W., Lee, C., Yang, H., Rübel, O., Bowen, B. P., Davin, L. B., Lewis, N. G. Dirigent protein-mediated lignan and cyanogenic glucoside formation in flax seed: Integrated omics and MALDI mass spectrometry imaging. *J. Nat. Prod.*, 2015, 78(6), 1231–1242.

Danzi, F., Pacchiana, R., Mafficini, A., Scupoli, M. T., Scarpa, A., Donadelli, M., Fiore, A. To metabolomics and beyond: A technological portfolio to investigate cancer metabolism. *Signal Transduct. Target. Ther.*, 2023, 8(1), 137. https://doi.org/10.1038/s41392-023-01380-0

de Oliveira, D. N., Ferreira, M. S., Catharino, R. R. Rapid and simultaneous in situ assessment of aflatoxins and stilbenes using silica plate imprinting mass spectrometry imaging. *PLoS One*, 2014, 9(3), e90901.

Dong, F., Yang, Z., Baldermann, S., Sato, Y., Asai, T., Watanabe, N. Herbivore-induced volatiles from tea (camellia sinensis) plants and their involvement in intraplant communication and changes in endogenous nonvolatile metabolites. *J. Agric. Food Chem.*, 2011, 59(24), 13131–13135. https://doi.org/10.1021/jf203396a

Dreisewerd, K. The desorption process in MALDI. *Chem. Rev.*, 2003, 103(2), 395–426.

Durai Ananda Kumar, T., Sai, C., Venkateswarlu, A., Supriya Reddy, K. Evolution of liquid chromatography: Technologies and applications. *Int. J. Res. Pharm. Sci.*, 2020, 11(3), 3204–3211. https://doi.org/10.26452/ijrps.v11i3.2449

Dutkiewicz, E. P., Su, C. H., Lee, H. J., Hsu, C. C., Yang, Y. L. Visualizing vinca alkaloids in the petal of Catharanthus roseus using functionalized titanium oxide nanowire substrate for surface-assisted laser desorption/ionization imaging mass spectrometry. *Plant J.*, 2021, 105(4), 1123–1133.

Engelberth, M. J., Engelberth, J. Monitoring plant hormones during stress responses. *J. Visual. Exp.*, 2009, 28, 1127. https://doi.org/10.3791/1127

Ernst, M., Silva, D. B., Silva, R. R., Vêncio, R. Z., Lopes, N. P. Mass spectrometry in plant metabolomics strategies: From analytical platforms to data acquisition and processing. *Nat. Prod. Rep.*, 2014, 31(6), 784–806.

Etalo, D. W., De Vos, R. C., Joosten, M. H., Hall, R. D. Spatially resolved plant metabolomics: Some potentials and limitations of laser-ablation electrospray ionization mass spectrometry metabolite imaging. *Plant Physiol.*, 2015, 169(3), 1424–1435.

Facchini, P. J., St-Pierre, B. Synthesis and trafficking of alkaloid biosynthetic enzymes. *Curr. Opin. Plant Biol.*, 2005, 8(6), 657–666.

Fang, M., Ivanisevic, J., Benton, H. P., Johnson, C. H., Patti, G. J., Hoang, L. T., Uritboonthai, W., Kurczy, M. E., Siuzdak, G. Thermal degradation of small molecules: A global metabolomic investigation. *Anal. Chem.*, 2015, 87(21), 10935–10941. https://doi.org/10.1021/acs.analchem.5b03003

Fiehn, O. Metabolomics—The link between genotypes and phenotypes. *Plant Mol. Biol.*, 2002, 48, 155–171.

Fiehn, O. Metabolomics by gas chromatography–mass spectrometry: Combined targeted and untargeted profiling. *Curr. Protocols Mol. Biol.*, 2016, 114(1), 30.4.1–30.4.32. https://doi.org/10.1002/0471142727.mb3004s114

Franceschi, P., Dong, Y., Strupat, K., Vrhovsek, U., Mattivi, F. Combining intensity correlation analysis and MALDI imaging to study the distribution of flavonols and dihydrochalcones in Golden Delicious apples. *J. Exp. Bot.*, 2012, 63(3), 1123–1133.

Franck, J., El Ayed, M., Wisztorski, M., Salzet, M., Fournier, I. On-tissue N-terminal peptide derivatizations for enhancing protein identification in MALDI mass spectrometric imaging strategies. *Anal. Chem.*, 2009, 81(20), 8305–8317.

Friesen, W. L., Schultz, B. J., Destino, J. F., Alivio, T. E., Steet, J. R., Banerjee, S., Wood, T. D. Two-dimensional graphene as a matrix for MALDI imaging mass spectrometry. *J. Am. Soc. Mass. Spectrom.*, 2015, 26(11), 1963–1966.

Gemperline, E., Jayaraman, D., Maeda, J., Ané, J. M., Li, L. Multifaceted investigation of metabolites during nitrogen fixation in Medicago via high resolution MALDI-MS imaging and ESI-MS. *J. Am. Soc. Mass. Spectrom.*, 2014, 26(1), 149–158.

Gohlke, R. S. Time-of-flight mass spectrometry and gas-liquid partition chromatography. *Anal. Chem.*, 1959, 31(4), 535–541. https://doi.org/10.1021/ac50164a024

Goodwin, R. J. Sample preparation for mass spectrometry imaging: Small mistakes can lead to big consequences. *J. Proteomics*, 2012, 75(16), 4893–4911.

Goodwin, R. J., Pitt, A. R. Mass spectrometry imaging of pharmacological compounds in tissue sections. *Bioanalysis*, 2010b, 2(2), 279–293.

Goodwin, R. J., Dungworth, J. C., Cobb, S. R., Pitt, A. R. Time-dependent evolution of tissue markers by MALDI-MS imaging. *Proteomics*, 2008, 8(18), 3801–3808.

Goodwin, R. J., Pitt, A. R., Harrison, D., Weidt, S. K., Langridge-Smith, P. R., Barrett, M. P., Logan Mackay, C. Matrix-free mass spectrometric imaging using laser desorption ionisation Fourier transform ion cyclotron resonance mass spectrometry. *Rapid Commun. Mass Spectrom.*, 2011, 25(7), 969–972.

Goodwin, R. J., Scullion, P., MacIntyre, L., Watson, D. G., Pitt, A. R. Use of a solvent-free dry matrix coating for quantitative matrix-assisted laser desorption ionization imaging of 4-bromophenyl-1, 4-diazabicyclo

(3.2. 2) nonane-4-carboxylate in rat brain and quantitative analysis of the drug from laser microdissected tissue regions. *Anal. Chem.*, 2010a, 82(9), 3868–3873.
Gorzolka, K., Bednarz, H., Niehaus, K. Detection and localization of novel hordatine-like compounds and glycosylated derivates of hordatines by imaging mass spectrometry of barley seeds. *Planta*, 2014, 239(6), 1321–1335.
Gowda, G. A., Djukovic, D. Overview of mass spectrometry-based metabolomics: Opportunities and challenges. *Methods Mol. Biol.*, 2014, 1198, 3–12. https://doi.org/10.1007/978-1-4939-1258-2_1
Gruhnert, C., Biehl, B., Selmar, D. Compartmentation of cyanogenic glucosides and their degrading enzymes. *Planta*, 1994, 195, 36–42.
Halkier, B. A., Møller, B. L. Biosynthesis of the cyanogenic glucoside dhurrin in seedlings of Sorghum bicolor (L.) Moench and partial purification of the enzyme system involved. *Plant Physiol.*, 1989, 90(4), 1552–1559.
Hertel Pereira, A. C., Auer, A. C., Biedel, L., de Almeida, C. M., Romão, W., Endringer, D. C. Analysis of gliricidia sepium leaves by MALDI mass spectrometry imaging. *J. Am. Soc. Mass. Spectrom.*, 2022, 33(3), 573–583.
Holmes, J. C., Morrell, F. A. Oscillographic mass spectrometric monitoring of gas chromatography. *Appl. Spectrosc.*, 1957, 11(2), 86–87. https://doi.org/10.1366/000370257774633394
Hölscher, D., Fuchser, J., Knop, K., Menezes, R. C., Buerkert, A., Svatoš, A., Schubert, U. S., Schneider, B. High resolution mass spectrometry imaging reveals the occurrence of phenylphenalenone-type compounds in red paracytic stomata and red epidermis tissue of Musa acuminata ssp. zebrina cv.'Rowe Red'. *Phytochemistry*, 2015, 116, 239–245.
Hölscher, D., Shroff, R., Knop, K., Gottschaldt, M., Crecelius, A., Schneider, B., Heckel, D. G., Schubert, U. S., Svatoš, A. Matrix-free UV-laser desorption/ionization (LDI) mass spectrometric imaging at the single-cell level: Distribution of secondary metabolites of Arabidopsis thaliana and Hypericum species. *Plant J.*, 2009, 60(5), 907–918.
Horn, P. J., Chapman, K. D. Metabolite Imager: Customized spatial analysis of metabolite distributions in mass spectrometry imaging. *Metabolomics*, 2014, 10, 337–348.
Houck, M. M., Siegel, J. A. Chapter 6—Separation methods. In: *Fundamentals of Forensic Science* (pp. 121–151). Essay, Academic Press, 2015.
Hübschmann, H.-J. *Handbook of GC-MS: Fundamentals and Applications*. Wiley-VCH Verlag GmbH & Co. KGaA, 2015.
Jun, J. H., Song, Z., Liu, Z., Nikolau, B. J., Yeung, E. S., Lee, Y. J. High-spatial and high-mass resolution imaging of surface metabolites of Arabidopsis thaliana by laser desorption-ionization mass spectrometry using colloidal silver. *Anal. Chem.*, 2010, 82(8), 3255–3265.
Jurchen, J. C., Rubakhin, S. S., Sweedler, J. V. MALDI-MS imaging of features smaller than the size of the laser beam. *J. Am. Soc. Mass. Spectrom.*, 2005, 16, 1654–1659.
Karban, R., Yang, L. H., Edwards, K. F. Volatile communication between plants that affects herbivory: A meta-analysis. *Ecol. Lett.*, 2013, 17(1), 44–52. https://doi.org/10.1111/ele.12205
Kaspar, S., Peukert, M., Svatos, A., Matros, A., Mock, H. P. MALDI-imaging mass spectrometry - An emerging technique in plant biology. *Proteomics*, 2011, 11(9), 1840–1850.
Kawamoto, T. Use of a new adhesive film for the preparation of multi-purpose fresh-frozen sections from hard tissues, whole-animals, insects and plants. *Arch. Histol. Cytol.*, 2003, 66(2), 123–143.
Kishimoto, K., Matsui, K., Ozawa, R., Takabayashi, J. Volatile C6-aldehydes and allo-ocimene activate defense genes and induce resistance against botrytis cinerea in Arabidopsis thaliana. *Plant Cell Physiol.*, 2005, 46(7), 1093–1102. https://doi.org/10.1093/pcp/pci122
Korte, A. R., Yandeau-Nelson, M. D., Nikolau, B. J., Lee, Y. J. Subcellular-level resolution MALDI-MS imaging of maize leaf metabolites by MALDI-linear ion trap-Orbitrap mass spectrometer. *Anal. Bioanal. Chem.*, 2015, 407, 2301–2309.
Kusari, S., Lamshöft, M., Kusari, P., Gottfried, S., Zühlke, S., Louven, K., Hentschel, U., Kayser, O., Spiteller, M. Endophytes are hidden producers of maytansine in Putterlickia roots. *J. Nat. Prod.*, 2014, 77(12), 2577–2584.
Kusari, S., Sezgin, S., Nigutova, K., Cellarova, E., Spiteller, M. Spatial chemo-profiling of hypericin and related phytochemicals in Hypericum species using MALDI-HRMS imaging. *Anal. Bioanal. Chem.*, 2015, 407, 4779–4791.

Lanekoff, I., Heath, B. S., Liyu, A., Thomas, M., Carson, J. P., Laskin, J. Automated platform for high-resolution tissue imaging using nanospray desorption electrospray ionization mass spectrometry. *Anal. Chem.*, 2012, 84(19), 8351–8356.

Lee, J. W., Ji, S. H., Lee, Y. S., Choi, D. J., Choi, B. R., Kim, G. S., Baek, N. I., Lee, D. Y. Mass spectrometry based profiling and imaging of various ginsenosides from Panax ginseng roots at different ages. *Int. J. Mol. Sci.*, 2017, 18(6), 1114.

Li, B., Bhandari, D. R., Janfelt, C., Römpp, A., Spengler, B. Natural products in Glycyrrhiza glabra (licorice) rhizome imaged at the cellular level by atmospheric pressure matrix-assisted laser desorption/ionization tandem mass spectrometry imaging. *Plant J.*, 2014a, 80(1), 161–171.

Li, B., Neumann, E. K., Ge, J., Gao, W., Yang, H., Li, P., Sweedler, J. V. Interrogation of spatial metabolome of Ginkgo biloba with high-resolution matrix-assisted laser desorption/ionization and laser desorption/ionization mass spectrometry imaging. *Plant, Cell Environ.*, 2018, 41(11), 2693–2703.

Li, C., Wang, Z., Jones, A. D. Chemical imaging of trichome specialized metabolites using contact printing and laser desorption/ionization mass spectrometry. *Anal. Bioanal.Chem.*, 2014b, 406, 171–182.

Li, L., Qiu, Z., Jiang, M., Zhang, B., Chen, Q., Zhang, C., Zheng, Z., Qiao, X. Visualizing the spatial distribution of *Arctium lappa* L. Root components by MALDI-TOF mass spectrometry imaging. *Foods*, 2022, 11(24), 3957.

Liu, J., Cai, Y. Z., Wong, R. N. S., Lee, C. K. F., Tang, S. C. W., Sze, S. C. W., Tong, Y., Zhang, Y. Comparative analysis of caffeoylquinic acids and lignans in roots and seeds among various burdock (*Arctium lappa*) genotypes with high antioxidant activity. *J. Agric. Food Chem.*, 2012, 60(16), 4067–4075.

Liu, X., Qiu, H., Lee, R. K., Chen, W., Li, J. Methylamidation for sialoglycomics by MALDI-MS: A facile derivatization strategy for both α2, 3-and α2, 6-linked sialic acids. *Anal. Chem.*, 2010, 82(19), 8300–8306.

López-Fernández, O., Domínguez, R., Pateiro, M., Munekata, P. E. S., Rocchetti, G., Lorenzo, J. M. Determination of polyphenols using liquid chromatography–tandem mass spectrometry technique (LC–MS/MS): A review. *Antioxidants*, 2020, 9(6), 479. https://doi.org/10.3390/antiox9060479

Manier, M. L., Reyzer, M. L., Goh, A., Dartois, V., Via, L. E., Barry III, C. E., Caprioli, R. M. Reagent precoated targets for rapid in-tissue derivatization of the anti-tuberculosis drug isoniazid followed by MALDI imaging mass spectrometry. *J. Am. Soc. Mass. Spectrom.*, 2011, 22(8), 1409–1419.

Marques, J. V., Dalisay, D. S., Yang, H., Lee, C., Davin, L. B., Lewis, N. G. A multi-omics strategy resolves the elusive nature of alkaloids in Podophyllum species. *Mol. Biosyst.*, 2014, 10(11), 2838–2849.

Martins, T. R., Berg, J. J., Blinka, S., Rausher, M. D., Baum, D. A. Precise spatio-temporal regulation of the anthocyanin biosynthetic pathway leads to petal spot formation in Clarkia gracilis (Onagraceae). *New Phytol.*, 2013, 197(3), 958–969.

McLafferty, F. W., Hertel, R. H., Villwock, R. D. Probability based matching of mass spectra. rapid identification of specific compounds in mixtures. *Org. Mass Spectrom.*, 1974, 9(7), 690–702. https://doi.org/10.1002/oms.1210090710

Mellon, F. A. Mass spectrometry | principles and instrumentation. *Encyclopedia Food Sci. Nutrit.*, 2003, 3739–3749. https://doi.org/10.1016/b0-12-227055-x/00746-x

Nielsen, K. A., Olsen, C. E., Pontoppidan, K., Møller, B. L. Leucine-derived cyano glucosides in barley. *Plant Physiol.*, 2002, 129(3), 1066–1075.

Pierik, R., Ballaré, C. L., Dicke, M. Ecology of plant volatiles: Taking a plant community perspective. *Plant, Cell Environ.*, 2014, 37(8), 1845–1853. https://doi.org/10.1111/pce.12330.

Piesik, D., Pańka, D., Jeske, M., Wenda-Piesik, A., Delaney, K. J., Weaver, D. K. Volatile induction of infected and neighbouring uninfected plants potentially influence attraction/repellence of a cereal herbivore. *J. Appl. Entomol.*, 2012, 137(4), 296–309. https://doi.org/10.1111/j.1439-0418.2012.01742.x

Piesik, Dariusz, Lemńczyk, G., Skoczek, A., Lamparski, R., Bocianowski, J., Kotwica, K., Delaney, K. J. Fusarium infection in maize: Volatile induction of infected and neighboring uninfected plants has the potential to attract a pest cereal leaf beetle, Oulema Melanopus. *J. Plant Physiol.*, 2011, 168(13), 1534–1542. https://doi.org/10.1016/j.jplph.2011.01.032

Ratnasari, J., Esyanti, R. R., Tan, M. I., Juliawaty, L. D., Shimma, S. Profile of cryptobrachytone C accumulation in Cryptocarya pulchrinervia leaves using MALDI-MSI. *Biodiversitas J. Biol. Div.*, 2021, 22(3).

Smith, J., Brown, A., Johnson, C. Derivatives of α-cyano-4-hydroxycinnamic acid: A comprehensive review. *J. Mass Spectrom.*, 2013, 48(5), 653–668.

Ren, J.-L., Zhang, A.-H., Kong, L., Wang, X.-J. Advances in mass spectrometry-based metabolomics for investigation of metabolites. *RSC Adv*., 2018, 8(40), 22335–22350. https://doi.org/10.1039/c8ra01574k

Sachdev, S., Ansari, S. A., Ansari, M. I., Fujita, M., Hasanuzzaman, M. Abiotic stress and reactive oxygen species: Generation, signaling, and defense mechanisms. *Antioxidants*, 2021, 10(2), 277. https://doi.org/10.3390/antiox10020277

Salem, M. A., Perez de Souza, L., Serag, A., Fernie, A. R., Farag, M. A., Ezzat, S. M., Alseekh, S. Metabolomics in the context of plant natural products research: From sample preparation to metabolite analysis. *Metabolites*, 2020, 10(1), 37. https://doi.org/10.3390/metabo10010037

Sánchez-Pérez, R., Jørgensen, K., Motawia, M. S., Dicenta, F., Møller, B. L. Tissue and cellular localization of individual β-glycosidases using a substrate-specific sugar reducing assay. *Plant J*., 2009, 60(5), 894–906.

Sarsby, J., Towers, M. W., Stain, C., Cramer, R., Koroleva, O. A. Mass spectrometry imaging of glucosinolates in Arabidopsis flowers and siliques. *Phytochemistry*, 2012, 77, 110–118.

Satyan, R. S., Aveek, N., Eganathan, P., Parida, A. Comparative histochemical localization of secondary metabolites in seed-raised and in vitro propagated plants of *Excoecaria agallocha* Linn. (Euphorbiaceae), the milky mangrove tree of historical significance. *Biotech. Histochem*., 2010, 85(5), 285–293.

Schmitt-Kopplin, P. *Comprehensive Foodomics* (Vol. 2), 1st edn., Elsevier, 2021.

Schwartz, S. A., Reyzer, M. L., Caprioli, R. M. Direct tissue analysis using matrix-assisted laser desorption/ionization mass spectrometry: Practical aspects of sample preparation. *J. Mass Spectrom*., 2003, 38(7), 699–708.

Seaman, C., Flinders, B., Eijkel, G., Heeren, R. M., Bricklebank, N., Clench, M. R. "Afterlife experiment": Use of MALDI-MS and SIMS imaging for the study of the nitrogen cycle within plants. *Anal. Chem*., 2014, 86(20), 10071–10077.

Seeley, E. H., Oppenheimer, S. R., Mi, D., Chaurand, P., Caprioli, R. M. Enhancement of protein sensitivity for MALDI imaging mass spectrometry after chemical treatment of tissue sections. *J. Am. Soc. Mass. Spectrom*., 2008, 19, 1069–1077.

Seneviratne, H. K., Dalisay, D. S., Kim, K. W., Moinuddin, S. G., Yang, H., Hartshorn, C. M., Davin, L. B., Lewis, N. G. Non-host disease resistance response in pea (Pisum sativum) pods: Biochemical function of DRR206 and phytoalexin pathway localization. *Phytochemistry*, 2015, 113, 140–148.

Shirasu, K., Nakajima, H., Rajasekhar, V. K., Dixon, R. A., Lamb, C. Salicylic acid potentiates an agonist-dependent gain control that amplifies pathogen signals in the activation of defense mechanisms. *Plant Cell*, 1997, 9(2), 261. https://doi.org/10.2307/3870546

Shroff, R., Schramm, K., Jeschke, V., Nemes, P., Vertes, A., Gershenzon, J., Svatoš, A. Quantification of plant surface metabolites by matrix-assisted laser desorption–ionization mass spectrometry imaging: Glucosinolates on Arabidopsis thaliana leaves. *Plant J.*, 2015, 81(6), 961–972.

Silva, D. B., Turatti, I. C. C., Gouveia, D. R., Ernst, M., Teixeira, S. P., Lopes, N. P. Mass spectrometry of flavonoid vicenin-2, based sunlight barriers in Lychnophora species. *Sci. Rep*., 2014, 4(1), 4309.

Soares, M. S., da Silva, D. F., Forim, M. R., Fernandes, J. B., Vieira, P. C., Silva, D. B., Lopes, N. P., de Carvalho, S. A., de Souza, A. A., Machado, M. A. Quantification and localization of hesperidin and rutin in Citrus sinensis grafted on C. limonia after Xylella fastidiosa infection by HPLC-UV and MALDI imaging mass spectrometry. *Phytochemistry*, 2015, 115, 161–170.

Soltwisch, J., Kettling, H., Vens-Cappell, S., Wiegelmann, M., Müthing, J., Dreisewerd, K. Mass spectrometry imaging with laser-induced postionization. *Science*, 2015, 348(6231), 211–215.

Sparkman, O. D., Penton, Z., Kitson, F. G. *Gas Chromatography and Mass Spectrometry: A Practical Guide*. Elsevier, Academic Press, 2011.

Stein, S. E., Scott, D. R. Optimization and testing of mass spectral library search algorithms for compound identification. *J. Am. Soc. Mass. Spectrom*., 1994, 5(9), 859–866. https://doi.org/10.1016/1044-0305(94)87009-8

Steven, R. T., Race, A. M., Bunch, J. para-Nitroaniline is a promising matrix for MALDI-MS imaging on intermediate pressure MS systems. *J. Am. Soc. Mass. Spectrom*., 2013, 24(5), 801–804. https://doi.org/10.1007/s13361-013-0586-0

Stoeckli, M., Chaurand, P., Hallahan, D. E., Caprioli, R. M. Imaging mass spectrometry: A new technology for the analysis of protein expression in mammalian tissues. *Nat. Med*., 2001, 7(4), 493–496.

Sturtevant, D., Lee, Y. J., Chapman, K. D. Matrix assisted laser desorption/ionization-mass spectrometry imaging (MALDI-MSI) for direct visualization of plant metabolites in situ. *Curr. Opin. Biotechnol*., 2016, 37, 53–60.

Sugiura, Y., Shimma, S., Setou, M. Thin sectioning improves the peak intensity and signal-to-noise ratio in direct tissue mass spectrometry. *J. Mass Spectrom. Soc. Japan*, 2006, 54(2), 45–48.

Sun, C., Cui, L., Zhou, B., Wang, X., Guo, L., Liu, W. Visualizing the spatial distribution and alteration of metabolites in continuously cropped Salvia miltiorrhiza Bge using MALDI-MSI. *J. Pharm. Anal.*, 2022, 12(5), 719–724.

Sun, C., Zhang, M., Dong, H., Liu, W., Guo, L., Wang, X. A spatially-resolved approach to visualize the distribution and biosynthesis of flavones in *Scutellaria baicalensis* Georgi. *J. Pharm. Biomed. Anal.*, 2020, 179, 113014.

Tarzi, O. I., Stéfano, L. D., Argüello, J. E., Oksdath-Mansilla, G., Erra-Balsells, R. Photochemical and thermal stability of some dihydroxyacetophenones used as UV-MALDI-MS matrices. *Photochem. Photobiol.*, 2013, 89(6), 1368–1374.

Thomas, S. N., French, D., Jannetto, P. J., Rappold, B. A., Clarke, W. A. Liquid chromatography–tandem mass spectrometry for clinical diagnostics. *Nat. Rev. Methods Primers*, 2022, 2(1), 96. https://doi.org/10.1038/s43586-022-00175-x

Ullmer, R., Plematl, A., Rizzi, A. Derivatization by 6-aminoquinolyl-N-hydroxysuccinimidyl carbamate for enhancing the ionization yield of small peptides and glycopeptides in matrix-assisted laser desorption/ionization and electrospray ionization mass spectrometry. *Rapid Commun. Mass Spectrom.*, 2006, 20(9), 1469–1479.

Wang, J., Yang, E., Chaurand, P., Raghavan, V. Visualizing the distribution of strawberry plant metabolites at different maturity stages by MALDI-TOF imaging mass spectrometry. *Food Chem.*, 2021, 345, 128838.

Wang, S., Blair, I. A., Mesaros, C. Analytical methods for mass spectrometry-based metabolomics studies. *Adv. Mass Spectrom. Biomed. Res.*, 2019, 1140, 635–647.

Wang, X., Han, J., Chou, A., Yang, J., Pan, J., Borchers, C. H. Hydroxyflavones as a new family of matrices for MALDI tissue imaging. *Anal. Chem.*, 2013, 85(15), 7566–7573.

Wang, Y., Xu, L., Shen, H., Wang, J., Liu, W., Zhu, X., Wang, R., Sun, X., Liu, L. Metabolomic analysis with GC-MS to reveal potential metabolites and biological pathways involved in PB & CD stress response of Radish Roots. *Sci. Rep.*, 2015, 5(1), 18296. https://doi.org/10.1038/srep18296

Wink, M. Evolution of secondary metabolites from an ecological and molecular phylogenetic perspective. *Phytochemistry*, 2003, 64(1), 3–19.

Yoshimura, K., Yamanashi, T., Takai, N., Takahashi, K. Spatial distribution of anthocyanins in black rice kernel and their antioxidant activity. *Biosci. Biotechnol. Biochem.*, 2012a, 76(6), 1185–1187. https://doi.org/10.1271/bbb.120056

Yoshimura, Y., Enomoto, H., Moriyama, T., Kawamura, Y., Setou, M., Zaima, N. Visualization of anthocyanin species in rabbiteye blueberry Vaccinium ashei by matrix-assisted laser desorption/ionization imaging mass spectrometry. *Anal. Bioanal. Chem.*, 2012b, 403, 1885–1895.

Zaima, N., Yoshimura, Y., Kawamura, Y., Moriyama, T. Distribution of lysophosphatidylcholine in the endosperm of Oryza sativa rice. *Rapid Commun. Mass Spectrom.*, 2014, 28(13), 1515–1520.

Zemski Berry, K. A., Lawson, L. M., Olsen, N. O. Advancements in MALDI matrices: A comparative study of 2-mercaptobenzothiazole and dithranol. *Anal. Bioanal. Chem.*, 2019, 411(18), 4095–4107.

Zhang, H., Cha, S., Yeung, E. S. Colloidal graphite-assisted laser desorption/ionization MS and MS n of small molecules. 2. Direct profiling and MS imaging of small metabolites from fruits. *Anal. Chem.*, 2007, 79(17), 6575–6584.

CHAPTER 14

Plant Metabolome Databases

SHABANA BEGUM[1], SIBASHISH KITYANIA[2], RAJAT NATH[3], DEEPA NATH[4], PRANAB BEHARI MAZUMDER[2], and ANUPAM DAS TALUKDAR[1*]

[1]*Department of Life Science and Bioinformatics, Assam University, Silchar, Assam, India*

[2]*Department of Biotechnology, Assam University, Silchar, Assam, India*

[3]*Department of Biotechnology and Microbiology, School of Natural Sciences, Techno India University Tripura (TIUT), Agartala, Tripura, India*

[4]*Department of Botany, Guru Charan University, Silchar, Assam, India*

*Corresponding authors

ABSTRACT

Plant metabolomics has come to the forefront as a crucial methodology for comprehending plant metabolism, physiology, and molecular biology. Investigating plant metabolic pathways in response to diverse environmental conditions is facilitated through the utilization of plant metabolomics databases. These information repositories provide comprehensive insight into a plant's metabolic processes, activities, and products, encompassing intricate details about the enzymes involved, metabolites generated, and the activated metabolic pathways that contribute to the plant's metabolic functionality. Plant metabolomic datasets have greatly facilitated the study of plant metabolism and the development of various innovative technologies. Some of the well-known plant metabolomics databases are PlantGDB, PMDB (plant metabolome databases), Plant Metabolic Network (PMN), MeRy-B, RIKEN Plant Metabolome Meta Database (RIKEN PMN), AraCyc, etc. The databases enable researchers to analyze the metabolic pathways of plants in response to various environmental situations and recognize and quantify metabolites and associated pathways. These databases provide helpful information on plant metabolism's activities, processes, and outcomes. The availability of these databases has helped develop a number of cutting-edge technologies, provided researchers with insight into plant evolution and can help researchers find new treatments and therapies and understand the mechanism underpinning plant adaptation.

14.1 INTRODUCTION

Plants produce compounds called metabolites as they grow and develop, and these molecules are crucial to the survival and general health of the plant. All components of the plant,

Exploring Plant Stress Tolerance: A Metabolomics Approach.
Shuvasish Choudhury, Debojyoti Moulick, and Mirza Hasanuzzaman (Eds.)

DOI: 10.1201/9781779641731-14

including the roots, stems, and leaves as well as the flowers, fruits, and seeds, contain these chemicals. Primary and secondary metabolites are the two main classifications of plant metabolites. The primary metabolites are those that are necessary for plant growth and to function normally examples include carbohydrates, proteins, lipids, vitamins, nucleic acids, and hormones, while secondary metabolites such as polyphenols, alkaloids, and terpenes are involved in the defense mechanism of the plant against infectious diseases and other predators and aid in its survival, as well as pollinator attractiveness, perception of color, scent, and plant flavors (Dias et al., 2012; Salam et al., 2023). It is estimated that there are about 200,000 plant metabolites, and they are detected using different techniques of analysis and extraction protocols (Fukushima and Kusano, 2013). Analyzing these small chemicals, or metabolites, is the focus of expanding the field of omics science known as metabolomics. It is the scientific examination of the distinctive chemical traces that particular biological processes generate, enabling the accurate characterization of cellular processes and pathways. A branch of metabolomics known as "plant metabolomics" focuses primarily on metabolites generated from plants, their functions in plant metabolism and physiology, and their environmental impacts (Camacho et al., 2005; Rochfort, 2005).

In the rapidly expanding field of study known as "metabolomics," small compounds (also known as "metabolites") found in plants are identified and measured (Monge et al., 2019). Metabolomics has gained more significance in recent years as a result of its potential to help with the study of disease etiology and in the discovery of biomarkers and hence are a valuable source of information for researchers studying plant metabolism The biochemical networks and pathways involved in disease processes, as well as the impact of pharmacological therapies on the metabolic pathways, can be better understood using metabolomics (Putri et al., 2013). Plant metabolomics databases are important because they provide in-depth and extensive knowledge concerning plant metabolism, among which are the recognition and description of metabolic routes, the patterns of gene expression, along with the measurement of metabolites (Marla et al., 2018). Recent years have seen a rise in the significance of plant metabolomics since it provides a thorough understanding of how a plant responds to stress, genetic change, environmental factors, and evaluate the biochemical differences between different plant species. The information in these databases can also be utilized to investigate the impact of genetic modification, medication, and environmental contaminants on plant metabolism by examining the metabolic profiles of plants subjected to varying environmental factors or genetic alterations, and hence can aid in the discovery of novel treatments and medications as well as the understanding of the mechanisms underlying plant adaptability (Patel et al., 2021). Furthermore, the information offered by these databases serves as vital both in plant breeding and genetic engineering as it makes it feasible to discover genes responsible for the synthesis of specific metabolites by analyzing the metabolic profiles of various plants. This knowledge provided by the database can be utilized to develop new plant varieties with superior characteristics, such as increased disease tolerance or higher crop yields (Hur et al., 2013; Staniek et al., 2013). Thus, plant metabolome databases (PMDB) have become a crucial step in the identification and examination of plant metabolites. As a result, plant metabolomics databases are important resources for plant biologists. They offer detailed and accurate information concerning the metabolites found in plants, which helps investigators discover metabolic processes, comprehend the ways plants react to their surroundings, and pinpoint possible sites for genetic engineering. Additionally, they can be utilized for predictive modeling and the discovery of new metabolites.

14.2 TYPES OF PMDB

To increase the efficiency of researchers, in identifying and exploring potential metabolites, their function in important metabolic pathways, and the therapeutic activities of these plant metabolites, several databases such as Golm Metabolome Database (GMD), RIKEN tandem mass spectral database (ReSpect), MetaCyc Metabolic Pathway Database, AraCyc Metabolic Pathway Databases, PlantCyc databases, PMDB, Metabolomic Repository Bordeaux (MeRy-B), PlantGDB, Plant Transcription Factor Database (PlantTFDB), TOMATOMET, Medicinal Plants Database for Drug Designing (MPD3), Plant trait Database (TRY), Transcriptome-Based Plant Database of Genes (EGENES), The Plant Metabolite Hub (PMhub), North East India Medical Plants Database (NEI-MPDB), and PlantPathMarks (PPMdb) play an important role in the advancement of plant research and development. A brief note on these databases is mentioned and listed in Table 14.1.

14.2.1 THE GMD

The GMD (http://gmd.mpimp-golm.mpg.de/) is an open access gas chromatography-mass spectrometry (GC–MS) database that combines mass spectral tags (MSTs) for both clean authentic standards and plant extracts with electron impact (EI) mass spectra that were all acquired under specific conditions. Biologically active metabolites that have been examined using GC–MS provide the basis for metabolite profiling analysis using the GMD reference library. Users can search their MSTs of their plant extract in the search area against GMD databases and match with the records of either recognized metabolites or distinctive identifiers for unidentified peaks in complex plant extracts. Even when the identity of the metabolites producing these peaks is undetermined, the database's inclusion of unidentified MSTs enables peak comparisons between samples. A web form or an application program interface (API) for automated scripting can be used to submit queries to GMD (Kopka et al., 2005). The decision tree search is an important aspect of GMD that predicts substructures that are likely to exist in an unknown metabolite, even if the entire structure has not been established (Hummel et al., 2010, 2013).

14.2.2 RESPECT

The ReSpect database, which is available online at http://spectra.psc.riken.jp/ website, is an open-access repository for MS/MS (MS2) data for phytochemicals that is focused on plant metabolites and currently has 9017 spectrum recordings. The majority of the MS2 data in

TABLE 14.1 List of Plant Metabolomic Databases

Sl. No.	Database	URL	Comments	Reference
1.	The Golm metabolome Database (GMD)	http://gmd.mpimp-golm.mpg.de/	A GC–MS database of plant metabolites	Hummel et al. (2010)
2.	RIKEN tandem mass spectral database (ReSpect)	http://spectra.psc.riken.jp/	An MS2 database for plant phytochemicals	Sawada et al. (2012)
3.	MetaCyc Metabolic Pathway Database	https://metacyc.org	A database of plant metabolic pathways	Krieger et al. (2004)
4.	AraCyc Metabolic Pathway Databases	https://www.arabidopsis.org/	A database of plant metabolic pathways	Mueller et al. (2003)
5.	PlantCyc databases	https://plantcyc.org/	A database of plant metabolic pathways	Zhang et al. (2010)
6.	Plant Metabolome Database (PMDB)	http://scbt.sastra.edu/pmdb/	A detailed database for metabolites	Udayakumar et al. (2012)
7.	Metabolomic Repository Bordeaux (MeRy-B)	http://services.cbib.u-bordeaux.fr/MERYB/index.php	An NMR database of plant metabolites	Ferry-Dumazet et al. (2011)
8.	PlantGDB	plantgdb.org	A database of plant genomes	Dong et al. (2004)
9.	PlantTFDB—Plant Transcription Factor Database	http://planttfdb.gao-lab.org/	A database of plant transcription factors	Guo et al. (2007)
10.	TOMATOMET	http://metabolites.in/tomato-fruits/	An LC–MS database of plant metabolites	Ara et al. (2021)
11.	Medicinal Plants Database for Drug Designing (MPD3)	https://mpd3.com/	A database of medicinal plant metabolites	Mumtaz et al. (2017)
12.	Plant Trait Database (TRY)	https://www.try-db.org/TryWeb/Home.php	A database for plant traits	Kattge et al. (2011)
13.	Transcriptome-Based Plant Database of Genes (EGENES)	http://www.genome.jp/kegg-bin/create_kegg_menu?category=plants_egenes	A database of plant genome	Masoudi-Nejad et al. (2007)
14.	The Plant Metabolite Hub (PMhub)	https://pmhub.org.cn/#/	A database for plant metabolites	Tian et al. (2023)
15.	North East India Medical Plants Database (NEI-MPDB)	https://neist.res.in/neimpdb/	A database for Northeastern medicinal plants and phytochemicals	Kiewhuo et al. (2022)
16.	PlantPathMarks (PPMdb)	http://ppmdb.easyomics.org/	A database for pathway-based markers in plant genome	Mokhtar et al. (2021)

the ReSpect database was compiled from the available scientific literature. In comparison to other MS2 databases, the ReSpect database differs from them in three unique characteristics that improve the efficacy of complex phytochemical structure annotation. The first characteristic is the fragment search technique, which searches adjusted MS2 data using the *m/z* values of fragment ions. ReSpect's second characteristic is the database's enumerated MS2 fragmentation association rules, which are used to identify intriguing connections between variables and the third characteristic that makes it unique from other databases is its plant-specific MS2 data resource. By searching the annotated untargeted MS2 data in the website's fragment, spectrum, and keyword search areas, users can narrow down potential structures of their untargeted MS2 data and also discover connections with other data (Matsuda et al., 2011; Sawada et al., 2012).

14.2.3 METACYC METABOLIC PATHWAY DATABASE

MetaCyc (https://metacyc.org) is an extensive, carefully selected database that includes complete information on enzymes, metabolites, and metabolic pathways that occur in a wide range of organisms, especially in microbes and plants. MetaCyc aims to categorize the whole field of metabolism by maintaining records of an accurate representation of every pathway that has been experimentally described. Approximately, 3105 metabolic pathways and comprehensive data on about 18,566 reactions and 18,973 metabolites are currently included in the database (Krieger et al., 2004). Every pathway in MetaCyc is represented as a sequence of chemical reactions, with complete information on the metabolic enzymes involved, the substrates and products of every reaction, and any cofactors or regulators that are likely to be involved in the process. Due to its user-friendly interface researchers can efficiently search for particular pathways, enzymes, or metabolites of interest using MetaCyc in association with the Pathway Tools software, to create new Pathway Genome Databases (PGDB) from annotated genes or genomes by using it as a reference database (Caspi et al., 2020; Karp et al., 2002).

14.2.4 ARACYC METABOLIC PATHWAY DATABASES

The AraCyc (https://www.arabidopsis.org/) metabolic pathway database is a comprehensive and expandable database of metabolic pathways and the enzymes involved, with a focus on the model organism *Arabidopsis thaliana*, sometimes known as thale cress. It was the first database of plant metabolism predicted computationally, generated by MetaCyc, and it has been further curated to improve data quality and broaden the coverage of pathways (Mueller et al., 2003). With 1418 distinct genes assigned to 204 pathways, AraCyc 2.0 offers robust functional annotations for these genes backed by evidence and citations imported from The *Arabidopsis* Information Resource (TAIR). AraCyc additionally provides the omics viewer, a user-friendly data visualization and analysis tools that enables the depiction of experimental enzymes, gene, or metabolite values and individual chemical structures onto a diagram of the whole AraCyc's pathway map (Paley and Karp, 2002; Rhee et al., 2006).

14.2.5 PLANTCYC DATABASE

PlantCyc databases (https://plantcyc.org/), created by the Plant Metabolic Network (PMN) is a broad network of databases of selected biochemical pathways, genes, enzymes, and compounds specific to different plant species that are involved in the primary and secondary metabolism of plants. These databases offer detailed and precise information concerning the metabolic pathways, enzymes, reactions, and compounds associated with plant growth and development and reactions to various environmental factors. About 500 varieties of plant species were covered by the current PlantCyc databases, including crop plants such as maize, rice, and soybeans and model organisms like *A. thaliana*. These databases are regularly updated and curated to reflect the most recent scientific results. These databases can be used to identify enzymes for specific metabolic pathways, analyze metabolite profiling data and deduce evolutionary data from metabolic connections. PlantCyc databases in collaboration with other databases such as MetaCyc (all-species reference metabolic pathway database) and reference enzyme sequence database can be used to develop particular metabolic pathway databases for plant species. Researchers can use the PlantCyc databases to annotate plant genomes and identify metabolic pathways in different plant species. They also enable comparative genomics studies, allowing researchers to compare pathway and enzyme data between plant species (Hawkins et al., 2021; Jaiswal and Usadel, 2016; Zhang et al., 2010).

14.2.6 PMDB

Extensive datasets on the three-dimensional structures of plant metabolites are available in PMDB (http://scbt.sastra.edu/pmdb/). Currently, the database includes detailed information on the common and metabolite name, as well as 2D and 3D structures with chemical formulas, canonical and isomeric forms of more than a thousand metabolites. The database can be used to draw, view, and work with metabolite structures using programs like JME Editor and Jmol. To provide more details on the metabolites, PMDB also has links both internally and externally to various databases such as KEGG, PubChem, and Cas number. PMDB is a user friendly and openly accessible online with a search engine and menu-driven links for exploring through metabolite names. PMDB seeks to support the scientific community in molecular and structural biology by providing researchers and scientists with instant access to all the information they need (Fiehn, 2002; Udayakumar et al., 2012).

14.2.7 MERY-B

MeRy-B Plant Databases (http://services.cbib.u-bordeaux.fr/MERYB/index.php) are user-friendly datasets of valuable in-depth information regarding different plant species such as their traits, habitats, geographical distribution, and applications. Plant metabolomics data can be stored and visualized using the MeRy-B platform, which is designed to facilitate the analysis and interpretation of NMR metabolic profiles from plants. Additionally, it also helps to identify biomarkers by detecting them through statistical methods and spectrum visualization. In addition to facilitating compound identification, researchers can explore and view the data, use statistical analysis and spectrum visualization tools to distinguish across profiles. MeRy-B further facilitates the overlay and visualization of NMR metabolomic profiles chosen via data or metadata queries. The easy-to-use interface of the Mery-B Plant Databases also provides advanced search tools making it simple for researchers to search for particular plants of interest, explore various categories, and swiftly obtain the needed information (Deborde and Jacob, 2014; Ferry-Dumazet et al., 2011).

14.2.8 PLANTGDB

PlantGDB (plantgdb.org) is an online library of plant genomes that incorporates genomic sequence information for every plant species. It annotates EST (expressed sequence tags) sequences and groups them into contigs, which stand for distinct genes. The database collects and evaluates Viridiplantae (green plants) sequences from UniProt and GenBank into distinct data sets according to the origin of the species or variant and the sequences are also updated and indexed every four months for GeneSeger or BLAST analysis (Kersey, 2019). The database by combining with bioinformatics technologies for predicting genes or cross-species examinations, aims to find gene sets that are either unique to a given species or shared by all plant species. For plant gene discovery, PlantGDB offers EST classification and assembly into putatively unique transcript segments in addition to GSS assembly. Researchers can carry out personalized sequence analysis procedures by using the database's other sequence retrieval and analysis tools, which include the web-based server BioExtract Server and the iterative alignment application Tracembler. Additionally, PlantGDB runs the Plant Genomics Research Outreach Portal, which provides users with a range of training and research resources (Dong et al., 2004; Duvick et al., 2007).

14.2.9 PLANTTFDB

PlantTFDB (http://planttfdb.gao-lab.org/) is an extensive library of putative transcription factors (TFs) predicted from different plant species. The most recent version, 5.0, includes 320,370 TFs from 165 species and is categorized into 58 families using the family assignment methods. The database provides extensive annotation for each identified TF both at the gene and family levels, along with cross-references to other reliable public sources like UniProt, RefSeq, STRING, and Entrez Gene, as well as functional domain data, three-dimensional structures, binding motifs, regulation details, and interaction. Additionally, researchers can query PlantTFDB by using distinct TF IDs provided to each TF, and common names in the quick search field or sequences in BLAST owing to its straightforward and user-friendly interface to explore specific individual TFs of various families for each species. The database also provides flexibility for complex searches. Researchers can search for specific TFs in one or more families, from the same or separate lineages within the species tree. Multiple query conditions, such as general descriptions, protein attributes like the range of sequence length, distinct tissues of gene expression, and multiple fields of annotation for TF entries, can be combined by users in a single search. Additionally, the database also allows to alter and store the search results in different forms for later use (Guo et al., 2007; Jin et al., 2014; Zhang et al., 2011).

14.2.10 TOMATOMET

TOMATOMET (http://metabolites.in/tomato-fruits/) is a database that was developed employing Java 8 (Oracle Corporation) and the Spring Boot (Pivotal Software, Inc.). The database utilizes liquid chromatography (LC)–MS to identify and preserve data on a collection of low-molecular-weight metabolites found in the red fruits of 25 mature tomato varieties. The database uses the PowerGet tool for the detection and alignment of peaks and comprises of 7118 peaks identified with electrospray ionization-positive method, with precise molecular weight, and every peak is labeled with chemical group information, resulting in the discovery of 1577 peaks corresponding to about 16 metabolite groups. The database emphasizes on the enormous chemical variety not just among plant species but also among tomato cultivars implying that the chemical analysis can help illuminate the metabolome and suggest desired biochemical targets to contribute to applications that involve breeding. The database also provides APIs to retrieve and explore the data for computational purposes (Ara et al., 2021).

14.2.11 MPD3

MPD3 (https://mpd3.com/) database is a freely available, database that has been developed as an open-access repository using computational methods (web servers such as Apache HTTP Server, MySQL, and PHP) for identifying and analyzing potential polyphenols from plants having therapeutic properties against a variety of pathogenic organisms, including bacteria, viruses, fungi, hormones, and enzymes. The database is manually compiled from reliable literature databases and comprises information on taxonomical classification, phytochemical activities, structures, test targets, and literature references of medicinal plants. The current MPD3 database comprises 632 Genus Records with 1022 Plant-based data (including 271 targets, 91 activities, along with 7062 entries for phytochemicals) and more than 2295 plant phytochemicals are available in three-dimensional structures for docking which are gathered from 986 references. Due to its straightforward and user-friendly interface, users can obtain information about medicinal plants from a simple query in the web browser (Hussain et al., 2021; Mumtaz et al., 2017).

14.2.12 TRY

TRY database (https://www.try-db.org/TryWeb/Home.php) is a database that is regarded as the primary database that compiles comprehensive records of plant characteristics or traits such as structural, anatomical, biological, biochemical processes, and phenological traits discovered across the world. The database is constructed, hosted, and administered in Jena, Germany, at the Max-Planck Institute for Biogeochemistry. The current TRY version 6 contains about 305,000 plant taxa from 6.24 million individual plants and 15,409,681 trait records. The database allows users to study across a number of areas, including communities and functional ecology, biogeography, and adaptive biology, and also to examine intra- and interspecific trait deviation, which offers new perspectives on the evolution of traits both temporal and spatial scales. Using the TRY database researchers can examine plant characteristics and their global environmental effects. Hence facilitating plant research (Fraser, 2020; Illa Bachs, 2019; Kattge et al., 2011).

14.2.13 EGENES

EGENES (http://www.genome.jp/kegg-bin/create_kegg_menu?category=plants_egenes) is a knowledge-driven freely available database of EST contigs that integrates plant genomic data with higher-level functional information to create a unified database that provides gene indices for every genome enabling researchers to obtain the whole library of genes that function in a plant organism. The database allows its users to identify and rebuild metabolic pathways based on the genome and their comparison with annotation based on ESTs, thereby enhancing the computational prediction of complex systems. The database by connecting genes and transcripts to a chain of interacting metabolites within the cell provides functional information for species for which only EST sequences are available. Furthermore, due to its user-friendly and simple interface, the database allows users to explore the important metabolic processes in plants in a biologically friendly manner so as to help the users study a single protein or gene, trying to make connections between large-scale datasets (Masoudi-Nejad et al., 2007).

14.2.14 PMHUB

The PMhub (https://pmhub.org.cn/#/) is an open-source database that contains comprehensive data on plant metabolites. The database currently contains in-depth information for each metabolite regarding the reference spectra, chemical reactions, genetic basis, important biological functions, and metabolic processes of a variety of plants, including *Oryza sativa*, *Zea mays*, *Triticum aestivum*, and *Lycopersicon esculentum*. The database also provides information on the chemical properties of plant metabolites for 188,837 through different sources, as well as 1,409,947 reaction spectra from standard high-resolution tandem MS spectra for these metabolites. The PMhub allows the reconstruction of a virtual network for each metabolite based on structural relationships and established routes of metabolism. These metabolites with profiled data are also assembled for genetic investigation, together with their related genomic and/or transcriptome information. The database allows researchers to automatically display potential genes for further functional studies when paired with the generated network. The PMhub's broad functionality makes it an invaluable tool for plant metabolomics researchers (Tian et al., 2023).

14.2.15 NEI-MPDB

NEI-MPDB (https://neist.res.in/neimpdb/) is a thorough resource designed to enhance the understanding of pharmacological indications of medicinal plants utilized in the north-eastern region of India that can aid in developing new drugs. The database contains 561 distinct medicinal plant species, along with 9225 phytochemicals. Several techniques, such as text mining, literature mining, and manual curation, have been used to thoroughly curate the database. The information provided by the database includes a list of medicinal plants peculiar to each state in Northeastern India, as well as details on the components, phyto-chemicals, uses, and ailments that the plants can treat. The database is freely available, with a user-friendly interface, and therefore offers researchers a variety of search options for plants, including those related to families, scientific names, common names, diseases, phytochemicals, and their availability in Northeast India's states (Kiewhuo et al., 2022).

14.2.16 PPMDB

PPMdb (http://ppmdb.easyomics.org/) is a comprehensive online database that offers straightforward, interactive access to hundreds of primers and markers mapped to pathways across 82 plant genomes to transform the field of molecular marker research. The database also developed two innovative marker approaches called pathway microsatellite-targeted markers (PMTMs) and pathway gene-targeted markers (PGTMs). The database consists of 2,690,742 pathway-based markers representing 9894 marker panels from 82 plant genomes including 691,555 PGTM and 1,999,187 PMTM. The database is freely accessible and provides users with three collaborative visualization features such as map browse, species comparison, and JBrowse, to assist users in visualizing, mapping, and contrasting the created markers with their KEGG reference route maps (Mokhtar et al., 2021).

14.3 FUTURE PROSPECTIVE OF DATABASES

Metabolomics is used significantly in both biotechnology and agricultural development. Combining metabolomics, transcriptomics, proteomics, and genomic data allows for a more comprehensive view of plant life. Researchers may utilize systems biology to understand the complicated links that exist throughout biological things (Manickam et al., 2023; Panda et al., 2018). Though metabolomics has demonstrated its usefulness, there are still significant obstacles to overcome, such as large-scale metabolite identifications. Metabolomics is also making it possible to better understand medicinal plants and identify significant metabolic QTLs for improved breeding (Sumner et al., 2015). Modern approaches have yielded significant insights into the biology of plants, made it easier to identify new bioactive substances, and created chances to enhance crop characteristics and address global agricultural issues. Plant metabolomics is expected to make a bigger contribution to our understanding of plant systems and the development of innovative solutions for a sustainable future as technology and analytical techniques advance. It will be possible to clarify the roles played by different metabolites in plant growth, development, and responses to environmental stimuli by connecting metabolite profiles with particular biological activities.

14.4 CONCLUSION

Plant sciences and systems biology are seeing rapid growth in the field of plant metabolomics. It comprises detailed analyses of the small molecules (metabolites) present in plant tissues and cells. Technological developments in plant biotechnology have significantly increased our understanding of the metabolic regulation of individual plants. Advanced molecular omics technologies have been widely applied throughout the last decades. High-throughput technologies such as GC–MS and LC–MS can be used to find new metabolic regulators in well-established pathways that affect cellular physiology and, eventually, plant phenotypic. With an emphasis on improving quality, yield-related variables have been given priority in recent metabolomics programs. Specifically, the integration of metabolomics with other methodologies such as transcriptomics, quantitative genetics, and genetic modification has demonstrated its critical significance in crop development.

KEYWORDS

- **metabolomics**
- **plant metabolism**
- **Plant Metabolome Databases**
- **metabolic pathways**
- **plant metabolic network**

REFERENCES

Ara, T., Sakurai, N., Takahashi, S., Waki, N., Suganuma, H., Aizawa, K., Matsumura, Y., Kawada, T., Shibata, D. TOMATOMET: A metabolome database consists of 7118 accurate mass values detected in mature fruits of 25 tomato cultivars. *Plant Direct*, 2021, 5(4), e00318.

Camacho, D., De La Fuente, A., Mendes, P. The origin of correlations in metabolomics data. *Metabolomics*, 2005, 1, 53–63.

Caspi, R., Billington, R., Keseler, I. M., Kothari, A., Krummenacker, M., Midford, P. E., Ong, W. K., Paley, S., Subhraveti, P., Karp, P. D. The MetaCyc database of metabolic pathways and enzymes—A 2019 update. *Nucleic Acids Res.*, 2020, 48(D1), D445–D453.

Deborde, C., Jacob, D. MeRy-B, a metabolomic database and knowledge base for exploring plant primary metabolism. *Plant Metab.: Methods Protocols*, 2014, 1083, 3–16.

Dias, D. A., Urban, S., Roessner, U. A historical overview of natural products in drug discovery. *Metabolites*, 2012, 2(2), 303–336.

Dong, Q., Schlueter, S. D., Brendel, V. PlantGDB, plant genome database and analysis tools. *Nucleic Acids Res.*, 2004, 32(suppl_1), D354–D359.

Duvick, J., Fu, A., Muppirala, U., Sabharwal, M., Wilkerson, M. D., Lawrence, C. J., Lushbough, C., Brendel, V. PlantGDB: A resource for comparative plant genomics. *Nucleic Acids Res.*, 2007, 36(suppl_1), D959–D965.

Ferry-Dumazet, H., Gil, L., Deborde, C., Moing, A., Bernillon, S., Rolin, D., Nikolski, M., de Daruvar, A., Jacob, D. MeRy-B: A web knowledgebase for the storage, visualization, analysis and annotation of plant NMR metabolomic profiles. *BMC Plant Biol.*, 2011, 11, 1–12.

Fiehn, O. Metabolomics–the link between genotypes and phenotypes. *Plant Mol. Biol.*, 2002, 48, 155–171.

Fraser, L. H. TRY—A plant trait database of databases. *Glob. Change Biol.*, 2020, 26(1), 189–190.

Fukushima, A., Kusano, M. Recent progress in the development of metabolome databases for plant systems biology. *Front. Plant Sci.*, 2013, 4, 73.

Guo, A.-Y., Chen, X., Gao, G., Zhang, H., Zhu, Q.-H., Liu, X.-C., Zhong, Y.-F., Gu, X., He, K., Luo, J. PlantTFDB: A comprehensive plant transcription factor database. *Nucleic Acids Res.*, 2007, 36(suppl_1), D966–D969.

Hawkins, C., Ginzburg, D., Zhao, K., Dwyer, W., Xue, B., Xu, A., Rice, S., Cole, B., Paley, S., Karp, P. Plant Metabolic Network 15: A resource of genome-wide metabolism databases for 126 plants and algae. *J. Integrat. Plant Biol.*, 2021, 63(11), 1888–1905.

Hummel, J., Strehmel, N., Bölling, C., Schmidt, S., Walther, D., Kopka, J. Mass spectral search and analysis using the golm metabolome database. In: *The Handbook of Plant Metabolomics* (pp. 321–343), 2013.

Hummel, J., Strehmel, N., Selbig, J., Walther, D., Kopka, J. Decision tree supported substructure prediction of metabolites from GC-MS profiles. *Metabolomics*, 2010, 6, 322–333.

Hur, M., Campbell, A. A., Almeida-de-Macedo, M., Li, L., Ransom, N., Jose, A., Crispin, M., Nikolau, B. J., Wurtele, E. S. A global approach to analysis and interpretation of metabolic data for plant natural product discovery. *Nat. Prod. Rep.*, 2013, 30(4), 565–583.

Hussain, N., Chanda, R., Abir, R. A., Mou, M. A., Hasan, M. K., Ashraf, M. A. MPDB 2.0: A large scale and integrated medicinal plant database of Bangladesh. *BMC Res. Notes*, 2021, 14(1), 1–6.

Illa Bachs, E. TRY plant trait database-enhanced coverage and open access. *Glob. Change Biol.*, 2019, 26(1), 119–188.

Jaiswal, P., Usadel, B. Plant pathway databases. *Plant Bioinf. Methods Protocols*, 2016, 1374, 71–87.

Jin, J., Zhang, H., Kong, L., Gao, G., Luo, J. PlantTFDB 3.0: A portal for the functional and evolutionary study of plant transcription factors. *Nucleic Acids Res.*, 2014, 42(D1), D1182–D1187.

Karp, P. D., Riley, M., Paley, S. M., Pellegrini-Toole, A. The metacyc database. *Nucleic Acids Res.*, 2002, 30(1), 59–61.

Kattge, J., Diaz, S., Lavorel, S., Prentice, I. C., Leadley, P., Bönisch, G., Garnier, E., Westoby, M., Reich, P. B., Wright, I. J. TRY—A global database of plant traits. *Glob. Change Biol.*, 2011, 17(9), 2905–2935.

Kersey, P. J. Plant genome sequences: Past, present, future. *Curr. Opin. Plant Biol.*, 2019, 48, 1–8.

Kiewhuo, K., Gogoi, D., Mahanta, H. J., Rawal, R. K., Das, D., Sastry, G. N. North East India medicinal plants database (NEI-MPDB). *Comput. Biol. Chem.*, 2022, 100, 107728.

Kopka, J., Schauer, N., Krueger, S., Birkemeyer, C., Usadel, B., Bergmüller, E., Dörmann, P., Weckwerth, W., Gibon, Y., Stitt, M. GMD@ CSB. DB: The Golm metabolome database. *Bioinformatics*, 2005, 21(8), 1635–1638.

Krieger, C. J., Zhang, P., Mueller, L. A., Wang, A., Paley, S., Arnaud, M., Pick, J., Rhee, S. Y., Karp, P. D. MetaCyc: A multiorganism database of metabolic pathways and enzymes. *Nucleic Acids Res.*, 2004, 32(suppl_1), D438–D442.

Manickam, S., Rajagopalan, V. R., Kambale, R., Rajasekaran, R., Kanagarajan, S., Muthurajan, R. Plant metabolomics: Current initiatives and future prospects. *Curr. Issues Mol. Biol.*, 2023, 45(11), 8894–8906. https://www.mdpi.com/1467-3045/45/11/558

Marla, S. S., Mirza, N., Nadella, K. Metabolic pathway analysis employing bioinformatic software. In: *Current Trends in Bioinformatics: An Insight* (pp. 175–192), 2018.

Masoudi-Nejad, A., Goto, S., Jauregui, R., Ito, M., Kawashima, S., Moriya, Y., Endo, T. R., Kanehisa, M. EGENES: Transcriptome-based plant database of genes with metabolic pathway information and expressed sequence tag indices in KEGG. *Plant Physiol.*, 2007, 144(2), 857–866.

Matsuda, F., Nakabayashi, R., Sawada, Y., Suzuki, M., Hirai, M. Y., Kanaya, S., Saito, K. Mass spectra-based framework for automated structural elucidation of metabolome data to explore phytochemical diversity. *Front. Plant Sci.*, 2011, 2, 40.

Mokhtar, M. M., El Allali, A., Hegazy, M.-E. F., Atia, M. A. PlantPathMarks (PPMdb): An interactive hub for pathways-based markers in plant genomes. *Sci. Rep.*, 2021, 11(1), 21300.

Monge, M. E., Dodds, J. N., Baker, E. S., Edison, A. S., Fernández, F. M. Challenges in identifying the dark molecules of life. *Ann. Rev. Anal. Chem.*, 2019, 12, 177–199.

Mueller, L. A., Zhang, P., Rhee, S. Y. AraCyc: A biochemical pathway database for Arabidopsis. *Plant Physiol.*, 2003, 132(2), 453–460.

Mumtaz, A., Ashfaq, U. A., ul Qamar, M. T., Anwar, F., Gulzar, F., Ali, M. A., Saari, N., Pervez, M. T. MPD3: A useful medicinal plants database for drug designing. *Nat. Prod. Res.*, 2017, 31(11), 1228–1236.

Paley, S. M., Karp, P. D. Evaluation of computational metabolic-pathway predictions for Helicobacter pylori. *Bioinformatics*, 2002, 18(5), 715–724.

Panda, A., Parida, A. K., Rangani, J. Advancement of metabolomics techniques and their applications in plant science: Current scenario and future prospective. In: *Plant Metabolites and Regulation Under Environmental Stress* (pp. 1–36). Elsevier, 2018.

Patel, M. K., Pandey, S., Kumar, M., Haque, M. I., Pal, S., Yadav, N. S. Plants metabolome study: Emerging tools and techniques. *Plants*, 2021, 10(11), 2409.

Putri, S. P., Nakayama, Y., Matsuda, F., Uchikata, T., Kobayashi, S., Matsubara, A., Fukusaki, E. Current metabolomics: Practical applications. *J. Biosci. Bioengineer.*, 2013, 115(6), 579–589.

Rhee, S., Zhang, P., Foerster, H., Tissier, C. AraCyc: Overview of an Arabidopsis metabolism database and its applications for plant research. *Plant Metabolomics* (pp. 141–154), 2006.

Rochfort, S. Metabolomics reviewed: A new "omics" platform technology for systems biology and implications for natural products research. *J. Nat. Prod.*, 2005, 68(12), 1813–1820.

Salam, U., Ullah, S., Tang, Z.-H., Elateeq, A. A., Khan, Y., Khan, J., Khan, A., Ali, S. Plant metabolomics: An overview of the role of primary and secondary metabolites against different environmental stress factors. *Life*, 2023, 13(3), 706.

Sawada, Y., Nakabayashi, R., Yamada, Y., Suzuki, M., Sato, M., Sakata, A., Akiyama, K., Sakurai, T., Matsuda, F., Aoki, T. RIKEN tandem mass spectral database (ReSpect) for phytochemicals: A plant-specific MS/MS-based data resource and database. *Phytochemistry*, 2012, 82, 38–45.

Staniek, A., Bouwmeester, H., Fraser, P. D., Kayser, O., Martens, S., Tissier, A., Van Der Krol, S., Wessjohann, L., Warzecha, H. Natural products-modifying metabolite pathways in plants. *Biotechnol. J.*, 2013, 8(10), 1159–1171.

Sumner, L. W., Lei, Z., Nikolau, B. J., Saito, K. Modern plant metabolomics: Advanced natural product gene discoveries, improved technologies, and future prospects. *Nat. Prod. Rep.*, 2015, 32(2), 212–229.

Tian, Z., Hu, X., Xu, Y., Liu, M., Liu, H., Li, D., Hu, L., Wei, G., Chen, W. PMhub 1.0: A comprehensive plant metabolome database. *Nucleic Acids Res.*, 2023, gkad811.

Udayakumar, M., Prem Chandar, D., Arun, N., Mathangi, J., Hemavathi, K., Seenivasagam, R. PMDB: Plant metabolome database—A metabolomic approach. *Med. Chem. Res.*, 2012, 21, 47–52.

Zhang, H., Jin, J., Tang, L., Zhao, Y., Gu, X., Gao, G., Luo, J. PlantTFDB 2.0: Update and improvement of the comprehensive plant transcription factor database. *Nucleic Acids Res.*, 2011, 39(suppl_1), D1114–D1117.

Zhang, P., Dreher, K., Karthikeyan, A., Chi, A., Pujar, A., Caspi, R., Karp, P., Kirkup, V., Latendresse, M., Lee, C. Creation of a genome-wide metabolic pathway database for Populus trichocarpa using a new approach for reconstruction and curation of metabolic pathways for plants. *Plant Physiol.*, 2010, 153(4), 1479–1491.

CHAPTER 15

Metabolomics Approaches to Regulating ROS in Plants

ABDUL MUKIT and SANJEEV KUMAR*

[1]*Department of Botany, Central University of Punjab, Bathinda, Punjab, India*

*Corresponding authors

ABSTRACT

Metabolomics is the detailed study of small molecules (called metabolites) within the biological systems that are basically the end-products of various cellular processes such as metabolism and signaling pathways. Metabolomics approaches involve various analytical techniques such as spectroscopy, mass spectrometry, chromatography, liquid chromatography–mass spectrometry, gas chromatography–mass spectrometry, high-performance liquid chromatography, and nuclear magnetic resonance. These analytical tools enable the identification, quantification, and characterization of metabolites in biological samples. When plants get succumbed to different stresses, cell membrane damages, nucleic acid, and protein degradations occur in cells due to the accumulation of reactive metabolites called reactive oxygen species (ROS). These are highly reactive molecules that include superoxide radicals, hydrogen peroxide, hydroxyl radicals, and singlet oxygen formed during the metabolism of plants. ROS in plants act as dwell-edged swords because, when present in low concentrations in the cells, they can mediate various signaling pathways and function as important signaling molecules in the regulation of several plant processes such as growth, development, and physiology; however, when present in excess, they become toxic and cause cellular damages. So, the ROS in plants need to be effectively regulated to maintain cellular homeostasis. Metabolomics approaches have proven to be valuable in understanding the regulating ROS levels in plants. This chapter gives an account of the approaches in which metabolomics can contribute to regulating ROS in plants. Here, various metabolomics approaches such as profiling of antioxidant metabolites, understanding ROS signaling pathways, identifying ROS producing metabolites, and targeted metabolomics have been discussed. These approaches can be employed to understand the regulating ROS levels in plants.

Exploring Plant Stress Tolerance: A Metabolomics Approach.
Shuvasish Choudhury, Debojyoti Moulick, and Mirza Hasanuzzaman (Eds.)

DOI: 10.1201/9781779641731-15

15.1 INTRODUCTION

15.1.1 METABOLOMICS

Metabolomics is the systemic identification and analysis of metabolites that are small molecules with low molecular weights and are the products of cellular metabolisms in an organism (Idle and Gonzalez, 2007). Metabolites take part in the metabolic reactions of the cells and are essential for its growth, maintenance, and normal functioning (Harrigan and Goodacre, 2003). Metabolites broadly include numerous small molecules such as sugars, organic acids, amino acids, and nucleotides (Idle and Gonzalez, 2007). The collection of all these metabolites both in terms of quality and quantity is referred to as metabolome (Harrigan and Goodacre, 2003). Metabolomics is the most recently developed omics technique and is generally used as a complementary approach to genomics, transcriptomics, and proteomics (Tan et al., 2016). Metabolomics has diverse applications in various branches of studies, including genetics, toxicology, botany, environmental analysis, forensics, and food sciences (Jones and Cheung, 2007). Various metabolomics tools such as gas chromatography–mass spectrometry (GCMS), liquid chromatography–mass spectrometry (LCMS), high-performance liquid chromatography (HPLC), and mass spectrometry (MS) help in the qualitative and quantitative identification and analysis of metabolites present in the entity under consideration. In the context of this chapter, metabolomics provides a comprehensive view of metabolic changes associated with reactive oxygen species (ROS) regulation in plants (Arbona et al., 2013). Various metabolomic approaches as discussed later help in the identification, quantification, and profiling of antioxidant metabolites as well as ROS producing metabolites. Such knowledge would be helpful in maintaining equilibrium status of the cells by regulating the excess ROS through antioxidant scavengers and thus protecting the cells from oxidative stress.

15.1.2 METABOLOMICS APPROACHES AND QUANTIFICATION TECHNIQUES

Metabolomics approaches are a set of techniques and methods used to study the metabolome. Based on the approach and scope, metabolomics can be differentiated broadly into two types: targeted metabolomics and nontargeted/untargeted metabolomics. Targeted metabolomics involves the quantification of specific metabolites based on the use of internal standards, whereas untargeted metabolomics involves the quantification of all metabolites present in a sample (Roberts et al., 2012). There are various techniques used for the quantification and analysis of metabolites and the most commonly used metabolomic tools are descripted in Table 15.1.

15.2 REACTIVE OXYGEN SPECIES

ROS are very reactive molecules of oxygen generated in an organism as an unavoidable consequence of normal metabolism involving oxygen and encompass various free radical and nonradical molecules such as superoxide anion, hydroxyl radical, hydrogen peroxide,

TABLE 15.1 Different Tools for Quantification and Analysis of Metabolites

Sr. No	Technique/Tool	Description	References
1	Chromatography	It is a technique used for separating a mixture into its individual components. The mixture to be separated is dissolved in a mobile phase that is fluid in nature (gas/liquid). This mobile phase carries the mixture through a system (column, plate, paper, etc.), on which a stationary phase (solid/liquid) is fixed. Based on the nature of mobile phase, chromatography is of two types, namely, gas chromatography and liquid chromatography.	(Poole and Poole, 2012; Rahman et al., 2015; Fanali et al., 2017; Dalvi, 2021)
2	Spectrophotometry	It is a technique to measure the amount of light absorbed or transmitted by a substance as a function of the wavelength or frequency of light and uses an instrument called spectrophotometer to measure the intensity of light at different wavelengths. It is a quantitative method where a sample is kept in the pathway of a beam of light, and the amount of light absorbed or transmitted by the sample is detected and measured. Such measurements would provide important insights about various properties such as chemical concentration and composition of the samples.	(Sunheimer et al., 2011; Yoon and Yoon, 2016; Kafle, 2019; Bachmann and Miller, 2020)
3	Mass spectrometry (MS)	This is a technique having high sensitivity and specificity and is used for determining the chemical compositions, structures, and various other properties of the samples. In this technique, samples are first ionized and then separated based on their mass-to-charge ratios and at last the ions are detected and measured based on their abundances.	(Van Bramer, 1998; Madeira and Florêncio, 2012; Patel et al., 2012; Awad et al., 2015)
4	Gas chromatography–mass spectrometry (GCMS)	It is a combination of two tools, that is, chromatography and MS. In this technique, samples that are usually volatile compounds are separated into individual components using a gas chromatography column and the separated compounds are then ionized, detected, and measured through the MS part. This tool has various advantages such as high resolution and sensitivity, and therefore, it is mostly used for the analysis of essential oils.	(Hites, 1997; Sparkman et al., 2011; Hussain and Maqbool, 2014; Stashenko and Martínez, 2014; Al-Rubaye et al., 2017)
5	Liquid chromatography–mass spectrometry	Like GCMS, this technique is also a combination of chromatography and MS. The only difference is it involves liquid chromatography in place of gas chromatography. Here, samples are dissolved in a liquid solvent and are separated into individual components using a liquid chromatography column and then like GCMS, the separated compounds are ionized, detected, and measured through MS.	(Ardrey, 2003; Allwood and Goodacre, 2010; Trufelli et al., 2011; López-Fernández et al., 2020; Pramod et al., 2020)

TABLE 15.1 *(Continued)*

Sr. No	Technique/Tool	Description	References
		It has several advantages over GCMS as it can analyze both small and large molecules with high sensitivity and is also suitable for complex sample mixtures. It has been preferably used for the analysis of nonvolatile and polar compounds such as peptides, proteins, and pharmaceuticals.	
6	High-performance liquid chromatography	It basically works on the principles of chromatography, more specifically liquid chromatography, where a solid constitutes the stationary phase and the mobile phase would be a liquid solvent or a mixture of solvents. It is a widely used analytical tool and has been used for the separation, identification, and quantification of components in complex mixtures, and is employed in various industries, including pharmaceuticals, environmental analysis, food science, forensics, and more. It is advantageous as it can handle a diverse range of sample volumes and concentrations and also provides high resolution and sensitivity.	(Lough and Wainer, 1995; Horváth, 2013; Meyer, 2013; Reuhs, 2017; Žuvela et al., 2019; Ali, 2022)
7	Nuclear magnetic resonance (NMR) spectroscopy	This technique is based on the interactions of magnetic field with the nucleus of atoms present in a molecule. It provides information about the dynamics and molecular structures of the compounds. In this technique, samples are first placed in strong magnetic fields. Then, by the application of radiofrequency pulse, the spin orientation of the nuclei is disturbed and later returned to original state by stopping the pulse. During this process of alterations in spin orientations, radiofrequency signals are emitted which are detected by a receiver. The detected frequency signals are used to obtain information about the connectivity and arrangement of the atoms in a molecule and also to know about the other properties such as molecular conformations and dynamic behavior. NMR spectroscopy is advantageous as it gives information about the structure–function relationships of molecules and plays a significant role in determining the 3D structures of proteins and nucleic acids. It is extensively used in fields of chemistry, biology, and material sciences.	(Wuethrich, 1989; Simpson et al., 2012; Günther, 2013; Elyashberg, 2015; Hatzakis, 2019; Lambert et al., 2019)

and singlet oxygen, respectively (Sharma et al., 2012). Table 15.2 shows the four main forms of ROS generated in plants along with their characteristic features and the cell organelle where they are mostly produced. These molecules contain at least one oxygen atom and unpaired electron(s) and have the capability to exist independently (Jakubczyk et al., 2020). Unlike other reactive species produced during metabolic processes, ROS can impact on cell function in specific or overlapping ways (Sies et al., 2017). In plants, ROS is generated mainly as products of metabolic processes during abiotic or biotic stress responses. Various stresses such as heat, cold, salt, UV rays, pollution, and pathogens instigate defense responses in the plants that produce ROS as byproducts at different organelles of the plant. ROS can be generated at various organelles depending on the site of reaction/response. In plants, the cell organelles that produce ROS include chloroplast, mitochondria, plasma membrane, apoplast, cell wall, peroxisomes, cytosol, and endoplasmic reticulum (Das and Roychoudhury, 2014). ROS are produced due to the unavoidable electrons leakage from the electron transport activities of the cell organelles such as mitochondria and chloroplasts (Sharma et al., 2012). ROS are highly reactive and cause oxidative stress resulting in damages to various biomolecules such as lipids, proteins and DNA, and ultimately can lead to cell death (Anjum et al., 2015). Thus, overproduction of ROS is toxic for living cells as it imbibes oxidative stress and may even lead to cell death. However, ROS in all is not bane for the plants as studies have evidently found that ROS accumulation below damaging levels would be a boon for plant growth and development.

Like ROS, reactive nitrogen species (RNS) are also highly reactive molecules and are generated from nitric oxides (NO) and oxidative metabolisms. The commonly occurring RNS in plants include free radicals such as –NO, nitrogen dioxide ($=NO_2$), and nonradicals like peroxynitrite ($ONOO^-$), nitroxyl anion (NO^-), nitrate (NO_3^-), nitrosonium cation (NO^+), dinitrogen trioxide (N_2O_3), dinitrogen tetroxide (N_2O_4), nitryl chloride (NO_2Cl), and nitrous acid (HNO_2) (Augusto et al., 2002; Halliwell, 2006; Airaki et al., 2015; Aranda-Caño et al., 2024). Plants being sessile are vulnerable to various biotic and abiotic stresses and such stresses lead to the instability in redox homeostasis. As a result, in consequence of activating defense mechanism, rapid formation of RNS and ROS occurs and acts as signaling molecules. These highly reactive species counter balance the in-built antioxidants system present in the cells. However, if the particular stress continues for longer period, the increasing accumulation of ROS and RNS can exceed the scavenging limit of the antioxidant systems and lead to cellular components and eventually to cell death (Ahmad et al., 2018; Chaki et al., 2021; Aranda-Caño et al., 2024). Among all the RNS that occur in plants, NO is the most important one because it acts as a signal molecule in defense responses as well as takes part in various physiological and developmental processes of the plants (Aranda-Caño et al., 2024). RNS in plants can be generated through various enzymatic and nonenzymatic processes. Enzymatic processes that generate RNS include L-arginine dependent nitric oxide synthase (NOS) activity, nitrate reductase activity, and other enzymatic systems (Saddhe et al., 2019; Gupta et al., 2022; Aranda-Caño et al., 2024). The nonenzymatic pathways of RNS generation include light-dependent NO production involving carotenoids in chloroplast (Cooney et al., 1994) or the reduction of NO_2^- to NO under low pH in highly reducing environments as observed in the apoplast of barley cells (Bethke et al., 2004; Aranda-Caño et al., 2024).

TABLE 15.2 Four Main Forms of ROS Generated in Plants during Stress Conditions

Sr. No	ROS Name	Characteristics	Cell Organelle	References
1	Singlet oxygen ($^1O_2$)	Short-lived molecule and can still cause damage to wide range of molecules. Causes oxidation of lipids and proteins.	Chloroplast	(Asada et al., 1974; Egorov et al., 1989; Telfer et al., 1994; Davies, 2004; Triantaphylides et al., 2008)
2	Superoxide (O_2^-)	The consumption of 1–2% of free oxygen in plants leads to generation of O_2^- and it is usually the first formed ROS. Its generation triggers the formation of more reactive ROS like ·OH and $^1O_2$. Reacts with Fe–S proteins (Mehler reaction). Dismutase to hydrogen peroxides.	Chloroplast, mitochondria, peroxisomes, cytosol	(Mehler, 1951; Asada et al., 1974; Sawyer et al., 1980; Elstner, 1982; Puntarulo et al., 1988, 1991; del Rio et al., 1989; Telfer et al., 1994; Møller, 2001; Bouché et al., 2003; Halliwell, 2006)
3	Hydrogen peroxides (H_2O_2)	Displays various beneficial functions at low concentrations, and causes severe cellular damages at increased concentrations. At low concentration, it performs as signal molecule and regulates various physiological process but at high concentration it induces oxidative stress and damages the biomolecules.	Chloroplast, mitochondria, peroxisomes, cytosol	(Asada et al., 1974; Telfer et al., 1994; Popov et al., 1997; Mittler and Berkowitz, 2001; Foyer et al., 2009)
4	Hydroxyl radicals (·OH)	Has single unpaired electron. Can lead to maximum cellular damage due to lack of its effective detoxification in plants.	Mitochondria, cytosol	(Haber and Weiss, 1934; Bors et al., 1979; Willson, 1979; Kehrer, 2000; Doudican et al., 2005)

15.2.1 ROS AS SIGNALING AND STRESS PRODUCTS

ROS are often referred as dual-edged swords because they can act as both beneficial and harmful for the plants depending on their concentrations. At lower or moderate concentrations in plant cells, ROS act as signal products (secondary messengers) and can mediate signaling pathways, such as closing of stomata (Yan et al., 2007) apoptosis (Mittler, 2002), and gravitropism of roots (Joo et al., 2001), which facilitate resistance against different environmental stresses and also play important roles in the overall growth and development (Torres et al., 2002; Miller et al., 2008). However, at higher concentration, ROS acts as stress products leading to oxidative stress that damage biomolecules and can eventually lead to cell death (Anjum et al., 2015). Metabolic ROS and signaling ROS are the two main sources of ROS production during stress, which are produced either as a result of alterations in metabolic activities or for signaling purpose as part of stress response (Choudhury et al., 2017). Thus, ROS can act as both signaling and stress products (Figure 15.1). Based on the equilibrium status between ROS generation and their regulation by antioxidant defense systems, ROS can act as damaging, protective, or signaling factors (Gratão et al., 2005). A balance between ROS generation and proper scavenging would protect the photosynthetic machinery, maintain membrane integrity, and prevent damages (Hasanuzzaman et al., 2020b). The concentration of ROS in plants can be monitored via the activities of different antioxidants (Ahmad et al., 2010).

15.2.2 PATHWAYS OF ROS SIGNALING AND STRESS REGULATIONS

Production of ROS is an obvious outcome in the metabolic processes during environmental stresses and plays a crucial role in cellular signaling and stress regulation. ROS were known as harmful products until when evidence was found regarding the roles of ROS as important signaling products (Das and Roychoudhury, 2014). Plants can sense ROS signals directly as well as indirectly. They perceive ROS signals directly with the help of key signaling proteins such as tyrosine phosphatase and indirectly through gene expressions, phosphorylation of protein, as well as using redox-sensitive proteins (Xiong et al., 2002). In the living cells, ROS sensors recognize the ROS produced during stress and then translate and transmit the signals to the nucleus via redox reactions in coordination with the mitogen-activated protein kinase (MAPK) pathway. Consequently, the stress responsive transcription factors get activated and bind to their specific cis-acting elements, resulting in the reprogramming of transcription, new protein translation, and the modification of old proteins (Foyer and Noctor, 2016). ROS are first sensed at apoplasts, after which signaling (which may be intercellular or intracellular) is activated which would lead to altered gene expressions (Singh et al., 2019). It is important to note that the strength, lifetime, and size of the ROS signaling pool is determined by the balance between ROS production and efficient scavenging by the antioxidants (Foyer and Shigeoka, 2011). The cell employ antioxidant defense mechanisms include various enzymatic and nonenzymatic antioxidants to counterbalance

ROS production and maintain redox homeostasis (Birben et al., 2012). There are several pathways involved in ROS signaling and stress regulation as discussed below.

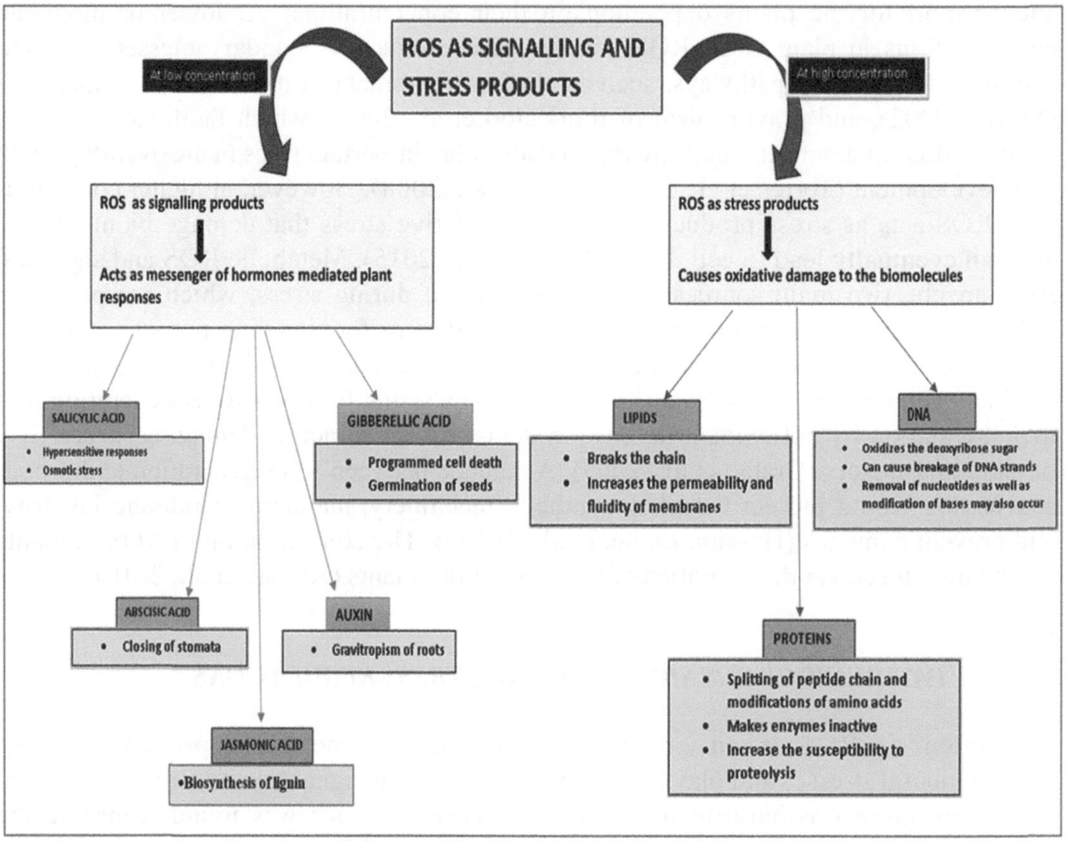

FIGURE 15.1 ROS at opposing concentrations act as signaling and stress products.

15.2.2.1 ACTIVATION OF NADPH OXIDASES

Nicotinamide adenine dinucleotide phosphate hydrogenoxidases (NADPH) are a family of enzymes responsible for generating ROS (Magnani and Mattevi, 2019). The production of ROS starts when specific signal transduction pathways activate these oxidases. This activation can be triggered by various stimuli, including growth factors, cytokines, and microbial pathogens (Bedard and Krause, 2007). NADPH oxidase is a plasma membrane-bound enzyme and member of the NOX family, and when activated, it catalyzes the production extracellular superoxide ($O_2^{\bullet -}$) (Foreman et al., 2003; Sagi and Fluhr, 2006). The NADPH oxidases of plants are also called as respiratory burst oxidase homologues (RBOHs) and they catalyze the formation of superoxides which are then involved in several essential processes in plants (Qu et al., 2017). In plants, the RBOHs play various roles in the overall growth and development of the cells, defenses against different environmental stimuli, and cellular signal transduction (Marino et al., 2012).

15.2.2.2 MITOCHONDRIAL ELECTRON TRANSPORT CHAIN

A key source of ROS generation in the cells is the electron transport chain operating within the mitochondria (Murphy, 2009). A small fraction of electrons leak from the electron transport chain (ETC) during normal oxidative phosphorylation, results in the formation of superoxide radicals and these radicals can act as signaling molecules under physiological conditions (Brand, 2016). Superoxide is produced by the electron transport chain when single electrons leak to oxygen (O_2) during the movement of electron pairs downward in the chain (Chance et al., 1979). The ROS generation in the mitochondria is an unavoidable consequence of the oxidative adenosine triphosphate (ATP) production (Harman, 1972). Mitochondria being the powerhouse of the cell during the normal process of ATP production produces ROS as byproducts of ETC which affect various processes in the cells (Huanget al., 2016). Recent evidence suggests that the ROS generated by mitochondria, called mitrochondrial ROS, may act as key intermediates in hormones-mediated plant signaling such as abscisic acid, jasmonic acid, salicylic acid, and auxins (Huang et al., 2016).

15.2.2.3 ROS-MEDIATED REDOX SIGNALING

All the metabolic processes in plants are sensitive to alterations in external conditions, and therefore, any imbalance in the metabolic processes might lead to the production and accumulation of ROS leading to oxidative stress. This oxidative stress causes oxidation of the biomolecules, thereby hampering normal cellular activities, affecting the integrity of the cell organelles, and thus hindering metabolic activities (Suzuki et al., 2012). ROS generated can serve as a secondary messenger in intracellular signaling and can also modify protein cysteine residues, leading to the formation of disulfide bonds and subsequent activation of specific signaling pathways (Moldogazieva et al., 2018). To maintain proper metabolic balance, redox signaling is a very important factor as it has the capability to regulate diverse cellular processes such as growth, proliferations, differentiations, and apoptosis of the cells (Matés et al., 2008). Studies have found that ROS metabolism is interlinked with redox regulation and this causes an optimized functioning of different cell organelles such as mitochondria and chloroplasts (Farooq et al., 2019). Further, it has also been established that network genes of ROS and the regulatory enzymes of redox signaling are expressed together in those cell organelles (Suzuki et al., 2012). According to recent studies, ROS are produced by the cells motives so that redox signals can be induced, and these redox signals are important as they can regulate various responses within the cells that are crucial for life (Finkel, 2011). In plants, redox signals trigger the expression of genes that would help in maintaining cellular redox homeostasis and progression of numerous physiological processes. The cell death is caused by an apoptosis pathway activated by ROS (Dangl and Jones, 2001). Overall, these studies reveal that redox signaling mediated by ROS can act as a key regulator of various responses in the cells that form part of different physiological processes.

15.2.2.4 STRESS-RESPONSIVE KINASES

In response to various stresses, the generated ROS can lead to the activation of various stress-responsive kinases such as MAPKs and other kinases (Son et al., 2011). MAPK signaling can regulate a number of cellular functions ranging from cell death to cell survival (Haagenson and Wu, 2010). ROS can lead tp activation of diverse transcription factors that regulate gene expressions involved in oxidative stress (Liu et al., 2017). MAPKs in plants are large families of proteins, especially involved in signaling. They have diversified roles in various external stimuli and form complex networks in cells. When a plant responds to any stress, the MAPK pathways get activated (Kim and Choi, 2010) and this MAPKs signaling modulates growth and development of plants during stress by various transcriptional as well as posttranscriptional regulations (Sharma et al., 2020). It has been found that MAPK signaling is induced by H_2O_2 and this signaling is important in the stress response as it helps to translate the environmental stimulus into intracellular alterations (Zhou et al., 2014).

15.2.2.5 NRF2 (NUCLEAR FACTOR-ERYTHROID 2-RELATED FACTOR 2) PATHWAY

This is an important protein involved in antioxidant responses of the cells and regulates the cellular defense against toxic and oxidative stimuli by specifically expressing those genes that are involved in oxidative stress response (He et al., 2020). In unstressed conditions, NRF2 protein is segregated in cytosol by KEAP1 protein (Kelch-like ECH-associated protein 1), which acts as its antagonist, and is maintained in low concentration due the proteasomal degradation (Taguchi and Yamamoto, 2017). However, on exposure to ROS, KEAP1 is modified leading to the stabilization and nuclear translocation of NRF2 (Nguyen et al., 2003). In the nucleus, the antioxidant response elements bind to NRF2 protein and thus promoting antioxidant enzymes expression (Hayes et al., 2009). Various electrophiles and quinones that have diverse sources as well as structures induce and activate the NRF2-mediated pathways (Hayes et al., 2009). Due to its involvement in diversified responses against broad range of substances, NRF2 protein has been regarded as the main regulator of defense mechanisms of cells against oxidative stress (Vomund et al., 2017).

15.2.2.6 HEAT SHOCK RESPONSE

When ROS is in high concentration, the resultant oxidative stress can induce heat shock response, a protective mechanism against various stressors including the oxidative stress (Lü et al., 2010). With the induction of heat shock response, the synthesis of special proteins called heat shock proteins (HSPs) is elevated due to the upregulation of special genes and simultaneously normal protein synthesis is minimized (ul Haq et al., 2019). These HSPs are multifunctional proteins that closely interact with the antioxidants (Reeg et al., 2016). HSPs are upregulated in response to ROS and act as molecular chaperones to maintain protein homeostasis and cellular integrity (Ahuja et al., 2010). The main functions of

HSPs in general include proper folding and unfolding of proteins, their transportation in the cells, and also degradation and discarding of alien proteins (Hartl et al., 2011). Under stress, HSPs are found to be present in almost all the compartments of the cells (Waters et al., 1996). Other than being just as stress responsive genes, HSPs also have roles in nonstressed conditions and contribute in the overall growth and development of the plant (Beck et al., 2007).

15.3 ANTIOXIDANTS

Antioxidants can be defined as the compounds that protect the cells from the damages of free radicals or ROS. When the plants are in normal unstressed conditions, the ROS metabolites are generated minimally and there exists a balance between the ROS production and subsequent quenching (Figure 15.2) (Sharma et al., 2012). Therefore, antioxidants can also be defined as ROS quenchers or scavengers that protect the cells from oxidative damages (Young and Woodside, 2001). Efficient functioning of antioxidant defense system helps the plants to be resistant and have better performance during environmental stresses. The antioxidant defense system is present in all the living organisms, although the mechanism of defenses may differ from species to species. Plants possess complex antioxidative defense systems and these antioxidants are present in different forms in the cellular environment (discussed in later sections). When in steady or equilibrium state, various antioxidants scavenge the ROS molecules in the cells (Foyer and Noctor, 2005). However, this equilibrium can be hindered by various environmental stress factors, such as heat, cold, and pathogens, and such disturbances cause overproduction of ROS in the cells, leading to oxidative stress. A study has found that O_2 consumption even at meager 1–2% can cause ROS formation in plants (Bhattachrjee, 2005). Higher plants elevate the levels of their in-built antioxidant defense systems during stresses to protect themselves from any oxidative damages that may occur due to the overproduction of ROS under stress conditions (Sharma et al., 2019).

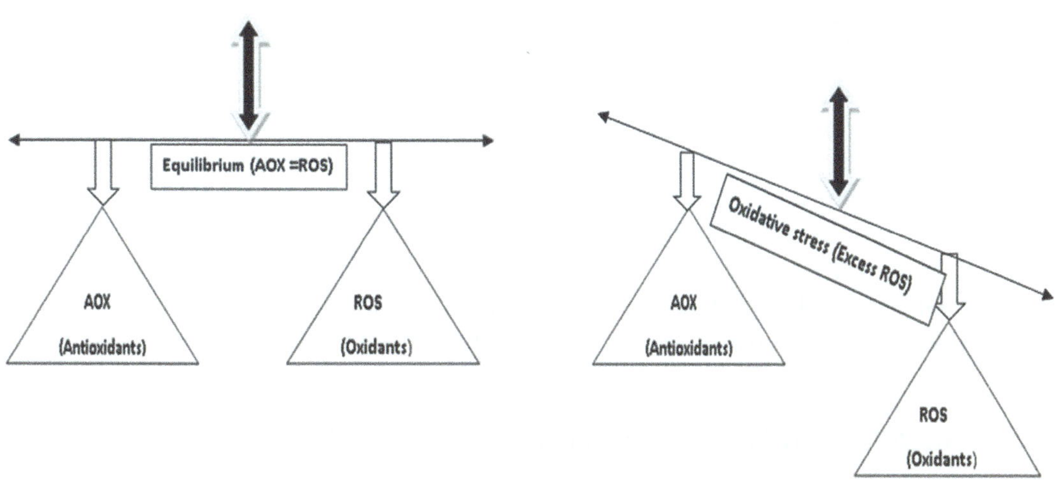

FIGURE 15.2 Diagrammatic representation of how antioxidant defense system and oxidative stress relate to ROS.

15.3.1 OXIDATIVE STRESS

The condition that arises within a cell when there is a lack of balance between the production of ROS and their scavenging by the antioxidants is referred to as oxidative stress (Persson et al., 2014). ROS are highly reactive molecules and their production beyond a limit in the cell creates an imbalance in redox status, thereby inducing an oxidative stress that may even proceed to cell death (Marino et al., 2012). These ROS are generally generated when the molecular oxygen undergoes incomplete reduction to water during aerobic metabolism, and when produced in excess they disturb the ROS–antioxidants equilibrium leading to oxidative stress and even to the death of cells in severe conditions (Hasanuzzaman et al., 2020a, 2020b). When a cell is under oxidative stress, various biomolecules such as lipids, proteins, DNA, and carbohydrates get modified due to the incurred oxidative damage (Gill and Tuteja, 2010). Besides this, during oxidative stress, due to the cellular damages, integral properties of membranes, such as fluidity, ion transport, and enzyme activity, also get modified (Sharma et al., 2012). Different environmental stresses contribute to ROS production and subsequent oxidative stress in different ways. For instance, drought stress causes closing of stomata and thus reduces carbon dioxide fixation creating a disequilibrium between the harvest and usage of light that creates a disbalance between the release of electrons and their acceptance causing overproduction of ROS from the leftover excess photons in the photosystem (Hasanuzzaman et al., 2020a, 2020b). Similarly, salinity also causes oxidative stress as factors such as osmotic stress, ion toxicity, genotoxicity, and nutritional deficiency that come along with salinity induce the production of ROS (Chawla et al., 2013). Further, metals/metalloids toxicity also cause oxidative stress resulting in imbalance of ROS–antioxidant equilibrium (Al Mahmud et al., 2018). Likewise, during heat and cold stress, ROS overproduction and oxidative stress occur (Guo et al., 2016; Pospíšil, 2016). Therefore, it is very important to strictly control ROS production and its proper scavenging in order to maintain equilibrium and avoid oxidative stress.

15.3.2 ANTIOXIDANT METABOLITES

In plants, the accumulation of excess ROS is regulated by the antioxidant defense system that includes a variety of scavengers which are either enzymatic or nonenzymatic low molecular metabolites (Gill et al., 2011). Such ROS scavenging entities are called as antioxidant metabolites. As ROS play diverse roles being a signaling as well as a stress product, it is important to maintain a balance or equilibrium rather than completely scavenging the ROS. In the cells, excess ROS are scavenged by antioxidant defense systems that consist of both enzymatic and nonenzymatic antioxidants that are discussed as follows. Table 15.3(a) and 15.3 (b) depict the different enzymatic and nonenzymatic antioxidants and their characteristic features, respectively.

15.4 METABOLOMICS AND ROS REGULATION IN PLANTS

Metabolomics approaches have proven valuable in understanding and regulating ROS levels in plants (Arbona et al., 2013). Metabolomics provides a comprehensive view of

TABLE 15.3 (a) Enzymatic Antioxidants

S no.	Antioxidant	Features	References
1	Superoxide dismutase (SOD)	Catalyze the dismutation of superoxide ($O_2^{\bullet -}$) to molecular oxygen (O_2) and water (H_2O_2). SOD activity increases in plants when exposed to various environmental stresses. Overproduction of SOD increases oxidative stress tolerance.	(Gupta et al., 1993; Racchi et al., 2001; Mishra et al., 2011)
2	Catalase (CAT)	Catalyze hydrogen peroxide (H_2O_2) dismutation to H_2O and O_2. In some plants, it has been observed that overproduction of CAT enhances tolerance against metal-induced oxidative stress.	(Scandalios et al., 1997; Guan et al., 2009)
3	Guaiacol peroxidase	Has roles in important cellular processes such as cell wall lignification, auxin degradation, ethylene biosynthesis, and responses against various environmental stresses. Functions as effective scavenger of different ROS under stressed conditions.	(Vangronsveld and Clijsters, 1994; Kobayashi et al., 1996)
4	Ascorbate peroxidase (APX)	It is found in various cell organelles and is regulated by redox signals and H_2O_2. APX present in the organelles scavenge H_2O_2 produced within the respective organelles, while the APX found in cytosols scavenges H_2O_2 present in cytosol.	(Mittler and Zilinskas, 1992; Patterson and Poulos, 1995)
5	Monodehydroascorbate reductase	Overexpression enhances tolerance against salt, poly(ethylene) glycol, and viologen-mediated oxidative stresses. It has the capability to reduce phenoxyl radicals	(Sakihama et al., 2000; Eltayeb et al., 2007; Li et al., 2010)
6	Dehydroascorbate reductase	Activity increased by environmental factors such as drought stress, metal toxicity, and cold stress. When expressed in excess, it infers increased tolerance to plants against herbicides and stress factors such as drought and salt.	(Sharma and Dubey, 2005; Eltayeb et al., 2011)
7	Glutathione reductase (GR)	Activity is correlated with oxidative stress resistance and this activity increases under stressed conditions. GR synthesis can be stimulated due to the oxidative stress caused by H_2O_2.	(Pastori and Trippi, 1992; Sharma and Dubey, 2007)

TABLE 15.3 (b) Nonenzymatic Antioxidants

S no.	Antioxidants	Features	References
1	Ascorbate	It is the most abundant antioxidant metabolite and plays important roles in defense against oxidative stresses Overexpression of enzymes associated with its biosynthesis process induces stress tolerances in plants.	(Zhang et al., 2011; Sharma et al., 2012)
2	Glutathione	It is found in almost all the compartments and organelles of the cell and plays key roles in intracellular defense against oxidative damages. It can directly function as ROS scavenger as it can react chemically with the free radicals. Higher level of reduced glutathione provides greater protection against oxidative stress.	(Foyer and Noctor, 2003; Eltayeb et al., 2010; Asada, 2019)
3	Tocopherols	These are lipid loving antioxidants and take part in scavenging of ROS such as oxygen free radicals, lipid peroxy radicals, and $^1O_2$. They protect the structure and function of photosystem by protecting the lipids and other components of membrane physically scavenging ROS and chemically reacting with oxygen in the chloroplasts.	(Diplock et al., 1989; Ivanov and Khorobrykh, 2003)
4	Carotenoids	These are also lipid loving antioxidants and have the ability to scavenge various forms of ROS Carotenoids scavenge $^1O_2$ (singlet oxygen) radical and protect the cells from damages of oxidative stress. Under salinity stress, higher contents of carotenoids can provide better adaptation to the plants.	(Young, 1991; Li et al., 2008; Gomathi and Rakkiyapan, 2011)
5	Phenolics	Secondary metabolites possess antioxidant properties and include flavonoids, tannins, hydroxycinnamate esters, and lignin. Can reduce the fluidity of membranes by modifying the packing orders of lipids. Increasing the concentration of flavonoids elevates the antioxidant capacity of the plants.	(Arora et al., 2000; Grace and Logan, 2000; Lukaszewicz et al., 2004)

metabolic changes associated with ROS regulation in plants which would help to understand the underlying mechanisms and develop strategies for ROS regulation and increase stress tolerance. Various approaches through which metabolomics can contribute to regulating ROS in the plants are discussed as follows.

15.4.1 PROFILING ANTIOXIDANT METABOLITES

The systematic analysis of metabolites to determine their quantity based on the observable features is termed as metabolic profiling (Clarke and Haselden, 2008). The most preferably used metabolomics tools include nuclear magnetic resonance (NMR) spectroscopy and MS because of their diverse advantages as discussed earlier (Clarke and Haselden, 2008). Among other techniques, GCMS is used for metabolic profiling as it is a multidimensional tool applied in different fields of biosciences due to its capability to identify and quantify diverse metabolites in a single sample (Lisec et al., 2006). Moreover, various metabolomics tools enable the quantification of important antioxidant compounds such as ascorbate, glutathione, and tocopherols simultaneously (Llanes et al., 2018). By determining the alterations in the concentrations of these metabolites under different stress conditions, important ROS scavenging antioxidants can be identified. The knowledge of these ROS scavenging antioxidants could be used to elevate their concentration by external supply to maintain the ROS–antioxidant equilibrium during stressed conditions.

15.4.2 UNRAVELLING ROS SIGNALING PATHWAYS

Various ROS signaling pathways have been discussed earlier. Metabolomics can reveal alterations in signaling molecules of these pathways linked to ROS. For example, metabolite profiling may identify changes in the levels of jasmonic acid, salicylic acid, or other secondary messengers involved in ROS signaling cascades. Understanding these pathways would help in uncovering how plants perceive and respond to ROS. So, in future, the specific signal molecules associated with specific ROS under different stress conditions could be emphasized to help plants overcome oxidative stress as well as ROS signaling.

15.4.3 IDENTIFICATION OF ROS PRODUCING METABOLITES

Metabolomics can aid in identifying specific metabolites that contribute to ROS formation (Noctor et al., 2015). Various metabolites such as NADPH oxidases, xanthine oxidases, NOS, and peroxisomal constituents produce ROS (Magnani and Mattevi, 2019). By studying changes in the metabolite levels before and after ROS induction, specific compounds and pathways involved in ROS production can be pointed out and this knowledge can guide the development of strategies to regulate the ROS levels in plants.

15.4.4 TARGETED METABOLOMICS

Targeted metabolomics involves the identification and analysis of specific metabolites of interest in a given sample (Roberts et al., 2012). Utilizing knowledge gained from metabolomics, targeted approaches can be developed to manipulate ROS equilibrium. For instance, by supplying plants with specific exogenous antioxidants or modifying metabolic pathways associated with ROS production, it would be possible to regulate ROS levels and mitigate oxidative stress.

15.5 METABOLOMICS DATA GENERATION AND ANALYSIS

The metabolomics analysis process involves a complex workflow starting from data detection and generation to the identification and characterization of the metabolites using different software. The different steps are discussed below.

15.5.1 METABOLOMICS DATA GENERATION

This is the initial step in the workflow where various techniques are used to isolate and characterize different groups of metabolites (Chen et al., 2022). MS and NMR spectroscopy are the two main basic techniques used for compound detections. Each has its own advantages and disadvantages (Beisken et al., 2015). The other metabolomic techniques such as GCMS and LCMS discussed earlier are mostly based on either of these basic two platforms. After the compounds are detected through a process called data preprocessing, raw signals are preprocessed to produce data in a suitable format for subsequent statistical analysis (Qiu et al., 2023). This generally involves activities such as noise reduction, retention time correction, and peak detection. The subsequent step is data processing that involves data normalization and naming of compounds (Chen et al., 2022).

15.5.2 STATISTICAL ANALYSIS

After data generation, the metabolomics data are analyzed using different types of data mining and statistical methods based on the experimental contexts. The analysis methods can broadly be categorized into two types, namely, univariate analysis and multivariate analysis. Univariate analysis gives a preliminary overview of the data characteristics that might be important in identifying the conditions under study (Chen et al., 2022). It involves statistical methods such as fold change analysis, *t*-tests and volcano plots, and one-way Analysis of Variance (ANOVA) and related correlation analysis for multigroup data. However, multivariate analysis involves the simultaneous observation and use of more than two statistical variables. Multivariate analysis includes multiple variance analysis (ANOVA), multiple regression analysis, factor analysis, principal component analysis, partial least squares discriminant analysis, cluster analysis, and machine learning (e.g., random forest and support vector machine) (Chen et al., 2022).

15.5.3 SOFTWARE TOOLS FOR DATA ANALYSIS

Metabolomics data are vast and to analyze such vast amount and variety of data, powerful software tools are needed. Various software tools for metabolomics data analyses are discussed in Table 15.4.

TABLE 15.4 Software Tools for Metabolomics Data Analysis

Sr. No	Software Tool	Description	References
1	*MS-DIAL*	• Free data preprocessing software. • Can process LCMS, GCMS, and NMR data. • Offers multiple data acquisition processing and also includes the spectra for compound ID.	(Tsugawa et al., 2015, 2020; Chen et al., 2022)
2	*MZmine3*	• Open-source software. • Mainly for MS data processing, specifically LCMS data. • Can also process GCMS and NMR data.	(Pluskal et al., 2010; Chen et al., 2022)
3	*EI-MAVEN*	• Open-source software. • Can process LCMS, GCMS, and NMR data.	(Agrawal et al., 2019; Chen et al., 2022)
4	*LipidMatch*	• Used for annotating lipids detected by LCMS.	(Koelmel et al., 2017)
5	*LipiDex*	• Identification of lipids using LCMS/MS. • Reduces biasness incurred in manual processing and thus improves the identification.	(Hutchins et al., 2018)
6	*MetFlow*	• Web-based tool, very simple to operate. • Enables users even with least idea about programming and statistics to carry out data analysis.	(Chen et al., 2022)
7	*MetaboAnalyst 5.0*	• Free and comprehensive web-based software. • Enables different functions such as data analysis, statistical analysis, functional annotation, and visualization of data.	(Xia et al., 2009; Pang et al., 2021)
8	*LipidSig*	• Web-based platform for the comprehensive analysis of lipidomic data. • Enables functions such as profiling, correlation analysis and networking	(Lin et al., 2021; Chen et al., 2022)
9	*LION*	• Web-based tool for the analysis of lipids.	(Molenaar et al., 2019)
10	*METLIN*	• A tandem MS/MS database for metabolites identification. • Free cloud-based platform.	(Smith et al., 2005; Tautenhahn et al., 2012)
11	*PaintOmics3*	• Web-based platform for the integrated visualization of multiomics data. • Accepts a wide variety of omics types, including transcriptomics, proteomics, and metabolomics.	(Hernández-de-Diego et al., 2018; Chen et al., 2022)
12	*3Omics*	• Web-based tool for analyzing, integrating and visualizing data. • Enables correlation analysis, phenotype mapping, pathway enrichment analysis, and coexpression analysis.	(Fernandez et al., 2007; Kuo et al., 2013)

TABLE 15.4 *(Continued)*

Sr. No	Software Tool	Description	References
13	*IMPala*	• Web tool for transcriptomics, proteomics, and metabolomics pathway analysis. • It is helpful to identify the pathways that are to be regulated at the transcriptional level, metabolic level, or both.	(Kamburov et al., 2011; Chen et al., 2022)
14	*MetPa*	• Web-based tool for the analysis and visualization of metabolomics data. • It helps to identify the most relevant metabolomic pathways by combining pathway topology feature analysis and pathway enrichment analysis programs.	(Xia and Wishart, 2010; Chen et al., 2022)
15	*MassTRIX*	• Web-based free tool for metabolomics pathway enrichment analysis.	(Suhre and Schmitt-Kopplin, 2008)
16	*MetaCore*™	• Web-based commercial tool. • Useful for analysis of high-throughput molecular data.	(Chen et al., 2022)
17	*OmicsNet*	• Used to annotate single nucleotide polymorphism, microbial taxa, or LC–MS peaks for network analysis. • Based on molecular interaction knowledge, this tool can be helpful in integrating different omics data.	(Zhou and Xia, 2018; Chen et al., 2022)

15.6 CONCLUSION

ROS are inevitably byproducts of cellular metabolisms and are generated by electron transport activities of cell organelles (Sharma et al., 2019). ROS play a double role in being stress products and agents of oxidative damages in plants as well as signaling products to prevent further damages (Das and Roychoudhury, 2014). ROS act as signaling molecules and facilitate various hormones-mediated plant responses, including responses toward stress and other growth and development (Joo et al., 2001; Torres et al., 2002; Miller et al., 2008). In contrast, when present in higher concentrations, ROS act as stress products and induce oxidative stress causing oxidative damage to the biomolecules, leading to their modifications along with the modifications of other membrane properties of the cell that may ultimately result in cell death (Das and Roychoudhury, 2014; Anjum et al., 2015; Sharma et al., 2019). To recuse themselves from such unavoidable oxidative damages, plants have complex antioxidant defense systems. The antioxidants scavenge the excess ROS generated in the cells and thus maintain the equilibrium status between the antioxidants and ROS. Metabolomics through the use of diverse metabolomic tools, such as LCMS, GCMS, HPLC, and NMR, would be very helpful in identifying and studying the metabolites involved in ROS production, signaling, and regulation pathways. Such knowledge would have potent impacts in maintaining equilibrium state in the cells through exogenous antioxidant supplements and thus protect the cells from oxidative damage during any environmental stress.

KEYWORDS

- **antioxidants**
- **ROS**
- **metabolites**
- **metabolite profiling**
- **signaling**
- **oxidative stress**
- **metabolomics**

REFERENCES

Agrawal, S., et al. El-MAVEN: A fast, robust, and user-friendly mass spectrometry data processing engine for metabolomics. In: *High-throughput Metabolomics: Methods and Protocols* (pp. 301–321), Springer, 2019.

Ahmad, P., et al. Exogenous application of nitric oxide modulates osmolyte metabolism, antioxidants, enzymes of ascorbate-glutathione cycle and promotes growth under cadmium stress in tomato. *Protoplasma*, 2018, 255, 79–93.

Ahmad, P., et al. Roles of enzymatic and nonenzymatic antioxidants in plants during abiotic stress. *Crit. Rev. Biotechnol.*, 2010, 30(3), 161–175.

Ahuja, I., et al. Plant molecular stress responses face climate change. *Trends Plant Sci.*, 2010, 15(12), 664–674.

Airaki, M., et al. Spatial and temporal regulation of the metabolism of reactive oxygen and nitrogen species during the early development of pepper (*Capsicum annuum*) seedlings. *Ann. Bot.*, 2015, 116(4), 679–693.

Al-Rubaye, A. F., et al. A review: uses of gas chromatography-mass spectrometry (GC-MS) technique for analysis of bioactive natural compounds of some plants. *Int. J. Toxicol. Pharm. Res.*, 2017, 9(1), 81–85.

Al Mahmud, J., et al. Insights into citric acid-induced cadmium tolerance and phytoremediation in Brassica juncea L.: Coordinated functions of metal chelation, antioxidant defense and glyoxalase systems. *Ecotoxicol. Environ. Saf.*, 2018, 147, 990–1001.

Ali, A. H. High-performance liquid chromatography (HPLC): A review. *Ann. Adv. Chem.*, 2022, 6, 010–020.

Allwood, J. W., Goodacre, R. An introduction to liquid chromatography–mass spectrometry instrumentation applied in plant metabolomic analyses. *Phytochem. Anal.*, 2010, 21(1), 33–47.

Anjum, N. A., et al. Lipids and proteins—Major targets of oxidative modifications in abiotic stressed plants. *Environ. Sci. Pollut. Res.*, 2015, 22, 4099–4121.

Aranda-Caño, L., et al. Reactive nitrogen species in plant metabolism. In: Lüttge, U., Cánovas, F. M., Risueño, M.-C., Leuschner, C., Pretzsch, H. (Eds.), *Progress in Botany* (Vol. 84. pp. 103–152). Springer Nature Switzerland: Cham, 2024.

Arbona, V., et al. Metabolomics as a tool to investigate abiotic stress tolerance in plants. *Int. J. Mol. Sci.*, 2013, 14(3), 4885–4911.

Ardrey, R. E. *Liquid Chromatography-mass Spectrometry: An Introduction.* John Wiley & Sons, 2003.

Arora, A., et al. Modulation of liposomal membrane fluidity by flavonoids and isoflavonoids. *Arch. Biochem. Biophys.*, 2000, 373(1), 102–109.

Asada, K. Production and action of active oxygen species in photosynthetic tissues. In: *Causes of Photooxidative Stress and Amelioration of Defense Systems in Plants* (pp. 77–104). CRC Press, 2019.

Asada, K., et al. Univalent reduction of molecular oxygen by spinach chloroplasts on illumination. *J. Biol. Chem.*, 1974, 249(7), 2175–2181.

Augusto, O., et al. Nitrogen dioxide and carbonate radical anion: Two emerging radicals in biology. *Free Radical Biol. Med.*, 2002, 32(9), 841–859.

Awad, H., et al. Mass spectrometry, review of the basics: Ionization. *Appl. Spectrosc. Rev.*, 2015, 50(2), 158–175.

Bachmann, L. M., Miller, W. G. Spectrophotometry. In: *Contemporary Practice in Clinical Chemistry* (pp. 19–133). Elsevier, 2020.

Beck, E. H., et al. Specific and unspecific responses of plants to cold and drought stress. *J. Biosci.*, 2007, 32, 501–510.

Bedard, K., Krause, K.-H. The NOX family of ROS-generating NADPH oxidases: Physiology and pathophysiology. *Physiol. Rev.*, 2007, 87(1), 245–313.

Beisken, S., et al. Getting the right answers: Understanding metabolomics challenges. *Expert Rev. Mol. Diagn.*, 2015, 15(1), 97–109.

Bethke, P. C., et al. Dormancy of Arabidopsis seeds and barley grains can be broken by nitric oxide. *Planta*, 2004, 219, 847–855.

Bhattachrjee, S. Reactive oxygen species and oxidative burst: Roles in stress, senescence and signal transduction in plant. *Curr. Sci.*, 2005, 89(7), 1113–1121.

Birben, E., et al. Oxidative stress and antioxidant defense. *World Allergy Organ. J.*, 2012, 5, 9–19.

Bors, W., et al. On the nature of biochemically generated hydroxyl radicals studies using the bleaching of p-nitrosodimethylaniline as a direct assay method: Studies using the bleaching of p-nitrosodimethylaniline as a direct assay method. *Eur. J. Biochem.*, 1979, 95(3), 621–627.

Bouché, N., et al. Mitochondrial succinic-semialdehyde dehydrogenase of the γ-aminobutyrate shunt is required to restrict levels of reactive oxygen intermediates in plants. *Proc. Natl. Acad. Sci.*, 2003, 100(11), 6843–6848.

Brand, M. D. Mitochondrial generation of superoxide and hydrogen peroxide as the source of mitochondrial redox signaling. *Free Radical Biol. Med.*, 2016, 100, 14–31.

Chaki, M., et al. New insights into the functional role of nitric oxide and reactive oxygen species in plant response to biotic and abiotic stress conditions. *Plant Growth and Stress Physiology* (pp. 215–235), 2021.

Chance, B., et al. Hydroperoxide metabolism in mammalian organs. *Physiol. Rev.*, 1979, 59(3), 527–605.

Chawla, S., et al. Salinity induced oxidative stress and antioxidant system in salt-tolerant and salt-sensitive cultivars of rice (*Oryza sativa* L.). *J. Plant Biochem. Biotechnol.*, 2013, 22, 27–34.

Chen, Y., et al. Guide to metabolomics analysis: A bioinformatics workflow. *Metabolites*, 2022, 12(4), 357.

Choudhury, F. K., et al. Reactive oxygen species, abiotic stress and stress combination. *Plant J.*, 2017, 90(5), 856–867.

Clarke, C. J., Haselden, J. N. Metabolic profiling as a tool for understanding mechanisms of toxicity. *Toxicol. Pathol.*, 2008, 36(1), 140–147.

Cooney, R. V., et al. Light-mediated conversion of nitrogen dioxide to nitric oxide by carotenoids. *Environ. Health Perspect.*, 1994, 102(5), 460–462.

Dalvi, S. B. *Chromatography in Dyestuff*, 2021.

Dangl, J. L., Jones, J. D. Plant pathogens and integrated defence responses to infection. *Nature*, 2001, 411(6839), 826–833.

Das, K., Roychoudhury, A. Reactive oxygen species (ROS) and response of antioxidants as ROS-scavengers during environmental stress in plants. *Front. Environ. Sci.*, 2014, 2, 53.

Davies, M. J. Reactive species formed on proteins exposed to singlet oxygen. *Photochem. Photobiol. Sci.*, 2004, 3, 17–25.

del Rio, L. A., et al. NADH induces the generation of superoxide radicals in leaf peroxisomes. *Plant Physiol.*, 1989, 89(3), 728–731.

Diplock, A. T., et al. Vitamin E: Biochemistry and health implications. *Ann. N. Y. Acad. Sci.*, 1989, 570, 1–535.

Doudican, N. A., et al. Oxidative DNA damage causes mitochondrial genomic instability in Saccharomyces cerevisiae. *Mol. Cell. Biol.*, 2005, 25(12), 5196–5204.

Egorov, S. Y., et al. Rise and decay kinetics of photosensitized singlet oxygen luminescence in water. Measurements with nanosecond time-correlated single photon counting technique. *Chem. Phys. Lett.*, 1989, 163(4–5), 421–424.

Elstner, E. F. Oxygen activation and oxygen toxicity. *Ann. Rev. Plant Physiol.*, 1982, 33(1), 73–96.

Eltayeb, A. E., et al. Overexpression of monodehydroascorbate reductase in transgenic tobacco confers enhanced tolerance to ozone, salt and polyethylene glycol stresses. *Planta*, 2007, 225, 1255–1264.

Eltayeb, A. E., et al. Greater protection against oxidative damages imposed by various environmental stresses in transgenic potato with higher level of reduced glutathione. *Breed. Sci.*, 2010, 60(2), 101–109.

Eltayeb, A. E., et al. Transgenic potato overexpressing Arabidopsis cytosolic AtDHAR1 showed higher tolerance to herbicide, drought and salt stresses. *Breed. Sci.*, 2011, 61(1), 3–10.
Elyashberg, M. Identification and structure elucidation by NMR spectroscopy. *TrAC Trends Analyt. Chem.*, 2015, 69, 88–97.
Fanali, S., et al. *Liquid Chromatography: Applications*. Elsevier, 2017.
Farooq, M. A., et al. Acquiring control: The evolution of ROS-Induced oxidative stress and redox signaling pathways in plant stress responses. *Plant Physiol. Biochem.*, 2019, 141, 353–369.
Fernandez, J. M., et al. iHOP web services. *Nucleic Acids Res.*, 2007, 35(suppl_2), W21–W26.
Finkel, T. Signal transduction by reactive oxygen species. *J. Cell Biol.*, 2011, 194(1), 7–15.
Foreman, J., et al. Reactive oxygen species produced by NADPH oxidase regulate plant cell growth. *Nature*, 2003, 422(6930), 442–446.
Foyer, C. H., et al. Photorespiratory metabolism: Genes, mutants, energetics, and redox signaling. *Ann. Rev. Plant Biol.*, 2009, 60, 455–484.
Foyer, C. H., Noctor, G. Redox sensing and signaling associated with reactive oxygen in chloroplasts, peroxisomes and mitochondria. *Physiol. Plant.*, 2003, 119(3), 355–364.
Foyer, C. H., Noctor, G. Redox homeostasis and antioxidant signaling: A metabolic interface between stress perception and physiological responses. *Plant Cell*, 2005, 17(7), 1866–1875.
Foyer, C. H., Noctor, G. Stress-triggered redox signaling: What's in pROSpect? *Plant, Cell Environ.*, 2016, 39(5), 951–964.
Foyer, C. H., Shigeoka, S. Understanding oxidative stress and antioxidant functions to enhance photosynthesis. *Plant Physiol.*, 2011, 155(1), 93–100.
Gill, S. S., et al. Amelioration of cadmium stress in crop plants by nutrients management: Morphological, physiological and biochemical aspects. *Plant Stress*, 2011, 5(1), 1–23.
Gill, S. S., Tuteja, N. Reactive oxygen species and antioxidant machinery in abiotic stress tolerance in crop plants. *Plant Physiol. Biochem.*, 2010, 48(12), 909–930.
Gomathi, R., Rakkiyapan, P. Comparative lipid peroxidation, leaf membrane thermostability, and antioxidant system in four sugarcane genotypes differing in salt tolerance. *Int. J. Plant Physiol. Biochem.*, 2011, 3(4), 67–74.
Grace, S. C., Logan, B. A. Energy dissipation and radical scavenging by the plant phenylpropanoid pathway. *Philos. Trans. R Soc. Lond. B Biol. Sci.*, 2000, 355(1402), 1499–1510.
Gratão, P. L., et al. Making the life of heavy metal-stressed plants a little easier. *Funct. Plant Biol.*, 2005, 32(6), 481–494.
Guan, Z., et al. Enhancement of Cd tolerance in transgenic tobacco plants overexpressing a Cd-induced catalase cDNA. *Chemosphere*, 2009, 76(5), 623–630.
Günther, H. *NMR Spectroscopy: Basic Principles, Concepts and Applications in Chemistry*. John Wiley & Sons, 2013.
Guo, Y., et al. Responses of unsaturated fatty acid in membrane lipid and antioxidant enzymes to chilling stress in sweet sorghum (*Sorghum bicolor* (L.) Moench) seedling. *J. Agric. Sci.*, 2016, 8(9), 71.
Gupta, A. S., et al. Increased resistance to oxidative stress in transgenic plants that overexpress chloroplastic Cu/Zn superoxide dismutase. *Proc. Natl. Acad. Sci.*, 1993, 90(4), 1629–1633.
Gupta, K. J., et al. Nitric oxide regulation of plant metabolism. *Mol. Plant*, 2022, 15(2), 228–242.
Haagenson, K. K., Wu, G. S. The role of MAP kinases and MAP kinase phosphatase-1 in resistance to breast cancer treatment. *Cancer Metastasis Rev.*, 2010, 29,143–149.
Haber, F., Weiss, J. The catalytic decomposition of hydrogen peroxide by iron salts. *Proc. R Soc. Lond. A Math. Phys. Sci.*, 1934, 147(861), 332–351.
Halliwell, B. Reactive species and antioxidants. Redox biology is a fundamental theme of aerobic life. *Plant Physiol.*, 2006, 141(2), 312–322.
Harman, D. The biologic clock: The mitochondria? *J. Am. Geriatr. Soc.*, 1972, 20(4), 145–147.
Harrigan, G. G., Goodacre, R. *Metabolic Profiling: Its Role in Biomarker Discovery and Gene Function Analysis: Its Role in Biomarker Discovery and Gene Function Analysis*. Springer Science & Business Media, 2003.
Hartl, F. U., et al. Molecular chaperones in protein folding and proteostasis. *Nature*, 2011, 475(7356), 324–332.
Hasanuzzaman, M., et al. Regulation of ROS metabolism in plants under environmental stress: A review of recent experimental evidence. *Int. J. Mol. Sci.*, 2020a, 21(22), 8695.

Hasanuzzaman, M., et al. Reactive oxygen species and antioxidant defense in plants under abiotic stress: Revisiting the crucial role of a universal defense regulator. *Antioxidants*, 2020b, 9(8), 681.

Hatzakis, E. Nuclear magnetic resonance (NMR) spectroscopy in food science: A comprehensive review. *Compr. Rev. Food Sci. Food Saf.*, 2019, 18(1), 189–220.

Hayes, J. D., et al. Cross-talk between transcription factors AhR and Nrf2: lessons for cancer chemoprevention from dioxin. *Toxicol. Sci.*, 2009, 111(2), 199–201.

He, F., et al. NRF2, a transcription factor for stress response and beyond. *Int. J. Mol. Sci.*, 2020, 21(13), 4777.

Hernández-de-Diego, R., et al. PaintOmics 3: A web resource for the pathway analysis and visualization of multi-omics data. *Nucleic Acids Res.*, 2018, 46(W1), W503–W509.

Hites, R. A. Gas chromatography mass spectrometry. *Handbook of Instrumental Techniques for Analytical Chemistry*, 1997, 1, 609–625.

Horváth, C. *High-performance Liquid Chromatography: Advances and Perspectives*, Elsevier, 2013.

Huang, S., et al. The roles of mitochondrial reactive oxygen species in cellular signaling and stress response in plants. *Plant Physiol.*, 2016, 171(3), 1551–1559.

Hussain, S. Z., Maqbool, K. GC-MS: Principle, technique and its application in food science. *Int. J. Curr. Sci.*, 2014, 13, 116–126.

Hutchins, P. D., et al. LipiDex: An integrated software package for high-confidence lipid identification. *Cell Syst.*, 2018, 6(5), 621–625. e625.

Idle, J. R., Gonzalez, F. J. Metabolomics. *Cell Metab.*, 2007, 6(5), 348–351.

Ivanov, B., Khorobrykh, S. Participation of photosynthetic electron transport in production and scavenging of reactive oxygen species. *Antioxid. Redox Signal.*, 2003, 5(1), 43–53.

Jakubczyk, K., et al. Reactive oxygen species - sources, functions, oxidative damage. *Pol Merkur Lekarski*, 2020, 48(284), 124–127.

Jones, O. A., Cheung, V. L. An introduction to metabolomics and its potential application in veterinary science. *Comp. Med.*, 2007, 57(5), 436–442.

Joo, J. H., et al. Role of auxin-induced reactive oxygen species in root gravitropism. *Plant Physiol.*, 2001, 126(3), 1055–1060.

Kafle, B. P. *Chemical Analysis and Material Characterization by Spectrophotometry*. Elsevier, 2019.

Kamburov, A., et al. Integrated pathway-level analysis of transcriptomics and metabolomics data with IMPaLA. *Bioinformatics*, 2011, 27(20), 2917–2918.

Kehrer, J. P. The Haber–Weiss reaction and mechanisms of toxicity. *Toxicology*, 2000, 149(1), 43–50.

Kim, E. K., Choi, E.-J. Pathological roles of MAPK signaling pathways in human diseases. *Biochim. Biophys. Acta (BBA)-Mol. Basis Dis.*, 2010, 1802(4), 396–405.

Kobayashi, K., et al. ε-(γ-Glutamyl) lysine cross-links of spore coat proteins and transglutaminase activity in Bacillus subtilis. *FEMS Microbiol. Lett.*, 1996, 144(2–3), 157–160.

Koelmel, J. P., et al. LipidMatch: An automated workflow for rule-based lipid identification using untargeted high-resolution tandem mass spectrometry data. *BMC Bioinf.*, 2017, 18, 1–11.

Kuo, T.-C., et al. 3Omics: A web-based systems biology tool for analysis, integration and visualization of human transcriptomic, proteomic and metabolomic data. *BMC Syst. Biol.*, 2013, 7, 1–15.

Lambert, J. B., et al. *Nuclear Magnetic Resonance Spectroscopy: An Introduction to Principles, Applications, and Experimental Methods*. John Wiley & Sons, 2019.

Li, F., et al. The maize phytoene synthase gene family: Overlapping roles for carotenogenesis in endosperm, photomorphogenesis, and thermal stress tolerance. *Plant Physiol.*, 2008, 147(3), 1334–1346.

Li, F., et al. Overexpression of chloroplastic monodehydroascorbate reductase enhanced tolerance to temperature and methyl viologen-mediated oxidative stresses. *Physiol. Plant.*, 2010, 139(4), 421–434.

Lin, W.-J., et al. LipidSig: A web-based tool for lipidomic data analysis. *Nucleic Acids Res.*, 2021, 49(W1), W336–W345.

Lisec, J., et al. Gas chromatography mass spectrometry–based metabolite profiling in plants. *Nat. Protoc.*, 2006, 1(1), 387–396.

Liu, Y., et al. Oxidative stress regulates mitogen-activated protein kinases and c-Jun activation involved in heat stress and lipopolysaccharide-induced intestinal epithelial cell apoptosis. *Mol. Med. Rep.*, 2017, 16(3), 2579–2587.

Llanes, A., et al. Metabolomic approach to understand plant adaptations to water and salt stress. *Plant Metabolites and Regulation Under Environmental Stress* (pp. 133–144). Elsevier, 2018.
López-Fernández, O., et al. Determination of polyphenols using liquid chromatography–tandem mass spectrometry technique (LC–MS/MS): A review. *Antioxidants*, 2020, 9(6), 479.
Lough, W. J., Wainer, I. W. *High Performance Liquid Chromatography: Fundamental Principles and Practice*. CRC Press, 1995.
Lü, J. M., et al. Chemical and molecular mechanisms of antioxidants: Experimental approaches and model systems. *J. Cell. Mol. Med.*, 2010, 14(4), 840–860.
Lukaszewicz, M., et al. Antioxidant capacity manipulation in transgenic potato tuber by changes in phenolic compounds content. *J. Agric. Food Chem.*, 2004, 52(6), 1526–1533.
Madeira, P. J. A., Florêncio, M. H. Applications of tandem mass spectrometry: From structural analysis to fundamental studies. In: *Tandem Mass Spectrometry—Applications and Principles* (pp. 1–32). 2012.
Magnani, F., Mattevi, A. Structure and mechanisms of ROS generation by NADPH oxidases. *Curr. Opini. Struct. Biol.*, 2019, 59, 91–97.
Marino, D., et al. A burst of plant NADPH oxidases. *Trends Plant Sci.*, 2012, 17(1), 9–15.
Matés, J. M., et al. Intracellular redox status and oxidative stress: Implications for cell proliferation, apoptosis, and carcinogenesis. *Arch. Toxicol.*, 2008, 82, 273–299.
Mehler, A. H. Studies on reactions of illuminated chloroplasts: I. Mechanism of the reduction of oxygen and other hill reagents. *Arch. Biochem. Biophys.*, 1951, 33(1), 65–77.
Meyer, V. R. *Practical High-performance Liquid Chromatography*. John Wiley & Sons, 2013.
Miller, G., et al. Reactive oxygen signaling and abiotic stress. *Physiol. Plant.*, 2008, 133(3), 481–489.
Mishra, S., et al. Arsenite treatment induces oxidative stress, upregulates antioxidant system, and causes phytochelatin synthesis in rice seedlings. *Protoplasma*, 2011, 248, 565–577.
Mittler, R. Oxidative stress, antioxidants and stress tolerance. *Trends Plant Sci.*, 2002, 7(9), 405–410.
Mittler, R., Berkowitz, G. Hydrogen peroxide, a messenger with too many roles? *Redox Rep.*, 2001, 6(2), 69–72.
Mittler, R., Zilinskas, B. Molecular cloning and characterization of a gene encoding pea cytosolic ascorbate peroxidase. *J. Biol. Chem.*, 1992, 267(30), 21802–21807.
Moldogazieva, N., et al. ROS and RNS signaling: Adaptive redox switches through oxidative/nitrosative protein modifications. *Free Radical Res.*, 2018, 52(5), 507–543.
Molenaar, M. R., et al. LION/web: A web-based ontology enrichment tool for lipidomic data analysis. *Gigascience*, 2019, 8(6), giz061.
Møller, I. M. Plant mitochondria and oxidative stress: Electron transport, NADPH turnover, and metabolism of reactive oxygen species. *Ann. Rev. Plant Biol.*, 2001, 52(1), 561–591.
Murphy, M. P. How mitochondria produce reactive oxygen species. *Biochem. J.*, 2009, 417(1), 1–13.
Nguyen, T., et al. Regulatory mechanisms controlling gene expression mediated by the antioxidant response element. *Ann. Rev. Pharm. Toxicol.*, 2003, 43(1), 233–260.
Noctor, G., et al. The metabolomics of oxidative stress. *Phytochemistry*, 2015, 112, 33–53.
Pang, Z., et al. MetaboAnalyst 5.0: Narrowing the gap between raw spectra and functional insights. *Nucleic Acids Res.*, 2021, 49(W1), W388–W396.
Pastori, G. M., Trippi, V. S. Oxidative stress induces high rate of glutathione reductase synthesis in a drought-resistant maize strain. *Plant Cell Physiol.*, 1992, 33(7), 957–961.
Patel, R., et al. Mass spectrometry-A review. *Vet. World*, 2012, 5(3).
Patterson, W. R., Poulos, T. L. Crystal structure of recombinant pea cytosolic ascorbate peroxidase. *Biochemistry*, 1995, 34(13), 4331–4341.
Persson, T., et al. Oxidative stress in Alzheimer's disease: Why did antioxidant therapy fail? *Oxid. Med. Cell. Longev.*, 2014, 2014, 427318.
Pluskal, T., et al. MZmine 2: Modular framework for processing, visualizing, and analyzing mass spectrometry-based molecular profile data. *BMC Bioinf.*, 2010, 11, 1–11.
Poole, C. F., Poole, S. K. *Chromatography Today*. Elsevier, 2012.
Popov, V., et al. Inhibition of the alternative oxidase stimulates H2O2 production in plant mitochondria. *FEBS Lett.*, 1997, 415(1), 87–90.
Pospíšil, P. Production of reactive oxygen species by photosystem II as a response to light and temperature stress. *Front. Plant Sci.*, 2016, 7, 1950.

Pramod, K., et al. An overview: LC-MS as tool of sample extraction and quantification in bioanalytical laboratories. *Asian J. Pharm. Anal.*, 2020, 10(3), 165–172.

Puntarulo, S., et al. Superoxide anion and hydrogen peroxide metabolism in soybean embryonic axes during germination. *Biochim. Biophys. Acta*, 1991, 1074(2), 277–283.

Puntarulo, S., et al. Hydrogen peroxide metabolism in soybean embryonic axes at the onset of germination. *Plant Physiol.*, 1988, 86(2), 626–630.

Qiu, S., et al. Small molecule metabolites: Discovery of biomarkers and therapeutic targets. *Signal. Transduct. Target. Ther.*, 2023, 8(1), 132.

Qu, Y., et al. Functional regulation of plant NADPH oxidase and its role in signaling. *Plant Signal. Behav.*, 2017, 12(8), e1356970.

Racchi, M., et al. Differential activity of catalase and superoxide dismutase in seedlings and in vitro micropropagated oak (*Quercus robur* L.). *Plant Cell Rep.*, 2001, 20, 169–174.

Rahman, M. M., et al. Basic overview on gas chromatography columns. In: *Analytical Separation Science* (pp. 823–834), 2015.

Reeg, S., et al. The molecular chaperone Hsp70 promotes the proteolytic removal of oxidatively damaged proteins by the proteasome. *Free Radical Biol. Med.*, 2016, 99, 153–166.

Reuhs, B. L. *High-Performance Liquid Chromatography*. Springer, 2017.

Roberts, L. D., et al. Targeted metabolomics. *Curr. Protocols Mol. Biol.*, 2012, 98(1), 30.32. 31–30.32. 24.

Saddhe, A. A., et al. Reactive nitrogen species: Paradigms of cellular signaling and regulation of salt stress in plants. *Environ. Exp. Bot.*, 2019, 161, 86–97.

Sagi, M., Fluhr, R. Production of reactive oxygen species by plant NADPH oxidases. *Plant Physiol.*, 2006, 141(2), 336–340.

Sakihama, Y., et al. Reduction of phenoxyl radicals mediated by monodehydroascorbate reductase. *Biochem. Biophys. Res. Commun.*, 2000, 279(3), 949–954.

Sawyer, D. T., et al. Redox reaction chemistry of superoxide ion. In: *Chemical and Biochemical Aspects of Superoxide and Superoxide Dismutase* (pp. 1–23). 1980.

Scandalios, J. G., et al. Catalases in plants: Gene structure, properties, regulation, and expression. *Cold Spring Harbor Monograph Series*, 1997, 34, 343–406.

Sharma, D., et al. MAP kinase as regulators for stress responses in plants: An overview. In: *Protein Kinases and Stress Signaling in Plants: Functional Genomic Perspective* (pp. 369–392), 2020.

Sharma, P., Dubey, R. Involvement of oxidative stress and role of antioxidative defense system in growing rice seedlings exposed to toxic concentrations of aluminum. *Plant Cell Rep.*, 2007, 26, 2027–2038.

Sharma, P., Dubey, R. S. Drought induces oxidative stress and enhances the activities of antioxidant enzymes in growing rice seedlings. *Plant Growth Regulation*, 2005, 46, 209–221.

Sharma, P., et al. Oxidative stress and antioxidative defense system in plants growing under abiotic stresses. *Handbook of Plant and Crop Stress* (pp. 93–136), 4th edn., CRC Press, 2019.

Sharma, P., et al. Reactive oxygen species, oxidative damage, and antioxidative defense mechanism in plants under stressful conditions. *J. Bot.*, 2012, 2012, 217037.

Sies, H., et al. Oxidative stress. *Ann. Rev. Biochem.*, 2017, 86(1), 715–748.

Simpson, A. J., et al. *Nuclear Magnetic Resonance Spectroscopy and its Key Role in Environmental Research*. ACS Publications, 2012.

Singh, A., et al. Reactive oxygen species-mediated signaling during abiotic stress. *Plant Gene*, 2019, 18, 100173.

Smith, C. A., et al. METLIN: A metabolite mass spectral database. *Ther. Drug Monit.*, 2005, 27(6), 747–751.

Son, Y., et al. Mitogen-activated protein kinases and reactive oxygen species: How can ROS activate MAPK pathways? *J. Signal Transduct.*, 2011, 2011, 792639.

Sparkman, O. D., et al. *Gas Chromatography and Mass Spectrometry: A Practical Guide*. Academic Press, 2011.

Stashenko, E., Martínez, J. R. Gas chromatography-mass spectrometry. In: *Advances in Gas Chromatography* (pp. 1–38). 2014.

Suhre, K., Schmitt-Kopplin, P. MassTRIX: Mass translator into pathways. *Nucleic Acids Res.*, 2008, 36(suppl_2), W481–W484.

Sunheimer, R. L., et al. Analysis: Principles of instrumentation. In: *Henry's Clinical Diagnosis and Management by Laboratory Methods E-Book* (p. 37), 2011.

Suzuki, N., et al. ROS and redox signaling in the response of plants to abiotic stress. *Plant, Cell Environ.*, 2012, 35(2), 259–270.
Taguchi, K., Yamamoto, M. The KEAP1–NRF2 system in cancer. *Front. Oncol.*, 2017, 7, 85.
Tan, S., et al. Introduction to metabolomics and its applications in ophthalmology. *Eye*, 2016, 30(6), 773–783.
Tautenhahn, R., et al. An accelerated workflow for untargeted metabolomics using the METLIN database. *Nat. Biotechnol.*, 2012, 30(9), 826–828.
Telfer, A., et al. Isolated photosynthetic reaction center of photosystem II as a sensitizer for the formation of singlet oxygen. Detection and quantum yield determination using a chemical trapping technique. *J. Biol. Chem.*, 1994, 269(18), 13244–13253.
Torres, M. A., et al. Arabidopsis gp91phox homologues AtrbohD and AtrbohF are required for accumulation of reactive oxygen intermediates in the plant defense response. *Proc. Natl. Acad. Sci.*, 2002, 99(1), 517–522.
Triantaphylides, C., et al. Singlet oxygen is the major reactive oxygen species involved in photooxidative damage to plants. *Plant Physiol.*, 2008, 148(2), 960–968.
Trufelli, H., et al. An overview of matrix effects in liquid chromatography–mass spectrometry. *Mass Spectrom. Rev.*, 2011, 30(3), 491–509.
Tsugawa, H., et al. MS-DIAL: Data-independent MS/MS deconvolution for comprehensive metabolome analysis. *Nat. Methods*, 2015, 12(6), 523–526.
Tsugawa, H., et al. A lipidome atlas in MS-DIAL 4. *Nat. Biotechnol.*, 2020, 38(10), 1159–1163.
ul Haq, S., et al. Heat shock proteins: Dynamic biomolecules to counter plant biotic and abiotic stresses. *Int. J. Mol. Sci.*, 2019, 20(21), 5321.
Van Bramer, S. E. *An Introduction to Mass Spectrometry* (p. 19013). Widener University, Department of Chemistry, One University Place: Chester, PA, 1998.
Vangronsveld, J., Clijsters, H. Toxic effects of metals. In: *Plants and the Chemical Elements: Biochemistry, Uptake, Tolerance and Toxicity* (pp. 149–177), 1994.
Vomund, S., et al. Nrf2, the master regulator of anti-oxidative responses. *Int. J. Mol. Sci.*, 2017, 18(12), 2772.
Waters, E. R., et al. Evolution, structure and function of the small heat shock proteins in plants. *J. Exp. Bot.*, 1996, 47(3), 325–338.
Willson, R. L. Hydroxyl radicals and biological damage in vitro: What relevance in vivo. *Oxygen Free Radicals and Tissue Damage,* Wiley Online Library, 1979, 65, 19–42.
Wuethrich, K. The development of nuclear magnetic resonance spectroscopy as a technique for protein structure determination. *Accounts Chem. Res.*, 1989, 22(1), 36–44.
Xia, J., et al. MetaboAnalyst: A web server for metabolomic data analysis and interpretation. *Nucleic Acids Res.*, 2009, 37(suppl_2), W652–W660.
Xia, J., Wishart, D. S. MetPA: A web-based metabolomics tool for pathway analysis and visualization. *Bioinformatics*, 2010, 26(18), 2342–2344.
Xiong, L., et al. Cell signaling during cold, drought, and salt stress. *Plant Cell*, 2002, 14(suppl_1), S165–S183.
Yan, J., et al. Reactive oxygen species and nitric oxide are involved in ABA inhibition of stomatal opening. *Plant, Cell Environ.*, 2007, 30(10), 1320–1325.
Yoon, J.-Y., Yoon, J.-Y. Spectrophotometry. In: *Introduction to Biosensors: From Electric Circuits to Immunosensors* (pp. 127–152), 2016.
Young, A. J. The photoprotective role of carotenoids in higher plants. *Physiol. Plant.*, 1991, 83(4), 702–708.
Young, I., Woodside, J. Antioxidants in health and disease. *Journal Clin. Pathol.*, 2001, 54(3), 176–186.
Zhang, C., et al. Overexpression of SlGMEs leads to ascorbate accumulation with enhanced oxidative stress, cold, and salt tolerance in tomato. *Plant Cell Rep.*, 2011, 30, 389–398.
Zhou, G., Xia, J. OmicsNet: A web-based tool for creation and visual analysis of biological networks in 3D space. *Nucleic Acids Res.*, 2018, 46(W1), W514–W522.
Zhou, J., et al. H_2O_2 mediates the crosstalk of brassinosteroid and abscisic acid in tomato responses to heat and oxidative stresses. *J. Exp. Bot.*, 2014, 65(15), 4371–4383.
Žuvela, P., et al. Column characterization and selection systems in reversed-phase high-performance liquid chromatography. *Chem. Rev.*, 2019, 119(6), 3674–3729.

CHAPTER 16

Metabolomics of Plant–Insect Interactions

USMAN ARIF, RABIAH ASHRAF, HAMMAD AHMED, MISBAH SALEEM, and ALLAH BAKHSH*

Seed Biotech Laboratories, Centre of Excellence in Molecular Biology, University of the Punjab, Lahore, Pakistan

*Corresponding authors

ABSTRACT

Metabolomics has emerged as a powerful tool for elucidating the complex biochemical interactions between plants and insects. These interactions involve a diverse array of chemical compounds that mediate defense, communication, and adaptation. Metabolomic profiling enables comprehensive analysis of metabolites involved in these processes, providing valuable insights into the molecular dialogue shaping plant–insect relationships. By examining metabolic changes triggered by insect herbivory or infestation, metabolomics helps identify key defense-related metabolites, including secondary metabolites, phytohormones, and stress-induced enzymes. It also facilitates the detection of insect-derived metabolites that influence plant physiology and modulate host defense mechanisms. Additionally, metabolomics reveals the crucial role of insect gut microbiota in producing bioactive compounds that impact plant–insect dynamics. Comparative metabolomic studies between resistant and susceptible plant genotypes further uncover pathways linked to plant resistance, guiding sustainable pest management and breeding of insect-resistant crops. Integration of metabolomics with genomics and transcriptomics promises to uncover novel defense networks and improve our understanding of plant metabolic plasticity under biotic stress. Overall, metabolomics offers a comprehensive framework for exploring the chemical ecology of plant–insect interactions and holds great potential for advancing knowledge of defense mechanisms and adaptation strategies.

16.1 INTRODUCTION

Plant interactions with members of their ecological community exhibit remarkable diversity that can vary from antagonistic to mutualistic relationship. Such interactions may trigger a significant metabolic rearrangement in plant physiology, which in turn may foster growth of beneficial organisms, while also limiting antagonistic forces such as herbivores and pathogens.

Exploring Plant Stress Tolerance: A Metabolomics Approach.
Shuvasish Choudhury, Debojyoti Moulick, and Mirza Hasanuzzaman (Eds.)

DOI: 10.1201/9781779641731-16

A better understanding of the molecular and cellular communication between plants, insects, and bacteria, as well as its evolutionary and ecological implications, is critical for comprehending how each partner adopts to its biotic environment. Moreover, recognizing how these biological interactions are influenced by environmental shifts and how scientists could mitigate any adverse effects will become increasingly crucial (Giron et al., 2018).

Considering these circumstances, metabolomics represents one of the most efficient bioanalytical tools for acquiring a picture of an organism's metabolites during a biological process, and it serves as a phenotyping tool. In its simplest form, the metabolomics approach includes recognizing and figuring out the set of metabolites (or particular metabolites) in biological specimens (tissues, cells, fluids, or organisms) under normal circumstances in contrast to altered conditions promoted by environmental modulations, disease, dietary intervention, or drug treatment (Klassen et al., 2017).

Biotic and abiotic stresses have a detrimental impact on crop production leading to a significant reduction in yearly crop output, qualitatively as well as quantitatively (Khan et al., 2022). Biologists, particularly agronomists have recently faced the challenge of finding an alternate strategy to deal with biotic and abiotic challenges such as insects, herbivores, and diseases, along with salt stress, drought, trace metals contamination, and extremely high temperatures (Hayat et al., 2021). All of these challenges have an impact on morphological and physiological characteristics, including inhibition of the functional groups of essential molecules, like enzymes.

Plants and microbes require primary metabolites to grow and develop properly. Secondary metabolites, on the other hand, occur near the stationary stage during growth and serve no direct part in the process of growth, development, or reproduction. Metabolic profiling of primary and secondary metabolites gives comprehensive information of biochemical processes occurring in plant metabolism (Ashraf and Ozturk, 2008). This chapter delves into the metabolomics of plant–insect interactions with a collection of studies focusing on the effects of specialized metabolites in modulating metabolic networks and rewiring genetic circuits resulting in coexistence of intricate plant–insect relationships.

16.2 METABOLITES ASSOCIATED WITH PLANT INTERACTIONS

Metabolomics is an exciting approach for interpreting both abiotic and biotic stress tolerance in crops. Metabolic profiling indicates various types of metabolites that are formed in plants, such as amino acids, sugars, polyamines, phenols, and terpenes. Primary metabolites, such as amino acids, carbohydrates, lipids, polyamines, as well as glycine betaine, have been proposed to be the primary contributors that function as plant's osmolytes as well as osmoprotectants against different environmental stress elements.

Plant-based secondary metabolites, such as terpenoids do not directly contribute to the growth process and development of plants. Nonetheless, these metabolites may contribute a crucial part in plant defense by defending against biotic factors such as herbivores, insects, and diseases. Furthermore, they can increase resistance to abiotic stress elements. Consequently, metabolomics techniques are becoming increasingly important in plants by discovering distinct phytochemicals involved in adaptive responses to a variety of stimuli (Salam et al., 2023).

16.3 CLASSIFICATION OF PLANTS SECONDARY METABOLITES

Secondary metabolites are classified into three major groups, terpenes, phenolic, and N-containing compounds, depending on plant biosynthetic pathway (Figure 16.1).

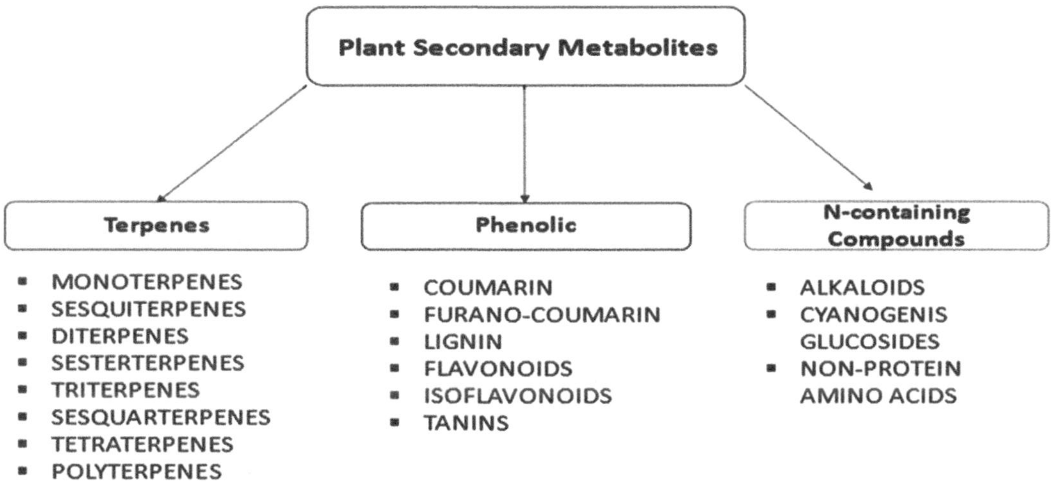

FIGURE 16.1 Classification of secondary metabolites.

16.3.1 TERPENES

Most secondary metabolites found in plants are called terpenes. Major components of biosynthesis are terpenes. For instance, steroids are derived from the triterpene squalene. The main components of the essential oils of many kinds of plants and flowers are also terpenes and terpenoids. There are around 25,000 different chemicals that make up terpenes (Divekar et al., 2022).

16.3.2 PHENOLIC COMPOUNDS

Almost 10,000 different chemicals make up the diverse class of secondary metabolites known as plant phenols. The most common type of secondary metabolites is phenolics. Phenolic compounds have hydroxyl groups on aromatic ring.

16.3.3 N-CONTAINING COMPOUNDS

Alkaloids are nonvolatile organic compound (VOC) having nitrogen in cyclic ring. Alkaloids are mainly present in plants and mostly found in flowering plants. Initially discovered 4000 years ago and primarily known for their therapeutic potential. They are primarily nitrogen containing secondary metabolites of plants, for example, caffeine, nicotine, cocaine, and morphine. Nicotine is the most well-known alkaloid. Many researches demonstrated that it is crucial for defending (wild) tobacco species against a range of native herbivorous insects, including *Diabrotica undecimpunctata*, *Spodoptera exigua*, and *Trimerotropis* spp. (Dey et al., 2020).

16.4 BIOSYNTHESIS OF SECONDARY PLANT METABOLITES

The secondary metabolites production in plants is complex and they act synergistically by specific mechanisms against different stresses to avoid the phytochemicals. Different enzymes and signaling pathways play a significant role in the formation of secondary metabolites. The biosynthesis of secondary metabolites in plants involves a number of constituents from primary metabolites, including sugars, fatty acids, amino acids, and acetyl CoA, which are by-products of shikimate pathway and tricarboxylic acid cycle (Wink and Schimmer, 2010).

A significant range of changes occur in the basic structures of metabolic processes including hydroxylation, oxidation, methylation, and phosphorylation due to modifications of enzymes. It is noticed that genes of secondary metabolites are activated and transcribed by a series of transcription factors that are involved in the synthesis of secondary metabolites, storage, and transport (Figure 16.2). Generally, biosynthetic pathways occur in the cell cytoplasm, but some compounds are biosynthesized in chloroplast (e.g., alkaloids and some in the endoplasmic reticulum includes sesquiterpenes and sterol (Divekar et al., 2022; Kutchan, 2005).

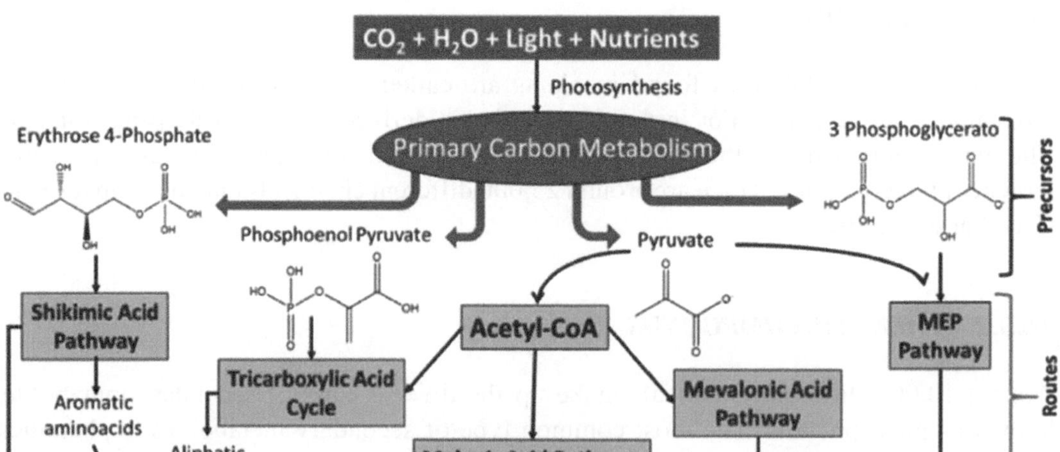

FIGURE 16.2 Secondary metabolite biosynthesis through different pathways and their relationship with primary metabolites.

Source: Reprinted from Twaij and Hasan, 2022. https://creativecommons.org/licenses/by/4.0/

16.5 INSECTS ADAPTATION AGAINST SECONDARY METABOLITES

Numerous tactics have been developed by insects to counteract the harmful effects of toxic chemical release by plants (Turlings and Erb, 2018). Since plants are sessile in nature, they require strong defensive mechanisms to keep herbivorous insects away from them. Large variety of plant secondary chemicals have been identified to date and these defense compounds are often chemicals in nature (Howe and Jander, 2008). Because they must endure the hazardous or chemically demanding conditions in which their host plants live, herbivores have evolved a variety of strategies for getting around plant defenses or even exploiting them for their own gain. Insects can store plant-derived xenobiotics to protect them from natural predators, or they can be quickly eliminated or broken down into nontoxic chemicals (Heckel et al., 2014). Here, we concentrate on the molecular processes that have been demonstrated to allow insects to successfully adapt to their host plants by overcoming chemical plant defenses (Figure 16.3).

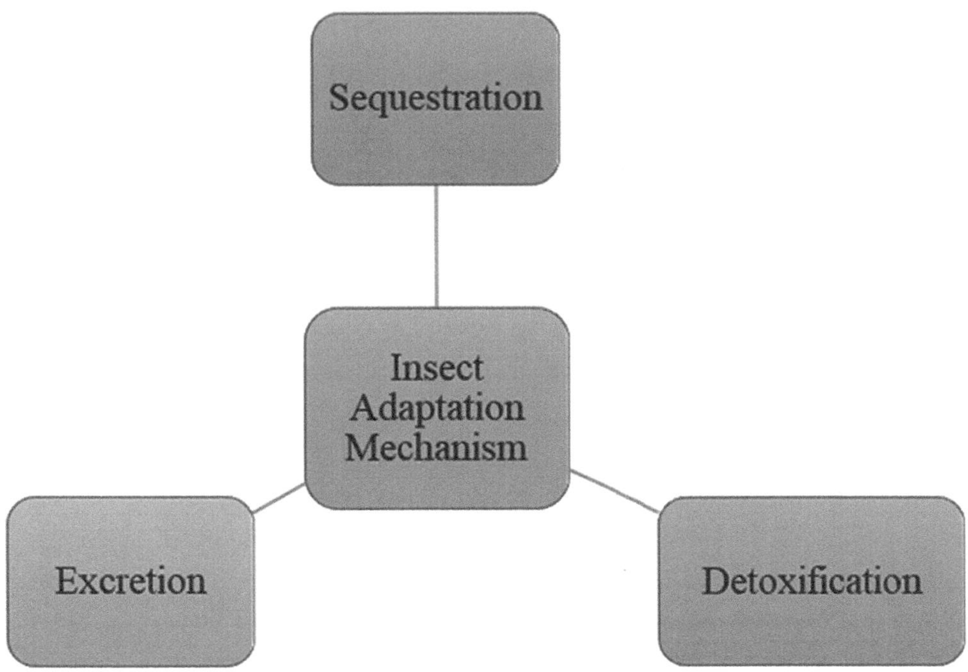

FIGURE 16.3 Mechanism of insect adaptations.

16.5.1 DETOXIFICATION

When a plant's poison is detoxified by insect pests, a variety of enzymes play a role in this process. Detoxifying enzymes aid in the detoxification or reduction of these hazardous secondary metabolites by insect pests following intake (Ferreres et al., 2008).

These enzymes' functions might change depending on the host plant's nutrition, and the kinds of insects that feed on them and their glycosylation, glutathionation, sulfation, and deacylation processes (Halkier and Gershenzon, 2006).

Enzymes involved in phase I detoxification, such as flavin monooxygenase, esterase, and cytochrome P450 monooxygenase (P450s), can reduce the biological activity of numerous exogenous and endogenous hazardous chemicals. Phase II enzymes primarily target hazardous metabolites of phase I metabolism and include UDP-glucuronyl transferases and glutathione transferases (Figure 16.4).

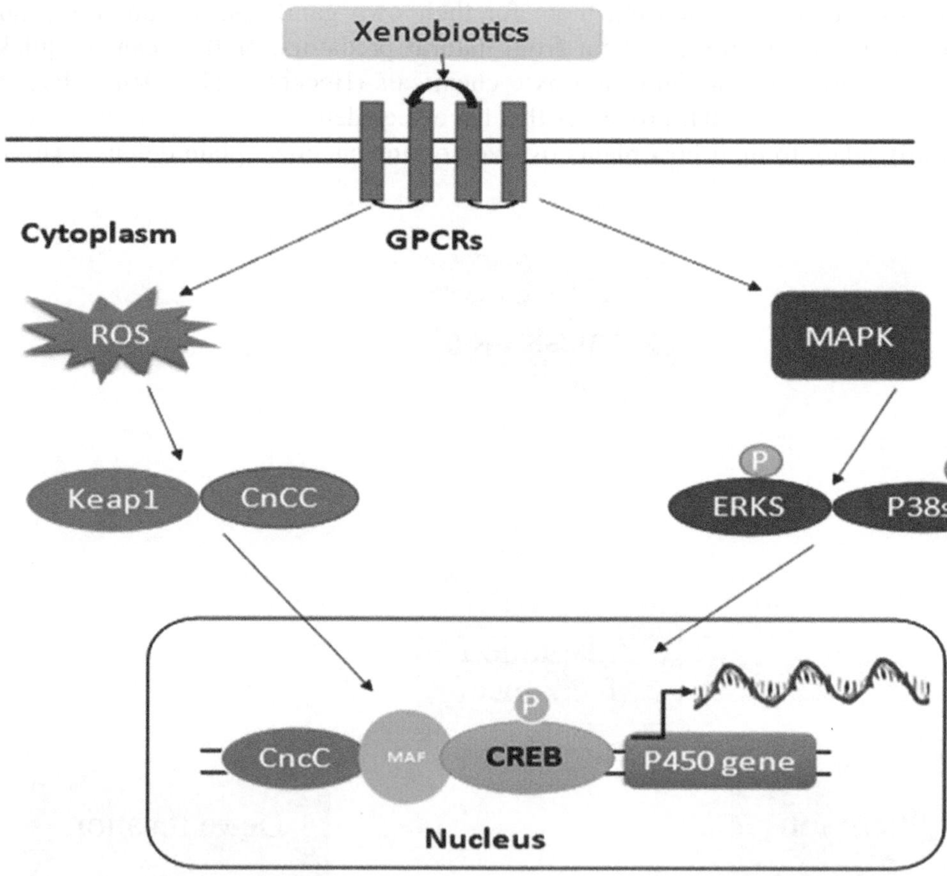

FIGURE 16.4 Regulation of P450 detoxification gene in response to xenobiotic stress, involving MAPK signaling pathway and reactive oxygen species (ROS). Activation of CnCC and ERKS phosphorylation activates CREB inside the nucleus. After phosphorylation, CREB attaches to the CRE-like sites situated in the promoter region of P450 genes, thereby stimulating their production.

In many insects, P450 genes play important roles in detoxification; pesticide resistance is mostly dependent on xenobiotic metabolism through CY P450s. In a similar manner, phytochemicals have the ability to alter the expression levels of *P450* genes linked to detoxification.

16.5.2 EXCRETION

Another class of metabolic enzymes involved in the breakdown of harmful substances is the esterase family. Through the creation of an enzyme pathway for cleavage or the sequestration of plant toxic metabolites, esterases detoxify harmful substances. As a result, these substances excrete out from insect's body.

Monarch caterpillars evolve specific mechanisms not only to tolerate the milkweed plant toxin cardiac glycosides but also use them defense against predator. Monarch is capable of excreting excess glycoside. Their malpighian tubules play an important role in filtering out and excreting any surplus glycosides, thereby regulating level within their bodies to avoid self-toxicity and maintaining enough to detect predators (Table 16.1).

TABLE 16.1 Insect's Adaptations Against Plant Secondary Metabolites

Insect Adaptation	Example	References
Cyanogenic glucosides	There is evidence that the high alkaline midgut lumen (pH 11) of the autumn webworm *Hyphantria cunea* protects the insect from negative effects by preventing the breakdown of cyanogenic glucosides in the gut	Chen et al. (2023)
Genetic adaptation	The monarch butterfly, *Danaus plexippus*, is the most well-known example of insensitivity. This insect can withstand cardenolides by avoiding the inhibition of Na^+/K^+-ATPase and the ensuing abnormalities in muscle contraction, brain function, and ion transport, now insect can tolerate cardenolides	Divekar et al. (2022)
Adaptation in *Zygena filipendulae* larvae	Another example is plant cyanogenic glycosides, which hydrolyze into toxic hydrogen cyanide (HCN) when tissue is disrupted. The larvae of the burnet moth *Zygaena filipendulae* restrict plant hydrolases by combining leaf-snipping with a very alkaline gut	Boter and Diaz (2023)

16.5.3 SEQUESTRATION

The above suggested process by which insects sequester pyrrolizidine alkaloids (PA). Most PA in plants are made and stored as nontoxic N-oxides. N-oxides are usually converted to (protoxic) tertiary lipophilic alkaloids in the digestive tracts of insects where they are passively absorbed into the body cavity through diffusion. These tertiary alkaloids become toxic to insects and vertebrates when they are bioactivated by cytochrome P450 enzymes. It has been demonstrated that the hemolymph of a number of insects contains flavin-dependent monooxygenases that change protoxic tertiary PAs back into nontoxic N-oxides. Since tertiary alkaloids are lipophilic, it was thought that passive sequestration (i.e., absorption) of these toxins would occur in *Creatonotos transiens* (Lepidoptera, Arctiidae) and *Zonocerus variegatus* (Orthoptera, Pyrgomorphidae) (Wang et al., 2012) (Figure 16.5).

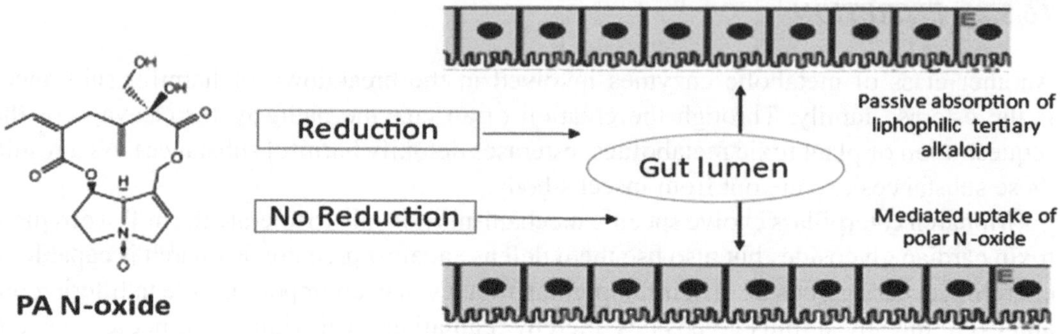

FIGURE 16.5 Physiological processes involved in sequestration.

16.6 METABOLIC SIGNALING AND COMMUNICATION BETWEEN INSECTS AND PLANTS

16.6.1 *ROS*

ROS are key redox-based signaling molecules in the plant defense response, according to mounting evidence (Skelly and Loake, 2013). Changes in redox homeostasis resulting in the alteration of cysteines in important regulatory proteins are emerging as a central responsive molecular switch that governs the activity of several regulators of plant defense responses (Yun et al., 2012).

16.6.2 *VOCS*

According to Dudareva et al. (2013), plants generate and release a wide range of VOCs to interact with their surroundings and as a defense against biotic and abiotic challenges. When plants are harmed by herbivores, they release VOCs that are created from scratch to fend off the invader. These VOCs can also serve as an early warning system to strengthen defenses in both the damaged and unharmed portions of the same plant (Heil and Silva Bueno, 2007).

16.6.3 *PHEROMONES*

Pheromones are defined as chemical signals unique to a species that facilitate communication among members of the same species. Insects release peptides called pheromones, which trigger particular reactions. Herbivorous insects synthesize chemicals from host plants and utilize them as precursors or sex pheromones. For instance, in order to promote leaking during mating, male orchid bees build terpenoid combinations from orchids and convey them as an aggregation pheromone (Wang et al., 2022).

16.6.4 SEMIOCHEMICALS

Semiochemicals are described as instructive molecules that, when produced from one creature, cause other members of the same species or distinct species to react physiologically or behaviorally. They are mostly employed as complimentary or alternative elements to pesticide treatments in various integrated pest management (IPM) strategies that address plant–insect or insect–insect interactions (Karlson and Butenandt, 1959). These substances primarily influence the behavior of different insect pests by chemical signals that travel between insects or between plants and insects. When it comes to IPM programs for managing insect pests, semiochemicals are thought to be a viable component (Quintana et al., 2003).

16.6.5 ALLELOCHEMICALS

Plants suppress competing plant species, act against herbivores and attacking microbes, specifically regulate the soil microbial community and encourage beneficial symbioses, and change below-ground ecological interactions through the production and release of allelochemicals and signaling chemicals (Abd El-Ghany, 2020).

Black walnut (*Juglans nigra*) is the source of juglone, a 1,4-naphthoquinone that was the first allelochemical to be validated. A woody species that impedes understory plants is the black walnut. Although the allelochemical juglone was isolated and identified only a few decades ago, this allelopathic occurrence was documented thousands of years ago.

16.7 PLANT ADAPTATIONS AGAINST INSECT PESTS

For decades, crop farmers have been mainly dependent on manmade chemicals for suppression of crop diseases. The use of large amounts of chemicals has varied side effects on the health of living organisms, pathogen resistance, and ecosystems. Several studies highlighted that hundreds of plants secondary metabolites are involved to play defensive role against biotic and abiotic stress. However, improving the production of different crops demands an understanding of the complex chemical interactions between plants and herbivores (War et al., 2012). Studying how plants resist herbivore aggression is similar to understanding how insect herbivores resist against plant reactions.

As a result of co-evolution, both plants and insects have discovered ways to get over the other's defense mechanisms. To response insect herbivory, plants have evolved an effective defense mechanism to recognize signals from damaged or attacked cells and trigger the plant's immune system to combat herbivores. To counter the pathogens attack plants have not only evolved morphological changes but also induced biochemical structures includes several secondary metabolites and defensive proteins that have toxic effects on the herbivores (Mostafa et al., 2022). This plant resistance behavior against insect herbivory is an important characteristic in crop protection and growth (War et al., 2012).

It is uncovered that formation and activation of defensive metabolites is facilitated by detection of stress through defense proteins or MAMPs recognition by recognition patterns.

These secondary metabolites modulate the corresponding molecular targets in pathogens including ion channels, neuroreceptors, or components of cytoskeleton (microtubules or tubulin) (Eid et al., 2015). In plants, the defensive role of secondary metabolites has also been confirmed by *in vitro* analysis of plants in which expression of metabolites was modified by modern techniques (War et al., 2018) (Table 16.2).

16.8 METABOLOMICS TECHNIQUES

In comparison to genetics (nucleic acids) and proteomics (peptides and proteins) and genomics (nucleic acids), metabolomics works with a broader range of chemically varied compounds. Metabolite analysis is challenging due to the huge variance (~106) in relative concentrations. As a result, extensive coverage is possible only by employing multiparallel complementary extraction and detection technologies with careful design of the experiment. Because no single technology currently exists (or probably in coming years) for identifying all compounds present in plants or other organisms, an array of multiple analytical methods such as mass spectrometry (MS), chromatography (gas chromatography [GC]–MS, liquid chromatography [LC]–MS), and nuclear magnetic resonance (NMR) are usually performed following general protocols. These metabolomics techniques offer unique strengths and limitations (Saito and Matsuda, 2010).

16.8.1 MS AND CHROMATOGRAPHY (GC–MS, LC–MS, CAPILLARY ELECTROPHORESIS [CE]–MS)

Metabolomics has grown dramatically in the field of natural product research, driven by an increasing fascination in comprehending the complex metabolic profiles of various biological systems (Sumner et al., 2015). In this data-driven landscape, MS appears out as the most prevalent preference for collecting data, due to its high-throughput capacities and sensitivity (Demarque et al., 2020) and, if combined with powerful chromatographic techniques (e.g., LC–MS, GC–MS, and CE–MS), enables separation and characterization of the tremendous compound diversity present in metabolome of a particular plant. Nevertheless, with the advancement in metabolomics research to decipher molecular frameworks essential for comprehending biological mechanisms, the fundamental limitations of MS in providing comprehensive structural information become evident (Letertre et al., 2020).

16.8.2 NMR

To solve MS limitations and increase the robustness of research, the incorporation of NMR spectroscopy appears as a critical complement to MS. In contrast to MS, NMR shines at providing immediate insights into the structure of molecules, permitting vital information for precise identification of the metabolites. Furthermore, NMR has intrinsic quantitative capabilities as a key method of measurement (Bharti and Roy, 2012).

TABLE 16.2 Role of Plant Secondary Metabolites in the Management of Various Plant Insect Pests

Type of Insect	Plant Species	Induced Defense	Protective Function	Reference
Brown planthopper (BPH)	*Oryza sativa* (rice)	Enriched phenylpropane, flavonoid and salicylic acid (SA) biosynthesis, and lower level of indole 3-acetic acid (IAA)	Reduced settling, feeding, and growth of BPH	Zhang et al. (2022)
White-backed planthopper (WBPH) *Sogatella furcifera*	Rice (*Oryza sativa*)	Elicited jasmonic acid pathway and suppressed SA (salicylic acid) pathway	Reduced WBPH feeding, weight, and survival by Silencing of mucin-like salivary protein (SFMLP) secretion	Liu et al. (2022)
Whitefly	Tobacco	Activate jasmonic acid pathway and suppressed SA (salicylic acid) pathway	Interrupts whitefly phloem feeding by silencing the whitefly *Bt56* gene	Xu et al. (2019)
Nematode	soybeans (*Glycine max*)	Production of glutathione (antioxidant and redox regulator) to combat ROS	Arrested nematode development	Chen et al. (2020)
Spodoptera littoralis (cotton leafworm)	*Artemisia granatensis* (royal chamomile)	Synthesis of sesquiterpenes, coumarin, and eudesmanolide	Antifeedant activity, aphid repellent	Barrero et al. (2013)
Myzocallis coryli	*Corylus avellana (hazel tree)*	Biosynthesis of flavonoids, tannins, phenolic acids, coumarins, and tannins	Inhibition of phytophages digestive process, decrease the availability of food uptake and permeability of the intestine	Gantner et al. (2019)
Myzus persicae (aphid)	*Arabidopsis* plants	Production of callose and ROS accumulation and activate salicylic acid (SA) signaling pathway	Reduce the aphid numbers and feeding performance of pathogens	Silva-Sanzana et al. (2022)
Anastrepha ludens (fruit fly)	*Persea americana*, Hass avocados	Production of phytoalexins, ROS, and polyphenols the induction of chitin receptors ROS	Potential ovicide metabolites counteract the insertion of eggs	Aluja et al. (2023)
Diaphorina citri (Asian citrus psyllid)	*Citrus sinensis* (sweet orange)	Emission of dimethyl disulfide (DMDS)	Antiherbivore activities activate defense responses in neighboring plants	Ling et al. (2022)
Plutella xylostella	*Brassica napus*	Production of glucosinolates and trypsin inhibitor	Reduced survivorship of *Plutella xylostella*	Nouri-Ganbalani et al. (2018)

Since NMR is crucial for structural elucidation, it is critical to recognize its own set of constraints. Sensitivity and database accuracy are significant challenges in NMR-based natural products and metabolomics. NMR is not as sensitive as MS, especially in detecting low-abundance metabolites (Borges and Teixeira, 2024).

Each metabolomics approach has advantages and disadvantages while analyzing plant–insect interactions. A blend of these strategies is frequently used to overcome these constraints and gain an extensive knowledge of the metabolic processes that underpin these intricate biological interactions. Integrating various analytical techniques can improve the accuracy, coverage, and comprehension of metabolomics data obtained from insect–plant interactions.

16.9 CONCLUSION

Although plant–insect interactions will persist and herbivores may coevolve in response to the genotypes of resistant plants, scientists are increasingly focused on eco-friendly, novel, and green approaches. One such approach is the use of plant secondary metabolites against pathogens. These metabolites have the potential to significantly reduce economic losses in stressed environments. Herbivore attacks and damage can be mitigated through modulating metabolic pathways by fluctuating carbon flux and rewiring genetic circuits linked to the biosynthesis of secondary metabolites. These metabolites mediate various defensive activities by stimulating the expression of genes involved in plant resistance mechanisms and enhancing the synthesis of specific enzymes required for secondary metabolite production.

Metabolomics is crucial in studying plant–insect interactions as it provides a comprehensive analysis of metabolites involved in these complex relationships. By profiling both primary and secondary metabolites, researchers can identify biochemical changes in plants in response to insect attacks. This insight helps uncover new secondary metabolites that play essential roles in plant defense mechanisms. Understanding these metabolic responses enables the development of crop varieties with enhanced resistance to pests, reducing reliance on chemical pesticides. Furthermore, metabolomics facilitates the discovery of novel bioactive compounds with potential applications in agriculture and pharmaceuticals, promoting sustainable and eco-friendly pest management strategies.

KEYWORDS

- **metabolomics**
- **plant-insect interactions**
- **secondary metabolites**
- **plant defense**
- **chemical ecology**
- **insect resistance**

REFERENCES

Abd El-Ghany, N. M. Pheromones and chemical communication in insects. In: *Pests, Weeds and Diseases in Agricultural Crop and Animal Husbandry Production* (pp. 1–13), 2020.

Aluja, M., Vázquez-Rosas-Landa, M., Cerqueda-García, D., Monribot-Villanueva, J. L., Altúzar-Molina, A., Ramírez-Vázquez, M., Velázquez-López, O., Rosas-Saito, G., Alonso-Sánchez, A. G., Ortega-Casas, R. Assessment of the molecular responses of an ancient angiosperm against atypical insect oviposition: The case of hass avocados and the Tephritid fly Anastrepha ludens. *Int. J. Mol. Sci.*, 2023, 24(3), 2060.

Ashraf, M., Ozturk, M. *Salinity and Water Stress: Improving Crop Efficiency* (Vol. 44). Springer Science & Business Media, 2008.

Barrero, A. F., del Pino, M. M. H., Portero, A. G., Burón, P. A., Arteaga, J. F., Alquézar, J. B., Díaz, C. E., Coloma, A. G. Terpenes and polyacetylenes from cultivated Artemisia granatensis boiss (Royal chamomile) and their defensive properties. *Phytochemistry*, 2013, 94, 192–197.

Bharti, S. K., Roy, R. Quantitative 1H NMR spectroscopy. *TrAC Trends Anal. Chem.*, 2012, 35, 5–26.

Borges, R. M., Teixeira, A. M. On the part that NMR should play in mass spectrometry metabolomics in natural products studies. *Front. Nat. Prod.*, 2024, 3, 1359151.

Boter, M., Diaz, I. Cyanogenesis, a plant defence strategy against herbivores. *Int. J. Mol. Sci.*, 2023, 24(8), 6982.

Chen, H., Hao, D., Chen, C., Sun, Y., Yu, X. Effects of midgut bacteria in Hyphantria cunea (Lepidoptera: Erebidae) on nuclear polyhedrosis virus and Bacillus thuringiensis (Bacillales: Bacillaceae). *J. Insect Sci.*, 2023, 23(2), 1.

Chen, X., Li, S., Zhao, X., Zhu, X., Wang, Y., Xuan, Y., Liu, X., Fan, H., Chen, L., Duan, Y. Modulation of (homo) glutathione metabolism and H_2O_2 accumulation during soybean cyst nematode infections in susceptible and resistant soybean cultivars. *Int. J. Mol. Sci.*, 2020, 21(2), 388.

Demarque, D. P., Dusi, R. G., de Sousa, F. D., Grossi, S. M., Silvério, M. R., Lopes, N. P., Espindola, L. S. Mass spectrometry-based metabolomics approach in the isolation of bioactive natural products. *Sci. Rep.*, 2020, 10(1), 1051.

Dey, P., Kundu, A., Kumar, A., Gupta, M., Lee, B. M., Bhakta, T., Dash, S., Kim, H. S. Analysis of alkaloids (indole alkaloids, isoquinoline alkaloids, tropane alkaloids). In: *Recent Advances in Natural Products Analysis* (pp. 505–567). Elsevier, 2020.

Divekar, P. A., Narayana, S., Divekar, B. A., Kumar, R., Gadratagi, B. G., Ray, A., Singh, A. K., Rani, V., Singh, V., Singh, A. K. Plant secondary metabolites as defense tools against herbivores for sustainable crop protection. *Int. J. Mol. Sci.*, 2022, 23(5), 2690.

Dudareva, N., Klempien, A., Muhlemann, J. K., Kaplan, I. Biosynthesis, function and metabolic engineering of plant volatile organic compounds. *New Phytologist.*, 2013, 198(1), 16–32.

Eid, S. Y., El-Readi, M. Z., Fatani, S. H., Eldin, E. E. M. N., Wink, M. Natural products modulate the multifactorial multidrug resistance of cancer. *Pharmacol. Pharm.*, 2015, 6(3), 146–176.

Ferreres, F., Valentão, P., Pereira, J. A., Bento, A., Noites, A., Seabra, R. M., Andrade, P. B. HPLC-DAD-MS/MS-ESI screening of phenolic compounds in Pieris brassicae L. reared on Brassica rapa var. rapa L. *J. Agric. Food Chem.*, 2008, 56(3), 844–853.

Gantner, M., Najda, A., Piesik, D. Effect of phenolic acid content on acceptance of hazel cultivars by filbert aphid. *Plant Prot. Sci.*, 2019, 55(2), 116–122.

Giron, D., Dubreuil, G., Bennett, A., Dedeine, F., Dicke, M., Dyer, L. A., Erb, M., Harris, M. O., Huguet, E., Kaloshian, I. Promises and challenges in insect–plant interactions. *Entomol. Exp. Appl.*, 2018, 166(5), 319–343.

Halkier, B. A., Gershenzon, J. Biology and biochemistry of glucosinolates. *Annu. Rev. Plant Biol.*, 2006, 57, 303–333.

Hayat, K., Khan, J., Khan, A., Ullah, S., Ali, S., Salahuddin, Fu, Y. Ameliorative effects of exogenous proline on photosynthetic attributes, nutrients uptake, and oxidative stresses under cadmium in pigeon pea (Cajanus cajan L.). *Plants*, 2021, 10(4), 796.

Heckel, D., Voelckel, C., Jander, G. *Annual Plant Reviews*, John Wiley & Sons, 2014.

Heil, M., Silva Bueno, J. C. Within-plant signaling by volatiles leads to induction and priming of an indirect plant defense in nature. *Proc. Natl. Acad. Sci.*, 2007, 104(13), 5467–5472.

Howe, G. A., Jander, G. Plant immunity to insect herbivores. *Annu. Rev. Plant Biol.*, 2008, 59, 41–66.

Karlson, P., Butenandt, A. Pheromones (ectohormones) in insects. *Ann. Rev. Entomol.*, 1959, 4(1), 39–58.
Khan, A., Ali, S., Khan, M., Hamayun, M., Moon, Y.-S. Parthenium hysterophorus's endophytes: The second layer of defense against biotic and abiotic stresses. *Microorganisms*, 2022, 10(11), 2217.
Klassen, A., Faccio, A. T., Canuto, G. A. B., da Cruz, P. L. R., Ribeiro, H. C., Tavares, M. F. M., Sussulini, A. Metabolomics: Definitions and significance in systems biology. *Metabolomics: From Fundamentals Clin. Appl.*, 2017, 3–17.
Kutchan, T. M. A role for intra-and intercellular translocation in natural product biosynthesis. *Cur. Opin. Plant Biol.*, 2005, 8(3), 292–300.
Letertre, M. P., Dervilly, G., Giraudeau, P. Combined nuclear magnetic resonance spectroscopy and mass spectrometry approaches for metabolomics. *Anal. Chem.*, 2020, 93(1), 500–518.
Ling, S., Qiu, H., Xu, J., Gu, Y., Yu, J., Wang, W., Liu, J., Zeng, X. Volatile dimethyl disulfide from guava plants regulate developmental performance of asian citrus psyllid through activation of defense responses in neighboring orange plants. *Int. J. Mol. Sci.*, 2022, 23(18), 10271.
Liu, Y., Yi, J., Jia, H., Miao, Y., Hou, M. Sogatella furcifera Saliva Mucin-like protein is required for feeding and induces rice defences. *Int. J. Mol. Sci.*, 2022, 23(15), 8239.
Mostafa, S., Wang, Y., Zeng, W., Jin, B. Plant responses to herbivory, wounding, and infection. *Int. J. Mol. Sci.*, 2022, 23(13), 7031.
Nouri-Ganbalani, G., Borzoui, E., Shahnavazi, M., Nouri, A. Induction of resistance against Plutella xylostella (L.) (Lep.: Plutellidae) by jasmonic acid and mealy cabbage aphid feeding in Brassica napus L. *Front. Physiol.*, 2018, 9, 347538.
Quintana, A., Reinhard, J., Faure, R., Uva, P., Bagnères, A.-G., Massiot, G., Clément, J.-L. Interspecific variation in terpenoid composition of defensive secretions of European Reticulitermes termites. *J. Chem. Ecol.*, 2003, 29, 639–652.
Saito, K., Matsuda, F. Metabolomics for functional genomics, systems biology, and biotechnology. *Annu. Rev. Plant Biol.*, 2010, 61, 463–489.
Salam, U., Ullah, S., Tang, Z.-H., Elateeq, A. A., Khan, Y., Khan, J., Khan, A., Ali, S. Plant metabolomics: An overview of the role of primary and secondary metabolites against different environmental stress factors. *Life*, 2023, 13(3), 706.
Silva-Sanzana, C., Zavala, D., Moraga, F., Herrera-Vásquez, A., Blanco-Herrera, F. Oligogalacturonides enhance resistance against aphids through pattern-triggered immunity and activation of salicylic acid signaling. *Int. J. Mol. Sci.*, 2022, 23(17), 9753.
Skelly, M. J., Loake, G. J. Synthesis of redox-active molecules and their signaling functions during the expression of plant disease resistance. *Antioxid. Redox Signal.*, 2013, 19(9), 990–997.
Sumner, L. W., Lei, Z., Nikolau, B. J., Saito, K. Modern plant metabolomics: Advanced natural product gene discoveries, improved technologies, and future prospects. *Nat. Prod. Rep.*, 2015, 32(2), 212–229.
Turlings, T. C., Erb, M. Tritrophic interactions mediated by herbivore-induced plant volatiles: Mechanisms, ecological relevance, and application potential. *Ann. Rev. Entomol.*, 2018, 63, 433–452.
Twaij, B. M., Hasan, M. N. Bioactive secondary metabolites from plant sources: Types, synthesis, and their therapeutic uses. *Int. J. Plant Biol.*, 2022, 13(1), 4–14.
Wang, H.-L., Ding, B.-J., Dai, J.-Q., Nazarenus, T. J., Borges, R., Mafra-Neto, A., Cahoon, E. B., Hofvander, P., Stymne, S., Löfstedt, C. Insect pest management with sex pheromone precursors from engineered oilseed plants. *Nat. Sustain.*, 2022, 5(11), 981–990.
Wang, L., Beuerle, T., Timbilla, J., Ober, D. Independent recruitment of a flavin-dependent monooxygenase for safe accumulation of sequestered pyrrolizidine alkaloids in grasshoppers and moths. *PLoS One*, 2012, 7(2), e31796.
War, A. R., Paulraj, M. G., Ahmad, T., Buhroo, A. A., Hussain, B., Ignacimuthu, S., Sharma, H. C. Mechanisms of plant defense against insect herbivores. *Plant Signal. Behav.*, 2012, 7(10), 1306–1320.
War, A. R., Taggar, G. K., Hussain, B., Taggar, M. S., Nair, R. M., Sharma, H. C. Plant defence against herbivory and insect adaptations. *AoB Plants*, 2018, 10(4), ply037.
Wink, M., Schimmer, O. Molecular modes of action of defensive secondary metabolites. *Annual Plant Reviews Volume 39: Functions and Biotechnology of Plant Secondary Metabolites*, 2010, 39, 21–161.

Xu, H.-X., Qian, L.-X., Wang, X.-W., Shao, R.-X., Hong, Y., Liu, S.-S., Wang, X.-W. A salivary effector enables whitefly to feed on host plants by eliciting salicylic acid-signaling pathway. *Proc. Natl. Acad. Sci.*, 2019, 116(2), 490–495.

Yun, B.-W., Spoel, S. H., Loake, G. J. Synthesis of and signalling by small, redox active molecules in the plant immune response. *Biochim. Biophys. Acta*, 2012, 1820(6), 770–776.

Zhang, Q., Li, T., Gao, M., Ye, M., Lin, M., Wu, D., Guo, J., Guan, W., Wang, J., Yang, K. Transcriptome and metabolome profiling reveal the resistance mechanisms of rice against brown planthopper. *Int. J. Mol. Sci.*, 2022, 23(8), 4083.

CHAPTER 17

Metabolomic Approaches to Regulate Reactive Oxygen Species: Unveiling Novel Insights for Plant Stress Tolerance

KABYASHREE BURAGOHAIN[1], RATUL NATH[1], DULUMONI TAMULY[2], JAY PRAKASH AWASTHI[3], BEDABRATA SAHA[4], and BHABEN CHOWARDHARA[5*]

[1]*Microbiology Laboratory, Department of Life Sciences, Dibrugarh University, Dibrugarh, Assam, India*

[2]*Department of Zoology, MDKG College, Dibrugarh, Assam, India*

[3]*Department of Botany, Government College Umariyapan, Katni, Madhya Pradesh, India*

[4]*Plant Pathology and Weed Research Department, Newe Ya'ar Research Centre, Agricultural Research Organization, Ramat Yishay, Israel*

[5]*Department of Botany, Faculty of Science and Technology, Arunachal University of Studies, Namsai, Arunachal Pradesh, India*

*Corresponding authors

ABSTRACT

Reactive oxygen species (ROS) are critical signaling molecules in plants involved in various physiological processes like growth, development, and stress responses. However, the excessive accumulation of ROS can lead to oxidative stress, causing cellular damage and compromising plant health and productivity. Metabolomics, a powerful systems biology approach, has emerged as a valuable tool to investigate the intricate metabolic networks that regulate ROS levels in plants. In the last 20 years, reverse genetic techniques have made it possible to directly assess the impact of a protein's abundance or lack on a plant's biochemistry. In parallel, top-down approaches have also been employed, which rely on sifting vast amounts of naturally occurring genetic diversity for metabolic diversity. This chapter provides a comprehensive analysis of recent advances in metabolomic approaches for understanding and manipulating ROS in plants, highlighting the potential of metabolites in enhancing plant stress tolerance and resilience, and the integration of metabolomics with other omics platforms to elucidate the complex cross-talk between ROS and cellular processes. Finally, we present future

Exploring Plant Stress Tolerance: A Metabolomics Approach.
Shuvasish Choudhury, Debojyoti Moulick, and Mirza Hasanuzzaman (Eds.)

DOI: 10.1201/9781779641731-17

perspectives on how metabolomics can revolutionize plant stress research, offering novel strategies for sustainable agriculture and global food security.

17.1 INTRODUCTION

All living organisms, including plants, encounter sudden and unfavorable environmental challenges. But unlike animals, plants cannot move to get away from stress because they are immobile. As such, plants have to deal with stress as soon as they are exposed to it and devise coping mechanisms immediately to survive (Singh et al., 2019). The widespread occurrence of challenging environmental conditions such as salinity, drought, extreme temperatures, toxic metals/metalloids, and flooding/waterlogging (WL) can be attributed to the significant and severe effects of climate change (Pereira, 2016; Raza et al., 2019). The augmentations of various abiotic stresses pose a significant challenge to the sustainable cultivation of crops. They must activate metabolic reactions and produce specific molecules to support repair and protection mechanisms, limiting the detrimental impacts of stress on their growth and reproductive abilities. Hence, in order to prevent the consequences of stress and a decline in crop yield, plants adapt to their physiological and biochemical conditions through molecular and epigenetic alterations (Singh et al., 2019). Furthermore, it results in the generation of oxidative stress due to the excessive buildup of reactive oxygen species (ROS), encompassing free radicals like superoxide anion ($O_2^{\bullet-}$), hydroperoxyl radical ($HO_2^{\bullet-}$), alkoxy radical ($RO^{\bullet-}$), and hydroxyl radical ($^{\bullet}OH$) as well as nonradical molecules like hydrogen peroxide (H_2O_2) and singlet oxygen ($^1O_2$) (Mehla et al., 2017; Hasanuzzaman et al., 2019). Initiation processes involving high-energy or electron-transfer reactions result in the conversion of atmospheric oxygen (O_2) into the aforementioned partially reduced or activated variations of molecular oxygen (Choudhury et al., 2017). The main locations within cells where ROS are primarily produced include chloroplasts, mitochondria, peroxisomes, apoplast, and plasma membranes (Singh et al., 2019). While ROS are naturally produced within plants as a result of regular cellular processes, their excessive accumulation under stressful conditions can lead to significant harm to essential cellular components such as carbohydrates, proteins, lipids, DNA, and more, owing to their extremely reactive features (Raja et al., 2017).

Plants primarily combat oxidative stress through an inherent protective system composed of various enzymatic components such as ascorbate peroxidase (APx), catalase (CAT), superoxide dismutase (SOD), glutathione reductase (GR), dehydroascorbate reductase (DHAR), glutathione peroxidase (GPX), monodehydroascorbate reductase (MDHAR), guaiacol peroxidase (GOPX), glutathione S-transferase (GST), ferritin, nicotinamide adenine dinucleotide phosphate (NADPH), oxidase-like alternative oxidase (AOX), peroxiredoxins (PRXs), thioredoxins (TRXs), and glutaredoxin (GRX), among others. Additionally, they employ nonenzymatic antioxidants like ascorbic acid (AsA), glutathione (GSH), phenolic acids, alkaloids, flavonoids, carotenoids, α-tocopherol, nonprotein amino acids, and more to address oxidative stress (Gill and Tuteja, 2010; Hasanuzzaman et al., 2012; Kaur et al., 2019). In plant cells, a constant equilibrium is maintained between the antioxidant defense system and the accumulation of ROS (Hasanuzzaman et al., 2012).

Sustaining an ideal ROS level within the cell is crucial for facilitating correct redox biological reactions and overseeing various vital processes in plants, including their growth and development (Mittler, 2017). The equilibrium of ROS generation and ROS scavenging sustains this intermediate stage (Hasanuzzaman et al., 2019). Nonetheless, during stress, an excessive production of ROS disrupts the balance and results in harm to the cells, triggering both programmed cell death (PCD) and a reduction in plant productivity (Raja et al., 2017).

In addition to their harmful effects, ROS are also recognized for their role as secondary messengers or signaling molecules. They transmit signals to the cell nucleus by means of redox reactions, primarily utilizing the mitogen-activated protein kinase (MAPK) pathway. This occurs in numerous cellular processes aimed at enhancing tolerance to various abiotic stresses (Singh et al., 2019). ROS play a crucial role in the adaptation of plants to environmental changes. They primarily function as signaling molecules, regulating various pathways during the plant's adjustment to stressful conditions (Antoniou et al., 2016; Choudhury et al., 2017). Numerous studies have demonstrated that ROS play a crucial role in the effective execution of various essential natural processes, such as cellular growth and specialization (Mittler, 2017). Furthermore, H_2O_2 plays a pivotal role in regulating stress responses in various crop plants, including rice (Sohag et al., 2020), wheat (Habib et al., 2020), maize (Terzi et al., 2014), mung bean (Fariduddin et al., 2014), soybean (Guler and Pehlivan, 2016), cucumber (Sun et al., 2016), sour orange (Tanou et al., 2012), strawberry (Christou et al., 2014), basil (Gohari et al., 2020), and rapeseed (Hasanuzzaman et al., 2017b). Hence, ROS fulfill a pivotal and twofold function in the field of plant biology, making them an intriguing subject of investigation for plant scientists.

The emergence of cutting-edge technologies has brought about a significant transformation in various facets of plant biology investigations and has paved the way for fresh opportunities to delve into the operational characteristics of cellular elements. One of these novel approaches is "metabolomics." Metabolites, which are the final outcomes of cellular activities, serve as direct indicators of how an organism reacts to a range of internal or external disturbances, be they inherent or environmental in nature (Choudhury et al., 2021). In the age of climate change, the repercussions of different abiotic stresses like droughts, floods, high temperatures, and the presence of harmful levels of heavy metals or metalloids resulting from human activities pose a significant global threat to agriculture. The occurrences of such events are expected to increase substantially in the near future. Therefore, it is crucial to gain insights into the elements that play a role in how plants react to, tolerate, and adapt to these nonliving stressors. Currently, the utilization of metabolomics to comprehend how plants respond to nonliving stressors is mainly propelled by advancements in molecular genetics and the enhancement and diversification of various analytical tools (Choudhury et al., 2021). This chapter provides a thorough overview of how metabolomics is becoming increasingly significant in the field of plant systems biology for investigating various facets of abiotic stress in plants.

17.2 SOURCES OF ROS PRODUCTION IN PLANTS

ROS are generated within the cell in both typical and demanding situations, occurring in multiple areas such as chloroplasts, mitochondria, peroxisomes, plasma membranes, endoplasmic reticulum (ER), and the cell wall. When exposed to light, chloroplasts and peroxisomes emerge as the primary contributors to ROS production, while in the absence of light, mitochondria take on the predominant role as ROS generation (Choudhury et al., 2013).

17.2.1 PEROXISOMES

Peroxisomes are small, spherical organelles enclosed by a single membrane, and they play a crucial role in generating hydrogen peroxide (H_2O_2) within cells through their essential oxidative processes (del Río et al., 2006; Das and Roychoudhury, 2014). Moreover, additional metabolic processes, such as the β-oxidation of fatty acids, the flavin oxidase pathway, and the conversion of $O^{\bullet 2-}$ radicals for generating peroxisomal ROS, exist (Das and Roychoudhury, 2014).

17.2.2 MITOCHONDRIA

Mitochondria also serve as the location where potentially damaging ROS, such as H_2O_2 and superoxide ($O^{\bullet 2-}$), are generated (Navrot et al., 2007). Mitochondria in plants have distinctive features compared to those in animals, as they operate within an oxygen-rich and carbohydrate-abundant environment, and they also play a role in the process of photorespiration (Rhoads et al., 2006). The primary offender in this scenario is the mitochondrial electron transport chain (mtETC), as it contains electrons with enough energy to convert oxygen into ROS. Complex I and Complex III within the mtETC are the key players responsible for generating ROS (Møller et al., 2007; Noctor et al., 2007). Complex I, also known as NADH dehydrogenase, is responsible for the direct reduction of oxygen (O_2) to form superoxide anion ($O^{\bullet 2-}$) within its flavoprotein region. This production of ROS at Complex I is intensified when there is a reversal of electron flow from Complex III to Complex I. This reversal occurs because of the absence of NAD^+-linked substrates (Turrens, 2003). A variety of enzymes present in the mitochondrial matrix are additional producers of ROS in the mitochondria. These enzymes include 1-galactono-γ-lactone dehydrogenase, which produces ROS indirectly by supplying electrons to the electron transport chain, and aconitase, which produces ROS directly (Rasmusson et al., 2008). Under typical circumstances, mitochondria typically generate ROS, but their production is significantly heightened when exposed to abiotic stressors (Pastore et al., 2007).

The chloroplast consists of a highly organized network of thylakoid membranes that serve as the home for both the machinery responsible for capturing light during photosynthesis and the structural components necessary for effective light absorption (Pfannschmidt, 2003). The primary generators of ROS within the thylakoid's light-harvesting system are the photosystems PSI and PSII (Das and Roychoudhury, 2014). Plants largely produce ROS because of the chloroplast. It is crucial to effectively manage and eliminate ROS within the chloroplast to ensure the ongoing survival of plants when they are under stress. This is evident in both genetically modified plants and stress-resistant varieties (Tseng et al., 2007).

17.2.3 APOPLAST

The apoplast, the space surrounding the plant cell membrane that allows for diffusion, plays a crucial role in transforming incoming carbon dioxide (CO_2) into a soluble, diffusible form. This processed CO_2 then enters the cytosol where it undergoes photosynthesis. During challenging environmental conditions, stress signals and the presence of abscisic acid (ABA) elevate the significance of the apoplast as a primary location for the generation of H_2O_2 (Hu et al., 2006). The NADPH oxidases found in both the guard cells and mesophyll cells of Arabidopsis, specifically AtRbohD and AtRbohF, are responsible for producing the apoplastic ROS needed to facilitate the closure of stomata induced by ABA (Kwak et al., 2003). In addition to these enzymes, there are other enzymes that generate ROS. These include peroxidases (POXs) sensitive to pH levels, oxidases linked to the cell wall, oxalate oxidases similar to germin, and polyamine oxidases. These enzymes primarily produce H_2O_2.

17.2.4 PLASMA MEMBRANES

The plant cell's outer plasma membrane serves a vital function by engaging with the constantly shifting environmental circumstances and conveying essential data to ensure the cell's ongoing survival. The NADPH-dependent oxidases found within the plasma membrane have garnered attention due to their gene expression patterns and the presence of various homologous forms under varying stress situations (Apel and Hirt, 2004). The generation of $O^{\cdot 2-}$ occurs in NADPH oxidase through the transfer of electrons from cytosolic NADPH to O_2. This results in either the spontaneous conversion of $O^{\cdot 2-}$ into H_2O_2 or its conversion facilitated by SOD. It is firmly established that NADPH oxidase plays a significant role in safeguarding plants against both pathogenic infections and adverse abiotic stress conditions (Kwak et al., 2003).

17.2.5 CELL WALLS

At the time of stress, the lipoxygenase (LOX) situated in the cell wall initiates the hydroperoxidation of polyunsaturated fatty acids (PUFA), turning thiol transform into a powerful source of ROS, including $OH^{\cdot}$, $O^{\cdot 2-}$, H_2O_2, and $^1O_2$. Furthermore, diamine oxidases in the cell wall use polyamines or diamines to produce ROS. When a pathogen attacks, the lignin precursors undergo substantial cross-linking through H_2O_2-mediated processes, strengthening the cell wall by incorporating lignin (Higuchi, 2006).

17.3 ROS GENERATION AND SCAVENGING MECHANISMS TO MAINTAIN CELLULAR REDOX HOMEOSTASIS

Cellular homeostasis within a plant is defined by a fundamental amount of ROS, which varies depending on factors like the plant's growth stage, its internal biological clock, the surrounding environment, its physiological state, and its interactions with the microbiomes in its roots and leaves. Various biological and environmental stressors have the potential to disturb this equilibrium, disengage metabolic processes, and result in the buildup of ROS in various cellular compartments (Mittler et al., 2022). The generation of ROS is a common occurrence in the cellular metabolism of living organisms and is prevalent in various cellular components such as chloroplasts, mitochondria, cytosol, peroxisomes, and so on (Han et al., 2013; Czarnocka and Karpiński, 2018; Hasanuzzaman et al., 2019).

Every plant cell strives to uphold an ever-changing equilibrium between ROS and antioxidants that scavenge ROS. When subjected to abiotic stress, this delicate cellular balance is disrupted, leading to an overwhelming increase in ROS production, tilting the scale toward oxidative reactions (Noctor et al., 2018). Inadequate energy engagement in the photosynthesis process under nonbiological stress conditions leads to a decrease in molecular oxygen levels and results in the generation of significant quantities of ROS, such as H_2O_2, O_2^-, $^1O_2$, OH^-, (Hasanuzzaman et al., 2017a; Singh et al., 2019). ROS are highly active molecules with the potential to harm a wide range of cellular biomolecules, such as carbohydrates, nucleic acids, lipids, proteins, and can disrupt their normal functions (Noctor et al., 2018; Soares et al., 2019). Furthermore, methylglyoxal (MG), a cytotoxic substance with reactive oxidizing properties, is naturally generated in small quantities within cells. However, when exposed to nonbiological stressors, its production escalates and plays a role in the development of oxidative stress. Like ROS, the generation of MG is heightened during abiotic stress conditions, which can harm the cellular ultrastructure and induce mutations, ultimately triggering PCD (Hasanuzzaman et al., 2017b).

In addition to inducing oxidative stress, ROS and MG serve as signaling molecules for stress resilience. They regulate the adaptation and defensive reactions by influencing certain antioxidants and their corresponding genetic pathways (Hasanuzzaman et al., 2017a; Singh et al., 2019). The overproduction of ROS and MG can also trigger a disruption in the balance of redox reactions within the cell. This disturbance can serve as a cue for cell death or the reduction of the plant's lifespan (Han et al., 2013; Hasanuzzaman et al., 2017).

Plant cells possess a robust antioxidant defense and glyoxalase system that effectively removes harmful ROS and MG. The antioxidant defense system comprises various elements, including both nonenzymatic constituents like AsA, GSH, alkaloids, alpha-tocopherol, nonprotein amino acids, and phenolic compounds, as well as enzymatic components (Gill and Tuteja, 2010; Hasanuzzaman et al., 2011). In the antioxidant defense system, the AsA–GSH pool plays a crucial and primary role in reducing the impact of stress by utilizing four key enzymes: APx, MDHAR, DHAR, and GR, to effectively remove ROS (Latowski et al., 2010; Hasanuzzaman et al., 2011; Hasanuzzaman et al., 2012). Typically, within the antioxidant defense mechanism, SOD serves as the initial defense against ROS by transforming superoxide (O_2^-) into H_2O_2. Following this, CAT and APX work to eliminate H_2O_2, converting it into water (H_2O). Additionally, GPX and GST also play a role in neutralizing H_2O_2 into H_2O with the assistance of GSH (Hasanuzzaman et al., 2011).

In situations of excessive light exposure, when the influx of photons surpasses the plant's energy requirements for carbon dioxide fixation, chloroplasts mainly generate $O_2^{\bullet-}$ and $^1O_2$ via photosystems I and II, respectively. In the event of photorespiration being triggered, as seen in C_3 plants, peroxisomes will additionally generate H_2O_2 (Galvez-Valdivieso et al., 2009; Kerchev et al., 2016; Wang et al., 2020). The generation of ROS may increase even more under drought conditions when there is a lack of CO_2 because stomata close, preventing excess energy absorbed by the photosynthetic system from being utilized for CO_2 fixation (Koffler et al., 2014; Zia et al., 2016; Hipsch et al., 2021). During heat stress, when the integrity of membrane complexes participating in various electron transfer pathways is compromised, mitochondria and chloroplasts generate $O_2^{\bullet-}$ and H_2O_2, leading to elevated levels of ROS in the cytoplasm and cell nucleus (Schwarzländer et al., 2009; Babbar et al., 2021).

The body's response to infections is characterized by a unique pattern of ROS accumulation. The primary source of $O_2^{\bullet-}$ and H_2O_2 in the apoplast is the activation of specific oxidases, such as RBOHs. Additionally, they are produced in chloroplasts as a result of disrupted and imbalanced metabolic pathways (Mittler et al., 2022). On the other hand, it has been demonstrated recently that viral infection leads to a reduction in peroxisomal ROS production as a result of viral protein interactions with glycolate oxidase (Yang et al., 2018).

Recent progress in the utilization of genetically engineered ROS detectors and fluorescent dyes has unveiled that distinct cellular compartments accumulate various forms of ROS during various stress conditions (Mittler et al., 2022). Hence, distinct stress-related mechanisms trigger varying patterns or indications of ROS buildup within cells. Furthermore, recent research has unveiled that ROS can be conveyed into or out of various cellular compartments, initiating diverse retrograde and anterograde signaling pathways linking different cellular compartments and the nucleus (Mittler et al., 2022). The various ROS and signals generated within distinct cellular compartments in reaction to diverse stimuli might initiate specialized signal transmission pathways tailored to specific stressors, thereby triggering specific adaptation and protection mechanisms against those particular stresses. The discovery that various types of stress lead to distinct patterns of ROS signatures can provide a foundation for future research on understanding how plants achieve specificity in their responses to stress.

17.4 ROS SIGNALING IN PLANT DEFENSE

Abiotic stress causes an overproduction of ROS due to disruptions in various metabolic processes and the occurrence of physiological abnormalities (Choudhury et al., 2017). The AsA–GSH pathway, which is a part of the antioxidant defense mechanisms, relies on NADPH as its source of energy. When NADPH levels are exhausted, these pathways become unable to protect against the harmful effects of ROS toxicity (Mittler et al., 2004; Choudhury et al., 2017). Nonetheless, it was only toward the conclusion of the 20th century and the onset of the 21st century that the roles of ROS, particularly H_2O_2, in how plants react to various stressors gained significant attention. A small number of scientific teams identified H_2O_2 as a signaling molecule, initiating adaptation mechanisms and bestowing resistance against a range of biotic and abiotic stress factors (Neill et al., 2002a, 2002b). ROS produced within the chloroplast under stress conditions may redirect electrons away from the photosynthetic apparatus, thereby averting excessive strain on the light-harvesting system and preventing subsequent harm. Similarly, ROS also serve to safeguard mitochondria in a comparable manner (Asada, 2006; Choudhury et al., 2017). Peroxidases in the cell wall could potentially play a role in generating ROS for signaling purposes. In this signaling process, H_2O_2 makes use of calcium ions (Ca^{2+}) and activates the MAPK pathway as a subsequent signaling cascade. Additionally, plant hormones, particularly ethylene (ET) and abscisic acid (ABA), participate in stress responses by interacting with ROS and bolstering stress tolerance. This underscores the dual function of ROS during stressful conditions (Kar, 2011). In addition to its role in signal transduction and its interaction with hormones, ROS can also oversee metabolic pathways in response to abiotic stress conditions. These pathways collectively govern the adaptation processes of plants, where redox reactions play a pivotal role in governing the transcription and translation of proteins and enzymes involved in stress adaptation. This ultimately safeguards plant cells from harm (Choudhury et al., 2017; Mittler, 2017). Furthermore, H_2O_2 influences the signaling pathways involving nitric oxide (NO) and calcium ions (Ca^{2+}), which regulate various aspects of plant growth, development, and a range of cellular and physiological responses when exposed to different abiotic stresses (Janicka et al., 2019; Niu and Liao, 2016). As internal H_2O_2 plays a role in enhancing resistance to nonliving environmental pressures, there is a growing focus on applying external H_2O_2, which has largely demonstrated its effectiveness (Savvides et al., 2016; Sun et al., 2016; Hasanuzzaman et al., 2017a; Sohag et al., 2020).

17.5 MECHANISMS OF ROS-MEDIATED STRESS RESPONSES

Numerous studies have provided compelling evidence indicating that various environmental conditions, including heat (Zhao et al., 2018), cold (Kawarazaki et al., 2013), drought, exposure to aluminum toxicity (Wu et al., 2017), the presence of organic pollutants (OPs) (Ahammed et al., 2017), and exposure to pathogens (Kim and Hwang, 2014; Yang et al., 2017), can trigger the production of ROS within plant cells (Table 17.1). ROS, functioning as signaling molecules, initiate signal transduction pathways when exposed to various stresses. Conversely, ROS also inflict permanent harm to cells due to their potent oxidative attributes, resulting in modifications to plant morphological structures that ultimately boost resistance (Bose et al., 2014; Frederickson Matika and Loake, 2014; Wahid et al., 2007). Due to the presence of multiple interchangeable ROS, it becomes challenging to distinguish between the harmful effects on cells and the signaling events triggered by a specific ROS. It's important to note that while ROS can lead to cell death, this process is essential for building resistance against stress. In essence, the activation of stress-induced responses by ROS needs to happen swiftly when stress appears and subside when the stress subsides. Interestingly, when exposed to higher temperatures (31°C under long-day conditions), plants lacking the ATP-dependent mitochondrial protease AtFtSH4 showed an interesting phenotype of premature growth cessation at both the shoot apical meristem (SAM) and root apical meristem (RAM). This phenomenon was linked to the buildup of internal oxidative stress and the gradual decline of mitochondrial function (Dolzblasz et al., 2016, 2018). These findings demonstrate the importance of sustaining proper mitochondrial function in the SAM and RAM, a process reliant on AtFtSH4, to maintain the stem cell's capabilities and its ability to cope with temperature stress during growth. The presence of ethylene triggers the activation of RBOHH, which belongs to the NADPH oxidase gene family. Using CRISPR/Cas9 to disable RBOHH results in a decrease in the buildup of ROS and inhibits the formation of inducible aerenchyma in rice roots, a crucial adaptation for rice to thrive in flooded and oxygen-deprived environments (Yamauchi et al., 2017). In periods of drought, ABA inhibits the buildup of H_2O_2 by stimulating the expression of CAT OsCATB, thus shielding cells from oxidative harm caused by ROS (Ye et al., 2011). When the wheat E3 ubiquitin ligase TaPUB1 (known as The Plant U-Box Proteins 1) is introduced into tobacco plants through transfection, it results in reduced levels of ROS and increased antioxidant capabilities in the transgenic plants. This enhancement leads to an improved survival rate of the transgenic tobacco plants when subjected to drought stress (DS) conditions (Zhang et al., 2017). In Arabidopsis, the transcription factor (TF) RRTF1 (redox-responsive transcription factor 1), which belongs to the APETALA2/ethylene response factor (AP2/ERF) family, plays a vital role in the central redox signaling network. Its expression is swiftly and briefly triggered by various ROS generated by both biotic and abiotic signals. Increased levels of RRTF1 in plants lead to the accumulation of ROS, implying that RRTF1 enhances the formation of ROS in reaction to stress. This augmentation of ROS production by RRTF1 may be crucial for the rapid and transient establishment of localized peaks in ROS, which are necessary to trigger appropriate downstream responses (Matsuo et al., 2015). When pathogens infiltrate, plants initiate the generation of ROS. This response is swiftly initiated upon pathogen detection and can potentially work in conjunction to activate the hypersensitive response (HR) (Huang et al., 2019). In the context of plant disease resistance, ROS have a dual beneficial function. They not only actively eliminate invading bacteria but also promote the reinforcement of neighboring cell walls, thereby impeding the further spread of pathogens (Huang et al., 2019). The rapid generation of ROS bursts serves as a shared signaling response in immunity across different domains of life. In order to defend against fungal infections, plants have evolved pattern-recognition receptors that can detect chitin, leading to the activation of intracellular MAPK cascades, ultimately resulting in the rapid synthesis of ROS (Kawasaki et al., 2017). In Arabidopsis, BIK1, a kinase associated with pattern recognition receptors, initiates a direct phosphorylation of RBOHD, a NADPH oxidase, triggering the burst of ROS induced by pathogen-associated molecular patterns and subsequently enhancing the plant's defense against bacterial infections (Kadota et al., 2014). Moreover, OPs cause environmental pollution, leading to the accumulation of both H_2O_2 and NO in the tips of plant roots. This accumulation results in an increase in malondialdehyde levels, which is an indicator of lipid peroxidation in cell membranes, as well as abnormal growth of the roots. However, the plant growth and stress tolerance regulator known as 24-epibrassinolide triggers both nonenzymatic and enzymatic antioxidant defense mechanisms in cucumber plants. This, in turn, elevates the levels of antioxidants, like SOD, CAT, and GSH, helping to maintain the balance of ROS within the cells. Consequently, this enhances cucumber's resistance to the harmful effects of OPs (Ahammed et al., 2017).

TABLE 17.1 ROS Responsible for Plant Stress Responses

ROS	Genes Involved	Stress Response	References
$O_2^{\cdot-}$, H_2O_2	ABA, AtNTL4	Drought	Huang et al. (2019)
H_2O_2	ABA, OsCATB	Water stress	Ye et al. (2011)
$O_2^{\cdot-}$, H_2O_2	Ethylene, OsRBOHH	Waterlogging	Yamauchi et al., 2017)
H_2O_2	Ethylene, OsEIN2	Disease resistance	Yang et al. (2017)
H_2O_2	AtPRX64	Al stress	Wu et al. (2017)
H_2O_2	SA, CaPAL1	Microbial pathogens	Kim and Hwang (2014)
$O_2^{\cdot-}$, H_2O_2	OsCATB	High temperature	Zhao et al. (2018)
$O_2^{\cdot-}$, H_2O_2	AtSRC	Cold temperature	Kawarazaki et al. (2013)
NO, H_2O_2	24-Epibrassinolide	Organic pollutants treatment	Ahammed et al. (2017)

17.6 METABOLOMICS AND MOLECULAR RESPONSES TO ABIOTIC STRESSES

Metabolic control plays a crucial role in preserving cellular osmotic balance when plants face abiotic stress, and the method of metabolite profiling has been extensively employed to examine how plants react at the molecular level to such stress conditions (Fàbregas and Fernie, 2019). Apart from its significance in cellular function, water plays a crucial role in plants because it consistently participates in transporting metabolites and vital nutrients to different parts of the plant. When there is not enough water available or when transpiration rates increase, it can lead to elevated DS (dehydration stress) and alterations in metabolite production (Jogawat et al., 2021). The diverse biological processes that empower plants to endure drought conditions can be observed throughout different phases of plant development, spanning cellular, organ-specific, and overall plant reactions. Drought stress affects plants at various levels, including the molecular scale (Anjum et al., 2011), leading to the heightened presence of proteins and metabolites associated with drought resistance (Reddy et al., 2004). Multiple molecular pathway sequences, such as the recognition of water shortage, initiation of signaling networks, and subsequent transcriptional, metabolic, and regulatory element reactions, enhance a plant's ability to withstand DS (Liu et al., 2014). The control of drought response at the molecular level is heavily influenced by regulatory components, including TFs and protein kinases. Families of TFs like MYB, NAC, bZIP, AP2/ERF, and AREB/ABF play a pivotal role in overseeing stomatal behavior and the activation of drought-responsive genes either before or after specific metabolic processes (Joshi et al., 2016; Kim et al., 2010) (Table 17.2).

The reaction of molecules to DS is a complex characteristic influenced by numerous genes. Various studies have explored multiple genes associated with DS at the transcriptional level using techniques like microarray analysis and real-time polymerase chain reaction (Kumar et al., 2021). Functional validation demonstrated that these genes offer defense against dehydration stress by engaging in stress perception, signal transduction, and regulating transcriptional networks to enhance drought tolerance (Umezawa et al., 2006; Wang et al., 2005). Hence, gaining knowledge about how plants respond at a molecular

TABLE 17.2 Major ROS Scavenging Enzymatic Antioxidants and Their Functions

Enzymatic Antioxidants	Location	Enzyme Code	Reactions Catalyzed	Functions	References
Dehydroascorbate reductase (DHAR)	Mitochondria, cytoplasm, and chloroplast	EC 1.8.5.1	DHA + 2GSH → AA + GSSG	DHAR reduces dehydroascorbate (DHA) to AA by using reduced glutathione (GSH) as an electron donor	
Monodehydroascorbate reductase (MDHAR)	Mitochondria, cytoplasm, and chloroplast	EC 1.6.5.4	MDHA + NAD(P)H → AA + $NAD(P)^+$	It regenerates AA from the short-lived MDHA	Mittler (2002)
Catalase (CAT)	Peroxisome and mitochondria	EC 1.11.1.6	$H_2O_2 \rightarrow H_2O + {}^1/_2O_2$	It catalyzes the dismutation of H_2O_2 into H_2O and O_2	Mittler (2002)
Superoxide dismutase (SOD)	Peroxisomes, cytosol, mitochondria, and chloroplast	EC 1.15.1.1	$O_2^- + O_2^- + 2H^+ \rightarrow 2H_2O_2 + O_2$	SOD forms the first line of defense against ROS-induced damages under environmental stresses	Das and Roychoudhury (2014)
Ascorbate peroxidase (APX)	Peroxisomes, cytosol, mitochondria, and chloroplast	EC 1.11.1.11	H_2O_2 + AA → $2H_2O$ + DHA	APX reduces H_2O_2 to H_2O and DHA, using ascorbic acid (AA) as a reducing agent	Das and Roychoudhury (2014)
Guaicol peroxidase (GPX)	Mitochondria, cytoplasm, ER, and chloroplast	EC 1.11.1.7	H_2O_2 + GSH → H_2O + GSSG	It eliminates excess H_2O_2 both during normal metabolism as well as during stress	Asada (1999)
Glutathione reductase (GR)	Mitochondria, cytoplasm, and chloroplast	EC 1.6.4.2	GSSG + NAD(P)H → 2GSH + $NAD(P)^+$	It catalyzes the formation of a disulfide bond in glutathione disulfide to maintain a high cellular GSH/GSSG ratio	Das and Roychoudhury (2014)

level to drought tolerance can offer valuable information for improving the ability to withstand drought in vulnerable plant species. Substantial research endeavors have been undertaken to investigate the molecular processes employed by plants to adapt to DS. When faced with DS, plants adapt by altering their transcriptional, proteomic, and metabolic pathways in order to safeguard cells from damage caused by stress (Kumar et al., 2021). Primary metabolites like glucose, sucrose, and trehalose serve as signaling molecules that control the expression of genes related to plant growth and the response to stress (Maruyama et al., 2014). Drought-induced stress increases the generation of ROS in various parts of cells, especially in chloroplasts and mitochondria. This process is regulated by a flexible and collaborative antioxidant mechanism that manages ROS levels within cells and establishes the redox state of plant cells (Kaur and Asthir, 2017).

17.7 ANALYTICAL METHODS IN PLANT METABOLOMICS

Achieving top-notch metabolic datasets does not necessarily demand the adoption of a particular approach. Instead, it typically hinges on the choice of the analytical tool to be employed. Metabolomic analysis primarily relies on nuclear magnetic resonance (NMR) and mass spectrometry (MS) methods. In the NMR-based approach, molecular properties are ascertained by assessing the magnetic characteristics of atomic nuclei, which aids in elucidating the chemical composition and ultimately the molecular structure of the compound (Liu and Locasale, 2017). The MS-based approaches surpass NMR-based techniques in terms of sensitivity and their ability to quantify numerous metabolites within a sample (Liu and Locasale, 2017). However, NMR's sensitivity is relatively lower, although it offers simplicity in quantifying metabolites since the molecule count corresponds directly to the number of nuclei. In contrast, MS necessitates the use of internal standards during extraction and sample preparation, which can be a laborious process, whether using gas chromatography (GC) or liquid chromatography (LC) for metabolite separation (Liu and Locasale, 2017).

Analytical chemistry uses the NMR spectroscopy as a potent instrument on a regular basis. The NMR spectrum provides a wealth of information about the identification of the chemical compounds, on the basis of which metabolites in a sample may be identified and measured, making it the preferred analytical instrument for metabolomics (Choudhury et al., 2021).

MS stands out as a widely embraced approach for plant metabolomics, possibly the most favored one. However, harnessing MS-based methods entails a multifaceted sequence of activities, commencing with experimental planning, sample preparation, preliminary procedures, analytical techniques, and culminating in data interpretation. Essentially, in metabolomics, a mass spectrometer is coupled with either LC or GC, which facilitates the effective separation of substances within a mixture. Subsequently, these compounds are detected through MS and their identification is achieved by comparing their mass spectra with the reference library established by the National Institute of Standards and Technology (Choudhury et al., 2021).

17.8 APPLICATION OF METABOLOMICS TO PLANT RESPONSES AGAINST STRESSES

Plant development and metabolic processes can be significantly influenced by a range of nonliving factors. These encompass environmental disruptions like drought, excessive water, extreme heat or cold, and the presence of harmful levels of metals or metalloids. These stresses have a substantial impact on multiple facets of plant metabolism and have a direct effect on how much a plant can produce and yield. In order to gain insight into how plants react to these nonliving stressors, metabolomics has emerged as a highly valuable approach (Choudhury et al., 2021). Plants have developed defense mechanisms to cope with various stressors and adapt to unfavorable environments. One of the most well-known defense responses in plants is the synthesis of specific metabolites. Examining these defense reactions using a metabolomics approach can provide a comprehensive understanding of how plants combat stressors. While sunlight is essential for most plants, except for a few parasitic species, ultraviolet (UV) light can be detrimental as it leads to the oxidation of DNA and cellular components through the generation of free radicals. Analyzing the metabolome in response to UV-B exposure in both wild-type Arabidopsis and mutants deficient in flavonoid or sinapoyl malate production revealed that early metabolic responses primarily involved primary metabolites. This suggests that these primary metabolites play a crucial role in priming the cells to later produce UV-B-absorbing secondary metabolites, such as flavonoids and sinapoyl malate (Kusano et al., 2011). Major research in the realm of plant responses to environmental stress primarily revolves around pinpointing genes linked to stress adaptation and tolerance. This is accomplished by analyzing the distinct gene expression patterns found in various plant or crop varieties. As previously discussed, metabolites represent the final products of various regulatory pathways, making metabolomic data invaluable for comprehending how plants react to stress. The alterations in a plant's metabolome have undergone thorough investigation in numerous plant species, even those exposed to diverse environmental stressors (Choudhury et al., 2021). Besides abiotic stress, plants also face the challenge of biotic stress, which can significantly impact their well-being. Ward and his colleagues conducted a study examining the alterations in Arabidopsis' metabolism when it comes to disease development. They highlighted distinct variations in the metabolites of Arabidopsis leaves when infected with the virulent *Pseudomonas syringae* compared to leaves that remained uninfected (Ward et al., 2010). Besides confirming modifications in phenolic and indolic compounds, they detected swift shifts in the quantities of amino acids and other nitrogen-containing compounds, distinct categories of glucosinolates, disaccharides, and molecules that impact the prevalence of ROS. These findings indicate that a comprehensive overhaul of metabolic pathways plays a role in how plants respond to stressors. Lately, numerous investigations have explored the broader metabolic alterations linked to both biotic and abiotic stressors through metabolome analysis (Urano et al., 2009; Widodo Patterson et al., 2009; Chaouch et al., 2010; Consonni et al., 2010; Mhamdi et al., 2010; Simon et al., 2010).

The identification of biomarkers is crucial for comprehending how plants respond to threats, and it has the potential to enable the assessment of disease status on a molecular level. This, in turn, can contribute to the creation of novel medications, early disease detection, and enhanced pest management. In the realm of human metabolomics, researchers have been actively engaged in discovering biomarkers using metabolomics techniques (Soga et al., 2006; Jansson et al., 2009; Koulman et al., 2009; Sreekumar et al., 2009). In studies, it may not be essential to definitively identify or label the majority of metabolites, as long as the biomarkers discovered prove to be valuable for early disease detection. Therefore, the effectiveness of the metabolomics technologies employed can have a direct impact in this domain, as a significant amount of time is still dedicated to analyzing data from numerous unidentified metabolites obtained through metabolome analysis (Okazaki and Saito, 2012).

Metabolomics has played a crucial role in aiding plant biologists and breeders in pinpointing crucial metabolic characteristics while also unveiling novel approaches and tactics for enhancing plant or crop yield. Numerous genes have been pinpointed, and their functional roles have been clarified. Additionally, research in the field of metabolomics has provided insights into the broader framework of how plants respond to and tolerate abiotic stress. Consequently, when combined with other functional genomics elements, metabolomics facilitates the establishment of a comprehensive network system that enables the understanding of nearly every conceivable aspect of how plants respond to abiotic stress (Choudhury et al., 2021). Due to recent advancements in metabolomics techniques, it is now possible to simultaneously detect several hundred metabolites in a tissue extract. This enables a more dependable comparison of metabolite composition between stressed and unstressed plants, whether through targeted or global (untargeted) approaches. While LC and GC-based MS methods and NMR spectrometry have been extensively used in plant metabolomics, accurately identifying certain metabolites still poses a significant challenge (Zivy et al., 2015). This mainly stems from the diverse array of physicochemical characteristics exhibited by metabolites. Consequently, the comprehensive analysis of the entire plant metabolome continues to pose a significant hurdle for plant scientists (Zivy et al., 2015). Plants, being stationary organisms, experience a wide array of environmental challenges known as abiotic stresses. These challenges encompass conditions like drought, excess water, high salinity in the soil, heavy metals, metalloids, extreme temperatures (both low and high), and variations in light intensity. These factors pose significant obstacles to plant growth and overall yield. When plants are exposed to such stresses, it disrupts their genetic functioning and leads to metabolic imbalances.

Research conducted on Arabidopsis plants during DS, for instance, has revealed the accumulation of specific metabolites. These metabolites include γ-amino butyric acid (GABA), raffinose family oligosaccharides (RFOs), various amino acids, and metabolites associated with the tricarboxylic acid cycle (Choudhury et al., 2021). Notably, GABA, classified as a nonprotein amino acid, plays a crucial role as a signaling molecule in regulating how plants respond to abiotic stresses (Cheng et al., 2018). GABA plays a crucial role in preserving the equilibrium between carbon and nitrogen levels, and it is linked to the promotion of cell growth as well as the regulation of amino acid and carbohydrate metabolism. GABA also serves as a potent osmolyte that helps counteract ROS generated due to cellular stress (Choudhury et al., 2021). The RFOs are acknowledged as compatible solutes that function as antioxidant compounds to mitigate the buildup of ROS under stress conditions and play a role in carbon allocation (ElSayed et al., 2014). In times of stress, RFOs play a role in supporting the stabilization of membranes and proteins. It has been observed to increase in concentration during periods of DS, where its function involves preserving cell turgor by serving as an osmolyte (Choudhury et al., 2021). Therefore, the primary reactions in plants to abiotic stresses involve the accumulation and changes in metabolites (Obata and Fernie, 2012). Numerous secondary metabolites tend to accumulate when plants experience stress. These compounds often serve a dual purpose: they help reduce the burden of stress by acting as effective scavengers of ROS, functioning as antioxidants, and also play a crucial role as signaling molecules. It is undeniable that these metabolic elements serve as crucial indicators for comprehending how plants respond to stress in a more comprehensive manner than ever before. The metabolome provides a comprehensive reflection of the plant's overall physiological condition during stressful periods, allowing us to deduce the various functions of individual metabolites in responding to and tolerating stress (Choudhury et al., 2021).

17.9 CONCLUSIONS AND FUTURE PROSPECTIVE

The ROS serves a dual purpose, arising as a natural byproduct of aerobic metabolism and serving as a marker during stressful conditions. They act not only as agents of harm to plants but also activate stress signaling pathways to mitigate further damage. ROS production is widespread, occurring in both intracellular and extracellular locations. Their detrimental effects encompass various biomolecules, including lipids, proteins, and DNA, compromising cell integrity and eventually causing cell death. However, plants have evolved a diverse array of defensive strategies at the morphological, metabolic, and genetic levels to adapt to adverse environmental conditions. This chapter provides insight into how both components of the antioxidant system, namely, antioxidant enzymes and nonantioxidant metabolites, collaborate to counteract ROS-induced damage and enhance tolerance to environmental stressors.

Despite significant recent progress, uncertainties and knowledge gaps persist regarding ROS formation and their impact on plants, largely due to their short lifespan and high reactivity. While the compartmentalization of antioxidants is well understood, further exploration is needed to elucidate the sensing and response mechanisms, as well as the delicate balance between ROS production and scavenging. Several questions remain unanswered, including the interplay between ROS and calcium signaling and the regulation of ROS during various environmental stressors. Future research may benefit from advanced imaging techniques, such as calcium imaging markers, to enhance our understanding of ROS metabolism. Additionally, advanced functional genomics, in conjunction with proteomics and metabolomics, will provide in-depth insights into the ROS network and associated responses. Compared to various scientific and experimental methods aimed at comprehending how plants respond to abiotic stress, metabolomics stands out as an area that has received relatively less attention. It is particularly intriguing to observe how metabolomic data can be applied in real-world plant breeding endeavors, transitioning from the laboratory to the field. It is undeniable that the transgenic approach, involving the overexpression of antioxidant genes, holds promise for enhancing tolerance to multiple stressors in the future.

KEYWORDS

- **metabolomics**
- **reactive oxygen species**
- **plant stress tolerance**
- **metabolites**
- **oxidative stress**

REFERENCES

Ahammed, G. J., He, B. B., Qian, X. J., Zhou, Y. H., Shi, K., Zhou, J., Yu, J. Q., Xia, X. J. 24-Epibrassinolide alleviates organic pollutants-retarded root elongation by promoting redox homeostasis and secondary metabolism in Cucumis sativus L. *Environ. Pollut.*, 2017, 229, 922–931. https://doi.org/10.1016/j.envpol.2017.07.076

Anjum, S. A., XiaoYu, X., LongChang, W., Saleem, M. F., Chen, M., Wang, L. Morphological, physiological and biochemical responses of plants to drought stress. *Afr. J. Agric Res.*, 2011, 6(9), 2026–2032.

Antoniou, C., Savvides, A., Christou, A., Fotopoulos, V. Unravelling chemical priming machinery in plants: The role of reactive oxygen–nitrogen–sulfur species in abiotic stress tolerance enhancement. *Curr. Opin. Plant. Biol.*, 2016, 33, 101–107. https://doi.org/10.1016/j.pbi.2016.06.020

Apel, K., Hirt, H. Reactive oxygen species: Metabolism, oxidative stress, and signal transduction. *Annu. Rev. Plant Biol.*, 2004, 55(1), 373–399. https://doi.org/10.1146/annurev.arplant.55.031903.141701

Asada, K. Production and scavenging of reactive oxygen species in chloroplasts and their functions. *Plant Physiol.*, 2006, 141(2), 391–396. https://doi.org/10.1104/pp.106.082040

Babbar, R., Karpinska, B., Grover, A., Foyer, C. H. Heat-induced oxidation of the nuclei and cytosol. *Front. Plant Sci.*, 2021, 11, 617779. https://www.frontiersin.org/articles/10.3389/fpls.2020.617779

Bose, J., Rodrigo-Moreno, A., Shabala, S. ROS homeostasis in halophytes in the context of salinity stress tolerance. *J. Exp. Bot.*, 2014, 65(5), 1241–1257. https://doi.org/10.1093/jxb/ert430

Chaouch, S., Queval, G., Vanderauwera, S., Mhamdi, A., Vandorpe, M., Langlois-Meurinne, M., Van Breusegem, F., Saindrenan, P., Noctor, G. Peroxisomal hydrogen peroxide is coupled to biotic defense responses by isochorismate synthase1 in a daylength-related manner. *Plant Physiol.*, 2010, 153(4), 1692–1705. https://doi.org/10.1104/pp.110.153957

Cheng, B., Li, Z., Liang, L., Cao, Y., Zeng, W., Zhang, X., Ma, X., Huang, L., Nie, G., Liu, W., Peng, Y. The γ-aminobutyric acid (GABA) alleviates salt stress damage during seeds germination of white clover associated with Na+/K+ transportation, dehydrins accumulation, and stress-related genes expression in white clover. *Int. J. Mol. Sci.*, 2018, 19(9), 9. https://doi.org/10.3390/ijms19092520

Choudhury, F. K., Rivero, R. M., Blumwald, E., Mittler, R. Reactive oxygen species, abiotic stress and stress combination. *Plant J.*, 2017, 90(5), 856–867. https://doi.org/10.1111/tpj.13299

Choudhury, S., Panda, P., Sahoo, L., Panda, S. K. Reactive oxygen species signaling in plants under abiotic stress. *Plant Signal. Behav.*, 2013, 8(4), e23681. https://doi.org/10.4161/psb.23681

Choudhury, S., Sharma, P., Moulick, D., Mazumder, M. K. Unrevealing metabolomics for abiotic stress adaptation and tolerance in plants. *J. Crop Sci. Biotechnol.*, 2021, 24(5), 479–493. https://doi.org/10.1007/s12892-021-00102-8

Christou, A., Filippou, P., Manganaris, G. A., Fotopoulos, V. Sodium hydrosulfide induces systemic thermotolerance to strawberry plants through transcriptional regulation of heat shock proteins and aquaporin. *BMC Plant Biol.*, 2014, 14(1), 42. https://doi.org/10.1186/1471-2229-14-42

Consonni, C., Bednarek, P., Humphry, M., Francocci, F., Ferrari, S., Harzen, A., Ver Loren van Themaat, E., Panstruga, R. Tryptophan-derived metabolites are required for antifungal defense in the arabidopsis mlo2 mutant. *Plant Physiol.*, 2010, 152(3), 1544–1561. https://doi.org/10.1104/pp.109.147660

Czarnocka, W., Karpiński, S. Friend or foe? Reactive oxygen species production, scavenging and signaling in plant response to environmental stresses. *Free Radic. Biol. Med.*, 2018, 122, 4–20. https://doi.org/10.1016/j.freeradbiomed.2018.01.011

Das, K., Roychoudhury, A. Reactive oxygen species (ROS) and response of antioxidants as ROS-scavengers during environmental stress in plants. *Front. Environ. Sci.*, 2014, 2. https://www.frontiersin.org/articles/10.3389/fenvs.2014.00053

del Río, L. A., Sandalio, L. M., Corpas, F. J., Palma, J. M., Barroso, J. B. Reactive oxygen species and reactive nitrogen species in peroxisomes. Production, scavenging, and role in cell signaling. *Plant Physiol.*, 2006, 141(2), 330–335. https://doi.org/10.1104/pp.106.078204

Dolzblasz, A., Gola, E. M., Sokołowska, K., Smakowska-Luzan, E., Twardawska, A., Janska, H. Impairment of meristem proliferation in plants lacking the mitochondrial protease AtFTSH4. *Int. J. Mol. Sci.*, 2018, 19(3), 3. https://doi.org/10.3390/ijms19030853

Dolzblasz, A., Smakowska, E., Gola, E. M., Sokołowska, K., Kicia, M., Janska, H. The mitochondrial protease AtFTSH4 safeguards Arabidopsis shoot apical meristem function. *Sci. Rep.*, 2016, 6(1), 1. https://doi.org/10.1038/srep28315

ElSayed, A. I., Rafudeen, M. S., Golldack, D. Physiological aspects of raffinose family oligosaccharides in plants: Protection against abiotic stress. *Plant Biol.*, 2014, 16(1), 1–8. https://doi.org/10.1111/plb.12053

Fàbregas, N., Fernie, A. R. The metabolic response to drought. *J. Exp. Bot.*, 2019, 70(4), 1077–1085. https://doi.org/10.1093/jxb/ery437

Fariduddin, Q., Khan, T. A., Yusuf, M. Hydrogen peroxide mediated tolerance to copper stress in the presence of 28-homobrassinolide in Vigna radiata. *Acta Physiol. Plant*, 2014, 36(10), 2767–2778. https://doi.org/10.1007/s11738-014-1647-0

Frederickson Matika, D. E., Loake, G. J. Redox regulation in plant immune function. *Antioxid. Redox Signal.*, 2014, 21(9), 1373–1388. https://doi.org/10.1089/ars.2013.5679

Galvez-Valdivieso, G., Fryer, M. J., Lawson, T., Slattery, K., Truman, W., Smirnoff, N., Asami, T., Davies, W. J., Jones, A. M., Baker, N. R., Mullineaux, P. M. The high light response in arabidopsis involves ABA signaling between vascular and bundle sheath cells. *Plant Cell*, 2009, 21(7), 2143–2162. https://doi.org/10.1105/tpc.108.061507

Gill, S. S., Tuteja, N. Reactive oxygen species and antioxidant machinery in abiotic stress tolerance in crop plants. *Plant Physiol. Biochem.*, 2010, 48(12), 909–930. https://doi.org/10.1016/j.plaphy.2010.08.016

Gohari, G., Alavi, Z., Esfandiari, E., Panahirad, S., Hajihoseinlou, S., Fotopoulos, V. Interaction between hydrogen peroxide and sodium nitroprusside following chemical priming of *Ocimum basilicum* L. against salt stress. *Physiol. Plant*, 2020, 168(2), 361–373. https://doi.org/10.1111/ppl.13020

Guler, N. S., Pehlivan, N. Exogenous low-dose hydrogen peroxide enhances drought tolerance of soybean (*Glycine max* L.) through Inducing antioxidant system. *Acta Biol. Hung.*, 2016, 67(2), 169–183. https://doi.org/10.1556/018.67.2016.2.5

Habib, N., Ali, Q., Ali, S., Javed, M. T., Zulqurnain Haider, M., Perveen, R., Shahid, M. R., Rizwan, M., Abdel-Daim, M. M., Elkelish, A., Bin-Jumah, M. Use of nitric oxide and hydrogen peroxide for better yield of wheat (*Triticum aestivum* L.) under water deficit conditions: Growth, osmoregulation, and antioxidative defense mechanism. *Plants*, 2020, 9(2), 2. https://doi.org/10.3390/plants9020285

Han, Y., Chaouch, S., Mhamdi, A., Queval, G., Zechmann, B., Noctor, G. Functional analysis of arabidopsis mutants points to novel roles for glutathione in coupling H_2O_2 to activation of salicylic acid accumulation and signaling. *Antioxid Redox Signal.*, 2013, 18(16), 2106–2121. https://doi.org/10.1089/ars.2012.5052

Hasanuzzaman, M., Bhuyan, M. H. M. B., Anee, T. I., Parvin, K., Nahar, K., Mahmud, J. A., Fujita, M. Regulation of ascorbate-glutathione pathway in mitigating oxidative damage in plants under abiotic stress. *Antioxidants*, 2019, 8(9), 9. https://doi.org/10.3390/antiox8090384

Hasanuzzaman, M., Hossain, M. A., da Silva, J. A. T., Fujita, M. Plant response and tolerance to abiotic oxidative stress: Antioxidant defense is a key factor. In: Venkateswarlu, B., Shanker, A. K., Shanker, C., Maheswari, M. (Eds.), *Crop Stress and its Management: Perspectives and Strategies* (pp. 261–315). Springer: the Netherlands, 2012a. https://doi.org/10.1007/978-94-007-2220-0_8

Hasanuzzaman, M., Hossain, M. A., Fujita, M. Nitric oxide modulates antioxidant defense and the methylglyoxal detoxification system and reduces salinity-induced damage of wheat seedlings. *Plant Biotechnol. Rep.*, 2011, 5(4), 353–365. https://doi.org/10.1007/s11816-011-0189-9

Hasanuzzaman, M., Hossain, M. A., Fujita, M. Exogenous selenium pretreatment protects rapeseed seedlings from cadmium-induced oxidative stress by upregulating antioxidant defense and methylglyoxal detoxification systems. *Biol. Trace Elem. Res.*, 2012b, 149(2), 248–261. https://doi.org/10.1007/s12011-012-9419-4

Hasanuzzaman, M., Nahar, K., Gill, S. S., Alharby, H. F., Razafindrabe, B. H. N., Fujita, M. Hydrogen peroxide pretreatment mitigates cadmium-induced oxidative stress in *Brassica napus* L.: An intrinsic study on antioxidant defense and glyoxalase systems. *Front. Plant Sci.*, 2017a, 8, 115. https://www.frontiersin.org/articles/10.3389/fpls.2017.00115

Hasanuzzaman, M., Nahar, K., Hossain, M. S., Mahmud, J. A., Rahman, A., Inafuku, M., Oku, H., Fujita, M. Coordinated actions of glyoxalase and antioxidant defense systems in conferring abiotic stress tolerance in plants. *Int. J. Mol. Sci.*, 2017b, 18(1), 1. https://doi.org/10.3390/ijms18010200

Higuchi, T. Look back over the studies of lignin biochemistry. *J. Wood Sci.*, 2006, 52(1), 2–8. https://doi.org/10.1007/s10086-005-0790-z

Hipsch, M., Lampl, N., Zelinger, E., Barda, O., Waiger, D., Rosenwasser, S. Sensing stress responses in potato with whole-plant redox imaging. *Plant Physiol.*, 2021, 187(2), 618–631. https://doi.org/10.1093/plphys/kiab159

Hu, X., Zhang, A., Zhang, J., Jiang, M. Abscisic acid is a key inducer of hydrogen peroxide production in leaves of maize plants exposed to water stress. *Plant Cell Physiol.*, 2006, 47(11), 1484–1495. https://doi.org/10.1093/pcp/pcl014

Huang, H., Ullah, F., Zhou, D. X., Yi, M., Zhao, Y. Mechanisms of ROS regulation of plant development and stress responses. *Front. Plant Sci.*, 2019, 10, 800. https://www.frontiersin.org/articles/10.3389/fpls.2019.00800

Janicka, M., Reda, M., Napieraj, N., Kabała, K. Plant abiotic stress: Function of nitric oxide and hydrogen peroxide. In: Gupta, D. K., Palma, J. M., Corpas, F. J. (Eds.), *Nitric Oxide and Hydrogen Peroxide Signaling in Higher Plants* (pp. 201–219). Springer International Publishing, 2019. https://doi.org/10.1007/978-3-030-11129-8_10

Jansson, J., Willing, B., Lucio, M., Fekete, A., Dicksved, J., Halfvarson, J., Tysk, C., Schmitt-Kopplin, P. Metabolomics reveals metabolic biomarkers of Crohn's disease. *PLoS One*, 2009, 4(7), e6386. https://doi.org/10.1371/journal.pone.0006386

Jogawat, A., Yadav, B., Chhaya Lakra, N., Singh, A. K., Narayan, O. P. Crosstalk between phytohormones and secondary metabolites in the drought stress tolerance of crop plants: A review. *Physiol. Plant*, 2021, 172(2), 1106–1132. https://doi.org/10.1111/ppl.13328

Joshi, R., Wani, S. H., Singh, B., Bohra, A., Dar, Z. A., Lone, A. A., Pareek, A., Singla-Pareek, S. L. Transcription factors and plants response to drought stress: Current understanding and future directions. *Front. Plant Sci.*, 2016, 7. https://www.frontiersin.org/articles/10.3389/fpls.2016.01029

Kadota, Y., Sklenar, J., Derbyshire, P., Stransfeld, L., Asai, S., Ntoukakis, V., Jones, J. D., Shirasu, K., Menke, F., Jones, A., Zipfel, C. Direct regulation of the NADPH oxidase RBOHD by the PRR-associated kinase BIK1 during plant immunity. *Mol. Cell*, 2014, 1(54), 43–55. https://doi.org/10.1016/j.molcel.2014.02.021

Kar, R. K. Plant responses to water stress: Role of reactive oxygen species. *Plant Signal. Behav.*, 2011, 6(11), 1741–1745. https://doi.org/10.4161/psb.6.11.17729

Kaur, G., Asthir, B. Molecular responses to drought stress in plants. *Biol Plant.*, 2017, 61(2), 201–209. https://doi.org/10.1007/s10535-016-0700-9

Kaur, N., Kaur, J., Grewal, S. K., Singh, I. Effect of heat stress on antioxidative defense system and its amelioration by heat acclimation and salicylic acid pre-treatments in three pigeonpea genotypes. *Indian J. Agric. Biochem.*, 2019, 32(1), 106. https://doi.org/10.5958/0974-4479.2019.00014.5

Kawarazaki, T., Kimura, S., Iizuka, A., Hanamata, S., Nibori, H., Michikawa, M., Imai, A., Abe, M., Kaya, H., Kuchitsu, K. A low temperature-inducible protein AtSRC2 enhances the ROS-producing activity of NADPH oxidase AtRbohF. *Biochim Biophys Acta (BBA) - Mol. Cell Res.*, 2013, 1833(12), 2775–2780. https://doi.org/10.1016/j.bbamcr.2013.06.024

Kawasaki, T., Yamada, K., Yoshimura, S., Yamaguchi, K. Chitin receptor-mediated activation of MAP kinases and ROS production in rice and Arabidopsis. *Plant Signal. Behav.*, 2017, 12(9), e1361076. https://doi.org/10.1080/15592324.2017.1361076

Kerchev, P., Waszczak, C., Lewandowska, A., Willems, P., Shapiguzov, A., Li, Z., Alseekh, S., Mühlenbock, P., Hoeberichts, F. A., Huang, J., Van Der Kelen, K., Kangasjärvi, J., Fernie, A. R., De Smet, R., Van de Peer, Y., Messens, J., Van Breusegem, F. Lack of glycolate oxidase1, but not glycolate oxidase2, attenuates the photorespiratory phenotype of CATALASE2-deficient arabidopsis. *Plant Physiol.*, 2016, 171(3), 1704–1719. https://doi.org/10.1104/pp.16.00359

Kim, D. S., Hwang, B. K. An important role of the pepper phenylalanine ammonia-lyase gene (PAL1) in salicylic acid-dependent signalling of the defence response to microbial pathogens. *J. Exp. Bot.*, 2014, 65(9), 2295–2306. https://doi.org/10.1093/jxb/eru109

Kim, T. H., Böhmer, M., Hu, H., Nishimura, N., Schroeder, J. I. Guard cell signal transduction network: Advances in understanding abscisic acid, CO_2, and Ca^{2+} signaling. *Annu. Rev. Plant. Biol.*, 2010, 61(1), 561–591. https://doi.org/10.1146/annurev-arplant-042809-112226

Koffler, B. E., Luschin-Ebengreuth, N., Stabentheiner, E., Müller, M., Zechmann, B. Compartment specific response of antioxidants to drought stress in Arabidopsis. *Plant Sci.*, 2014, 227, 133–144. https://doi.org/10.1016/j.plantsci.2014.08.002

Koulman, A., Lane, G. A., Harrison, S. J., Volmer, D. A. From differentiating metabolites to biomarkers. *Anal. Bioanal. Chem.*, 2009, 394(3), 663–670. https://doi.org/10.1007/s00216-009-2690-3

Kumar, M., Kumar Patel, M., Kumar, N., Bajpai, A. B., Siddique, K. H. M. Metabolomics and molecular approaches reveal drought stress tolerance in plants. *Int. J. Mol. Sci.*, 2021, 22(17), 17. https://doi.org/10.3390/ijms22179108

Kusano, M., Tohge, T., Fukushima, A., Kobayashi, M., Hayashi, N., Otsuki, H., Kondou, Y., Goto, H., Kawashima, M., Matsuda, F., Niida, R., Matsui, M., Saito, K., Fernie, A. R. Metabolomics reveals comprehensive reprogramming involving two independent metabolic responses of Arabidopsis to UV-B light. *Plant J.*, 2011, 67(2), 354–369. https://doi.org/10.1111/j.1365-313X.2011.04599.x

Kwak, J. M., Mori, I. C., Pei, Z. M., Leonhardt, N., Torres, M. A., Dangl, J. L., Bloom, R. E., Bodde, S., Jones, J. D. G., Schroeder, J. I. NADPH oxidase AtrbohD and AtrbohF genes function in ROS-dependent ABA signaling in Arabidopsis. *EMBO J.*, 2003, 22(11), 2623–2633. https://doi.org/10.1093/emboj/cdg277

Latowski, D., Surówka, E., Strzałka, K. Regulatory role of components of ascorbate–Glutathione pathway in plant stress tolerance. In: Anjum, N. A., Chan, M.-T., Umar, S. (Eds.), *Ascorbate-Glutathione Pathway and Stress Tolerance in Plants* (pp. 1–53). Springer: the Netherlands, 2010. https://doi.org/10.1007/978-90-481-9404-9_1

Liu, J., Chen, N., Chen, F., Cai, B., Dal Santo, S., Tornielli, G. B., Pezzotti, M., Cheng Z M (Max). Genome-wide analysis and expression profile of the bZIP transcription factor gene family in grapevine (Vitis vinifera). *BMC Genom.*, 2014, 15(1), 281. https://doi.org/10.1186/1471-2164-15-281

Liu, X., Locasale, J. W. Metabolomics: A primer. *Trends Biochem. Sci.*, 2017, 42(4), 274–284. https://doi.org/10.1016/j.tibs.2017.01.004

Maruyama, K., Urano, K., Yoshiwara, K., Morishita, Y., Sakurai, N., Suzuki, H., Kojima, M., Sakakibara, H., Shibata, D., Saito, K., Shinozaki, K., Yamaguchi-Shinozaki, K. Integrated analysis of the effects of cold and dehydration on rice metabolites, phytohormones, and gene transcripts. *Plant Physiol.*, 2014, 164(4), 1759–1771. https://doi.org/10.1104/pp.113.231720

Matsuo, M., Johnson, J. M., Hieno, A., Tokizawa, M., Nomoto, M., Tada, Y., Godfrey, R., Obokata, J., Sherameti, I., Yamamoto, Y. Y., Böhmer, F. D., Oelmüller, R. High Redox responsive transcription factor1 levels result in accumulation of reactive oxygen species in Arabidopsis thaliana shoots and roots. *Mol. Plant*, 2015, 8(8), 1253–1273.

Mehla, N., Sindhi, V., Josula, D., Bisht, P., Wani, S. H. An introduction to antioxidants and their roles in plant stress tolerance. In: Khan, M. I. R., Khan, N. A. (Eds.), *Reactive Oxygen Species and Antioxidant Systems in Plants: Role and Regulation under Abiotic Stress* (pp. 1–23). Springer, 2017. https://doi.org/10.1007/978-981-10-5254-5_1

Mhamdi, A., Mauve, C., Gouia, H., Saindrenan, P., Hodges, M., Noctor, G. Cytosolic NADP-dependent isocitrate dehydrogenase contributes to redox homeostasis and the regulation of pathogen responses in Arabidopsis leaves. *Plant Cell Environ.*, 2010, 33(7), 1112–1123. https://doi.org/10.1111/j.1365-3040.2010.02133.x

Mittler, R. ROS are good. *Trends Plant Sci.*, 2017, 22(1), 11–19. https://doi.org/10.1016/j.tplants.2016.08.002

Mittler, R., Vanderauwera, S., Gollery, M., Breusegem, F. V. Reactive oxygen gene network of plants. *Trends Plant Sci.*, 2004, 9(10), 490–498. https://doi.org/10.1016/j.tplants.2004.08.009

Mittler, R., Zandalinas, S. I., Fichman, Y., Van Breusegem, F. Reactive oxygen species signalling in plant stress responses. *Nat. Rev. Mol. Cell Biol.*, 2022, 23(10), 10. https://doi.org/10.1038/s41580-022-00499-2

Møller, I. M., Jensen, P. E., Hansson, A. Oxidative modifications to cellular components in plants. *Annu. Rev. Plant Biol.*, 2007, 58(1), 459–481. https://doi.org/10.1146/annurev.arplant.58.032806.103946

Navrot, N., Rouhier, N., Gelhaye, E., Jacquot, J. P. Reactive oxygen species generation and antioxidant systems in plant mitochondria. *Physiol. Plant*, 2007, 129(1), 185–195. https://doi.org/10.1111/j.1399-3054.2006.00777.x

Neill, S., Desikan, R., Hancock, J. Hydrogen peroxide signalling. *Curr. Opin. Plant Biol.*, 2002a, 5(5), 388–395. https://doi.org/10.1016/S1369-5266(02)00282-0

Neill, S. J., Desikan, R., Clarke, A., Hurst, R. D., Hancock, J. T. Hydrogen peroxide and nitric oxide as signalling molecules in plants. *J. Exp. Bot.*, 2002b, 53(372), 1237–1247. https://doi.org/10.1093/jexbot/53.372.1237

Niu, L., Liao, W. Hydrogen peroxide signaling in plant development and abiotic responses: Crosstalk with nitric oxide and calcium. *Front. Plant Sci.*, 2016, 7, 230. https://www.frontiersin.org/articles/10.3389/fpls.2016.00230

Noctor, G., Paepe, R. D., Foyer, C. H. Mitochondrial redox biology and homeostasis in plants. *Trends Plant Sci.*, 2007, 12(3), 125–134. https://doi.org/10.1016/j.tplants.2007.01.005

Noctor, G., Reichheld, J. P., Foyer, C. H. ROS-related redox regulation and signaling in plants. *Semin. Cell Dev. Biol.*, 2018, 80, 3–12. https://doi.org/10.1016/j.semcdb.2017.07.013

Obata, T., Fernie, A. R. The use of metabolomics to dissect plant responses to abiotic stresses. *Cell Mol. Life Sci.*, 2012, 69(19), 3225–3243. https://doi.org/10.1007/s00018-012-1091-5

Okazaki, Y., Saito, K. Recent advances of metabolomics in plant biotechnology. *Plant Biotec. Rep.*, 2012, 6(1), 1–15. https://doi.org/10.1007/s11816-011-0191-2

Pastore, D., Trono, D., Laus, M. N., Di Fonzo, N., Flagella, Z. Possible plant mitochondria involvement in cell adaptation to drought stress: A case study: Durum wheat mitochondria. *J. Exp. Bot.*, 2007, 58(2), 195–210. https://doi.org/10.1093/jxb/erl273

Pereira, A. Plant abiotic stress challenges from the changing environment. *Front. Plant Sci.*, 2016, 7. https://www.frontiersin.org/articles/10.3389/fpls.2016.01123

Pfannschmidt, T. Chloroplast redox signals: How photosynthesis controls its own genes. *Trends Plant Sci.*, 2003, 8(1), 33–41. https://doi.org/10.1016/S1360-1385(02)00005-5

Raja, V., Majeed, U., Kang, H., Andrabi, K. I., John, R. Abiotic stress: Interplay between ROS, hormones and MAPKs. *Environ. Exp. Bot.*, 2017, 137, 142–157. https://doi.org/10.1016/j.envexpbot.2017.02.010

Rasmusson, A. G., Geisler, D. A., Møller, I. M. The multiplicity of dehydrogenases in the electron transport chain of plant mitochondria. *Mitochondrion*, 2008, 8(1), 47–60. https://doi.org/10.1016/j.mito.2007.10.004

Raza, A., Razzaq, A., Mehmood, S. S., Zou, X., Zhang, X. L. Y., Xu, J. Impact of climate change on crops adaptation and strategies to tackle its outcome: A review. *Plants*, 2019, 8(2), 2. https://doi.org/10.3390/plants8020034

Reddy, A. R., Chaitanya, K. V., Vivekanandan, M. Drought-induced responses of photosynthesis and antioxidant metabolism in higher plants. *J. Plant Physiol.*, 2004, 161(11), 1189–1202. https://doi.org/10.1016/j.jplph.2004.01.013

Rhoads, D. M., Umbach, A. L., Subbaiah, C. C., Siedow, J. N. Mitochondrial reactive oxygen species. Contribution to oxidative stress and interorganellar signaling. *Plant Physiol.*, 2006, 141(2), 357–366. https://doi.org/10.1104/pp.106.079129

Savvides, A., Ali, S., Tester, M., Fotopoulos, V. Chemical priming of plants against multiple abiotic stresses: Mission possible? *Trends Plant Sci.*, 2016, 21(4), 329–340. https://doi.org/10.1016/j.tplants.2015.11.003

Schwarzländer, M., Fricker, M. D., Sweetlove, L. J. Monitoring the in vivo redox state of plant mitochondria: Effect of respiratory inhibitors, abiotic stress and assessment of recovery from oxidative challenge. *Biochim. Biophys. Acta(BBA)—Bioenergetics*, 2009, 1787(5), 468–475. https://doi.org/10.1016/j.bbabio.2009.01.020

Simon, C., Langlois-Meurinne, M., Bellvert, F., Garmier, M., Didierlaurent, L., Massoud, K., Chaouch, S., Marie, A., Bodo, B., Kauffmann, S., Noctor, G., Saindrenan, P. The differential spatial distribution of secondary metabolites in Arabidopsis leaves reacting hypersensitively to *Pseudomonas syringae* pv. Tomato is dependent on the oxidative burst. *J. Exp. Bot.*, 2010, 61(12), 3355–3370. https://doi.org/10.1093/jxb/erq157

Singh, A., Kumar, A., Yadav, S., Singh, I. K. Reactive oxygen species-mediated signaling during abiotic stress. *Plant Gene*, 2019, 18, 100173. https://doi.org/10.1016/j.plgene.2019.100173

Soares, C., Carvalho, M. E. A., Azevedo, R. A., Fidalgo, F. Plants facing oxidative challenges—A little help from the antioxidant networks. *Environ. Exp. Bot.*, 2019, 161, 4–25. https://doi.org/10.1016/j.envexpbot.2018.12.009

Soga, T., Baran, R., Suematsu, M., Ueno, Y., Ikeda, S., Sakurakawa, T., Kakazu, Y., Ishikawa, T., Robert, M., Nishioka, T., Tomita, M. Differential metabolomics reveals ophthalmic acid as an oxidative stress biomarker indicating hepatic glutathione consumption*. *J. Biol. Chem.*, 2006, 281(24), 16768–16776. https://doi.org/10.1074/jbc.M601876200

Sohag, A. A. M., Tahjib-Ul-Arif, M., Brestic, M., Afrin, S., Sakil, M. A., Hossain, M. T., Hossain, M. A., Hossain, M. A. Exogenous salicylic acid and hydrogen peroxide attenuate drought stress in rice. *Plant Soil Environ.*, 2020, 66 (2020)(No. 1), 7–13. https://doi.org/10.17221/472/2019-PSE

Sreekumar, A., Poisson, L. M., Rajendiran, T. M., Khan, A. P., Cao, Q., Yu, J., Laxman, B., Mehra, R., Lonigro, R. J., Li, Y., Nyati, M. K., Ahsan, A., Kalyana-Sundaram, S., Han, B., Cao, X., Byun, J., Omenn, G. S., Ghosh, D., Pennathur, S., Alexander, D. C., Chinnaiyan, A. M. Metabolomic profiles delineate potential role for sarcosine in prostate cancer progression. *Nature*, 2009, 457(7231), 7231. https://doi.org/10.1038/nature07762

Sun, Y., Wang, H., Liu, S., Peng, X. Exogenous application of hydrogen peroxide alleviates drought stress in cucumber seedlings. *S. Afr. J. Bot.*, 2016, 106, 23–28. https://doi.org/10.1016/j.sajb.2016.05.008

Tanou, G., Filippou, P., Belghazi, M., Job, D., Diamantidis, G., Fotopoulos, V., Molassiotis, A. Oxidative and nitrosative-based signaling and associated post-translational modifications orchestrate the acclimation of citrus plants to salinity stress. *Plant J.*, 2012, 72(4), 585–599. https://doi.org/10.1111/j.1365-313X.2012.05100.x

Terzi, R., Kadioglu, A., Kalaycioglu, E., Saglam, A. Hydrogen peroxide pretreatment induces osmotic stress tolerance by influencing osmolyte and abscisic acid levels in maize leaves. *J. Plant Interact.*, 2014, 9(1), 559–565. https://doi.org/10.1080/17429145.2013.871077

Tseng, M. J., Liu, C. W., Yiu, J. C. Enhanced tolerance to sulfur dioxide and salt stress of transgenic Chinese cabbage plants expressing both superoxide dismutase and catalase in chloroplasts. *Plant Physiol. Biochem.*, 2007, 45(10), 822–833. https://doi.org/10.1016/j.plaphy.2007.07.011

Turrens, J. F. Mitochondrial formation of reactive oxygen species. *J. Physiol.*, 2003, 552(2), 335–344. https://doi.org/10.1111/j.1469-7793.2003.00335.x

Umezawa, T., Fujita, M., Fujita, Y., Yamaguchi-Shinozaki, K., Shinozaki, K. Engineering drought tolerance in plants: Discovering and tailoring genes to unlock the future. *Curr. Opin. Biotechnol.*, 2006, 17(2), 113–122. https://doi.org/10.1016/j.copbio.2006.02.002

Urano, K., Maruyama, K., Ogata, Y., Morishita, Y., Takeda, M., Sakurai, N., Suzuki, H., Saito, K., Shibata, D., Kobayashi, M., Yamaguchi-Shinozaki, K., Shinozaki, K. Characterization of the ABA-regulated global responses to dehydration in Arabidopsis by metabolomics. *Plant J*, 2009, 57(6), 1065–1078. https://doi.org/10.1111/j.1365-313X.2008.03748.x

Wahid, A., Gelani, S., Ashraf, M., Foolad, M. R. Heat tolerance in plants: An overview. *Environ. Exp. Bot.*, 2007, 61(3), 199–223. https://doi.org/10.1016/j.envexpbot.2007.05.011

Wang, F. Z., Wang, Q. B., Kwon, S. Y., Kwak, S. S., Su, W. A. Enhanced drought tolerance of transgenic rice plants expressing a pea manganese superoxide dismutase. *J. Plant Physiol.*, 2005, 162(4), 465–472. https://doi.org/10.1016/j.jplph.2004.09.009

Wang, L., Leister, D., Guan, L., Zheng, Y., Schneider, K., Lehmann, M., Apel, K., Kleine, T. The Arabidopsis SAFEGUARD1 suppresses singlet oxygen-induced stress responses by protecting grana margins. *Proc. Natl. Acad. Sci.*, 2020, 117(12), 6918–6927. https://doi.org/10.1073/pnas.1918640117

Ward, J. L., Forcat, S., Beckmann, M., Bennett, M., Miller, S. J., Baker, J. M., Hawkins, N. D., Vermeer, C. P., Lu, C., Lin, W., Truman, W. M., Beale, M. H., Draper, J., Mansfield, J. W., Grant, M. The metabolic transition during disease following infection of Arabidopsis thaliana by *Pseudomonas syringae* pv. Tomato. *Plant J.*, 2010, 63(3), 443–457. https://doi.org/10.1111/j.1365-313X.2010.04254.x

Widodo Patterson, J. H., Newbigin, E., Tester, M., Bacic, A., Roessner, U. Metabolic responses to salt stress of barley (*Hordeum vulgare* L.) cultivars, Sahara and Clipper, which differ in salinity tolerance. *J. Exp. Bot.*, 2009, 60(14), 4089–4103. https://doi.org/10.1093/jxb/erp243

Wu, Y., Yang, Z., How, J., Xu, H., Chen, L., Li, K. Overexpression of a peroxidase gene (AtPrx64) of Arabidopsis thaliana in tobacco improves plant's tolerance to aluminum stress. *Plant Mol. Biol.*, 2017, 95(1), 157–168. https://doi.org/10.1007/s11103-017-0644-2

Yamauchi, T., Yoshioka, M., Fukazawa, A., Mori, H., Nishizawa, N. K., Tsutsumi, N., Yoshiok, H., Nakazono, M. An NADPH oxidase RBOH functions in rice roots during lysigenous aerenchyma formation under oxygen-deficient conditions. *Plant Cell*, 2017, 29(4), 775–790. https://doi.org/10.1105/tpc.16.00976

Yang, C., Li, W., Cao, J., Meng, F., Yu, Y., Huang, J., Jiang, L., Liu, M., Zhang, Z., Chen, X., Miyamoto, K., Yamane, H., Zhang, J., Chen, S., Liu, J. Activation of ethylene signaling pathways enhances disease resistance by regulating ROS and phytoalexin production in rice. *Plant J*, 2017, 89(2), 338–353. https://doi.org/10.1111/tpj.13388

Yang, M., Li, Z., Zhang, K., Zhang, X., Zhang, Y., Wang, X., Han, C., Yu, J., Xu, K., Li, D. Barley stripe mosaic virus γb interacts with glycolate oxidase and inhibits peroxisomal ROS production to facilitate virus infection. *Mol. Plant*, 2018, 11(2), 338–341. https://doi.org/10.1016/j.molp.2017.10.007

Ye, N., Zhu, G., Liu, Y., Li, Y., Zhang, J. ABA controls H2O2 accumulation through the induction of OsCATB in rice leaves under water stress. *Plant Cell Physiol.*, 2011, 52(4), 689–698. https://doi.org/10.1093/pcp/pcr028

Zhang, G., Zhang, M., Zhao, Z., Ren, Y., Li Q Wang, W. Wheat TaPUB1 modulates plant drought stress resistance by improving antioxidant capability. *Sci. Rep.*, 2017, 7(1), 1. https://doi.org/10.1038/s41598-017-08181-w

Zhao, Q., Zhou, L., Liu, J., Cao, Z., Du, X., Huang, F., Pan, G., Cheng, F. Involvement of CAT in the detoxification of HT-induced ROS burst in rice anther and its relation to pollen fertility. *Plant Cell Rep.*, 2018, 37(5), 741–757. https://doi.org/10.1007/s00299-018-2264-y

Zia, A., Walker, B. J., Oung, H. M. O., Charuvi, D., Jahns, P., Cousins, A. B., Farrant, J. M., Reich, Z., Kirchhoff, H. Protection of the photosynthetic apparatus against dehydration stress in the resurrection plant *Craterostigma pumilum*. *Plant J.*, 2016, 87(6), 664–680. https://doi.org/10.1111/tpj.13227

Zivy, M., Wienkoop, S., Renaut, J., Pinheiro, C., Goulas, E., Carpentier, S. The quest for tolerant varieties: The importance of integrating "omics" techniques to phenotyping. *Front. Plant Sci.*, 2015, 6, 448. https://www.frontiersin.org/articles/10.3389/fpls.2015.00448

Index

E

F

G

H

Q

R

S

T